Algal Functional Foods and Nutraceuticals: Benefits, Opportunities and Challenges

Edited by

Avinash Mishra

Applied Phycology and Biotechnology
Central Salt and Marine Chemicals Research Institute
Bhavnagar (Gujarat)
India

Algal Functional Foods and Nutraceuticals: Benefits, Opportunities and Challenges

Editor: Avinash Mishra

ISBN (Online): 978-981-5051-87-2

ISBN (Print): 978-981-5051-88-9

ISBN (Paperback): 978-981-5051-89-6

First published in 2022.

Bentham Science Publishers Pte. Ltd.
80 Robinson Road #02-00
Singapore 068898
Singapore
Email: subscriptions@benthamscience.net

BENTHAM SCIENCE

CONTENTS

FOREWORD

The algae, including seaweeds, have gained significant global importance largely due to their extensive utilization in human foods, feed, nutraceuticals, personal care products and several other industrial products of commercial value. In the COVID pandemic, there is a growing awareness among the consumers to eat healthy diets with functional foods, otherwise known as nutraceuticals, that not only collectively boost the immune system by preventing viral and bacterial infections but also contribute to overall wellness. In recent times, considerable effort has been directed to unlock and harness the potential of algae for functional foods in conventional and modified formats. Since the functional foods represent a tiny fraction of the biomass, the remaining portion is utilized sustainably to produce beneficial commodity value products in the biorefinery model. The primary components produced from biorefinery have attracted innovations in the downstream processing industry for developing high-value niche products to create additional revenue streams and promotet6 inclusive sectoral growth. Also, algae farming is a relatively low carbon footprint activity with many positive environmental impacts.

This book gives a most updated account of the global trends and developments in the cultivation of edible algae along with their environmental impacts, trends, potentials, application and health benefits of algal-based nutraceuticals, probiotics and prebiotics, bioactive properties of algal foods, secondary metabolites, culture-induced nutraceutical compounds, algae-based bio packaging materials, biosafety aspects, etc. Further, the methods for extraction of algal metabolites, pigments and bioactive compounds have also been included and presented in this book. Also, how do algal-based foods impact human health, challenges in consumer acceptance of algal food, and industry needs besides a roadmap ahead are also covered. This book comprises 20 chapters by eminent experts in the field, providing deeper insights and newer perspectives of algae-based functional foods, their health benefits to consumers, bioprospecting of primary and secondary metabolites, and extraction methods.

I am sure the contents of this book provide the reader with a great deal of information, further advancing the knowledge in the area of algal functional foods. The researchers working on algal foods and human health will find this book a very useful reference resource.

CRK Reddy
Chief Executive Officer
Indian Centre for Climate and Societal Impacts Research
Mandvi-Katchch, 370465, India
Formerly Divisional Chair and
Chief Scientist at CSIR-CSMCRI

PREFACE

"Let food be thy medicine and medicine be thy food," a quote by the Father of Medicine, Hippocrates, extended our quest towards the use of algal resources as a functional food for human health. Algae have been used as food and food ingredients from ancient times. The objective of the book is to present the current knowledge and recent advances in the field of Algal Functional Foods and Nutraceuticals to stimulate further research and to sensitize the exploration of algal-food for novel applications. The benefits of algae-based foods, opportunities in the exploration of value-added functional foods from algae, and challenges to be encountered during nutraceuticals or product development are key topics, which are the genesis of this book. The concept of algae as a source of biologically active ingredients for the formulation of functional foods and nutraceuticals has also been discussed. The book is organized into twenty-one chapters, covering information on algal cultivation, habitat, nutraceuticals, bioactive compounds, metabolites, pigments, food packaging materials, and toxicity.

Chapters introduce the cultivation, habitat and environmental impact of the edible algae. Techniques available for cultivation, different types of habitat, and how the environment influences food quality have been discussed in detail. The next few chapters discuss the potential, trends, application in the present scenario, and prospects of algal-based nutraceuticals. Subsequent chapters discuss the extraction of bioactive compounds, their potential, application, and chemical structures of algal metabolites. Various biological properties, such as antioxidant, anticancer, anti-inflammatory, anti-obesity and anti-diabetic are also discussed in the book. The role of algae as prebiotic and probiotic, and their nutritional values have enlightened their wide range of scope. The role of algal-based value-added products in food packaging, biodegradable film and food coating are also key highlights of the book. Bio-refinery concept and role of different algal metabolites and pigments are discussed in detail. Finally, the toxicity of algal foods to human health is also discussed.

The book covers basic concepts as well as advances in Algal Functional Foods. The book is valuable for algal-farmers, scientists, students, industries and consumers to understand the potential and prospects of algae for Foods and Nutraceuticals. The book will strengthen the concept of readers and also allow them to select an appropriate approach for their research. The book will also enable readers to think differently from the conventional approaches.

With the grace of Lord Shiva, and the blessing of my parents and family members, I have completed this book. The compilation was made possible only through the voluntary contributions of international experts who provided a valuable and global vision of the edible algae. I express my deep appreciation and gratitude to all the authors for their valuable contributions. My gratitude is also extended to Dr. Kannan Srinivasan, Director CSMCRI, Prof. Bhavanath Jha, former Head of the Division, and all students associated with me. I acknowledge Dr. CRK Reddy, an eminent seaweed-biotechnologist, for forwarding this book. This publication is approved with CSIR-CSMCRI PRIS registration number 02/2021.

I thank my wife, Mrs. Chetna Mishra, for being my strength and daughters Ekisha and Aaral for unconditional love.

Avinash Mishra
Principal Scientist, Molecular Biology and Biotechnology
Applied Phycology and Biotechnology Division
CSIR-Central Salt and Marine Chemicals Research Institute
Bhavnagar (Gujarat), India

SUMMARY

Marine ecosystem is highly diversified, and edible algae including seaweeds have been considered as a potential source of functional food, dietary supplements, metabolites and bioactive compounds. Algal-based functional foods have potential health benefits, and their commercial value depends on their applications in the food and nutraceutical industries. Seaweed based foods are preferentially selected over other types of marine foods but still they are considered underutilize marine resources. In this book, different aspect of algal based functional foods has been discussed. In brief, cultivation techniques, environmental impact, habitat, nutraceutical potential, extraction of bioactive metabolites, functional-food composition, bio-prospection, culture-induced nutraceutical compounds, algae-based bio-packaging, algal-biorefinery, toxicity, trends and future prospects have been discussed in detail.

List of Contributors

A. Catarina Guedes	CIIMAR/CIMAR-LA – Interdisciplinary Centre of Marine and Environmental Research, University of Porto, Novo Edifício do Terminal de Cruzeiros do Porto de Leixões, Av. General Norton de Matos, s/n, 4450-208, Matosinhos, Portugal
Aditya Velidandi	Department of Biotechnology, National Institute of Technology, Warangal, Telangana, India
Ana M.M. Gonçalves	Department of Biology, University of Aveiro, Campus Universitário de Santiago, 3810-193 Aveiro, Portugal CESAM, University of Aveiro, Campus Universitário de Santiago, 3810-193 Aveiro, Portugal University of Coimbra, MARE – Marine and Environmental Sciences Centre/ ARNET - Aquatic Research Network, Department of Life Sciences, Calçada Martim de Freitas, 3000-456, Coimbra, Portugal
Ángel J. Gutiérrez	Area of Toxicology, School of Medicine, Universidad de La Laguna, 38071 Santa Cruz de Tenerife, Tenerife, Canary Islands, Spain
Antony V. Samrot	School of Bioscience, Faculty of Medicine, Bioscience and Nursing, MAHSA University, Jalan SP2, Bandar Saujana Putra, 42610, Jenjarom, Selangor, Malaysia
Anuruddhika Udayangani Rathnayake	Department of Marine Biotechnology, Gangneung-Wonju National University, Gangneung 25457, Republic of Korea Department of Science and Technology, Faculty of Applied Sciences, Uva Wellassa University, Badulla, Sri Lanka
Arturo Hardisson	Area of Toxicology, School of Medicine, Universidad de La Laguna, 38071 Santa Cruz de Tenerife, Tenerife, Canary Islands, Spain
C. Kurinjimalar	Centre for Advanced Studies in Botany, University of Madras, Chennai-600 025, India
Carmen Rubio	Area of Toxicology, School of Medicine, Universidad de La Laguna, 38071 Santa Cruz de Tenerife, Spain
Chandrashekar Mootapally	School of Applied Sciences & Technology (SAST-GTU), Gujarat Technological University, Chandkheda Campus, Ahmedabad, Gujarat, 382424, India
Cláudia Nunes	CICECO, Department of Chemistry, University of Aveiro, Campus Universitário de Santiago, 3810-193 Aveiro, Portugal LAQV-REQUIMTE, Department of Chemistry, University of Aveiro, Campus Universitário de Santiago, 3810-193 Aveiro, Portugal
D. Inbakandan	Centre for Ocean Research (DST-FIST Sponsored Centre), MoES - Earth Science & Technology Cell (Marine Biotechnological Studies), Col. Dr. Jeppiaar Research Park, Sathyabama Institute of Science and Technology, Chennai 600 119, India
Dailos González-Weller	Area of Toxicology, School of Medicine, Universidad de La Laguna, 38071 Santa Cruz de Tenerife, Tenerife, Canary Islands, Spain Health Inspection and Laboratory Service, Canary Health Service, 38006 Santa Cruz de Tenerife, Tenerife, Canary Islands, Spain

Danilo B. Largo	Department of Biology, University of San Carlos, Cebu City, Philippines
Darshita K. Pandya	Division of Applied Phycology and Biotechnology, CSIR-Central Salt and Marine Chemicals Research Institute, Bhavnagar-364002, Gujarat, India
Emilin Renitta R.	Department of Food Processing, Karunya Institute of Technology and Science, Coimbatore, Tamil Nadu, India
F. Xavier Malcata	LEPABE – Laboratory for Process Engineering, Environment, Biotechnology and Energy, University of Porto, Rua Dr. Roberto Frias, s/n, 4200-465 Porto, Portugal FEUP – Department of Chemical Engineering, Faculty of Engineering, University of Porto, Rua Dr. Roberto Frias, s/n, 4200-465 Porto, Portugal ALICE - Associate Laboratory in Chemical Engineering, Faculty of Engineering, University of Porto, Rua Dr. Roberto Frias, 4200-465 Porto, Portugal
Fernando J.M. Gonçalves	Department of Biology, University of Aveiro, Campus Universitário de Santiago, 3810-193 Aveiro, Portugal CESAM, University of Aveiro, Campus Universitário de Santiago, 3810-193 Aveiro, Portugal
G. Kavitha Aravind	Providence College for Women, Bandishola, Spring Field, Coonoor, The Nilgiris, Tamil Nadu 643104, India
Hee-Guk Byun	Department of Marine Biotechnology, Gangneung-Wonju National University, Gangneung 25457, Republic of Korea
Helena M. Amaro	CIIMAR/CIMAR-LA – Interdisciplinary Centre of Marine and Environmental Research, University of Porto, Novo Edifício do Terminal de Cruzeiros do Porto de Leixões, Av. General Norton de Matos, s/n, 4450-208, Matosinhos, Portugal
Indira Wickramasinghe	Department of Food Science & Technology, Faculty of Applied Sciences, University of Sri Jayewardenepura, Nugegoda, Sri Lanka
Isuru Wijesekara	Department of Food Science and Technology, Faculty of Applied Sciences, University of Sri Jayewardenepura, Gangodawila, Nugegoda, Sri Lanka
Jayani Samarathunga	Department of Food Science and Technology, Faculty of Applied Sciences, University of Sri Jayewardenepura, Gangodawila, Nugegoda, Sri Lanka
Jayappriyan K.R.	Centre for Advanced Studies in Botany, University of Madras, Chennai-600 025, India
João C. Marques	University of Coimbra, MARE – Marine and Environmental Sciences Centre/ARNET - Aquatic Research Network, Department of Life Sciences, Calçada Martim de Freitas, 3000-456, Coimbra, Portugal
Joana Assunção	CIIMAR/CIMAR-LA – Interdisciplinary Centre of Marine and Environmental Research, University of Porto, Novo Edifício do Terminal de Cruzeiros do Porto de Leixões, Av. General Norton de Matos, s/n, 4450-208, Matosinhos, Portugal LEPABE – Laboratory for Process Engineering, Environment, Biotechnology and Energy, University of Porto, Rua Dr. Roberto Frias, s/n, 4200-465 Porto, Portugal

K. Sundar	Department of Biotechnology, Kalasalingam of Academy of Research and Education, Virudhunagar, Tamil Nadu, India
K. Vasumathi	Department of Biotechnology, Kalasalingam of Academy of Research and Education, Virudhunagar, Tamil Nadu, India
K.G. Kaushani	Department of Food Science and Technology, Faculty of Applied Sciences, University of Sri Jayewardenepura, Gangodawila, Nugegoda, Sri Lanka
Luke Webster	Algae Biotechnology Laboratory, School of Agriculture and Food Sciences, The University of Queensland, Brisbane, Queensland 4072, Australia
M. Kaviraj	Centre for Advanced Studies in Botany, University of Madras, Chennai-600 025, India
M.M. Jayakody	Department of Food Science and Technology, Faculty of Applied Sciences, University of Sri Jayewardenepura, Gangodawila, Nugegoda, Sri Lanka
M.P.G Vanniarachchy	Department of Food Science and Technology, Faculty of Applied Sciences, University of Sri Jayewardenepura, Gangodawila, Nugegoda, Sri Lanka
M. Premalatha	Department of Energy and Environment, National Institute of Technology Tiruchirappalli-620015, Tamil Nadu, India
M. Vijayakumar	Centre for Advanced Studies in Botany, University of Madras, Chennai-600 025, India
Madhura Jayasinghe	Department of Food Science and Technology, Faculty of Applied Sciences, University of Sri Jayewardenepura, Gangodawila, Nugegoda, Sri Lanka
Mangal S. Rathore	Division of Applied Phycology and Biotechnology, CSIR-Central Salt and Marine Chemicals Research Institute, Bhavnagar-364002, Gujarat, India Academy of Scientific and Innovative Research (AcSIR), Ghaziabad-201002, India
Manuel A. Coimbra	LAQV-REQUIMTE, Department of Chemistry, University of Aveiro, Campus Universitário de Santiago, 3810-193 Aveiro, Portugal
Marlene A. Trindade	Department of Biology, University of Aveiro, Campus Universitário de Santiago, 3810-193 Aveiro, Portugal
Mona M. Ismail	National Institute of Oceanography and Fisheries, NIOF, Egypt
Monica G. Kavale	Academy of Scientific and Innovative Research, Ghaziabad-201002, Uttar Pradesh, India CSIR-Central Salt and Marine Chemicals Research Institute, Bhavnagar-364002, Gujarat, India
Neelam M. Nathani	School of Applied Sciences & Technology (SAST-GTU), Gujarat Technological University, Chandkheda Campus, Ahmedabad, Gujarat, 382424, India
Neha Kumari	Laboratory of Photobiology and Molecular Microbiology, Centre of Advanced Study in Botany, Institute of Science, Banaras Hindu University, Varanasi-221005, India
Ninian P.P. Pabbathi	Department of Biotechnology, National Institute of Technology, Warangal, Telangana, India
Peer M. Schenk	Algae Biotechnology Laboratory, School of Agriculture and Food Sciences, The University of Queensland, Brisbane, Queensland 4072, Australia

Prashakha J. Shukla	Parul Institute of Applied Sciences, Parul University, Vadodara - 391760, Gujarat, India
R. Rajkumar	Department of Environmental Sciences, Bharathiar University, Coimbatore-641 046, India
Rajeshwar P. Sinha	Laboratory of Photobiology and Molecular Microbiology, Centre of Advanced Study in Botany, Institute of Science, Banaras Hindu University, Varanasi-221005, India
Rathinam Raja	Research and Development Wing, Sree Balaji Medical College and Hospital (SBMCH), Bharath Institute of Higher Education and Research (BIHER), Chromepet, Chennai - 600 044, India
Saigeetha S.	Department of Biotechnology, School of Bio and Chemical Engineering, Sathyabama Institute of Science and Technology, Chennai, Tamil Nadu, 600119, India
Sandunika Senarathna	Department of Food Science & Technology, Faculty of Applied Sciences, University of Sri Jayewardenepura, Nugegoda, Sri Lanka
Seneviratne Navaratne	Department of Food Science & Technology, Faculty of Applied Sciences, University of Sri Jayewardenepura, Nugegoda, Sri Lanka
Shahrukh A. Siddiqui	Division of Applied Phycology and Biotechnology, CSIR-Central Salt and Marine Chemicals Research Institute, Bhavnagar-364002, Gujarat, India Academy of Scientific and Innovative Research (AcSIR), Ghaziabad-201002, India
Sonal Mishra	Laboratory of Photobiology and Molecular Microbiology, Centre of Advanced Study in Botany, Institute of Science, Banaras Hindu University, Varanasi-221005, India
Soraya Paz	Area of Toxicology, School of Medicine, Universidad de La Laguna, 38071 Santa Cruz de Tenerife, Tenerife, Canary Islands, Spain
Surabhi Agrawal	Division of Applied Phycology and Biotechnology, CSIR-Central Salt and Marine Chemicals Research Institute, Bhavnagar-364002, Gujarat, India Academy of Scientific and Innovative Research (AcSIR), Ghaziabad-201002, India
Taylor J. Wass	Algae Biotechnology Laboratory, School of Agriculture and Food Sciences, The University of Queensland, Brisbane, Queensland 4072, Australia
Tejal K. Gajaria	Academy of Scientific and Innovative Research, Ghaziabad-201002, Uttar Pradesh, India CSIR-Central Salt and Marine Chemicals Research Institute, Bhavnagar-364002, Gujarat, India
Xuan Ji	Algae Biotechnology Laboratory, School of Agriculture and Food Sciences, The University of Queensland, Brisbane, Queensland 4072, Australia

CHAPTER 1

Cultivation of Edible Algae: Present and Future

Danilo B. Largo[1,*]

[1] *Department of Biology, University of San Carlos, Cebu City, Cebu 6000, Philippines*

Abstract: The use of algae as food by humans started in pre-recorded history and is most widespread in Asia, where algae are used as part of the peoples' daily diet. Of more than ten thousand species of algae that have been described taxonomically, only about a hundred species are known to be edible and are generally recognized as safe (GRAS). Algae for human consumption come from both natural population and open-sea aquaculture, but with increasing issues of marine pollution, global warming and marine use conflict, the production of some algal species from aquaculture is shifting towards the more controlled condition of land-based production systems or in sea areas far from sources of pollutants. The preparation of edible algae comes in various forms that are either consumed directly as fresh salad or pickled in vinegar for species with foliose, delicate or succulent nature or as blanched or cooked recipes for species with fleshy, rubbery or firm texture, or they are consumed indirectly as an ingredient or additive of some food recipes as gelling, hardening, or thickening agent. Many species of micro- and macroalgae have nutritional profiles that make them a perfect food for individuals who are on a diet or are health conscious. This chapter describes some of the most common algal genera whose biomass is mainly produced from culture systems that involve a land-based culture facility (*e.g.*, photobioreactor) and or seedling collection procedure prior to open sea cultivation.

Keywords: Aquaculture , Edible Algae, Eucheumatoids, Kelps, Macroalgae, Mariculture, Marine Agronomy, Microalgae, Nori, Nutraceutical, Outdoor cultivation, Photobioreactor, Raceways, Seaweed aquaculture, Seaweed cultivation, Seaweed farming, Tank culture.

INTRODUCTION

Man has used algae since the beginning of recorded history, known as "manna from heaven," in biblical times, which could be an edible blue-green alga [1]. Algae is mainly consumed by people in countries located along or near the sea, with east Asians most known to gather algae from the sea (seaweeds). The most consumed edible algae are from cultivated species, either through land-based facilities or cultivation systems on the coastal and open seas.

* **Corresponding author Danilo B. Largo:** Department of Biology, University of San Carlos, Cebu City, Philippines; Tel: +63 (32) 230-0100; Fax: +63 (32) 255-4341; E-mail: dblargo@usc.edu.ph

Avinash Mishra (Ed.)

The term "edible" means "something that is suitable or safe to eat", but some species of algae contain compounds that can be harmful to health, such as *Sargassum fusiforme* (formerly *Hizikia fusiformis*), which contains a high amount of arsenic acid – a heavy metal [2]. Even the most popular among the green algae being consumed in many Oriental and Pacific Island countries, referred to as 'green caviar' (*Caulerpa racemosa* complex), contain a toxic chemical compound called caulerpin, and caulerpic, when eaten in large quantities, may cause health issues, although the review by Rengasamy *et al.* [3] says otherwise. Indeed, many seaweed species possess natural defense systems against potential grazers and colonizers. Having utilized these species as food for centuries, coastal people may have developed tolerance to mild amounts of these compounds or may have learned and used techniques to remove these toxins to safer levels. The phlorotannins found in the brown algae, including the edible ones, not only afford the seaweeds with a natural defense against grazers and microbial pathogens, but are important compounds with health benefits to humans if taken in a safe amount [2].

COLOR MATTERS

Macroalgae are classified, regardless of size, into four major color groupings based on their dominant pigment – *blue-green* (cyanobacteria), *green* (chlorophytes), *brown* (phaeophytes) and *red* (rhodophytes), and color variations in terms of intensities and hues depending on their habitat and amount of light received by the thallus. It is not uncommon to find a green red seaweed or a red green seaweed in the seafood market. The intensity and hues of their natural colors make some species attractive as "sidings" or garnishings in some food recipes.

ALGAE AS FUNCTIONAL FOODS

The most commonly appreciated algae in the seafood market are macroalgae, or seaweeds, that are most popular in oriental (Japanese, Chinese, Korean) cooking, and are presented in a wide variety of preparations. They could range from fresh, blanched, to boiled or deeply cooked, depending on the texture of the seaweed thallus. Their uses also depend on the species and include applications as taste enhancers (*e.g.*, 'umami' made from *Saccharina*) or as garnishing in various cuisines and recipes [4] (Mouritsen *et al.* 2018). The microalgae, which are mostly members of cyanobacteria (*e.g.*, *Spirulina, Aphanizomenon flos-aquae*) and green algae (*e.g.*, *Chlorella, Dunaliella*), are mainly popular in the health supplement market due to their high content of protein (as high as 70% of dry wt.), antioxidants, vitamins, amino acids, and micronutrients taken in by health conscious individuals or those who are on a controlled diet. Conversely, macroalgae are becoming popular in the culinary industry and are popular as functional foods associated with the body and mental health and wellness [5]. This chapter describes selected edible seaweeds that are utilized in the modern food industry, either as food for direct human consumption or as raw materials to derive nutraceuticals and seaweed extract-based edible products.

SPECIES OF EDIBLE ALGAE AND THEIR CULTIVATION

Of the more than a hundred species of algae documented to be used as low-calorie food for human consumption or as raw materials for extracting phycocolloids for food applications, only a few (probably less than 30) are cultivated using various agronomic systems (Table **1**).

Table 1. List of edible micro- and macroalgae utilized as food for direct consumption and or sources of food-grade extracts.

-	Popular Name(s)	Uses	Source of Biomass	References
MICROALGAE				
A. Cyanobacteria	-	-	-	-
1. *Spirulina* (*Arthrospira*)	-	as food supplement	cultured	[6 - 9]
B. Chlorophyta	-			-
2. *Chlorella*	-		cultured	[14, 15]
3. *Dunaliella*	-		cultured	[12, 16]
MACROALGAE				
C. Chlorophyceae (green algae)		-	-	-
1. *Capsosiphon fulvescens*	Mesangi (Korean)	as an additive to foods (rice noodles)	cultured	[17, 18]
2. *Caulerpa (C. lentillifera,* 3. *C. racermosa, C. peltata,* 4. *C. sertularioides)*	Sea grapes, green caviar	as seaweed salad in Philippines	wild, cultured	[19]
5. *Codium (C. bartlettii,* 6. *C. edule,* 7. *C. fragile,* 8. *C. intricatum,* 9. *C. tomentosum)*	-	as seaweed salad in Philippines	wild, cultured	[20, 21]
10. *Monostroma nitidum*	Aonori, aosa (Japanese)	as condiment, paste or jam	wild, cultured	[22]
11. *Ulva (U. compressa,* 12. *U. lactuca,* 13. *U. linza,* 14. *U. prolifera)*			wild, cultured	[23, 24]

(Table 1) cont.....

-	Popular Name(s)	Uses	Source of Biomass	References
D. Phaeophyceae (Brown algae		-	-	-
15. *Ascophyllum nodosum*	wrack	as source of food-grade alginate	wild	[25]
16. *Cladosiphon okamuranus*	Okinawa mozuku	as seaweed salad in Japan	cultured	[26]
17. *Ecklonia (E. cava,* 18. *E. stolonifera)*	-	as source of food-grade alginate	wild, cultured	[27]
19. *Hydroclathrus clathratus*	-	as seaweed salad in Philippines	wild	[19]
20. *Nemacystis decipiens*	itomozuku	as seaweed salad in Japan	-	[28]
21. *Pelvetia siliquosa*	-	as source of food-grade alginate	-	[29]
22. *Saccharina (S. acgotensis,* 23. *S. angustata,* 24. *S. diabolica,* 25. *S. digitata,* 26. *S. japonica,* 27. *S. hyperborea,* 28. *S. longissimima,* 29. *S. religiosa)*	Konbu (Japanese) Kaidai (Chinese)	as source of food-grade alginate; as taste enhancer	cultured	[28, 30, 31]
30. *Sargassum (S. fusiforme = Hizikia fusiformis,* 31. *S. fulvellum,* 32. *S. horneri,* 33. *S. naozhouense*)	Hiziki (Japanese)	as boiled seaweed served with rice in Japan, China and Korea	wild, cultured	[28, 32 - 36]
34. *Undaria* (*U. peterseniana,* 35. *U. pinnatifida*)	Wakame (Japanese)	prepared as soup in Japan; as source of food-grade alginate	cultured	[28, 30, 31, 37, 38]
E. Rhodophyceae (red algae)		-	-	-
36. *Acanthopeltis japonica*	-	used in making noodle-like gel dish tokoroten	wild	[28, 30, 42]
37. *Ahnfeltia plicata*	-	as source of food-grade agar	wild	[30]
38. *Asparagopsis taxiformis*	-	Boiled into soup in China	wild	[39, 42]
39. *Ceramium kondoi*	-	Steamed with flour in China	wild	[30, 42]
40. *Chondracanthus chamissoi*	-	used as salad in Peru	wild, cultured	[40]
41. *Chondrus (C. crispus,* 42. *C. crassicaulis)*	Irish moss, dulse	direct human food	wild	[41]

(Table 1) cont.....

-	Popular Name(s)	Uses	Source of Biomass	References
43. *Eucheuma (E. denticulatum,* 44. *E. arnoldii,* 45. *E. gelatinae)*	'Goso', agal-agal	direct human food as pickled salad; source of food-grade carrageenan	cultured	[19]
46. *Gelidiella acerosa*	-	source of food-grade agar	wild	[19]
47. *Gelidium (G. amansii,* 48. *G elegans,* 49. *G. japonicum,* 50. *G. pacificum)*	-	direct human food as salad or as garnishing for sashimi; source of food-grade agar	wild	[28, 38, 42]
51. *Gloiopeltis (G. furcata,* 52. *G tenax)*	-	as source of food-grade agar	wild	[38, 42]
53. *Gracilaria (G. asiatica,* 54. *G. chilensis,* 55. *G. coronopifolia,* 56. *G. lemaneiformis,* 57. *G. verrucosa,* 58. *G. tenustipitata)*	Limu manauea (Hawaii); agar-agar (Philippines)	direct human food; source of food-grade agar; used in the production of tokoroten	wild, cultured	[30, 31, 39, 42, 43]
59. *Grateloupia turuturu*	-	stir-fried with meat in China	wild	[38, 42]
60. *Halymenia (H. durvillaei,* 61. *H. dilatata,* 62. *H. microcarpa)*	-	source of food-grade carrageenan	wild	[42, 98]
63. *Himanthalia elongata*	-	potential source of dietary fucoxanthin	-	[44]
64. *Kappaphycus (K. alvarezii,* 65. *K. cottonii,* 66. *K. procrusteanum,* 67. *K. striatum*	'Goso', agal-agal (Phil.)	food as pickled salad in Philippines	cultured	[39]
68. *Meristotheca* (*M. papulosa,* 69. *M. procumbens,* 70. *M. senegalensis*)	Tosaka (Japanese)	prepared as salad	wild	[45, 46]
71. *Nemalion vermiculare*	-	prepared as salad in Korea	wild	[25, 27]
72. *Palmaria palmata*	dulse	prepared as salad	wild	[41]
73. *Pterocladia capillacea*	-	used to produce noodle-like gel dish tokoroten	-	[38]
74. *Pyropia* (formerly *Porphyra*) *(P. haitanensis,* 75. *P. umbilicalis,* 76. *P. yezoensis)*	Nori (Japanese), kim or gim (Korean), purple laver	direct food either as raw dried or prepared as stir-fried in Japan, Korea and China	wild, cultured	[30, 31, 38, 41, 42, 47]

(Table 1) cont.....

-	Popular Name(s)	Uses	Source of Biomass	References
77. *Scinaia moniliformis*	-	as seaweed salad in Philippines; source of food-grade carrageenan	wild	[19]

Microalgae

Food-grade microalgae are mainly artificially produced through culture, either under controlled environmental conditions in indoor tanks, in close (tube or flat) bioreactors, or in open outdoor conditions under the natural source of light using open pond systems and raceways (Fig. **1**). There are advantages and disadvantages in both systems depending on the purpose and desired quality of the products. In an indoor system, since environmental factors can be controlled (irradiance, photoperiod, temperature, salinity and nutrients), products are of high quality – i.e., free of contaminants and, often, with customized nutritional content. Its main disadvantage is in terms of cost, especially in scaling up to a production volume that meets the demand of certain markets. Open culture systems, on the other hand, have issues of quality due to airborne contaminants (*e.g.*, bird wastes, dust), quality of water supply, and influence of climatic conditions being common problems. Open culture systems are commonly used in tropical countries where the sun is abundant, while indoor culture systems are common in countries in higher latitudes where solar radiation is limited, thus requiring a great amount of energy and, therefore, cost.

Only a few species of microalgae are utilized as food for direct human consumption, which is mainly manufactured as ingestible tablets, pills or capsules. The three most common microalgal genera are discussed in this chapter:

Spirulina

Spirulina (*Arthrospira*) is a blue-green algal genus with unbranched, spiral-shaped filaments found in both marine, brackish and fresh waters, usually high in carbonate and bicarbonate content with pH of up to 11. *Spirulina* is rich in nutrients – high in protein (60-70% of dry wt.), low fat, high content of vitamins (esp. B12) and essential fatty acid gamma-linoleic acid [6]. Food-grade *Spirulina* is commercially produced using both open ponds and bioreactors and raceways, which, in the 1980s, produced about 18 and 33-tonne dry wt ha^{-1} yr^{-1}, respectively, and culture systems have been continuously improved to produce harvestable biomass at a lesser cost with the use of better technology and use of improved culture media (*e.g.*, Zarrouk's media) [7, 8]. At the currently available culture technology, the biomass of 10 g m^{-2} day^{-1} that could translate to about 30 tonnes ha^{-1} yr^{-1} is achievable, but the industry is looking into a more efficient culture system that could produce quantities meeting the increasing market demand for healthy foods. *Spirulina* is commonly sold in food/health supplement shops.

Fig. (1). The traditional **(A, B)** and thin culture **(C, D)** systems used in microalgal production. **A** – open pond/raceways, **B** – close tube bioreactor, **C** – thin flat plate bioreactor, and **D** – thin cascade raceways. (Photo source: Trends in Biotechnology).

Dunaliella

Dunaliella is a bi-flagellated unicellular, naked (absence of cell wall) green algal genus considered the richest natural source of β-carotene, so that species of these algae appear as orange-red instead of green [10]. Aside from β-carotene, an antioxidant, *Dunaliella* also contains high amounts of fatty acids and protein used in the nutraceutical, pharmaceutical and health food industries. Many species of *Dunaliella* (*e.g.*, *D. salina*, *D. tertiolecta, D. parva*) are produced in large quantities in areas with high solar output, minimal cloudiness, warm climate, and hypersaline waters [11], such as those in Andalusian coast of Spain and Texcoco, Mexico. The downside of the open-cultivation system is product quality and issues of biomass versus quantity of β-carotene, where carotenogenesis is high when growth is least [12].

Chlorella

Chlorella is a green algal genus with species made up of unicells, occurring either singly or in colonies, and are found in both fresh and marine waters. Members of this genus are high in quality protein (40-70% content) and are commonly sold in health supplement shops in the form of tablets. A number of species (*e.g.*, *C. vulgaris, C. pyrenoidosa*, *C. ellipsoidea*) initially maintained in agar medium as tube culture, are cultured mainly in earthen ponds in places like Hawaii, Thailand, and Mexico. *Chlorella* is one of the most important algae with high nutritional value, making it the first 'superfood' brought by astronauts to outer space.

Future Prospects for Microalgae

The need for healthy food with a wide range of medical benefits makes microalgae the food of the future. Due to their simple growth requirements abundant in nature – sunlight, water and nutrients, they are easy to grow for mass production. However, harnessing these requirements entails an innovative approach to obtain biomass of commercial quantity per square meter of surface available in an area. Whilst microalgae are mainly waterborne organisms, their mass production relies mainly on the land-based facility in a standard system that meets food-grade quality generally recognized as safe (GRAS). The prospect of microalgae as a source of nutritionally-rich edible oils, including triglycerides, and as animal protein analogues, make the microalgal culture industry more promising than ever before. New technological innovations in the cultural system are making it possible to increase production in a more efficient and sustainable way. For instance, the introduction of biorefinery technology that is being integrated into the circular bioeconomy is slowly resolving the bottlenecks and challenges of microalgal production through the new techniques of recovering and separating the biomass component, as well as minimizing the production of waste products. However, the downstream processes of harvesting and fractionating organic compounds, which account for 40% of the total cost, remain a challenge [13]. Once these processes are resolved, the single-product capacity of biorefinery could gradually shift towards synthesizing multiple high- and medium-value products. This is taking advantage of the fact that a metabolic imbalance during times of stressful culture conditions, such as nutrient depletion, high irradiance, temperature, salinity, pH, *etc*., can cause the cessation of growth which could redirect the electron flux during the photosystem stage. This then results in the synthesis and hyperaccumulation of certain compounds (*e.g*., lipids, carbohydrates and carotenoids) that are deemed as commercially important. Moreover, certain metabolites, such as antioxidants (*e.g.* phycobiliproteins, carotenoids or astaxanthin), can co-accumulate and be harvested at certain stages and conditions during culture. Multiple products from a single biorefinery system make the biorefinery approach a more sustainable and cost-effective.

Macroalgae (Seaweeds)

Macroalgae are abundant in the marine environment, and thus are also better known as seaweeds. As many species are good sources of food for direct human consumption and are sources of many bioactive compounds for various industrial applications, the term "seaweed" is inappropriate as it connotes undesirable plants ('weeds') that are better burned or disposed of. However, the Chinese character of the English word "sea" (海), which is pronounced as 'kai' or 'umi' in Japanese, depicts it as "mother", while the English word "weed" (藻) pronounced as 'so' in Japanese, which means "tree", is a life-giving creature. Thus, seaweed being used as food for man in various forms can be better called into a much kinder word as "sea-vegetable" [47]. Some species are harvested from the wild (*e.g.*, *Chondrus crispus*, *Meristotheca papulosa*), while some others are produced through cultivation (Table **1**), either through the vegetative method or from spores or gametes. The following species of seaweeds have been cultivated using well-established techniques for a number of years:

Pyropia (formerly *Porphyra*) spp.

Pyropia, or *nori* in Japanese, is one of the most important edible seaweeds in the world, with a well-established cultural technology that was developed in Japan early in the 1940's and well-adopted in China and Korea [30, 31, 38, 42]. The culture of *Pyropia* begins with collecting mature fronds of the seaweed, which produce both female and male gametes at the frond margins. In a concrete tank, the non-motile spermatia, which are produced in large numbers, passively move towards the female gametes and the zygotes that form after fertilization divide to produce carpospores. The carpospores are made to settle and penetrate within oyster shells spread across the bottom of the tank containing sterilized seawater. The carpospores germinate into the Conchocelis stage (first described by British phycologist Kathleen Drew-Baker). This filamentous stage produces conchosporangia from late summer to early autumn. By late September to early October, the conchosporangia start to liberate conchospores that are then made to settle into seeding nets as artificial substrates placed inside the tank. The liberation of conchospores is promoted by treating the cultures with low-temperature seawater (18-20 °C) 5-7 days before seeding. Regulating light intensity to 800-1000 lux with low seawater temperature further promotes conchospore settlement and adhesion into nets. Several nets can accommodate thousands of conchospores that develop into young thalli, which, after two weeks or so, will form monospores that also develop into new thalli unto the seeding nets. Before transplanting, the nets bearing the conchospores are exposed to the air for 15-20 days to discourage the growth of diatoms that mainly affects the growth of the young thalli. The modern cultivation system of *Pyropia* is either a system where nets are hung between poles (pole system) that allows emersion of nets during low tides to maintain foul-free cultivars, or a floating system where nets are kept afloat using buoys. The floating method makes it possible to culture *Pyropia* in deeper waters. As young thalli grow to 1-3 cm in length, the nets are removed from the field and dried on land to reduce the water content of thalli to 20-40% before placing the nets temporarily in cold storage at a temperature of -20°C. These can be returned to the field to continue the cultivation as required, making harvesting of *Pyropia* possible all throughout the year. *Pyropia* is harvested using a harvesting boat equipped with a mechanical scraper and the harvested thalli brought and processed further into a land-based facility. Here the seaweeds are washed in washing machines and the minced product poured into wooden frame, dried into thin sheets in an oven dryer and packed. By automating the production of *Pyropia*, from the nursery to the post-harvest stages of sheeting and drying, Japan has been able to increase its production volume [28, 47]. The strict quality standard being implemented by the nori-producing countries based on the crispiness and "greenness" of *nori* products determines the qualities that are desired by the market.

Pyropia can be eaten raw or used as a wrapper in molded rice (*sushi*), or in some cases, shredded and used as a rice topping. Korea has a unique way of adding value to *Pyropia* (known there as *kim*) by adding chili powder to give it a spicy taste [38].

Saccharina spp.

There are at least seven (7) species of the genus *Saccharina* (formerly *Laminaria*) that are utilized as food and are mainly produced from culture farms in the temperate cold waters of East Asia (China, Japan and Korea) [30, 31, 38, 48, 49] and in some production areas of Europe (Spain, U.K., Sweden) [50 - 52] and North America (U.S., Canada) [53] (Fig. **2**). *Laminaria*, a.k.a. sugar kelp, is cultivated starting with fertile sporophytic thalli collected from the wild. The specialized reproductive branch called sporophyll is excised from the sporophyte and placed into a flask maintained under temperature-controlled conditions. Thousands of spores shed from the sporophyll could then develop into filamentous gametophytes. These microscopic structures release male and female gametes, which can then form zygotes that can be seeded by pouring the contents of the flask into seeding ropes placed inside a tank filled with seawater. The seeding rope bearing the young sporophytes is then cut into short pieces, which are then individually inserted into the twine of culture ropes. They are then deployed in the field to further grow the sporophytes into maturity. This method, developed and referred to as *force-cultivation* in Japan [49], shortens the time of culturing *Saccharina* from two years to only a year. After harvesting, the crops are washed with freshwater, boiled, dried, and the fronds cut into short pieces and packed. The product is known as *kombu* in Japanese, which is used as a roll wrap (usually with tuna chunk inside) and as a source of a taste-enhancing ingredient called 'dashi' used in much Japanese cooking. However, a high amount of iodine is an issue that prevents the consumption of *Saccharina* in western countries [51].

Fig. (2). Harvesting of sugar kelp (*Saccharina*) from a culture ground in a U.S. seaweed farm. (Photocredit: GreenWave/Ron Gautreau/Seaweed Aquaculture | NOAA Fisheries).

Undaria pinnatifida

Undaria is a cold-water brown algal genus that grows on rocky substratum from 1 to 8 m depth. Considered an invasive seaweed in some regions of the world (*e.g.*, New Zealand, Europe, U.S. west coast) [54, 55], *Undaria* is cultivated in Japan as a source of food using methods quite similar to *Saccharina* production but with an additional genetic tweak that enables the culture of this cold-water species (a.k.a. "northern type") into warm waters (a.k.a. "southern type") [37]. The new technique of artificial seeding of zygotes is done by brushing the zygotes directly into a seeding rope which is wound around a PVC frame instead of pouring the zygote suspension into a tank. This takes 6 months to produce young sporophytes that can be grown out into the field, which in turn takes another 2-3 months to reach harvestable size [28]. This method can produce harvestable *Undaria* crops in less than a year. *Undaria* (*wakame* in Japanese) is used as the main ingredient in Japanese *miso* soup and Korean *miyeok-guk* and in other food preparations.

Sargassum spp.

There are at least four species of edible *Sargassum* that are utilized as food in Japan, China, and Korea, namely *S. fusiforme* (formerly *Hizikia fusiformis*), *S. horneri, S. fulvellum,* and *S. naozhouense* [56]. Different species of *Sargassum* are known to have anti-inflammatory [57], antioxidant [58], anti-diabetic, and anti-obesity [59] activities. The culture of *Sargassum* has been developed only relatively recently using techniques adopted from *Saccharina* and *Undaria* culture, where zygotes from gamete-producing thalli collected from the field are seeded onto seeding ropes or seedling collectors placed in the tank and allowed to grow until they develop into young plantlets ready for out-planting in the field [60 - 62] (Fig. **3**).

Fig. (3). A cultivar of *Sargassum siliquosum* in a floating long line to produce harvestable crop. (Photo credit: This author).

Early development of *Sargassum* is very slow and may take a couple of months in the tank, but the hatchery can be shortened by manipulating the temperature and light conditions (irradiance and photoperiod) and addition of fertilizer [63]. The technique for out-planting seedlings of *Sargassum* involves inserting seedling-bearing seeding rope into the culture rope similar to those of *Saccharina* and *Undaria*. Sexual reproduction is a common method used by most aquaculturists, but vegetative techniques have also been attempted [33] involving the use of young plants collected from the natural population that is clipped between the twine of culture ropes. Out-planting takes an additional 3-9 months depending on the species wherein thalli grow following their annual pattern.

Edible *Sargassum,* such as *S. fusiforme* is rich in nutrients but has to be pre-cooked to remove toxic arsenic the seaweed is known to produce in high amounts.

Cladosiphon Okamuranus

Also known as Okinawa *mozuku*, this brown seaweed (under Family Chordariaceae, Order Ectocarpales) which has a soft and delicate thallus, is only produced in the subtropical to tropical waters of southern Japan through mariculture at depths of not more than 3 meters. Its artificial culture begins with sporophytes (macroscopic phase) collected during low-temperature months (October-June). Thalli are placed in seed tanks (6-8 kg/1 tonne tank) made of polycarbonate provided with strong agitation to discharge a substantial amount of zoospores while maintaining water temperature at 24°C [26]. Released zoospores adhere to surfaces of the inner walls of the plastic tank to develop into gametophytes (microscopic phase) that, later on, produce gametes over the summer period (June-October). Towards the autumn period, PVC seed plates (5cm x 10cm, at 300 plates per tank) are introduced into the tanks while the water temperature is maintained between 7.0-30.5°C and culture medium density of 1.032-1.230 (at 15°C) and seawater medium that remains unchanged during the nursery period. This is to monitor adhered gametophytes (seeds) until they are ready to develop into discoidal sporophytes (apothecia). During this period, seed nets are placed into the tanks where microthallic gametophytes begin to grow into macrothallic sporophytes beginning in the autumn period. Seeded nets (arranged in stacks of up to 12 nets) are then transferred into intermediate nursery ground, usually over seagrass beds at less than 1-meter depth. The microthalli develop into juvenile thalli until they are suitable or large enough to be transferred to the main cultivation ground, ideally where there is good water exchange. Here, the direction of water current is important to consider on how the culture nets should be oriented to avoid losses of cultivars. Harvesting is done using a vacuum pump operated from aboard a boat after cultivars reach lengths of about 30 cm (after 80-90 days cultivation). Harvested crops are washed first on shore and again at the processing factory to maintain a high quality that is meant to be consumed as fresh seaweed salad or as a source of phycocolloid fucoidan. The latter is used as an additive in health foods, drinks and cosmetics.

Eucheuma Denticulatum, *Kappaphycus Alvarezii* and *K. striatum*

Not just as a source of carrageenan, but these red seaweeds are also a source of food for direct human consumption in many Southeast Asian countries such as the Philippines and Indonesia from which the bulk of the cultured crops are produced. These seaweeds are cultivated over shallow reefs where vegetative cuttings are tied to fixed or floating lines, producing harvestable crops after 30-45 days. The culture of these species is well described by several authors [39, 64]. Today, eucheumatoids are cultivated in almost all tropical waters of the world from cultivars mostly originating from the Philippines. These relatively hard or cartilaginous seaweeds are prepared into fresh salad or pickles by cutting them into short fragments; and to make the seaweeds softer, they are blanched in hot water for a few minutes, the water is removed, and the seaweeds are then added with vinegar and garnished with dried fish and spices such as tomato, onion, and ginger.

Caulerpa spp.

The green seaweed, *Caulerpa*, where balloon-like vesicles characterize some species, thus the popular name 'sea grapes' or 'green caviar', is known for thallus plasticity, diverse thallus morphologies, including the shape of vesicles. The thallus can be modified by manipulating irradiance, resulting in the seaweed having several varietal names. There are two species of *Caulerpa* known to be utilized as food: *Caulerpa lentillifera* and *C. racemosa* (Fig. **4**), although a third one (*C. peltata*) is considered a subspecies of *C. racemosa*. *C. lentillifera* has been cultivated in the Philippines since the 1960's in earthen ponds with humus soil used for growing milkfish and shrimp. It is also cultured using vegetative cuttings in land-based tanks supplied with clean, filtered seawater. *C. racemosa*, although abundant in the wild in some areas, is also grown in tanks. *Caulerpa* is usually eaten fresh after dipping in vinegar and is also used as decorative sidings in several recipes.

Monostroma (M. nitidum and *M. latissimum*) and *Ulva* spp.

Monostroma and related ulvoid taxa represent a group of foliose green algae that are utilized as food for direct human consumption and for other purposes (like animal fodder, organic fertilizer, *etc*.). The harvesting of *Monostroma* started in Japan, mainly from the natural population growing in brackish waters of river mouths (*e.g*., Shimanto River, in Shikoku Island) and inner bays. The culture of *Monostroma* has been going on for some time, but it is yet to develop into a commercial scale [65] (Kida 1990, as cited in [22] Ohno 1993). Mature vegetative fronds of *Monostroma* undergo gametogenesis in spring to early summer to produce motile haploid, elongated biflagellate gametes. Gametes undergo conjugation to form zygotes (as diploid sporophyte generation) which then increase in size to develop into a spherical body (*Codiolum* stage) in summer. At maturation in autumn, the sporophyte produces quadriflagellate zoospores, which germinate and develop into the foliose fronds. The culture of *Monostroma* follows this cycle involving, first, the collection of spores using spore collectors made of nets placed in the natural grounds where the seaweed is expected to appear in early autumn. Due to the increasing incidence of pollution, this practice of spore collection from natural grounds has been replaced by artificially seeding gametes into seed collectors during neap tides in April [22].

Fig. (4). *Caulerpa racemosa* (left) and *Caulerpa lentillifera* (right) – two of the most common species utilized as food for direct human consumption.

On the other hand, Ulva can be differentiated from *Monostroma* in having two-cell layered thalli (distromatic) and thus appear darker in color than the latter, where the thalli are composed of a single cell layer and thus are lighter green. *Ulva* is also harvested from a natural population in eutrophied waters where loosely attached, if not, floating thalli occur in abundance, sometimes in a proportion that creates environmental problems referred to as 'green tides'. Provided *Ulva* comes from unpolluted water, it is considered as GRAS and is fit for human consumption. The seaweed, however, can be cultivated in controlled environmental conditions to produce food-grade products, such as in land-based tank facilities supplied with nutrient-rich deep ocean water [23] (Fig. **5**). Having an isomorphic alternation of generation, *Ulva* can be artificially cultured using either sporophytic or gametophytic thalli to obtain spores or gametes, respectively. Mature fronds produce the reproductive structures when artificially stimulated using warm water temperature, releasing them to settle on artificial substrates like nets and ropes.

Together, with *Monostroma*, *Ulva* is harvested, cleaned by washing in fresh water and dried to crispiness that makes them suitable for making into fine powder or flakes, to be used as condiments in various food recipes such as the popular 'okonomiyake' or snack food like crackers and cakes (Fig. **6**).

Fig. (5). Land-based culture facility in Cape Muroto, Kochi Prefecture, Japan using nutrient-rich deep ocean seawater supplied to tanks for the production of high-value marine species including seaweeds (*e.g. Ulva prolifera* as shown here). (Photocredit: Alvin Monotilla).

Fig. (6). Mixed *Monostroma* and *Ulva* used as condiments (flakes) applied to popular Japanese snack foods such as 'okonomiyake', chips and cakes.

Gracilaria spp. and *Gelidium* spp.

The red seaweed *Gracilaria*, aside from being one of the main sources of the phycocolloid agar [42, 43], is an edible seaweed consumed directly by coastal people in many parts of the world, either prepared in pickled form (especially in Asia) or as an extract. Agar from *Gracilaria* is used in food and drinks as a gelling agent in desserts, a thickening agent in juices and beverages, and an extender in canned meat and other food products. The seaweed is harvested mainly from natural population in brackish to marine waters, especially in protected bays and lagoons, or cultured vegetatively in earthen ponds and land-based tanks, or propagated using bottom culture techniques in shallow waters (either by tying fragments in sand-filled plastic tubes or inserted directly into loose sandy or sandy-muddy bottom, as practiced in Chile and other parts of the world, or simply spread at the bottom of ponds (Fig. **7**). *Gracilaria* is also used as a component in integrated aquaculture of fish and shrimp to serve as a biofilter of inorganic wastes. High-value product from *Gracilaria* includes bacteriological-grade agar and agarose used for biotechnology applications. Food-grade agar from *Gracilaria* and *Gelidium* is processed in Japan using a traditional method of freeze-thawing the agar extract in the open air (Fig. **8**), after which they are dried and further processed into powder, in flakes, or in strips.

Fig. (7). Typical earthen culture ponds where *Gracilaria* is cultured by spreading fragments directly into the bottom made up of soft substrate.

Fig. (8). Traditional way of drying agar after extraction process in the freezing open air in the mountains of Nagano Prefecture, Japan. (Photo credit: Prof. Masao Ohno).

Capsosiphon Fulvescens

This green algal species is a newly cultivated species in Korea (Sohn 2004), similar in principle to *Saccharina* and *Undaria* culture. The species has been assessed to be high in protein retinol and ascorbic acid [18], which are found to have several health benefits. The polysaccharides extracted from *C. fulvescens* (both crude and fractionated) have been found to strongly stimulate macrophage cell lines, producing considerable amounts of nitric oxide (NO) and prostaglandin E2 (PGE2), suggesting strong immunostimulatory properties, with potential application as immunostimulant [66]. Made into powder (*mesangi* in Korean), the use of this seaweed as an additive in food products is continuously explored in Korea [67 - 69].

Codium spp.

The green algal genus *Codium* has been utilized as food in some countries, especially in Asia. The seaweed is known to be rich in vitamins, amino acids and minerals. The experimental culture of *Codium* (*C. fragile*) has been pioneered in Korea using both sexual and asexual method of reproduction [70]. The seaweed is used as an additive to *kimchi*, a traditional fermented vegetable [20].

GLOBAL PROSPECT OF ALGAL UTILIZATION IN THE FOOD INDUSTRY

Seaweeds and microalgae will continue to be a major source of food and food ingredients for human consumption, which are not only low in calories but high in nutritional content with a lot of health benefits. In 2018, the global macroalgal production was placed at 32.39 million tonnes, with 8 of the top ten major producing countries coming from Asia (China, Indonesia, Republic of Korea, Philippines, People's Democratic Republic of Korea, Japan, Malaysia, Vietnam) [71]. These are mainly seaweeds grown in different agronomic systems with a fast turnover of biomass used as food or as raw materials for the extraction of phycocolloids such as agar, carrageenan and alginates. In addition, global microalgal production, has been estimated to be less than 20,000 tonnes, mainly from raceway ponds [72]. Of the more than 11,000 species listed in Algaebase (www.algaebase.org), only a few hundreds have been recognized as edible species, leaving the majority of species in the list still unexplored and unexploited for their usefulness as either food for humans or as sources of bioactive compounds that have health benefits, such as those found to have anticancer, anti-inflammatory, antimicrobial, antihelminthic, antiobesity, antidiabetic, antilipidemic, anticholesterolemic, and so many other health benefits, not to mention that some species are gaining popularity in the health and wellness industry. On the other hand, some algae have been recognized as sources of feedstock with high nutritional content used in animal production (including aquatic species in aquaculture), while some other algal species are also being recognized as having the convertible biomass to be used as organic fertilizer (*e.g.*, *Ulva*, *Asparagopsis*, kelps) or biostimulants, because of their hormone-like activity (*e.g. Eucheuma/Kappaphycus*, *Sargassum*, *Ascophyllum, Ecklonia*) for agricultural crop production – both contributing to the world's food security. With the human population continuing to expand, especially in less developed countries, and the terrestrial spaces used for agriculture becoming smaller and smaller everyday due to conversion of spaces for housing and industrial uses, algae from the ocean will be the only remaining sustainable food reserve that mankind can exploit in the future. The availability of these algal resources will depend, to some extent, on the health of our oceans and water bodies and on the continuous improvement and innovation in algal culture techniques for the different species to be produced and utilized. For instance, improvements in photobioreactors, such as the development of biorefineries for microalgal production of high-value products, offer a brighter prospect for more microalgal species to be exploited as sources of functional foods. The increasing demand for polysaccharides from seaweeds for food and other applications requires the production of seaweed biomass that is not only of high quality but also of quantity that can be possible in the wider oceanic space. Besides, with warming seas due to global warming and climate change, it is now imperative that seaweed farming be done in areas where water temperature and other conditions are within an optimal range for growth. Indeed, most phycologists agree (*e.g.*, discussion forum of Phyconomy Coalition, www.phyconomy.org) that, to overcome high water temperature in the shallow coastal waters, is to move the farming of seaweeds to the deeper offshore waters where water temperature and salinity conditions are more stable. Of course, this should be carried out within the Exclusive Economic Zone of a particular maritime country or a defined zone within the EEZ such as a 3-mile distance from the shore in the case of the United States [73]. However, the challenge of often rough seas in offshore areas during stormy weather conditions that could destroy culture lines, thus to crop loss, needs to be overcome by constructing a more durable and robust floating structures in the open sea. This has been demonstrated in kelp farming by designs developed in the U.S. (*e.g.* Marine Biomass Program), northern Spain, North Sea, Netherlands (SPAR Buoys and H-frame), Norway (*e.g.* Seaweed Career), Republic of Korea (*e.g.* Tension-Led Plantform), Chile (*e.g.* BioArchitecture Lab), and Faroe Islands (*e.g.* Macroalgal Cultivation Rig) [73]. The high installation cost that hindered the development and execution of offshore seaweed cultivation will have to be offset by the value of the products developed from the cultivated seaweeds. On the other hand, the option of cultivating seaweeds that produce food-grade quality and or high value products can be done in land-based aquaculture tanks supplied with nutrient-rich seawater. Such is the use in Japan of deep nutrient-rich ocean water pumped from more than 300 meters below the surface [74], in the culture of desirable marine species including, among others, seaweeds such as *Ulva prolifera*, *Saccharina* spp. and *Undaria* spp. Using open rectangular tanks supplied with the deep ocean water, growth of *Ulva*, for instance, has increased growth performance of cultivars with the desirable quality of products.

THE VALUE OF ALGAE AS FUNCTIONAL FOOD

The diversity of algae found in the world's water bodies is yet to be fully exploited for their potential as an edible food source. Tropical regions with high algal species diversity could be sources of a large number of edible algae with functionalities that relate to human health (*e.g* [75 - 77].). The literature is full of information showing that algae are excellent sources of functional food with high nutritional properties [78]. Many species have been investigated in experiments with animal models and human cell lines to have strong immunostimulatory (*e.g* [75].), antioxidant (*e.g* [58, 76].), antihypertensive [79, 80], antiobesity [59, 79], antidiabetic [80, 81], anti-inflammatory [82, 83], and antimicrobial properties [76, 84, 85] that address various diseases and health deficiencies. The worldwide increase in lifestyle diseases brought about by a boom in ready-to-eat convenient food which are mostly lacking in nutritional elements, makes algae and seaweeds a better alternative food and food additives for people who are health conscious.

SUSTAINABLE SEAWEED INDUSTRY THROUGH INTELLECTUAL PROPERTY PROTECTION FOR NEW STRAINS OF EDIBLE SEAWEEDS

High on the agenda among countries with an active seaweed industry is the continuous improvement of their genetic stock to support a sustainable industry. Some seaweed producing countries (*e.g.* China, Japan, Republic of Korea) have joined the International Union for the Protection of New Varieties of Plants (UPOV) which, to date, has 77 member countries. For example, Republic of Korea significantly increased its seaweed production after it joined the UPOV on January 7, 2002 (UPOV Publication 423) and enacted a law exercising its intellectual property rights over 19 new varieties of seaweeds [86]. These consist of 13 *Pyropia*, 5 *Undaria*, and one *Saccharina* – three of the most commercially important seaweeds in the world for their use as source of food and polysaccharide extracts. Also known as "plant patent", new plant variety protection requires that the new variety has distinctness, uniformity and stability (DUS) and by affording exclusive rights to the developer of these new varieties encourages further development of more seaweed varieties that could lead to better quality crops. Efforts to develop fast-growing seaweed varieties, disease and pest resistant, and producing high-value products are ongoing at various levels of development in many seaweed-producing countries. Indeed, Korea became a major player in the seaweed industry and is now the third top seaweed producers in just a decade or so, next only to China and Indonesia [71]. New strains of the main source of carrageenan, species of *Kappaphycus*, have been developed both in the Philippines [87] and Brazil [88, 89] to address the carrageenophyte's deteriorating productivity and increasing susceptibility to 'ice-ice' disease. While genetic improvement has been made in microalgae such as *Hematococcus* [90] and *Dunaliella* [91], genetic improvement has also been undertaken for macroalgae, such as in edible seaweeds *Pyropia* (as *Porphyra*) and *Gracilaria,* with the aim at producing high-quality seaweeds with improved productivity, and species that avoid inbreeding and loss of genetic variability [92]. Modern techniques to genetically improve cell lines are available [93] for various seaweed crops to sustain mankind's future source of healthy food.

CONCLUSION

The ocean is indeed a rich source of edible algae, but of the more than a hundred species known to be fit for human consumption, only a few species (*e.g.*, *Spirulina, Chlorella, Pyropia, Saccharina, Undaria*) have been utilized on a commercial scale that has established methods for culture and mass production, either in a land-based facility or in the open sea. Many other species that are known for their gastronomic functionalities are yet to be fully developed to be well appreciated by culinary enthusiasts. I hope that this chapter is able to ignite the curiosity of the uninitiated and help explore further the many uses of algae for all their health benefits for healthy living.

CONSENT FOR PUBLICATION

Not applicable.

CONFLICT OF INTEREST

The author(s) declare no conflict of interest, financial or otherwise.

ACKNOWLEDGMENTS

The author is grateful to Dr. Avinash Mishra for the opportunity in contributing to this this book. Author is also thankful to the anonymous reviewers who helped improve this paper.

REFERENCES

[1] Belasco W. Algae burgers for a hungry world? The rise and fall of *Chlorella* cuisine. Technol Cult 1997; 38(3): 608-34. https://www.jstor.org/stable/3106856 [Internet].

[http://dx.doi.org/10.2307/3106856]

[2] Cherry P, O'Hara C, Magee PJ, McSorley EM, Allsopp PJ. Risks and benefits of consuming edible seaweeds. Nutr Rev 2019; 77(5): 307-29.
[http://dx.doi.org/10.1093/nutrit/nuy066] [PMID: 30840077]

[3] Rengasamy KRR, Mahomoodally MF, Aumeeruddy MZ, Zengin G, Xiao J, Kim DH. Bioactive compounds in seaweeds: An overview of their biological properties and safety. Food Chem Toxicol 2020; 135: 111013.
[http://dx.doi.org/10.1016/j.fct.2019.111013] [PMID: 31794803]

[4] Mouritsen OG, Rhatigan P, Pérez-Lloréns JL. The rise of seaweed gastronomy: phycogastronomy. Bot Mar 2019; 62(3): 195-209.
[http://dx.doi.org/10.1515/bot-2018-0041]

[5] Pereira L. A review of the nutrient composition of selected edible seaweeds.Seaweed: Ecology, nutrient composition and medicinal uses 2011. https://www.researchgate.net/publication/235767788 [Internet]

[6] Vonshak A, Richmond A. Mass production of the blue-green alga Spirulina: An overview. Biomass 1988; 15(4): 233-47. [Internet].
[http://dx.doi.org/10.1016/0144-4565(88)90059-5]

[7] Shi W, Li S, Li G, *et al.* Investigation of main factors affecting the growth rate of *Spirulina.* Optik (Stuttg) 2016; 127(16): 6688-94. [Internet].
[http://dx.doi.org/10.1016/j.ijleo.2016.04.125]

[8] Andrade BB, Cardoso LG, Assis DJ, Costa JAV, Druzian JI, da Cunha Lima ST. Production and characterization of *Spirulina* sp. LEB 18 cultured in reused Zarrouk's medium in a raceway-type bioreactor. Bioresour Technol 2019; 284: 340-8.
[http://dx.doi.org/10.1016/j.biortech.2019.03.144] [PMID: 30954902]

[9] Soni RA, Sudhakar K, Rana RS. Comparative study on the growth performance of *Spirulina platensis* on modifying culture media. Energy Rep 2019; 5: 327-36. [Internet].
[http://dx.doi.org/10.1016/j.egyr.2019.02.009]

[10] Borowitzka MA. Micro-algal biotechnology 1988; 456-65. https://agris.fao.org/agris-search/search.do?recordID=US201300607458 [Internet]

[11] Ben-Amotz A. *Dunaliella* β-Carotene. In: Seckbach J. editor. Enigmatic microorganisms and life in extreme environments. Cellular origin and life in extreme habitats. Dordrecht: Springer. Available from.
[http://dx.doi.org/10.1007/978-94-011-4838-2_31]

[12] Borowitzka MA. The mass culture of *Dunaliella salina.* 1990.http://www.fao.org/3/ab728e/AB728E06.htm [Internet]

[13] Gifuni I, Pollio A, Safi C, Marzocchellia A, Olivieri G. Current bottlenecks and challenges of the microalgal biorefinery. Trends Biotechnol [Internet]. 2019 Oct 20 2018; 37: (3).
[http://dx.doi.org/10.1016/j.tibtech.2018.09.006]

[14] Travieso L, Benítez F, Sánchez E, Borja R, Martín A, Colmenarejo MF. Batch mixed culture of *Chlorella vulgaris* using settled and diluted piggery waste. Ecol Eng 2006; 28(2): 158-65. [Internet].
[http://dx.doi.org/10.1016/j.ecoleng.2006.06.001]

[15] Hadj-Romdhane F, Zheng X, Jaouen P, *et al.* The culture of *Chlorella vulgaris* in a recycled supernatant: Effects on biomass production and medium quality. Bioresour Technol 2013; 132: 285-92.
[http://dx.doi.org/10.1016/j.biortech.2013.01.025] [PMID: 23411460]

[16] Doucha J, Lívanský K. Production of high-density *Chlorella* culture grown in fermenters. J Appl Phycol 2012; 24(1): 35-43. [Internet].
[http://dx.doi.org/10.1007/s10811-010-9643-2]

[17] Moreira A, Cruz S, Marques R, Cartaxana P. The unexplored potential of green macroalgae in aquaculture 2021. [Internet]
[http://dx.doi.org/10.1111/raq.12580]

[18] Hwang EK, Amano H, Park CS. Assessment of the nutritional value of *Capsosiphon fulvescens* (Chlorophyta): developing a new species of marine macroalgae for cultivation in Korea. J Appl Phycol 2008; 20(2): 147-51. [Internet].
[http://dx.doi.org/10.1007/s10811-007-9198-z]

[19] Trono G. The seaweed resources of the Philippines. In: Ohno M, Critchley, Largo DB, Gillespie R, editors. The seaweed resources of the world. Japan: Japan International Cooperation Agency. 1998; pp. 47-61.

[20] Hwang EK, Baek JM, Park CS. Cultivation of the green alga, Codium fragile (Suringar) Hariot, by artificial seed production in Korea. In: Borowitzka MA, Critchley AT, Kraan S, Peters A, Sjøtun K, Notoya M, Editors. Nineteenth International Seaweed Symposium. Developments in Applied Phycology 2007; 2. Dordrecht: Springer.
[http://dx.doi.org/10.1007/978-1-4020-9619-8_3]

[21] Marques R, Cruz S, Calado R, *et al.* Effects of photoperiod and light spectra on growth and pigment composition of the green macroalga *Codium tomentosum*. J Appl Phycol 2021; 33(1): 471-80. [Internet].
[http://dx.doi.org/10.1007/s10811-020-02289-9]

[22] Ohno M. Cultivation of the green alga Monostroma and Enteromorpha "aonori".Seaweed Cultivation and Marine Ranching 1993; 7-15.

[23] Hiraoka M, Oka N. Hiraoka M, Oka N. Tank cultivation of Ulva prolifera in deep seawater using a new "germling cluster" method. J Appl Phycol [Internet]. 2008; 20: 97-102.
[http://dx.doi.org/10.1007/s10811-007-9186-3]

[24] Carl C, de Nys R, Paul NA. Carl C, de Nys R, Paul NA. The seeding and cultivation of a tropical species of filamentous Ulva for algal biomass production. PLOS One [Internet]. 2014; 9(6.)
[http://dx.doi.org/10.1371/journal.pone.0098700]

[25] Peinado I, Girón J, Koutsidis G, Ames JM. Chemical composition, antioxidant activity and sensory evaluation of five different species of brown edible seaweeds. Food Res Int 2014; 66: 36-44. [Internet].
[http://dx.doi.org/10.1016/j.foodres.2014.08.035]

[26] Toma T. Cultivation of the brown alga, Cladosiphon okamuranus "Okinawa mozuku".Seaweed Cultivation and Marine Ranching 1993; 51-6.

[27] Hwang EK, Park CS. Seaweed cultivation and utilization of Korea. Algae 2020; 35(2): 107-21. [Internet].
[http://dx.doi.org/10.4490/algae.2020.35.5.15]

[28] Tanaka K, Ohno M, Largo D. An update on the seaweed resources of Japan. In: Critchley AT, Hurtado A, Pereira L, Cornish M, Largo D, Paul N, editors. Seaweed Resources of the World: A 2020 Vision. Part 4. Botanica Marina 2020; 63(1): 105-17.
[http://dx.doi.org/10.1515/bot-2018-0100]

[29] Sanjeewa KKA, Jeon YJ. Edible brown seaweeds: a review. J Food Bioact 2018; 2: 37-50.
[http://dx.doi.org/10.31665/JFB.2018.2139]

[30] Ohno M, Largo DB. The seaweed resources of Japan. In: Ohno M, Critchley, Largo DB, Gillespie R, editors. The seaweed resources of the world. Japan: Japan International Cooperation Agency, 1998; 1-14.

[31] Wu CY. The seaweed resources of China. In: Ohno M, Critchley, Largo DB, Gillespie R, editors. The seaweed resources of the world. Japan: Japan International Cooperation Agency, 1998; 34-46.

[32] Hwang EK, Park CS, Baek JM. Artificial seed production and cultivation of the edible brown alga, *Sargassum fulvellum* (Turner) *C. Agardh*: developing a new species for seaweed cultivation in Korea. Eighteenth International Seaweed Symposium. 25-31. [http://dx.doi.org/10.1007/s10811-006-9021-2]

[33] Ito R, Terawaki T, Satuita CG, Kitamura H. Cultivation of hiziki, *Sargassum fusiforme*, on culture rope using naturally growing plants. Aquat Sci 2008; 56(1): 97-103. https://www.jstage.jst.go.jp/article/aquaculturesci/56/1/56_97/_pdf [Internet].

[34] Matanjun P, Mohamed S, Mustapha NM, Muhammad K. Nutrient content of tropical edible seaweeds, *Eucheuma cottonii, Caulerpa lentillifera* and *Sargassum polycystum.* J Appl Phycol 2009; 21(1): 75-80. [Internet]. [http://dx.doi.org/10.1007/s10811-008-9326-4]

[35] Hu ZM, Li JJ, Sun ZM, *et al.* Hidden diversity and phylogeographic history provide conservation insights for the edible seaweed *Sargassum fusiforme* in the Northwest Pacific. Evol Appl 2017; 10(4): 366-78. [http://dx.doi.org/10.1111/eva.12455] [PMID: 28352296]

[36] Wang L, Oh JY, Yang HW, Fu X, Kim JI, Jeon YJ. Fucoidan isolated from the popular edible brown seaweed *Sargassum fusiforme* suppresses lipopolysaccharide-induced inflammation by blocking NF-κB signal pathway. J Appl Phycol 2021; 33(3): 1845-52. [Internet]. [http://dx.doi.org/10.1007/s10811-021-02390-7]

[37] Ohno M, Matsuoka M. Undaria cultivation "Wakame".Seaweed cultivation and marine ranching 1993; 41-9.

[38] Sohn CH. The seaweed resources of Korea. In: Ohno M, Critchley, Largo DB, Gillespie R, editors. The seaweed resources of the world. Japan: Japan International Cooperation Agency, 1998; pp. 15-33.

[39] Trono GT, Largo DB. Seaweed Resources of the Philippines. In: Critchley A, Hurtado A, Pereira L, Cornish M, Largo D, Paul N, editors. Seaweed resources of the world: a 2020 vision. Botanica Marina 2019; 62: pp. (5)483-98. [http://dx.doi.org/10.1515/bot-2018-0069]

[40] Avila M, Piel MI, Caceres JH. Cultivation of the red alga *Condracanthus chamissoi: sexual reproduction and seedling production in culture under controlled conditions. J Appl Phycol* 2010; 23: pp. 529-36. [http://dx.doi.org/10.1007/s10811-010-9628-1]

[41] MacArtain P, Gill CIR, Brooks M, Campbell R, Rowland IR. Nutritional value of edible seaweeds. Nutr Rev 2007; 65(12): 535-43. [http://dx.doi.org/10.1111/j.1753-4887.2007.tb00278.x] [PMID: 18236692]

[42] Wang X, He L, Ma Y, *et al.* Economically important red algae resources along the Chinese coast: History, status, and prospects for their utilization. Algal Res 2020; 46: 101817. [Internet]. [http://dx.doi.org/10.1016/j.algal.2020.101817]

[43] Critchley AT. Gracilaria (Gracilariales, Rhodophyta): An economically important agarophyte.Seaweed Cultivation and Marine Ranching 1993; 89-112.

[44] Rajauria G, Foley B, Abu-Ghannam N. Characterization of dietary fucoxanthin from *Himanthalia elongata* brown seaweed. Food Res Int 2017; 99(Pt 3): 995-1001. [http://dx.doi.org/10.1016/j.foodres.2016.09.023] [PMID: 28865626]

[45] Borlongan IA, Terada R, Hurtado A. Concise review of the genus Meristotheca (Rhodophyta: Solieriaceae). J Appl Phycol [Internet]. 2021 Oct. 27, 2020; 33: 167-81. [http://dx.doi.org/10.1007/s10811-020-02288-w]

[46] N'Yeurt ADR. Meristotheca procumbens P. Gabrielsonet et Kraft (Gigartinales, Solieriaceae): an edible seaweed from Rotuma Island. S Pac J Nat Sci 1995; 14: 243-50. https://repository.usp.ac.fj/5277/1/Meristotheca_procumbens_P._Gabrielson_et_kraft_(Gigartinales-S

olieriaceae)-an_edible_seaweed_from_Rotuma_Island.pdf [Internet].

[47] Oohusa T. The cultivation of Porphyra "Nori". In: Ohno M, Critchley AT, editors. Seaweed Cultivation and Marine Ranching. Japan: Japan International Cooperation Agency (JICA), 1993; pp. 57-73.

[48] Nisizawa K. Seaweeds kaiso Bountiful harvest from the seas 2002.

[49] Kawashima S. Cultivation of the brown alga, Laminaria "Kombu".Seaweed Cultivation and Marine Ranching 1993; 25-40.

[50] Peteiro C, Salinas JM, Freire O, Fuertes C. Cultivation of the autoctonous seaweed *Laminaria saccharina* off the Galician coast (NW Spain): production and features of the sporophytes for the annual and biennial harvest. Thalassas 2006; 22: 45-53.

[51] Lüning K, Mortensen L. European aquaculture of sugar kelp (*Saccharina latissima*) for food industries: iodine content and epiphytic animals as major problems. Bot Mar 2015; 58(6): 449-55. [Internet].
[http://dx.doi.org/10.1515/bot-2015-0036]

[52] Forbord S, Skjermo J, Arff J, *et al.* Development of *Saccharina latissima* (Phaeophyceae) kelp hatcheries with year-round production of zoospores and juvenile sporophytes on culture ropes for kelp aquaculture. J Appl Phycol 2012; 24(3): 393-9. [Internet].
[http://dx.doi.org/10.1007/s10811-011-9784-y]

[53] Kim J, Stekoll M, Yarish C. Opportunities, challenges and future directions of open-water seaweed aquaculture in the United States. Phycologia 2019; 58(5): 446-61. [Internet].
[http://dx.doi.org/10.1080/00318884.2019.1625611]

[54] Hay CH, Luckens PA. The Asian kelp *Undaria pinnatifida* (Phaeophyta: Laminariales) found in a New Zealand harbour. N Z J Bot 1987; 25(2): 329-32. [Internet].
[http://dx.doi.org/10.1080/0028825X.1987.10410079]

[55] Minchin D, Nunn J. The invasive brown alga *Undaria pinnatifida* (Harvey) Suringar, 1873 (Laminariales: Alariaceae), spreads northwards in Europe. BioInvasions Rec 2014; 3(2): 57-63. [Internet].
[http://dx.doi.org/10.3391/bir.2014.3.2.01]

[56] Xie EY, Liu DC, Jia C, Chen XL, Yang B. Artificial seed production and cultivation of the edible brown alga *Sargassum naozhouense* Tseng et Lu. J Appl Phycol 2013; 25(2): 513-22. [Internet].
[http://dx.doi.org/10.1007/s10811-012-9885-2]

[57] Wang L, Yang HW, Ahn G, *et al. In vitro* and *in vivo* anti-inflammatory effects of sulfated polysaccharides isolated from the edible brown seaweed, *Sargassum fulvellum.* Mar Drugs 2021; 19(5): 277.
[http://dx.doi.org/10.3390/md19050277] [PMID: 34063550]

[58] Park PJ, Shahidi F, Jeon YJ. Antioxidant activities of enzymatic extracts from an edible seaweed *Sargassum horneri* using ESR spectrometry. J Food Lipids 2004; 11(1): 15-27. [Internet].
[http://dx.doi.org/10.1111/j.1745-4522.2004.tb00257.x]

[59] Murakami S, Hirazawa C, Ohya T, *et al.* The edible brown seaweed *Sargassum horneri* (Turner) *C. Agardh* ameliorates high-fat diet-induced obesity, diabetes, and hepatic steatosis in mice. Nutrients 2021; 13(2): 551.
[http://dx.doi.org/10.3390/nu13020551] [PMID: 33567531]

[60] Le HN, Hughes AD, Kerrison PD. Early development and substrate twine selection for the cultivation of *Sargassum muticum* (Yendo) Fensholt under laboratory conditions. J Appl Phycol 2018; 30(4): 2475-83.
[http://dx.doi.org/10.1007/s10811-018-1459-5] [PMID: 30147239]

[61] Ko SJ, Kim YK, Hong SW, *et al.* Artificial seed production and cultivation of *Sargassum macrocarpum* (Fucales, Phaeophyta). Algae 2020; 35(2): 123-31. [Internet].

[http://dx.doi.org/10.4490/algae.2020.35.5.27]

[62] Largo DB, Diola AG, Rance GMS. Culture of the brown seaweed *Sargassum siliquosum* J. Agardh (Phaeophyceae, Ochrophyta): from hatchery to out-planting. J Appl Phycol 2020; 32(6): 4081-98. [Internet].
[http://dx.doi.org/10.1007/s10811-020-02228-8]

[63] Redmond S, Kim JK, Yarish C, Pietrak M, Bricknell I. Culture of Sargassum in Korea: techniques and potential for culture in the US Orono 2014.seagrant.umaine.edu/extension/korea-aquaculture [Internet]

[64] Dawes CJ, Trono GC Jr, Lluisma AO. Clonal propagation of *Eucheuma denticulatum* and *Kappaphycus alvarezii* for Philippine seaweed farms. Hydrobiologia 1993; 260-261(1): 379-83. [Internet].
[http://dx.doi.org/10.1007/BF00049044]

[65] Kida W. Culture of seaweeds *monostroma*. Mar Behav Physiol 1990; 16(2): 109-31. [Internet].
[http://dx.doi.org/10.1080/10236249009378745]

[66] Karnjanapratum S, Tabarsa M, Cho M, You S. Characterization and immunomodulatory activities of sulfated polysaccharides from *Capsosiphon fulvescens*. Int J Biol Macromol 2012; 51(5): 720-9.
[http://dx.doi.org/10.1016/j.ijbiomac.2012.07.006] [PMID: 22796642]

[67] Jung B-M, Park S-O, Shin T-S. Development and quality characteristics of rice noodles made with added Capsosiphon fulvescens powder. Kor J Cookery Sci [Internet] 2009; 25: 180-1188. https://www.koreascience.or.kr/article/JAKO200916638023874.page

[68] Park BH, You MJ, Cho HS. Quality characteristics of dried noodle containing *Capsosiphon fulvescens* powder. Journal of the East Asian Society of Dietary Life 2015; 25(2): 300-8.
[http://dx.doi.org/10.17495/easdl.2015.4.25.2.300]

[69] Park ID. Quality properties of fish paste containing *Capsosiphon fulvescens* powder. J Korean Soc Food Cult 2016; 31(4): 357-63.
[http://dx.doi.org/10.7318/KJFC/2016.31.4.357]

[70] Hwang EK, Baek JM, Park CS. Artificial seed production using the reproduction methods in *Codium fragile* (Chlorophyta). Hangug Susan Haghoi Ji 2005; 38(3): 164-71.
[http://dx.doi.org/10.5657/kfas.2005.38.3.164]

[71] FAO. The state of world fisheries and aquaculture. Sustainability in action, 2020.
[http://dx.doi.org/10.4060/ca9229en]

[72] Tredici MR, Rodolfi L, Biondi N, Bassi N, Sampietro G. Techno-economic analysis of microalgal biomass production in a 1-ha Green Wall Panel (GWP®) plant. Algal Res 2016; 19: 253-63. [Internet].
[http://dx.doi.org/10.1016/j.algal.2016.09.005]

[73] Bak UG, Gregersen O, Infante J. Technical challenges for offshore cultivation of kelp species: lessons learned and future directions. In: Critchley AT, Hurtado A, Pereira L, Cornish M, Largo D, Paul N, editors. Seaweed Resources of the World: A 2020 Vision. Part 4. Botanica Marina 2020; 63(4): 341-53.
[http://dx.doi.org/10.1515/bot-2019-0005]

[74] Nakasone T, Akeda S. The application of deep sea water in Japan. In: Proceedings of the twenty-eighth UJNR Aquaculture Panel Symposium. Kihei, Hawai'i [Internet]. 1999 Nov pp. 10-2. http://www.aquamaris.org/wp-content/uploads/Usos-de-agua-de-mar-profunda-en-Japon.pdf

[75] Hao H, Han Y, Yang L, *et al.* Structural characterization and immunostimulatory activity of a novel polysaccharide from green alga *Caulerpa racemosa* var *peltata*. Int J Biol Macromol 2019; 134: 891-900.
[http://dx.doi.org/10.1016/j.ijbiomac.2019.05.084] [PMID: 31100398]

[76] Dixit DC, Reddy CRK, Balar N, Suthar P, Gajaria T, Gadhavi DK. Assessment of the nutritive, biochemical, antioxidant and antibacterial potential of eight tropical macro algae along Kachchh coast, India as human food supplements. J Aquat Food Prod Technol 2018; 27(1): 61-79. [Internet].

[http://dx.doi.org/10.1080/10498850.2017.1396274]

[77] Kelman D, Posner EK, McDermid KJ, Tabandera NK, Wright PR, Wright AD. Antioxidant activity of Hawaiian marine algae. Mar Drugs 2012; 10(12): 403-16. [http://dx.doi.org/10.3390/md10020403] [PMID: 22412808]

[78] Shannon E, Abu-Ghannam N. Seaweeds as nutraceuticals for health and nutrition. Phycologia 2019; 58(5): 563-77. [Internet]. [http://dx.doi.org/10.1080/00318884.2019.1640533]

[79] Seca A, Pinto D. Overview on the antihypertensive and anti-obesity effects of secondary metabolites from seaweeds. Mar Drugs 2018; 16(7): 237. [http://dx.doi.org/10.3390/md16070237] [PMID: 30011911]

[80] Awang AN, Ng JL, Matanjun P, Sulaiman MR, Tan TS, Ooi YBH. Anti-obesity property of the brown seaweed, *Sargassum polycystum* using an *in vivo* animal model. J Appl Phycol 2014; 26(2): 1043-8. [Internet]. [http://dx.doi.org/10.1007/s10811-013-0149-6]

[81] Unnikrishnan PS, Jayasri MA. Marine algae as a prospective source for antidiabetic compounds–a brief review. Curr Diabetes Rev 2018; 14(3): 237-45. [http://dx.doi.org/10.2174/1573399812666161229151407] [PMID: 28034290]

[82] Lee JC, Hou MF, Huang HW, *et al.* Marine algal natural products with anti-oxidative, anti-inflammatory, and anti-cancer properties. Cancer Cell Int 2013; 13(1): 55. [http://dx.doi.org/10.1186/1475-2867-13-55] [PMID: 23724847]

[83] Barbalace MC, Malaguti M, Giusti L, Lucacchini A, Hrelia S, Angeloni C. Anti-inflammatory activities of marine algae in neurodegenerative diseases. Int J Mol Sci 2019; 20(12): 3061. [http://dx.doi.org/10.3390/ijms20123061] [PMID: 31234555]

[84] Murugan AC, Karim MR, Yusoff MBM, Tan SH, Asras MFBF, Rashid SS. New insights into seaweed polyphenols on glucose homeostasis. Pharm Biol 2015; 53(8): 1087-97. [http://dx.doi.org/10.3109/13880209.2014.959615] [PMID: 25630358]

[85] Abu-Ghannam N, Rajauria G. Antimicrobial activity of compounds isolated from algae.Functional ingredients from algae for foods and nutraceuticals 2013; 287-306. Internet [http://dx.doi.org/10.1533/9780857098689.2.287]

[86] Hwang EK, Choi HG, Kim JK. Seaweed resources of Korea. In: Critchley AT, Hurtado A, Pereira L, Cornish M, Largo D, Paul N, editors. Seaweed Resources of the World: A 2020 Vision. Part 4. Botanica Marina 2020; 63(4): 395-405. [http://dx.doi.org/10.1515/bot-2020-0007]

[87] Hinaloc LAR, Roleda MY. Phenotypic diversity, growth and sexual differentiation in the progeny of wild *Kappaphycus* (Gigartinales, Florideophyceae). Phycologia [Internet]; 2021. Available from:. [http://dx.doi.org/10.1080/00318884.2021.1946307]

[88] de Paula EJ, Pereira RTL, Ohno M. Strain selection in *Kappaphycus alvarezii* var. *alvarezii* (Solieriaceae, Rhodophyta) using tetraspore progeny. J Appl Phycol 1999; 11: 111-21. [Internet]. [http://dx.doi.org/10.1023/A:1008085614360]

[89] Bulboa C, de Paula EJ, Chow F. Germination and survival of tetraspores of *Kappaphycus alvarezii* var. *alvarezii* (Solieriaceae, Rhodophyta) introduced in subtropical waters of Brazil. Phycol Res 2008; 56(1): 39-45. [Internet]. [http://dx.doi.org/10.1111/j.1440-1835.2008.00483.x]

[90] Kathiresan S, Sarada R. Towards genetic improvement of commercially important microalga *Haematococcus pluvialis* for biotech applications. J Appl Phycol 2009; 21(5): 553-8. [Internet]. [http://dx.doi.org/10.1007/s10811-009-9414-0]

[91] Feng S, Xue L, Liu H, Lu P. Improvement of efficiency of genetic transformation for *Dunaliella salina* by glass beads method. Mol Biol Rep 2009; 36(6): 1433-9.

[http://dx.doi.org/10.1007/s11033-008-9333-1] [PMID: 18792804]

[92] Robinson N, Winberg P, Kirkendale L. Genetic improvement of macroalgae: status to date and needs for the future. J Appl Phycol 2013; 25(3): 703-16. [Internet]. [http://dx.doi.org/10.1007/s10811-012-9950-x]

[93] Charrier BÃ, Rolland E, Gupta V, Reddy CRK. Production of genetically and developmentally modified seaweeds: exploiting the potential of artificial selection techniques. Front Plant Sci 2015; 6: 127. https://www.frontiersin.org/article/10.3389/fpls.2015.00127 [http://dx.doi.org/10.3389/fpls.2015.00127] [PMID: 25852700]

CHAPTER 2

Cultivation Techniques to Induce High-Value Nutraceuticals in Microalgae

Xuan Ji[1], **Luke Webster**[1], **Taylor J. Wass**[1] and **Peer M. Schenk**[1,*]

[1] *Algae Biotechnology Laboratory, School of Agriculture and Food Sciences, The University of Queensland, Brisbane, Queensland 4072, Australia*

Abstract: Microalgae are unicellular autotrophic and heterotrophic microorganisms that can exceed the areal productivity of land plants by order of magnitude. Microalgae are producers of food, medicines, high-value bioactive substances and biofuels. They are highly adaptable, allowing them to thrive in freshwater and saltwater, and can be cultivated on non-arable land. Different cultivation methods have varying effects on the yield of bioactive substances produced by microalgae. As such, selecting the appropriate cultivation conditions is imperative for efficient compound production. This chapter summarizes the current state of microalgae cultivation, techniques for overproduction of high-value nutraceuticals, and future prospects, with the aim of providing the reader with fresh ideas for the cultivation of microalgae for human health.

Keywords: Bioactives, Bioreactor, Compound Production, Cultivation, Microalgae, Microbiome, Nutraceuticals, Omega-3 Fatty Acid, Raceway Pond, Vegan Protein.

INTRODUCTION

Light, carbon dioxide, and water give life to plants. Through photosynthesis, plants convert these three components into important carbohydrates for growth and reproduction. However, traditional agriculture requires fertilizers to promote plant growth during cropping, which has led to the loss of soil nutrients and greenhouse gas emissions from fertilizer production [1]. Due to the scarcity of arable land, human demand for protein and oils has led to extensive deforestation in the Amazon and other tropical rainforests [2]. The substantial increase in demand is likely to cause more damage, and the environment, in turn, affects animal or plant production, creating a negative feedback loop. Nowadays, various extreme weather events and natural disasters show that Earth urgently needs to

* **Corresponding author Peer M. Schenk:** Algae Biotechnology Laboratory, School of Agriculture and Food Sciences, The University of Queensland, Brisbane, Queensland, Australia; Tel: +61 7 33651142; Fax: +61-7-33651699; E-mail: p.schenk@uq.edu.au

restore and maintain the functionality of the ecosystem as soon as possible. Microalgae are likely to have a reduced impact on the environment as they are much more productive (approx. 70 times more protein yields than soybean), and their cultivation does not depend on arable land or freshwater resources [3]. As the materials and equipment for production are simple and easy to obtain, microalgae can be grown in large quantities [4]. Microalgae-derived compounds are considered environmentally friendly agricultural products, and, therefore, the microalgae industry has been developed extensively to help achieve sustainability and production goals.

According to data from the United Nations, more than 2 billion people are affected by micronutrient deficiency, of which 159 million are children [5]. Healthy people must not only meet the recommended daily calorific intake but also ensure a balanced diet [6]. Presently, fish such as salmon are the primary source of dietary omega-3 polyunsaturated fatty acids (PUFAs). However, many people find it difficult to access fish due to geographical location or costs. Fisheries are one of the hardest-hit industries due to unsustainable harvesting practices. Over-fishing has led to a decrease in fish populations and a decrease in diversity [7]. At the same time, marine pollution reduces fish production and quality [8]. Fish provides fish oil (which contains PUFAs) and high-quality protein. These nutrients can also be provided by edible microalgae, which contain a large amount of omega-3 fatty acids and amino acids. Indeed, fish acquire these compounds from microalgae through the food chain. Microalgae can therefore be seen as a food substitute for fish. *Nannochloropsis* spp. and *Chlorella minutissima* are rich in the omega-3 unsaturated fatty acid EPA (eicosapentaenoic acid; 20:5,n-3), while *Cryptomonas* spp. is high in levels of DHA (docosahexaenoic acid; 22:6,n−3). Microalgae also produce polysaccharides, protein, folic acid and vitamins [9]. Cultivation and extraction of these metabolites with high efficiency and low cost as dietary supplements or nutraceuticals is a promising aspect for future research.

MICROALGAE AND THEIR METABOLITES

Microalgae are microscopic organisms that typically source their energy from photosynthesis, living in both fresh and marine water. Microalgae include a wide range of prokaryotic or eukaryotic microorganisms, such as cyanobacteria, diatoms, protists, and unicellular plants [10, 11]. Microalgae have a simple cell structure. They exist in aquatic conditions as single cells, chains or groups. Unlike plants, algae do not possess specialized tissue types, such as roots and stems, which results in reduced growth rates of plants compared to microalgae.

Consistent with their simple cell structure, microalgae growth conditions are usually simple and easy to establish, requiring only light, carbon dioxide, water and nutrients (mainly phosphorus and nitrogen). Microalgae can use photosynthesis to convert light energy into chemical energy for cell development or organic synthesis. The main chemical components produced by microalgae are a variety of lipids, high-quality proteins, and carbohydrates.

As microalgae mature, they gradually accumulate lipids and carbohydrates as energy reserves. The production of PUFAs is typically highest during exponential growth, as they occur predominately in the form of polar lipids in membranes. PUFA content reaches its peak at the end of the exponential phase, before nitrogen deprivation starts to occur, after which saturated fatty acids in the form of triacylglycerides dominate [12]. In some microalgae, such as *Nannochloropsis*, the mass fraction of EPA in the cell increases with time, but the relative proportion of EPA within the lipids remains unchanged unless culture conditions are varied [13].

In order to adapt to the environment, microalgae will quickly adjust the components in the cells during the growth process. Many of these components have multiple biological activities and can be used as food additives, nutritional health products and even pharmaceutical products [14, 15]. Therefore, microalgae are regarded as feedstocks for nutraceuticals with great market potential.

Among the various metabolites obtained from microalgae, fatty acids have received special attention due to their nutritional importance and rich content. Some microalgae species can produce a large amount of PUFAs and have the potential to replace animal oil and vegetable oil. The two most important omega-3 PUFAs, EPA and DHA, have been implicated in the prevention of cardiovascular diseases and are important for infants' brain function, nervous system development and eye development [16, 17]. Regular consumption of omega-3 fatty acids can prevent high blood pressure, hyperlipidemia, arrhythmia, myocardial infarction, and thrombosis by increasing the high-density lipoprotein/low-density lipoprotein (HDL/LDL) ratio and reducing the total cholesterol/HDL ratio. For example, oil extracted from *Crypthecodinium cohnii* contains 40% to 50% of DHA, and from *Nannochloropsis oceanica* contains 25% EPA which, when cultivated at a large scale, can be used as sustainable sources for omega-3 fatty acids [12].

Carotenoids, especially β-carotene and astaxanthin, inositol, folic acid, calcium pantothenate, niacin, and vitamins A, B2, B6, B12, C, and E are common nutraceuticals [18]. The β-carotene in the green alga *Dunaliella salina* is as high as 14% of its dry weight. It is considered a safe vitamin A supplement product, which can maintain the health of the eyes and skin and prevent night blindness. Inositol is extracted from *Nannochloropsis oculata* with a content of about 20.32% in monosaccharide extracts, which has a positive effect on the female reproductive system [19, 20]. Long-term use of folic acid can reduce the risk of stroke, and cardiovascular and cerebrovascular diseases, which is essential for the development of the baby's neural tube. *Picochlorum* sp. contains high concentrations of folic acid, with a total folate content of up to 6.5 mg/100 g [21]. Alkaloids, flavonoids, pigments, phenols, saponins, steroids, tannins, and terpenes are also secondary metabolites of microalgae. They are bioactives for medical treatment, pharmaceuticals, plant growth, insect resistance, and disease resistance [22].

The edible aspect of microalgae has a long history. The earliest record shows that *Spirulina* spp. was used as food in the 16th century [23]. Later, due to food shortages caused by the Second World War, people used microalgae as a source of protein [24]. Today, modern microalgae production has a history of more than 30 years.

MICROALGAE CULTIVATION TECHNIQUES

The cultivation conditions of microalgae are one of the most important factors affecting the quantity and quality of algal nutraceuticals. Cultivation technology can be divided into biological factors and non-biological factors. Biological factors include the incorporation of bacteria, rotifers, *etc*. Non-biological factors include the cultivation system, culture concentration, light, temperature, pH, salinity, nutrients, and plant hormones [25, 26].

Recently, knowledge of microorganisms as biological fertilizers to stimulate the growth and metabolite productivity of microalgae has made substantial progress [13]. Added probiotic bacteria can mutually influence the physiology and metabolism of photosynthetic microalgae. Bacteria can provide nutrients for algal growth and can improve the productivity of some nutraceuticals *via* the secretion of plant growth hormones [27]. This relationship is described in Fig (**1**).

The selection or breeding of bacteria beneficial to the growth of microalgae where bacteria from nature are adapted to co-culture with algae, is a field that has only recently begun to be explored with promising results [28]. Its basic steps are to collect bacteria in nature, eliminate pathogens harmful to the human body, and co-culture microalgae with plant growth promoting bacteria (PGPB) that have the characteristics of promoting plant growth. Bacteria promote the growth of microalgae generally in the following aspects; firstly, symbiotic nitrogen-fixing bacteria such as *Rhizobium, Azorhizobium, Azospirillum, and Enterobacter* can convert the inert nitrogen in the air into ionic nitrogen that can be directly absorbed by algae/plants [29]. Under mixed nutrient growth, the biomass obtained of co-cultured *Chlorella vulgaris* and *Rhizobium* was three times higher than that of axenic algal cultures, and the fatty acid content was 13 times higher [30]. Additionally, phosphate solubilizing bacteria can release the ineffective phosphorus in minerals into a bioavailable form that is absorbed by microalgae under the catalysis of alkaline phosphatase [31]. Some bacteria, such as *Azospirillum,* can induce cell division and differentiation by producing plant hormones, like auxins or gibberellins [32]. Auxins such as indole-3-acetic acid (IAA) can also trigger microalgal lipid synthesis [33]. Moreover, siderophores are an important mediator of the interaction between microalgae and bacteria, which can cause interactions such as cooperation, utilization, and competition between the two. Siderophores are small-molecule iron chelates secreted by bacteria at low intracellular iron concentrations [34]. They can be divided into four categories: catecholate, phenolate, hydroxamate, and carboxylate [35]. Insufficient iron will also reduce the cell's chlorophyll concentration and carotenoid composition and limit photosynthesis, but excessive iron may be toxic to the growth of algae. Bacteria, *Idiomarina loihiensis* RS14, increased the fat content of *Chlorella variabilis* by 7% under iron-deficient conditions [36].

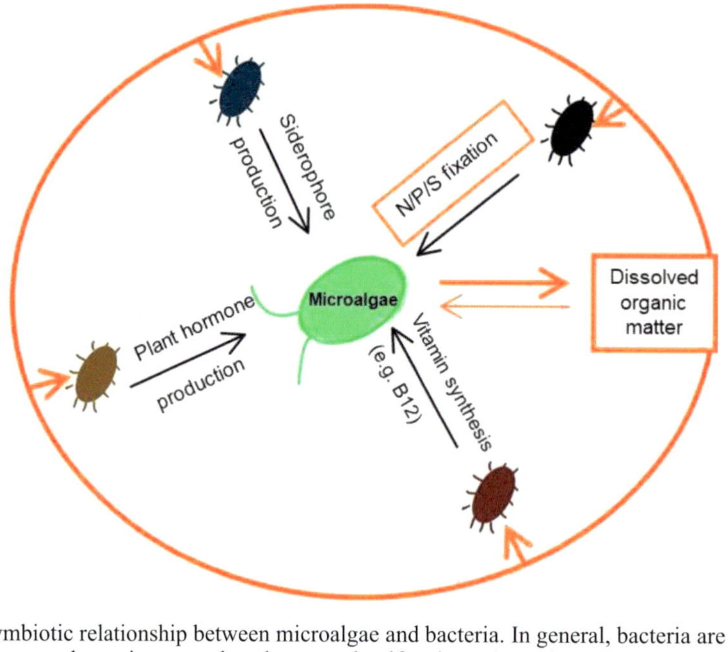

Fig. (1). The symbiotic relationship between microalgae and bacteria. In general, bacteria are able to provide inorganic nutrients such as nitrogen, phosphorus and sulfur through remineralization. Microalgae provide dissolved organic matter in return, which is produced by photosynthesis. Some specific bacteria engage in the secretion of cofactors such as vitamin B or use siderophores to bind iron in the water, which is usually limiting in seawater. At the same time, microalgae support nutrient resources through the turnover of dissolved organic carbon and debris of old cells.

In 1994, Fuqua proposed quorum sensing, that is, during the growth of microbial populations, the increase in population density leads to changes in their physiological and biochemical characteristics, and regulates the specific expression of certain genes as well as physiological functions and lifestyle habits [37]. Some evidence suggests that microbes and bacteria affect each other through the secretion of recognition of quorum sensing molecules [38]. In one study, quorum sensing molecules extracted from wastewater improved the growth of *Chlorella* sp. and had a positive impact on lipid production. To be more specific, the expression of genes encoding enzymes responsible for fatty acid synthesis was significantly up-regulated, and the lipid content, lipid productivity and production all increased [39].

Photobioreactor Systems

One of the most important parameters in algae cultivation is the type of photobioreactor (PBR) system used. The systems can be divided into two kinds: a closed system and an open system, with respective advantages and disadvantages. The most optimal culture system varies depending upon the microalgae used and the desired product.

Closed PBR Systems

Closed PBRs come in various styles, such as vertical tube PBRs, horizontal tube PBRs, stirred tank PBRs and flat panel PBRs. The vertical tube PBR has two forms, bubble column and airlift column. The air or CO_2 pump is located at the bottom of the vertical tube PBR, which mixes algae and culture fluid through the power of bubbles rising and bursting [40]. The net photosynthetic activity of the microalgae can be improved by the "flashing light effect", that is, the light with changing brightness, thereby increasing the productivity of microalgae biomass and overall photosynthesis effectiveness [41]. Because of the air injection, the surface area of the interface between the culture medium and carbon dioxide increases. The top of the column usually has an outlet for air exhaust, which reduces the negative effect of oxygen on the growth of algae. However, the bottom of the bubble column will cause algal cells to settle if insufficient pressure is used [42]. Vertical tube PBRs have a high average photosynthetic efficiency and volumetric productivity because they have a low incident photon flux density on the surface of the reactor. Therefore, the biomass yield per unit area is high, which is suitable for large-scale cultivation, but considering the cost, it is only suitable for the production of high-value products.

Horizontal Tube PBRs

Horizontal tube cultivation systems usually comprise a long horizontal tube stack or ring. The tubes are generally thinner, and the light source has higher permeability. Its shape gives advantages to light-loving microalgae, resulting in high light conversion efficiency. An artificial light source changes the parameters to stimulate the growth and metabolism of algae [43]. Horizontal tube PBRs generally use an automatic evaporative cooling system to control the temperature. In order to avoid the precipitation of algal cells, some PBR tubes require high flow rates, which may cause the rupture of some thin-walled microalgae such as *Spirulina*. Installing an internal mixer can prevent this from happening but at a higher cost. This PBR covers a large area and is suitable for large-scale cultivation in factories. Compared with horizontal tube PBR, the dissolved oxygen content ratio is higher in vertical tube PBRs because the pressure in the system is higher and thus, the solubility is increased [44].

Flat Panel PBRs

The flat panel PBR is the most common PBR used to cultivate microalgae. It is made of transparent materials such as glass, in a flat cube form, and can be inclined at an angle. It is equipped with a culture fluid inlet, outlet, aeration and an exhaust port. The diversion devices divide the PBR shell into an algae liquid rising area and a descending area [45]. The aeration port provides a suitable concentration of carbon dioxide and removes oxygen in time. Under the action of the air-lift driving force, the culture fluid circulates in the reactor [46]. Obviously, the photosynthetic efficiency of this device is very high, and the biomass production efficiency is very impressive. However, it also has some disadvantages, for example, algae can easily adhere to the transparent wall, which negatively affects light penetration. Expansion of cultivation requires materials of larger specifications, which is inconvenient for fabrication, installation and transportation. In addition, PBRs are generally hard to clean and it is also a problem that it is difficult to control the culture temperature. If the surface area is too large, heat dissipation becomes a design constraint, as microalgae typically grow optimally in warmer temperatures.

Stirred Tank PBRs

A stirred tank PBR has the most traditional stirring method, which is similar to a fermentation tank. It is stirred by the mechanical movement of impellers of different sizes and shapes [47]. It can mix algae well and ensure uniform light, nutrient absorption and heat transfer. The use of baffles can increase turbulence which promotes gas exchange. The air introduced at the bottom provides a carbon source for the algae, and artificial light sources are added [44]. The light exposure area is not as large as the first three photobioreactors, which affects the efficiency and absorption of light energy into chemical energy. Currently, this equipment is only used in small-scale cultivation or laboratories probably because of the suboptimal light exposure.

Open-Pond Cultivation Systems

Open-pond microalgae cultivation systems cannot ensure contamination-free operation and are greatly affected by the environment, but it is cheap and suitable for large-scale, low-cost cultivation. As large volumes can be used, i.e., a full-scale 1-acre raceway pond usually contains > 1.2 million L, the water is less prone to temperature fluctuations. Common open-air cultivation methods include unstirred and raceway-type ponds where water circulation is assisted by paddle wheels. Rapid growth of dominant microalgae is needed if algae are to be grown continuously (*e.g.*, *Nannochloropsis* spp.). Alternatively, a batch process can be used where open ponds are inoculated from clean cultures grown in closed PBRs (*e.g.*, *Haematococcus pluvia*lis). To ensure optimal growth, raceway ponds are typically less than 30 cm deep and require frequent nutrient supply, pH regulation by addition of CO_2, and control of grazers (*e.g.*, by ammonia or pH changes). As open raceway ponds are typically quite large, it is important that microalgae in continuous systems are monitored for cell culture density and harvested within the window of exponential growth (i.e., avoid over- and under harvesting). Open raceway ponds are by far the cheapest way of growing microalgae and are popular within the industry for the growth of *Spirulina*, *Chlorella* and *Nannochloropsis*.

Attached Microalgae Cultivation

Attached microalgae cultivation is a relatively new concept. It means that microalgae grow in the form of a biofilm, attached to the surface of a supporting matrix. It is different in structure from the closed systems and open systems above, but the basic growth elements are the same (all need light, water, nutrition, carbon dioxide). This system is easier to harvest and consumes less energy. For example, *Chlorella vulgaris* grown on biofilms saved 99.7% of energy required for dewatering and drying [48]. Since microalgae adhere to the support matrix, the density is very high, with about 15% w/w solids and the water flow cannot take it away, so it is easy to change the concentration and composition of the cultivation medium. For example, attached microalgae cultivation can more easily provide a nitrogen starvation environment that is conducive to promoting lipid synthesis. It is more water-saving, relatively compact, and has higher light energy conversion efficiency and biomass production efficiency [49]. This mode is suitable for spherical or slightly oblong cells of approximately 10–12 μm, such as *Chlorococcum* sp [50]. Some articles show that the protein content of microalgae cultivated in the attachment system is higher, but the results are influenced by various conditions [51]. The cultivation support material of the system is a big problem as it cannot meet the demand of large-scale cultivation, due to its high cost and not being recyclable. In conclusion, attached microalgae cultivation as a novel cultivation system has the potential for large-scale cultivation, and it also meets the needs of contemporary industrial energy conservation and emission reduction.

Light

Light provides the requisite energy for the growth of microalgae. Light intensity, wavelength and light duty cycle will affect the efficiency of microalgae photosynthesis and nutrient production. Different types of microalgae respond differently to light intensity and photoperiod [52]. First of all, different algae have different light compensation points and saturation points, and light intensity has different effects on biomass yields. When light intensity increases, the microalgae will respond to it, increasing productivity and accelerating reproduction. However, when the light intensity exceeds the light saturation point, the algae will be photo-inhibited under stress, and photosynthetic efficiency will decrease. In order to obtain more expensive nutraceuticals at the same time, it is very important to master the light compensation and saturation point of microalgae, as well as their correct mix of colors that make up the "light quality". For example, in an experiment on the influence of light intensity on *Chlorella vulgaris*, an increase of light intensity from 37.5 μmol photon m^{-2} s^{-1} to 62.5 μmol photon m^{-2} s^{-1} resulted in an increase in biomass, but at 100 μmol photon m^{-2} s^{-1}, the phenomenon of light suppression occurred at times [53].

An increase in light duration is related to an increase in specific growth rates. The light and dark responses of photosynthesis have different functions. Light reactions synthesize ATP and NADPH, and dark reactions are responsible for the formation of the photosynthates (sugars). Therefore, the length of illumination greatly influences the content of organic matter in microalgae, such as protein, carbohydrate, lipid, and vitamins. An increase of the illumination time from 8 h to 16 h positively affected the production of *C. vulgaris* biomass [54]. Moreover, with the increase in light time, saturated fatty acids showed an overall upward trend, while monounsaturated fatty acids and polyunsaturated fatty acids showed a downward trend. The synthesis of saturated fatty acids requires a large amount of ATP and NADPH produced by photosynthesis. It is possible that microalgae convert the excessive light energy absorbed into saturated fatty acids in order to protect themselves. As many nutraceuticals are unsaturated fatty acids, attention should be paid to the relationship between growth and the efficiency of unsaturated fatty acid synthesis. Different wavelengths also can be used to stimulate the biosynthesis of specific compounds. For example, blue light was found to be advantageous to stimulate astaxanthin production in *H. pluvia*lis [55].

Flashing Light

High-frequency flashing of 10 kHz LEDs has been used in the cultivation of microalgae in order to enhance photosynthetic efficiency [41]. This occurs through the prevention of non-photochemical quenching, which is a system whereby microalgae dissipate excess light as heat. This operation is to avoid damage caused by light saturation and the accumulation of unsaturated fatty acids, which are diminished by oxidative stress triggered by light saturation [56]. By pulsing the light, the photosynthetic apparatus allows time for the reaction centre to return to a basal state, and thus can accept more photons per unit time compared to continuous illumination. Further, pulsed light is more cost-effective than continuous illumination as the duty cycle (i.e., the period of time a light is on and consuming electricity) is reduced [57, 58]. It has been reported that compared with a constant light source, flashing light can promote biomass accumulation and growth of the microalga *Dunaliella salina* [59]. The vertical photoreactor uses this principle.

Dark Treatment

Microalgae in an organic carbon-rich culture medium do not need to undergo photosynthesis and grow heterotrophically, thereby increasing the per unit area yield of microalgae biomass and lipid. It is reported that lipid contents of *Scenedesmus* sp. increased more than 60%, and *Chlorella* sp. produced 80% more lipids in 11 days of dark growth [60] compared to control under normal conditions. In a recent study by Chua *et al.*, 2020, the EPA content of *N. oceanica* grown under light followed by a dark period increased by 85.9% compared with control cultures, as thylakoids membranes (that harbor phospholipids with EPA) were more abundant [61].

Temperature

The temperature has a great influence on the growth of microalgae and the biochemical composition of their products. Temperature increase, in general, is associated with a significant decrease in protein content and an increase in lipids and carbohydrates, and vice versa. But there are some differences depending on the species. For example, low temperature positively affected the protein content and quality of the algae *Chlorella* UMACC 234, and *Chlorella* UMACC 237. However, the protein quality of *Navicula* UMACC 231 and *Stichococcus* UMACC 238 did not increase at 6-9°C. The percentage of PUFA in *Navicula* UMACC 231, especially the percentage of EPA, also decreased [62].

Cold Treatment

A cold shock is a common approach to stimulating cold-adapted microalgae to produce high biomass products. The optimum growth temperature of most microalgae is generally 25°C, and cold treatment is generally to control the temperature at 5, 10, 15, and 20°C [63]. For example, after 3 days of 15°C cold treatment, the EPA content of *N. oceanica* increased from 26% to 36%, which was higher than that at 30°C. At the same time, the carotenoid content was higher, which may be caused by the combined oxidative stresses caused by changes in the ambient temperature [61]. For *Nannochloropsis salina*, the optimum temperature for producing EPA was found to be 20°C [64].

CONCLUSION

In summary, microalgae cultivation is strongly influenced by different cultivation techniques, and appropriate conditions may induce the best yield of the desired product. Table **1** summarizes examples of how valuable metabolite biosynthesis in microalgae can be induced by various cultivation techniques. With the deepening of research on microalgae, low-cost and environmentally friendly technologies will positively impact the production of microalgae. For industrial cultivation, the main dilemma lies in the high cost of harvesting and processing. In the future, significant cost reductions can be expected from automation and artificial intelligence that learn to stimulate and maximize nutraceutical biosynthesis from microalgae under varying cultural conditions.

Table 1. Examples of induction treatments for the biosynthesis of valuable compounds in microalgae.

Microalgae	Desirable compound	Induction conditions	Reference
Most marine microalgae and some freshwater microalgae (*e.g. Tetraselmis* spp., *Chlorella* spp.)	Triacylglycerides	Nitrogen starvation in the light	[65]
Dunaliella salina	Beta-carotene	Nitrogen starvation in bright light under high salinity; plant hormone methyl jasmonate	[66]
Nannochloropsis spp.	Eicosapentaenoic acid	Rapid cell division (exponential growth) followed by cold and/or dark treatment; growth-promoting bacteria	[12, 13]
*Haematococcus pluvia*lis	Astaxanthin	Nitrogen starvation in bright light (especially blue light)	[55]

CONSENT FOR PUBLICATION

Not applicable.

CONFLICT OF INTEREST

The author(s) declares no conflict of interest, financial or otherwise.

ACKNOWLEDGEMENT

We wish to acknowledge Dr. Skye Thomas-Hall for providing his expertise in useful discussions on the topic of induction of high-value compounds in microalgae.

REFERENCES

[1] Hunter MC, Smith RG, Schipanski ME, Atwood LW, Mortensen DA. agriculture in 2050: recalibrating targets for sustainable intensification. Bioscience 2017; 67(4): 386-91. [http://dx.doi.org/10.1093/biosci/bix010]

[2] de Area Leão Pereira EJ, de Santana Ribeiro LC, da Silva Freitas LF, de Barros Pereira HB. Brazilian policy and agribusiness damage the Amazon rainforest. Land Use Policy 2020; 92: 104491. [http://dx.doi.org/10.1016/j.landusepol.2020.104491]

[3] Correa DF, Beyer HL, Possingham HP, Fargione JE, Hill JD, Schenk PM. Microalgal biofuel production at national scales: Reducing conflicts with agricultural lands and biodiversity within countries. Energy 2021; 215: 119033. [http://dx.doi.org/10.1016/j.energy.2020.119033]

[4] Travieso L, Benitez F, Dupeiron R. Sewage treatment using immobilied microalgae. Bioresour Technol 1992; 40(2): 183-7. [http://dx.doi.org/10.1016/0960-8524(92)90207-E]

[5] Sidhu AVAPS. Securing healthy diets for the billions who cannot afford them would save trillions in costs 2020.https://www.who.int/news/item /13-07-2020-as-mor -e-go-hungry-and- malnutrition-persists-achieving -zero-hunger-by-2030-in-doubt-un-report-warns

[6] Welch RM, Graham RD. A new paradigm for world agriculture: meeting human needs. Field Crops Res 1999; 60(1-2): 1-10. [http://dx.doi.org/10.1016/S0378-4290(98)00129-4]

[7] Rice J. Achieving and Maintaining Sustainable Fisheries 2016. https://www.un.org/zh/ chronicle/ article/20642

[8] Sarker PK, Gamble MM, Kelson S, Kapuscinski AR. Nile tilapia (*Oreochromis niloticus*) show high digestibility of lipid and fatty acids from marine *Schizochytrium* sp. and of protein and essential amino acids from freshwater *Spirulina* sp. feed ingredients. Aquacult Nutr 2016; 22(1): 109-19. [http://dx.doi.org/10.1111/anu.12230]

[9] Edelmann M, Aalto S, Chamlagain B, Kariluoto S, Piironen V. Riboflavin, niacin, folate and vitamin B12 in commercial microalgae powders. J Food Compos Anal 2019; 82: 103226. [http://dx.doi.org/10.1016/j.jfca.2019.05.009]

[10] Zullaikah S, Utomo AT, Yasmin M, Ong LK, Ju YH. 9 - Ecofuel conversion technology of inedible lipid feedstocks to renewable fuel.Advances in Eco-Fuels for a Sustainable Environment. Woodhead Publishing 2019; pp. 237-76. [http://dx.doi.org/10.1016/B978-0-08-102728-8.00009-7]

[11] Bahadar A, Bilal Khan M. Progress in energy from microalgae: A review. Renew Sustain Energy Rev 2013; 27: 128-48.
[http://dx.doi.org/10.1016/j.rser.2013.06.029]

[12] Chua ET, Schenk PM. A biorefinery for Nannochloropsis: Induction, harvesting, and extraction of EPA-rich oil and high-value protein. Bioresour Technol 2017; 244(Pt 2): 1416-24.
[http://dx.doi.org/10.1016/j.biortech.2017.05.124] [PMID: 28624245]

[13] Liu B, Eltanahy EE, Liu H, *et al.* Growth-promoting bacteria double eicosapentaenoic acid yield in microalgae. Bioresour Technol 2020; 316: 123916.
[http://dx.doi.org/10.1016/j.biortech.2020.123916] [PMID: 32768998]

[14] Wells ML, Potin P, Craigie JS, *et al.* Algae as nutritional and functional food sources: revisiting our understanding. J Appl Phycol 2017; 29(2): 949-82.
[http://dx.doi.org/10.1007/s10811-016-0974-5] [PMID: 28458464]

[15] Chew KW, Yap JY, Show PL, *et al.* Microalgae biorefinery: High value products perspectives. Bioresour Technol 2017; 229: 53-62.
[http://dx.doi.org/10.1016/j.biortech.2017.01.006] [PMID: 28107722]

[16] Weiser M, Butt C, Mohajeri M. Docosahexaenoic acid and cognition throughout the lifespan. Nutrients 2016; 8(2): 99.
[http://dx.doi.org/10.3390/nu8020099] [PMID: 26901223]

[17] Chilton F, Dutta R, Reynolds L, Sergeant S, Mathias R, Seeds M. Precision nutrition and omega-3 polyunsaturated fatty acids: A case for personalized supplementation approaches for the prevention and management of human diseases. Nutrients 2017; 9(11): 1165.
[http://dx.doi.org/10.3390/nu9111165] [PMID: 29068398]

[18] Fabregas J, Herrero C. Vitamin content of four marine microalgae. Potential use as source of vitamins in nutrition. J Ind Microbiol Biotechnol 1990; 5(4): 259-63.

[19] Ben H, Ben I, Garrab M, Aly R, Gagnon J, Naghmouchi K. Antimicrobial, antioxidant, cytotoxic and anticholinesterase activities of water-soluble polysaccharides extracted from microalgae Isochrysis galbana and Nannochloropsis oculata. J Serb Chem Soc 2017; 82(5): 509-22.
[http://dx.doi.org/10.2298/JSC161016036B]

[20] Monastra G, Unfer V, Harrath AH, Bizzarri M. Combining treatment with myo-inositol and D -chir- -inositol (40:1) is effective in restoring ovary function and metabolic balance in PCOS patients. Gynecol Endocrinol 2017; 33(1): 1-9.
[http://dx.doi.org/10.1080/09513590.2016.1247797] [PMID: 27898267]

[21] Woortman DV, Fuchs T, Striegel L, *et al.* Microalgae a Superior Source of Folates: Quantification of Folates in Halophile Microalgae by Stable Isotope Dilution Assay. Front Bioeng Biotechnol 2020; 7(481): 481.
[http://dx.doi.org/10.3389/fbioe.2019.00481] [PMID: 32039182]

[22] Singh R, Parihar P, Singh M, *et al.* Uncovering Potential Applications of Cyanobacteria and Algal Metabolites in Biology, Agriculture and Medicine: Current Status and Future Prospects. Front Microbiol 2017; 8(515): 515.
[http://dx.doi.org/10.3389/fmicb.2017.00515] [PMID: 28487674]

[23] Marles RJ, Barrett ML, Barnes J, *et al.* United States pharmacopeia safety evaluation of spirulina. Crit Rev Food Sci Nutr 2011; 51(7): 593-604.
[http://dx.doi.org/10.1080/10408391003721719] [PMID: 21793723]

[24] Bule MH, Ahmed I, Maqbool F, Bilal M, Iqbal HMN. Microalgae as a source of high-value bioactive compounds. Front Biosci (Schol Ed) 2018; 10(2): 197-216.
[PMID: 28930527]

[25] Zhuang LL, Yu D, Zhang J, *et al.* The characteristics and influencing factors of the attached microalgae cultivation: A review. Renew Sustain Energy Rev 2018; 94: 1110-9.

[http://dx.doi.org/10.1016/j.rser.2018.06.006]

[26] Suparmaniam U, Lam MK, Uemura Y, Lim JW, Lee KT, Shuit SH. Insights into the microalgae cultivation technology and harvesting process for biofuel production: A review. Renew Sustain Energy Rev 2019; 115: 109361.
[http://dx.doi.org/10.1016/j.rser.2019.109361]

[27] Lu Y, Xu J. Phytohormones in microalgae: a new opportunity for microalgal biotechnology? Trends Plant Sci 2015; 20(5): 273-82.
[http://dx.doi.org/10.1016/j.tplants.2015.01.006] [PMID: 25697753]

[28] Yao S, Lyu S, An Y, Lu J, Gjermansen C, Schramm A. Microalgae-bacteria symbiosis in microalgal growth and biofuel production: a review. J Appl Microbiol 2019; 126(2): 359-68.
[http://dx.doi.org/10.1111/jam.14095] [PMID: 30168644]

[29] Peccia J, Haznedaroglu B, Gutierrez J, Zimmerman JB. Nitrogen supply is an important driver of sustainable microalgae biofuel production. Trends Biotechnol 2013; 31(3): 134-8.
[http://dx.doi.org/10.1016/j.tibtech.2013.01.010] [PMID: 23414785]

[30] Ferro L, Colombo M, Posadas E, Funk C, Muñoz R. Elucidating the symbiotic interactions between a locally isolated microalga *Chlorella vulgaris* and its co-occurring bacterium Rhizobium sp. in synthetic municipal wastewater. J Appl Phycol 2019; 31(4): 2299-310.
[http://dx.doi.org/10.1007/s10811-019-1741-1]

[31] Prentice MJ, Hamilton DP, Willis A, O'Brien KR, Burford MA. Quantifying the role of organic phosphorus mineralisation on phytoplankton communities in a warm-monomictic lake. Inland Waters 2019; 9(1): 10-24.
[http://dx.doi.org/10.1080/20442041.2018.1538717]

[32] Dao GH, Wu GX, Wang XX, Zhang TY, Zhan XM, Hu HY. Enhanced microalgae growth through stimulated secretion of indole acetic acid by symbiotic bacteria. Algal Res 2018; 33: 345-51.
[http://dx.doi.org/10.1016/j.algal.2018.06.006]

[33] Salama ES, Kabra AN, Ji MK, Kim JR, Min B, Jeon BH. Enhancement of microalgae growth and fatty acid content under the influence of phytohormones. Bioresour Technol 2014; 172: 97-103.
[http://dx.doi.org/10.1016/j.biortech.2014.09.002] [PMID: 25247249]

[34] Naito K, Imai I, Nakahara H. Complexation of iron by microbial siderophores and effects of iron chelates on the growth of marine microalgae causing red tides. Phycol Res 2008; 56(1): 58-67.
[http://dx.doi.org/10.1111/j.1440-1835.2008.00485.x]

[35] Ghavami N, Alikhani HA, Pourbabaei AA, Besharati H. Effects of two new siderophore-producing rhizobacteria on growth and iron content of maize and canola plants. J Plant Nutr 2017; 40(5): 736-46.
[http://dx.doi.org/10.1080/01904167.2016.1262409]

[36] Rajapitamahuni S, Bachani P, Sardar RK, Mishra S. Co-cultivation of siderophore-producing bacteria Idiomarina loihiensis RS14 with Chlorella variabilis ATCC 12198, evaluation of micro-algal growth, lipid, and protein content under iron starvation. J Appl Phycol 2019; 31(1): 29-39.
[http://dx.doi.org/10.1007/s10811-018-1591-2]

[37] Zhou D, Zhang C, Fu L, *et al.* Responses of the microalga Chlorophyta sp. to bacterial quorum sensing molecules (N-acylhomoserine lactones): aromatic protein-induced self-aggregation. Environ Sci Technol 2017; 51(6): 3490-8.
[http://dx.doi.org/10.1021/acs.est.7b00355] [PMID: 28233977]

[38] Borowitzka MA. Chemically-mediated interactions in microalgae The physiology of microalgae. Springer 2016; pp. 321-57.

[39] Zhang C, Li Q, Fu L, Zhou D, Crittenden JC. Quorum sensing molecules in activated sludge could trigger microalgae lipid synthesis. Bioresour Technol 2018; 263: 576-82.
[http://dx.doi.org/10.1016/j.biortech.2018.05.045] [PMID: 29783193]

[40] Brindley Alías C, García-Malea López MC, Acién Fernández FG, Ferníndez Sevilla JM, García

Sánchez JL, Molina Grima E. Influence of power supply in the feasibility of *Phaeodactylum tricornutum* cultures. Biotechnol Bioeng 2004; 87(6): 723-33. [http://dx.doi.org/10.1002/bit.20179] [PMID: 15329930]

[41] Martín-Girela I, Curt MD, Fernández J. Flashing light effects on CO2 absorption by microalgae grown on a biofilm photobioreactor. Algal Res 2017; 25: 421-30. [http://dx.doi.org/10.1016/j.algal.2017.06.008]

[42] Torzillo G, Chini Zittelli G. Tubular Photobioreactors. In: Prokop A, Bajpai RK, Zappi ME, editors. Algal Biorefineries: Volume 2: Products and Refinery Design. Cham: Springer International Publishing 2015; pp. 187-212. [http://dx.doi.org/10.1007/978-3-319-20200-6_5]

[43] Molina E, Fernández J, Acién FG, Chisti Y. Tubular photobioreactor design for algal cultures. J Biotechnol 2001; 92(2): 113-31. [http://dx.doi.org/10.1016/S0168-1656(01)00353-4] [PMID: 11640983]

[44] Singh RN, Sharma S. Development of suitable photobioreactor for algae production – A review. Renew Sustain Energy Rev 2012; 16(4): 2347-53. [http://dx.doi.org/10.1016/j.rser.2012.01.026]

[45] Spier MR, Vandenberghe L, Medeiros ABP, Soccol CR. Application of different types of bioreactors in bioprocesses. Bioreactors: design, properties and applications Nova Science Publishers Inc: New York 2011; pp. 55-90.

[46] Sierra E, Acién FG, Fernández JM, García JL, González C, Molina E. Characterization of a flat plate photobioreactor for the production of microalgae. Chem Eng J 2008; 138(1-3): 136-47. [http://dx.doi.org/10.1016/j.cej.2007.06.004]

[47] Gupta PL, Lee SM, Choi HJ. A mini review: photobioreactors for large scale algal cultivation. World J Microbiol Biotechnol 2015; 31(9): 1409-17. [http://dx.doi.org/10.1007/s11274-015-1892-4] [PMID: 26085485]

[48] Huang Y, Xiong W, Liao Q, *et al.* Comparison of *Chlorella vulgaris* biomass productivity cultivated in biofilm and suspension from the aspect of light transmission and microalgae affinity to carbon dioxide. Bioresour Technol 2016; 222: 367-73. [http://dx.doi.org/10.1016/j.biortech.2016.09.099] [PMID: 27741475]

[49] Ji B, Zhang W, Zhang N, Wang J, Lutzu GA, Liu T. Biofilm cultivation of the oleaginous microalgae Pseudochlorococcum sp. Bioprocess Biosyst Eng 2014; 37(7): 1369-75. [http://dx.doi.org/10.1007/s00449-013-1109-x] [PMID: 24362561]

[50] Cheng P, Ji B, Gao L, Zhang W, Wang J, Liu T. The growth, lipid and hydrocarbon production of *Botryococcus braunii* with attached cultivation. Bioresour Technol 2013; 138: 95-100. [http://dx.doi.org/10.1016/j.biortech.2013.03.150] [PMID: 23612166]

[51] Chew KW, Chia SR, Show PL, Yap YJ, Ling TC, Chang JS. Effects of water culture medium, cultivation systems and growth modes for microalgae cultivation: A review. J Taiwan Inst Chem Eng 2018; 91: 332-44. [http://dx.doi.org/10.1016/j.jtice.2018.05.039]

[52] Bouterfas R, Belkoura M, Dauta A. The effects of irradiance and photoperiod on the growth rate of three freshwater green algae isolated from a eutrophic lake. Limnetica 2006; 25(2): 647-56. [http://dx.doi.org/10.23818/limn.25.43]

[53] Amini Khoeyi Z, Seyfabadi J, Ramezanpour Z. Effect of light intensity and photoperiod on biomass and fatty acid composition of the microalgae, *Chlorella vulgaris*. Aquacult Int 2012; 20(1): 41-9. [http://dx.doi.org/10.1007/s10499-011-9440-1]

[54] Seyfabadi J, Ramezanpour Z, Amini Khoeyi Z. Protein, fatty acid, and pigment content of *Chlorella vulgaris* under different light regimes. J Appl Phycol 2011; 23(4): 721-6. [http://dx.doi.org/10.1007/s10811-010-9569-8]

[55] Ma R, Thomas-Hall SR, Chua ET, *et al.* Gene expression profiling of astaxanthin and fatty acid pathways in *Haematococcus pluvialis* in response to different LED lighting conditions. Bioresour Technol 2018; 250: 591-602.
[http://dx.doi.org/10.1016/j.biortech.2017.11.094] [PMID: 29216572]

[56] Sforza E, Simionato D, Giacometti GM, Bertucco A, Morosinotto T. Adjusted light and dark cycles can optimize photosynthetic efficiency in algae growing in photobioreactors. PLoS One 2012; 7(6): e38975.
[http://dx.doi.org/10.1371/journal.pone.0038975] [PMID: 22745696]

[57] Park KH, Lee CG. Optimization of algal photobioreactors using flashing lights. Biotechnol Bioprocess Eng; BBE 2000; 5(3): 186-90.
[http://dx.doi.org/10.1007/BF02936592]

[58] Matthijs HCP, Balke H, van Hes UM, Kroon BMA, Mur LR, Binot RA. Application of light-emitting diodes in bioreactors: Flashing light effects and energy economy in algal culture *(Chlorella pyrenoidosa)*. Biotechnol Bioeng 1996; 50(1): 98-107.
[http://dx.doi.org/10.1002/(SICI)1097-0290(19960405)50:1<98::AID-BIT11>3.0.CO;2-3] [PMID: 18626903]

[59] Abu-Ghosh S, Fixler D, Dubinsky Z, *et al.* Flashing light enhancement of photosynthesis and growth occurs when photochemistry and photoprotection are balanced in *Dunaliella salina*. Eur J Phycol 2015; 50(4): 469-80.
[http://dx.doi.org/10.1080/09670262.2015.1069404]

[60] Zhang TY, Wu YH, Zhu S, Li FM, Hu HY. Isolation and heterotrophic cultivation of mixotrophic microalgae strains for domestic wastewater treatment and lipid production under dark condition. Bioresour Technol 2013; 149: 586-9.
[http://dx.doi.org/10.1016/j.biortech.2013.09.106] [PMID: 24140357]

[61] Chua ET, Dal'Molin C, Thomas-Hall S, Netzel ME, Netzel G, Schenk PM. Cold and dark treatments induce omega-3 fatty acid and carotenoid production in *Nannochloropsis oceanica*. Algal Res 2020; 51: 102059.
[http://dx.doi.org/10.1016/j.algal.2020.102059]

[62] Teoh ML, Chu WL, Marchant H, Phang SM. Influence of culture temperature on the growth, biochemical composition and fatty acid profiles of six Antarctic microalgae. J Appl Phycol 2004; 16(6): 421-30.
[http://dx.doi.org/10.1007/s10811-004-5502-3]

[63] Renaud SM, Thinh LV, Lambrinidis G, Parry DL. Effect of temperature on growth, chemical composition and fatty acid composition of tropical Australian microalgae grown in batch cultures. Aquaculture 2002; 211(1-4): 195-214.
[http://dx.doi.org/10.1016/S0044-8486(01)00875-4]

[64] Willette S, Gill SS, Dungan B, *et al.* Alterations in lipidome and metabolome profiles of *Nannochloropsis salina* in response to reduced culture temperature during sinusoidal temperature and light. Algal Res 2018; 32: 79-92.
[http://dx.doi.org/10.1016/j.algal.2018.03.001]

[65] Lim DKY, Schuhmann H, Thomas-Hall SR, *et al.* RNA-Seq and metabolic flux analysis of Tetraselmis sp. M8 during nitrogen starvation reveals a two-stage lipid accumulation mechanism. Bioresour Technol 2017; 244(Pt 2): 1281-93.
[http://dx.doi.org/10.1016/j.biortech.2017.06.003] [PMID: 28625352]

[66] Ahmed F, Fanning K, Netzel M, Schenk PM. Induced carotenoid accumulation in *Dunaliella salina* and *Tetraselmis suecica* by plant hormones and UV-C radiation. Appl Microbiol Biotechnol 2015; 99(22): 9407-16.
[http://dx.doi.org/10.1007/s00253-015-6792-x] [PMID: 26201492]

CHAPTER 3

Environment Impact on Algal Food Quality

K. Vasumathi[1,*], K. Sundar[1] and **M. Premalatha[2]**

[1] *Department of Biotechnology, Kalasalingam Academy of Research and Education, Tamil Nadu, India*

[2] *Department of Energy and Environment, National Institute of Technology, Tamil Nadu, India*

Abstract: The demand for nutrient-rich foods is steadily rising. Algal foods have gained a lot of attention lately because of their health benefits and nutritional worth. To overcome the challenges caused by environmental, seasonal, and nutritional changes, algae use a variety of routes and create a variety of phytochemicals. The combination of macro and micro nutrients also changes, which can result in the development of hazardous metabolites in some cases. Proteins, polysaccharides, lipids, and vitamins are the key components produced by algae. Similarly, because of their superior nutritional value, many microalgae are used as a whole-cell food source. The environmental influence on the production of these macronutrients and species employed as functional foods will be discussed in this chapter.

Keywords: Bioaccessibility, Bioactive Compounds, Bioavailability, *Chlorella*, *Dunaliella*, Environmental Conditions, Food Safety, Functional Food, Light Intensity, Lipids, Macroalgae, Microalgae, Micronutrients, Pigments, Polysaccharides, Proteins, Seasonal Variation, *Spirulina*, Temperature, Vitamins.

INTRODUCTION

Algae are a type of organism found in various environments, such as ponds, freshwater, and even seawater. They are capable of photosynthesis because they need sunlight, water, and carbon to grow. Algae are divided into two major categories, microalgae (which can only be seen using a microscope) and macroalgae (which can be seen without a microscope). They are used as food supplements, additives, poultry and animal feed, fertilizers, and bioactive substances [1]. Seaweeds are traditionally used as food in China and Japan. Next in line, the Republic of Korea is found to be the largest consumer of seaweed (22.41 kg/capita/year in 2013) [2] now, it is used in many countries [3].

* **Corresponding author K. Vasumathi:** Department of Biotechnology, Kalasalingam of Academy of Research and Education, Virudhunagar, Tamil Nadu, India; Tel: +91-4563-289042; Fax: +91-4563-289322; E-mail: mathikrishna@gmail.com

The most widely used species are Japanese kelp, *Laminaria japonica, Kappaphycus alvarezii*, and *Eucheuma* spp. *Gracilaria* spp., *Undaria* sp., and *Porphyra* spp [4].

The pigments present in the algae are mainly responsible for the color of the algae. Fucoxanthin pigment is responsible for the coloration of brown algae [5]. So far, 2000 species have been reported worldwide and are multicellular [6, 7]. The species used for human consumption are *Macrocystis pyrifera Undaria, pinnatifida* (wakame), *Hizikia fusiformis* (hijiki), and *Laminaria japonica* (kombu) [8, 9]. *Laminaria digitata* (oarweed), *Ascophyllum nodosum* (rockweed), and *Fucus vesiculosus* (bladder wrack) are used as animal feed [10].

Chlorophyll, β-carotene, and xanthophylls are the pigments responsible for the coloration of green algae [6]. Chlorophyta and Charophyta are the major groups of microalgae; 8000 species have been reported so far under this division [11]. They are widely distributed in nature. Chlorophyta is unicellular and multicellular, whereas Charophyta is only multicellular [12]. *Undaria lactuca, Undaria intestinalis*, *Undaria compressa*, *Chlorella* and *Spirulina* are the most commonly consumed species of green algae.

The pigments Phycobilin are responsible for the coloration of red algae. Around 6000 species have been reported under this category and are multicellular [6]. *Chondrus crispus, Mastocarpus stellatus*, *Porphyra tenera* (nori), *Porphyra tenera*, and *Palmaria palmata* (dulse) are the majorly used seaweeds for the production of carrageenan, sushi, and food [13, 14].

Micro and macroalgae have gained increased attention in recent years as an alternative source of nutrition, health benefits, and medicine. These are rich in essential nutrients, such as proteins, aminoacids, vitamins, lipids, polysaccharides, bioactive compounds, and others [15]. Bioactive compounds are found to reduce the risk of heart disease, hyperglycemia, hyperlipidemic, malnutrition, deficiencies, *etc* [16 - 19]. Algae are regarded as alternate sources of nutrition due to their numerous advantages. The consumption of these products is influenced by a lack of knowledge about the nutritional makeup of algae species, geographic areas, and seasons. In addition, the bioavailability of products and their metabolism in humans also influence the health benefits of algal products. The algal diet is also highly influenced by the growing conditions, harvesting, extraction processes, and storage. As a result, this chapter aims to learn about the elements that determine the nutritional value of algae.

BIOAVAILABILITY OF NUTRITIVE PRODUCTS

Bioavailability is a term that refers to a product's nutritional value. It is characterised as a mix of bioaccessibility (extraction to intestinal transit) and bioactivity (uptake in various tissues to its physiological effects) [20]. *in vitro* research aid in the comprehension of physiological impacts. Nonetheless, there is a deficiency in the intake of many nutritional values. As a result, absorption, distribution, metabolism and excretion (ADME) play an important role in the human body. The digestive process is aided by a variety of enzymes found in the small intestine, including trypsin, chymotrypsin, carboxypeptidases, α-amylase, lipase, and other enzymes [21]. Cooking aids in the digestion of starch-based foods in general [22]. Protein digestion is aided by hydrochloric acid. The lipases break down triacylglycerols. Vitamin B12 absorption is aided by the intrinsic factor generated in the stomach. Although there is still no research on algal starch absorption. It has been reported that fucoxanthin pigment is absorbed in a considerably higher rate than other lipids [23].

Cian *et al.* [24] concluded that undigested protein and carbohydrates help to stimulate the immune responses and induction of microbes. The major impact of the colonial flora on health benefits is reported [25]. Other prevalent issues reported are the characteristics of the algae, harvesting season, and formulation may mainly affect the food properties. The intact cell wall, solubility, and presence of toxic metabolites, biomass and metabolite composition, cultivation conditions, anthropogenic factors, extraction, and preparation are the major causes of the above-said issues [26]. The various biological techniques and simulated gastrointestinal digestion help to understand the bioavailability of many foods, but still, algal foods need more attention [17, 27].

PROTEINS

The aminoacids are classified into essential and non-essential ones. The human body cannot able to synthesize essential amino acids. It will be obtained only from dietary sources. The composition and quality vary; hence, the bioavailability and further digestion become complex [28, 29]. The plant-based proteins lack some of the essential aminoacids, such as histidine, isoleucine, leucine, lysine, methionine, phenylalanine, threonine, tryptophan, and valine [30]. They are harder to digest since they are rich in insoluble polysaccharides. Whereas animal origin proteins are rich in saturated fats and cholesterol, they lead to cardiovascular disease and diabetes. Hence, FAO recommends the usage of plant-based proteins [31]. *Spirulina* is called SCP and is found to be GRAS and *Chlorella*, and many other seaweeds are found to be the richest source of proteins. The other algae are also rich sources of protein with essential aminoacids. FAO reported that the protein composition and quality are on par with the soybean and egg [10, 32]. In most species, tryptophan and lysine are absent [33 - 35], whereas in red algae, leucine and isoleucine are found in low concentration, and in brown algae, methionine, cysteine, and lysine are absent [33, 36]. In most seaweeds, aspartic acid and glutamic acid are found in higher concentrations, whereas cysteine is found in nil or very low concentrations [37]. *Fucus* sp. contains 20 – 40%, and *Ulva* sp. contains 20 – 30% of aspartic acid and glutamic acid [32].

Spirulina platensis and *Chlorella*, which are commercially produced, contain 40 – 70% dry wt protein [38]. Recently, these algae are used as flavor foods to increase their nutritional value, for example, included in salads, baking items, *etc* [38 - 44]. The other algae, such as *Porphyra* sp., *Pyropia* sp., and *Palmaria palmata* are also rich in proteins [13, 33, 45 - 47].

The production of proteins is decreased due to the nutrient limitation in summer stratification [48 - 50]. Seasonal variation in protein content is very less documented. It has been concluded that the harvesting could be done at a high protein concentration [51 - 53]. The sampling and extraction protocols also affect the quality and quantity of proteins. The protein was found to be in ~15% of dry mass in June blade, while in November blade was found as ~23% protein in the Brittany Coast [48]. The protein concentration of four kelps was found to be higher in winter and polysaccharide content in summer than other macroalgae [50]. Shuuluka et al. [54] concluded that estimation procedures must be replicated to quantify the proteins. Generally, a conversion factor of 6.25 for total nitrogen-to-protein concentration will be used for the estimation, *i.e.*, the N present in the sample, is wholly considered as a nitrogen source. It has been reported that the usage of this conversion factor will either over/under estimate the availability of proteins [47, 55 - 57]. They concluded that the seasonal variations may lead to these, hence the usage of the same conversion factor around the year will lead to misinterpretations.

Glutamic acid and aspartic acid are predominant in marine algae [13, 32, 55]. Glutamate is responsible for the taste called umami in kelp [58, 59], and they were found to decrease after harvesting of *Pyropia yezoensis* [60]. Alanine and glycine also add flavor [13]. Taurine, a non-protein found in marine algae, is present in bile acids, and forms complex and lower cholesterol levels in the blood [60, 61]. The protein is less digested due to the presence of inhibitory soluble fibers [10, 62 - 67]. Pretreatments help to overcome this indigestion. Freezing, milling, digesting enzymes, and cell disruption techniques have been used to improve digestion [44, 68, 69]. It has been reported that cooking has increased the bioaccessibility of proteins [70]. Further research is required for the complete digestion of proteins in the human gut.

LIPIDS

Lipids are the main constituents of the cell wall, storage, and signaling [71]. Different types of lipids play different functions, for example, phospholipids and glycolipids are present in the cell membrane, whereas triacylglycerols are used as storage.

Microalgae constitute 40-60% of TAG, which gained more attention to produce biodiesel [72, 73], whereas macroalgae contain phospholipids and glycolipids [13]. Long-chain unsaturated fatty acids are classified into two families: the linoleic acids (n- 6 or omega 6 fatty acids) and the α-linolenic acids (n-3 or Omega 3 fatty acids). Other than that, carotenoids are also used as functional foods [13]. The edible microalgae contain essential fatty acids such as eicosapentaenoic acid and docosahexaenoic acid, have health benefits and are an alternative to fish oils [74]. Generally, humans and animals cant able to produce EPA, hence supplementation is required [75]. Other fatty acid includes stearidonic acid, arachidonic acid is also present in marine algae. The bioactivity of DHA and EPA ensures the supplementation of the nutritional benefits of these fatty acids [74, 76, 77]. It constitutes about ~50 to 100% depending on the matrix [78, 79]. Cardioprotective effects are reported in TAG from *Crypthecodinium cohnii* [80] and *Schizochytrium* sp. [75, 81]. and incorporated in infant products [82 - 84]. reported improvement in cognitive and visual activity, especially in infants. Similarly, DHA from *Schizochytrium* sp. increased the plasma phospholipids and erythrocytes [85, 86]. Algal oil, salmon, and fortified snacks commercially available are reported to have required DHA content [85, 74]. Literature reports that the variation in dosage, duration, and population size resulted in ambiguous results. Both EPA and DHA reported a reduction in TAG levels [87, 88]. DHA also helps to maintain BP, heart rate-lowering effect similar to fish oils, uptake by cells, and platelet function [89]. It has been suggested that the algal DHA supplements help the vegetarian obtain the required essential fatty acids [74, 90]. No side effects have been reported for neither EPA nor DHA when compared to the fish oils [74, 91, 92].

The primary sources for the production of DHA and EPA are the microalgae compared to other zooplankton, fish, and other multicellular organisms and a sustainable way for the production of these essential fatty acids, hence incorporated in the food web [93, 94]. A major factor that affects the production of DHA and EPA is found to be ocean acidification. Increasing CO2 concentrations in the atmosphere severely affects the production of these essential fatty acids [95 - 97]. *Eustigmatophyceae, Trachydiscus minutus* and *Phaeodactylum tricornutum* are reported to produce these essential fatty acids [96]. Recently these have been commercially produced by fermentation process

[98]. Different regions of polar, tropical, and temperature habitats are reported to produce these essential fatty acids [99 - 103]. Among the Red, brown, and green algae, red and brown algae are reported to produce EPA, whereas green algae are reported to produce less EPA but more DHA [104].

Jiang and Gao [105] reported that the DHA and EPA are produced at low temperatures. High temperature is required for biomass production. Hence two-stage cultivation practices seem to be best for the production of these PUFA, a minimum of 12 hours is required to induce the EPA production. The Winter season is best suited to obtain the enhanced yield of EPA. Generally, Marine algae are reported to have low lipid content; hence an increased amount of supplementation of these algae is required to meet the requirement [13, 17, 106]. The composition of the fatty acid has to be quantified to meet the requirements. *Porphyra umbilicalis* and *Palmaria palmata* have some good EPA but still do not meet the body requirements [99, 100, 106, 107]. Hence the researchers are required to concentrate growth conditions to enhance the production of these fatty acids. On the other hand, green algae have been reported to produce more EPA in the Gulf of Finland [108]. Metabolic studies are required to overproduce these fatty acids [109, 110]. reported genes involved in the pathway for the production of PUFA in *Emiliania huxleyi*. The formulation studies have to be concentrated based on the composition to achieve the required dose. These may help us to explore the commercialization of these microalgae.

POLYSACCHARIDES

Generally, the excess energy is stored in the form of polysaccharides in both micro and macroalgae. Humans cannot digest all kindly of polysaccharides [111], and these are called dietary fibers. These may be fermented by the enzymes produced by the microbes [24, 112 - 114]. The cell wall is made up of polyuronides and methylated, acetylated, pyruvylated, and sulfated polysaccharides [115 - 118]. Unknowingly these polysaccharides are consumed as food. These polysaccharides are incorporated as food ingredients based on the recommendation of the regulatory agents [13, 37, 119 - 126]. Dietary fiber constitutes about 20–65% dry weight in *Gracilaria* sp. [127 - 129]. These dietary fibers are reported to have physiological effects (functional value) in addition to nutritional values [61]. Red algae are reported to have 60–85% soluble fibers, whereas 50-55% have green algae [130]. Upon breakdown, these polysaccharides produce acetate, propionate, and butyrate and are oxidized/utilized in signaling pathways [112, 113, 131, 132]. These may enhance/modify the microbiomes [24, 114, 131, 133 - 135] and complexity in their interaction for functional property

[136, 137]. Many *in vitro* studies on animal models proved the bioactivity of these polymers and their ingestion by whole cells or extracts as food supplements [138 - 142].

Commercially available microalgae are *Chlorella*, Arthrospira (*Spirulina*), Dunaliella, Haematococcus, *Scenedesmus*, Aphanizomenon, Odontella, and Porphyridium. Out of which, *Chlorella* is rich in polymers [140], *Spirulina* being rich in proteins. These possess cardioprotective, antidiabetic, antihypertensive, anticancer, anti-inflammatory, *etc* [143 - 147]. They exhibit immunomodulating effects when fed with 5g/day supplement of *Chlorella* [148 - 151]. In addition, extracts of acidic polysaccharides possess anti-tumor properties [145, 152 - 154]. Very limited species are reported to exhibit functional property [155]. Brown algae are the major sources of alginates, whereas sulfated polysaccharides and esterified ones are common among brown, red, and green algae [156, 157]. Recently, Sulphated polysaccharides from red algae have been reported to eliminate the entry of viruses, which can be externally used. This paves the way for researchers to explore the potential of various algae and their bioactive compounds. The structural and functional analysis is yet to be derived for the already reported bioactive compounds, and also, the variation of bioactive with seasonal or environmental factors is to be witnessed [158, 159].

VITAMINS

Vitamins are micronutrients only obtained through diet since they cannot be synthesized in sufficient quantities in the body. Different classes of vitamins have different functions; thus, the deficiencies of each class cause various diseases [160, 161]. These are the starting materials for enzyme production and thus play a critical role in metabolic activities. Literature reports that algae are the richest source of vitamins. *Porphyra umbilicalis, Gracilaria changii,* and *Eisenia arborea* produces vitamin C equivalent to fruits and vegetables [162 - 164].

Marine algae are rich in vitamin B1, B12, A, and E. Kelp is reported to be the richest source of vitamin E [165]. Similarly, *Codium fragile* and *Gracilaria chilensis* are reported to have β-carotene, *Tetraselmis suecica, Isochrysis galbana, Dunaliella tertiolecta*, and *Chlorella stigmatophora* are rich in vitamin A, E, and B and they have a very high concentration compared to conventional foods [166]. They are also rich in other micronutrients which are essentially required for the body.

The pretreatment, extraction procedures, and conditions resulted in variation of these phytochemicals [164, 166, 167]. The quantity may also vary due to seasonal and environmental variations. This is depicted by their experiments showed that the samples collected from different sites resulted in different compositions of β-carotene and vitamin C. The seasonal variation of these vitamins has been reported [164]. He concluded that vitamin C, B2, B1, and A were high during April and low in summer (June, July, August), and provitamin A was found high in January. The reason behind the variation is yet to be derived. Vitamin B12 cannot be synthesized by plants, hence it has to be supplemented for the vegetarians [168 - 171]. Most algae cant able to produce vitamin B12 [168, 172]. Only, prokaryotes can synthesize vitamin B12 [173], whereas the eukaryotes, which have been associated with B12 synthesizing bacteria, can produce it [168]. The concentration may vary with cultivation, harvesting, and processing methods [174, 175].

Various studies reported the bioaccessibility of vitamin B12 in animal models and humans [176, 177], and [178] reported the non-bioaccessibility of these compounds in humans. It has been reported that these are absorbed in the gastrointestinal tract [179]. Whereas the precursors of vitamin B12 (pseudovitamin) bind with intrinsic factors, reducing the absorption in the intestine [180, 181]. *Spirulina* fractions consist majorly of pseudovitamins, thus reducing the function value of these vitamins [174, 182]. *Ulva* and *Pyropia* enhance the concentration of B12 in rats. Hence, using marine algae helps overcome the vitamin B12 deficiency rather than the green algae [182]. concluded that the radioisotope dilution assay (RIDA) helps to differentiate between vitamin and pseudo, whereas the normal assays cannot able to differentiate these. The processing methods would denature/alter the chemical nature and thus affects the uptake. Vitamin B12 of *Pyropia tenera* is denatured by the air-drying process [183]. Lyophilization helps to overcome these problems [176]. The cultivation conditions such as temperature, light, and moisture content also heavily influence the chemical nature of the vitamins [184 - 186]. So far, no detailed studies describe the cause of the factors affecting the vitamins. The bioavailability of Vitamin E is comparatively less than that of vitamin B12.

Vitamin E is available in 8 different forms, in which α- and γ-tocopherols are the most common and active precursors [165]. Lipid-rich foods help to enhance the uptake of fat-soluble vitamins. Many kinds of literature reported on the concentration [163 - 165, 187] or depicted the usage as functional foods [176 - 178]. But more studies are required on the interaction of these above two factors, which gives more insights into the functional foods based on composition [176]. Ecological variations also affect the concentration and type of vitamins [168, 172, 188]. More studies are required to decode the pathways and genes responsible for vitamin production and the complexity of involvement of the environmental parameters of the same.

FOOD SAFETY

Algae are found to be major sources of proteins, carbohydrates, vitamins, minerals, and bioactive compounds [189]. They also accumulate heavy metals. FAO/WHO released legislation on heavy metal concentration in foods that are used for human consumption. The acceptable limit of methylmercury, arsenic, lead, and cadmium was found to be 1.6, 15, 25, and 25 μg/kg body weight. The above-said levels of lead were found to be low in foodstuff and arsenic should be lowered since it causes cancer and other adverse effects based on the CONTAM panel (EFSA). In algae, the concentration of lead and mercury are considered to be safe for human consumption. Arsenic and cadmium were reported to be more in algae [190, 191]. Microalgae were reported to have desired concentration of heavy metals [192]. The nutritional value varies between different batches due to many factors such as seasonal, environmental, nutritional, *etc* [193]. Thus, the quality and quantity are consecutively varied from production to processing of foods [194]. The usage of food-grade reagents also plays an implication in the downstream steps for human consumption, while this is not required in the case of animal feed [195]. And also, the processes and reagents should be environmentally friendly [196]. Downstream processing, such as purification, membrane technologies, *etc*., plays a major role in the extraction of ingredients, which is more prevalently applied in many food applications [197, 198].

ENVIRONMENTAL CONDITIONS

The nature of growth, protein, polysaccharides and fatty acid composition depends on physiology and environmental factors [199]. Algae can able to utilize photosynthetically active radiation of 400-700 nm. Each alga will be specific saturation light intensity beyond which it is determinantal to the pigments as well as the algae, especially under low temperatures [200, 201]. PUFA continues to accumulate in the early phase of growth [202 - 205]. The PUFA is negatively affected by high irradiance [204]. Similarly, higher uptake of carbon was reported with higher irradiances and vice versa at low irradiances. Temperature also affects the biochemical composition of the macromolecules [206 - 208]. Higher temperature and irradiance are found to accumulate more starch rather than proteins and lipids. The fixed carbon is converted to lipids towards the end of ice algal growth. The storage lipids enhanced during the end of the ice age compared to structural lipids [207, 208]. Similarly, nutrient limitation also plays a major role in the accumulation of lipids [209]. Literature reports that food quality and its isotopes are greatly affected by biochemical changes. The influence of conditions on the growth of various microalgae species is discussed as follows.

SPIRULINA

Blue-green algae have been used as traditional food due to their higher nutritive values especially proteins. Among the various species, *S. platensis* and *S. maxima* are majorly used as food. Around 900 tons of food-grade *Spirulina* is produced in many parts of the world. Very high salinity conditions are required for the growth of *Spirulina*; this helps eliminate the contamination of other species. Hence, this could be cultivated under open pond conditions easily [210, 211]. However, the low salinity conditions result in contamination. Generally, pond depth is limited to 30 cm due to the limitation in light penetration. The optimum flow velocity has to be maintained in the ponds to avoid the aggregation of cells and physical damage to the cells. Zarrouk medium and CFTRI medium are employed for the cultivation of *Spirulina*. It requires a high bicarbonate concentration, *Spirulina*. The suitable temperature was found to be 35 – 37 C [212]. Photoinhibition was reported in the outdoor culture during the peak sunlight hours [213]. The best time for harvesting is in the evening hours; this could avoid the loss of biomass due to respiration [214]. reported that there is a constant change in size and shape of the cell due to seasonal variations. The whole cell is used as a food supplement; hence quality is required for human consumption (IS-12895, Table **1**). Phaeophorbides are reported in the culture, *i.e.*, the destruction of pigments due to which nutritive value reduces and thus causes health defects. This may be due to the increased light conditions and decrease in pH. Further food safety studies have been extensively studied in *Spirulina* [215 - 219]. They are rich in vitamins, minerals, and essential aminoacids except for methionine and cysteine [220].

Table 1. Quality of *spirulina* as per indian standard 12895.

-	Indian Standard IS 12895: 1990 (Amendment No.1, June 1991)
Microbial Contamination	-
Coliform (or E.Coli)	Negative
Salmonella	Negative
Shigella	Negative
Staphylococcus	Negative
Mould	-
Heavy metal contaminants	-
Lead	<2.5 μg/g
Mercury	<0.1 μg/g
Cadmium	<1.0 μg/g
Arsenic	< 1.1 μg/g
Other contaminants	-

(Table 1) cont.....

-	Indian Standard IS 12895: 1990 (Amendment No.1, June 1991)
Unidentified insect fragments	-
Rodent hairs	-
Appearance and texture	-
Moisture	<9%
Total ash (on a dry basis)	<9%
Acid-insoluble ash (on dry basis)	<0.5%
Protein (on dry basis)	>55%
Phaeophorbide (total)	-
Appearance	-
Color	-
Taste	-

DUNALIELLA SALINA

Dunaliella Salina is the potential source for the production of β-carotene under large-scale cultivation. Since, it requires high salinity for growth, accumulates high pigments, and thus provides higher returns. The USA and Australia are the largest producers of *DUNALIELLA SALINA*. Australian conditions are well suited for the production of β-carotene. Fed-batch mode of cultivation has opted for the production. The low salinity only leads to contamination. Rain will lead to the low salinity mode, hence should be avoided.

CHLORELLA

The single-cell green alga *Chlorella* has been used as a nutritional supplement. Mainly *C. vulgaris, C. ellipsoidea,* and *C. pyrenoidosa* were cultivated on a large scale. It can be cultivated in any mode photoautotrophically, mixotrophically and heterotrophically [221]. L-Ascorbic acid is produced heterotrophically using C. *pyrenoidosa* [222]. The major drawback in the cultivation of *Chlorella* is the cost involved in harvesting. Centrifugation is the only best technique, but handling large volumes is a problem.

OTHER ALGAE

Scientists are involved in screening algae to be used as food sources or as sources of nutritional supplements. Becker and Venkataraman [223, 224] have extensively researched *Scenedesmus* sp. as an alternative for the protein source. But it requires more cost. *Scenedesmus* is cultivated on a large scale in the Czech Republic at Trebon. They use the slope for uniformly distributing the culture to expose the cells to light, which is collected at the tank. The cultures are circulated only during the daytime, and during the night, it is stored in tanks. An increased yield was obtained in this system than in the conventional paddle wheel.

Scenedesmus was reported to produce PUFAs. Marine diatoms, dinoflagellates produce these fatty acids [225 - 227]. Heterotopic cultivation enhances the yield and helps to overcome the light requirements. But the macromolecule composition varies [228]. *Crypthecodinium cohnii* is one of the prominent species for the production of DHA [229]. Thraustochytrids (organisms with a close affinity to the red and brown algae), *Schizochytridium* spp [230]. and *Thraustochytrium* spp [231, 232]. also, produce DHA.

CULTIVATION CONDITIONS

The major drawback is the usage of the cost involved in the cultivation of microalgae for food and nutritional supplement. Open ponds are cheaper, but it has major disadvantages. Limited for few species, contamination, can not able to maintain condition which leads to the destruction of nutritional value, carbon dioxide loss due to wind and low productivity [233]. Hence, closed photobioreactors are required for the cultivation of these algae [234]. The basic requirements for algae are light, nutrients, and carbon sources. The food-grade chemicals for nutrients and carbon sources can be supplemented easily. The major cost is involved in achieving the required light intensity. Recently, heterotrophic cultivation has been used to overcome these problems [235]. However, all the species could not be grown under heterotrophic conditions, and the nutritive composition varies with the cultivation conditions.

The various designs have been employed for the cultivation of microalgae, and the following are commercially available in the market:

1. Flat plate reactors of various designs [236 - 238].

2. Stirred tank reactors with internal illumination [239 - 241]

3. Tower reactors with internal fiber-optic illumination [242, 243].

4. Tubular reactors [244, 245].

Out of these, flat plate reactors have resulted in greater yield but require more land area. Whereas the tubular reactor is widely used. A decrease in the path length enhances the yield in the reactor. Generally, optimum pathlength was found to be 500 and 250 mm. PVC, glass, and silicon materials are used for the construction of these reactors. The cultures are continuously circulated; hence, adhesions are also avoided in these reactors. Heat exchangers, Acid-base pumps are connected to control the temperature and pH of the medium. Gas exchange systems are incorporated to remove the oxygen produced during photosynthesis. Mass flow meters are used for CO_2 supplementation. Algae can able to fix NOx and Sox. Harvesting at the right concentration also helps to enhance productivity [246]. *Spirulina platensis, Chlorella* spp., *Porphyridium cruentum, Phaeodactylum tricornutum*, Tetraselmis spp., *Isochrysis* spp., *Chaetoceros muelleri, C. calcitrans, Pavlova lutheri, Chroomonas salina, Haematococcus pluvialis,* and *Anabaena* sp. were the species successfully cultivated in the closed bioreactors. The major advantages of closed systems include avoidance of contamination, higher biomass, and nutrition value of food. Due to the controlled conditions, the biochemical composition quality and quantity could be easily achieved. Closed photobioreactors help to achieve better food quality along with specific bioactive compounds.

CONCLUSION

Researchers have paid attention to very few species and explored their nutritive value, which creates the evidence to use as functional foods. The above chapter depicts the variation of these nutritive values due to seasonal, environmental, nutrient levels, and cultivation conditions. In many cases, uncertain results are obtained *In vivo* studies. This could be overcome with a better assessment of these functional foods. The major challenges prevailing are the quantification of bioavailability and bioaccessibility in the digestive system. Furthermore, research is required on the ADMET process of these algae in human models, from absorption in epithelial cells to elimination, structural and functional elucidation, and interaction in the human metabolism has to be derived.

CONSENT FOR PUBLICATION

Not applicable.

CONFLICT OF INTEREST

The authors declare declare no conflict of interest, financial or otherwise.

ACKNOWLEDGEMENT

Declared none.

REFERENCES

[1] Henrikson R. Earth food spirulina. Laguna Beach, CA: Ronore Enterprises, Inc, 187: 1989.

[2] Food and Agriculture Organization. Food Balance Sheets; Food and Agriculture Organisation of the United Nations: Rome, Italy, 2013.

[3] FAO Inland Water Resources and Aquaculture Service—Review of the state of world aquaculture., 2003. FAO Fisheries Department. Circular no. 886. Rev. no. 2. Rome, 95.

[4] Food and Agriculture Organization. The State of World Fisheries and Aquaculture: Opportunties and Challenges; Food and Agriculture Organisation of the United Nations: Rome, Italy, 2014.

[5] Kadam SU, Tiwari BK, O'Donnell CP. Application of novel extraction technologies for bioactives from marine algae. J Agric Food Chem 2013; 61(20): 4667-75. [http://dx.doi.org/10.1021/jf400819p] [PMID: 23634989]

[6] Cock JM, Sterck L, Rouzé P, *et al.* The Ectocarpus genome and the independent evolution of multicellularity in brown algae. Nature 2010; 465(7298): 617-21. [http://dx.doi.org/10.1038/nature09016] [PMID: 20520714]

[7] Hoek C, Mann D, Jahns HM, Jahns M. Algae: an introduction to phycology. Cambridge University Press 1995.

[8] Foster MS, Schiel DR. The ecology of giant kelp forests in California: a community profile. 2017.

[9] Fitzgerald C, Gallagher E, Tasdemir D, Hayes M. Heart health peptides from macroalgae and their potential use in functional foods. J Agric Food Chem 2011; 59(13): 6829-36. [http://dx.doi.org/10.1021/jf201114d] [PMID: 21574559]

[10] Fleurence J. Seaweed proteins. Trends Food Sci Technol 1999; 10(1): 25-8. [http://dx.doi.org/10.1016/S0924-2244(99)00015-1]

[11] Karol KG, McCourt RM, Cimino MT, Delwiche CF. The closest living relatives of land plants. Science 2001; 294(5550): 2351-3. [http://dx.doi.org/10.1126/science.1065156] [PMID: 11743201]

[12] McCourt RM, Delwiche CF, Karol KG. Charophyte algae and land plant origins. Trends Ecol Evol 2004; 19(12): 661-6. [http://dx.doi.org/10.1016/j.tree.2004.09.013] [PMID: 16701329]

[13] Holdt SL, Kraan S. Bioactive compounds in seaweed: functional food applications and legislation. J Appl Phycol 2011; 23(3): 543-97. [http://dx.doi.org/10.1007/s10811-010-9632-5]

[14] Mouritsen OG, Dawczynski C, Duelund L, Jahreis G, Vetter W, Schröder M. On the human consumption of the red seaweed dulse (*Palmaria palmata* (L.) Weber & Mohr). J Appl Phycol 2013; 25(6): 1777-91. [http://dx.doi.org/10.1007/s10811-013-0014-7]

[15] Stévant P, Rebours C, Chapman A. Seaweed aquaculture in Norway: recent industrial developments and future perspectives. Aquacult Int 2017; 25(4): 1373-90. [http://dx.doi.org/10.1007/s10499-017-0120-7]

[16] Khotimchenko SV, Kulikova IV. Lipids of different parts of the lamina of *Laminaria japonica* Aresch 2000. [http://dx.doi.org/10.1515/BOT.2000.008]

[17] Maehre HK, Malde MK, Eilertsen KE, Elvevoll EO. Characterization of protein, lipid and mineral contents in common Norwegian seaweeds and evaluation of their potential as food and feed. J Sci Food Agric 2014; 94(15): 3281-90. [http://dx.doi.org/10.1002/jsfa.6681] [PMID: 24700148]

[18] Sánchez-Machado DI, López-Cervantes J, López-Hernández J, Paseiro-Losada P. Fatty acids, total lipid, protein and ash contents of processed edible seaweeds. Food Chem 2004; 85(3): 439-44. [http://dx.doi.org/10.1016/j.foodchem.2003.08.001]

[19] Cordain L, Eaton SB, Miller JB, Mann N, Hill K. The paradoxical nature of hunter-gatherer diets: meat-based, yet non-atherogenic. Eur J Clin Nutr 2002; 56(S1) (Suppl. 1): S42-52. [http://dx.doi.org/10.1038/sj.ejcn.1601353] [PMID: 11965522]

[20] Carbonell-Capella JM, Buniowska M, Barba FJ, Esteve MJ, Frígola A. Analytical methods for determining bioavailability and bioaccessibility of bioactive compounds from fruits and vegetables: A review. Compr Rev Food Sci Food Saf 2014; 13(2): 155-71. [http://dx.doi.org/10.1111/1541-4337.12049] [PMID: 33412647]

[21] Gropper SS, Smith JL. Advanced nutrition and human metabolism Cengage Learning 2012.

[22] Hardy K, Brand-Miller J, Brown KD, Thomas MG, Copeland L. The importance of dietary carbohydrate in human evolution. Q Rev Biol 2015; 90(3): 251-68. [http://dx.doi.org/10.1086/682587] [PMID: 26591850]

[23] Peng J, Yuan JP, Wu CF, Wang JH. Fucoxanthin, a marine carotenoid present in brown seaweeds and diatoms: metabolism and bioactivities relevant to human health. Mar Drugs 2011; 9(10): 1806-28. [http://dx.doi.org/10.3390/md9101806] [PMID: 22072997]

[24] Cian R, Drago S, de Medina F, Martínez-Augustin O. Proteins and carbohydrates from red seaweeds: evidence for beneficial effects on gut function and microbiota. Mar Drugs 2015; 13(8): 5358-83. [http://dx.doi.org/10.3390/md13085358] [PMID: 26308006]

[25] Duffy LC, Raiten DJ, Hubbard VS, Starke-Reed P. Progress and challenges in developing metabolic footprints from diet in human gut microbial cometabolism. J Nutr 2015; 145(5): 1123S-30S. [http://dx.doi.org/10.3945/jn.114.194936] [PMID: 25833886]

[26] Sensoy I. A review on the relationship between food structure, processing, and bioavailability. Crit Rev Food Sci Nutr 2014; 54(7): 902-9. [http://dx.doi.org/10.1080/10408398.2011.619016] [PMID: 24499069]

[27] Moreda-Piñeiro J, Moreda-Piñeiro A, Romarís-Hortas V, *et al. In-vivo* and *in-vitro* testing to assess the bioaccessibility and the bioavailability of arsenic, selenium and mercury species in food samples. Trends Analyt Chem 2011; 30(2): 324-45. [http://dx.doi.org/10.1016/j.trac.2010.09.008]

[28] Boisen S, Eggum BO. Critical evaluation of *in vitro* methods for estimating digestibility in simple-stomach animals. Nutr Res Rev 1991; 4(1): 141-62. [http://dx.doi.org/10.1079/NRR19910012] [PMID: 19094329]

[29] Wells ML, Potin P, Craigie JS, *et al.* Algae as nutritional and functional food sources: revisiting our understanding. J Appl Phycol 2017; 29(2): 949-82. [http://dx.doi.org/10.1007/s10811-016-0974-5] [PMID: 28458464]

[30] Young VR, Pellett PL. Plant proteins in relation to human protein and amino acid nutrition. Am J Clin Nutr 1994; 59(5) (Suppl.): 1203S-12S. [http://dx.doi.org/10.1093/ajcn/59.5.1203S] [PMID: 8172124]

[31] Food and Agriculture Organization; World Health Organization. Protein Quality Evaluation—Report of Joint FAO/WHO Expert Consultation 1991.

[32] Fleurence J. The enzymatic degradation of algal cell walls: a useful approach for improving protein accessibility? J Appl Phycol 1999; 11(3): 313-4. [http://dx.doi.org/10.1023/A:1008183704389]

[33] Dawczynski C, Schubert R, Jahreis G. Amino acids, fatty acids, and dietary fibre in edible seaweed products. Food Chem 2007; 103(3): 891-9. [http://dx.doi.org/10.1016/j.foodchem.2006.09.041] [PMID: 26065750]

[34] Kolb N, Vallorani L, Stocchi V. Chemical composition and evaluation of protein quality by amino acid score method of edible brown marine algae arame (*eisenia bicyclis*) and hijiki (*hijikia fusiforme*). Acta Aliment 1999; 28(3): 213-22. [http://dx.doi.org/10.1556/AAlim.28.1999.3.1]

[35] Volkmann H, Imianovsky U, Oliveira JLB, Sant'Anna ES. Cultivation of Arthrospira (spirulina) platensis in desalinator wastewater and salinated synthetic medium: protein content and amino-acid

profile. Braz J Microbiol 2008; 39(1): 98-101.
[http://dx.doi.org/10.1590/S1517-83822008000100022] [PMID: 24031187]

[36] Mišurcová L, Buňka F, Vávra Ambrožová J, Machů L, Samek D, Kráčmar S. Amino acid composition of algal products and its contribution to RDI. Food Chem 2014; 151: 120-5.
[http://dx.doi.org/10.1016/j.foodchem.2013.11.040] [PMID: 24423510]

[37] MacArtain P, Gill CIR, Brooks M, Campbell R, Rowland IR. Nutritional value of edible seaweeds. Nutr Rev 2007; 65(12): 535-43.
[http://dx.doi.org/10.1111/j.1753-4887.2007.tb00278.x] [PMID: 18236692]

[38] Gantar M, Svirčev Z. Microalgae and cyanobacteria: food for thought 1. J Phycol 2008; 44(2): 260-8.
[http://dx.doi.org/10.1111/j.1529-8817.2008.00469.x] [PMID: 27041182]

[39] Lubitz JA. The Protein Quality, Digestibility, and Composition of Algae, *Chlorella* 71105. J Food Sci 1963; 28(2): 229-32.
[http://dx.doi.org/10.1111/j.1365-2621.1963.tb00189.x]

[40] Ciferri O. Spirulina, the edible microorganism. Microbiol Rev 1983; 47(4): 551-78.
[http://dx.doi.org/10.1128/mr.47.4.551-578.1983] [PMID: 6420655]

[41] Jassby A. Spirulina: a model for microalgae as human food. Algae and human affairs 1988; 149-79.

[42] Belay A. Mass culture of Spirulina outdoors—the Earthrise Farms experience. Spirulina platensis 1997; 1(1): 131-58.

[43] Szabo NJ, Matulka RA, Chan T. Safety evaluation of whole algalin protein (wap) from *Chlorella* protothecoides. Food Chem Toxicol 2013; 59: 34-45.
[http://dx.doi.org/10.1016/j.fct.2013.05.035] [PMID: 23733102]

[44] Safi C, Zebib B, Merah O, Pontalier PY, Vaca-Garcia C. Morphology, composition, production, processing and applications of *Chlorella* vulgaris: A review. Renew Sustain Energy Rev 2014; 35: 265-78.
[http://dx.doi.org/10.1016/j.rser.2014.04.007]

[45] Pereira L. A review of the nutrient composition of selected edible seaweeds. Seaweed: Ecology, Nutrient Composition and Medicinal Uses. Nova Science Publishers, Inc., Coimbra 2011; 15-47.

[46] Taboada MC, Millán R, Miguez MI. Nutritional value of the marine algae wakame (*Undaria pinnatifida*) and nori (Porphyra purpurea) as food supplements. J Appl Phycol 2013; 25(5): 1271-6.
[http://dx.doi.org/10.1007/s10811-012-9951-9]

[47] Angell AR, Mata L, de Nys R, Paul NA. The protein content of seaweeds: a universal nitrogen-t--protein conversion factor of five. J Appl Phycol 2016; 28(1): 511-24.
[http://dx.doi.org/10.1007/s10811-015-0650-1]

[48] Galland-Irmouli AV, Fleurence J, Lamghari R, *et al.* Nutritional value of proteins from edible seaweed *Palmaria palmata* (dulse). J Nutr Biochem 1999; 10(6): 353-9.
[http://dx.doi.org/10.1016/S0955-2863(99)00014-5] [PMID: 15539310]

[49] Johnson RB, Kim JK, Armbruster LC, Yarish C. Nitrogen allocation of Gracilaria tikvahiae grown in urbanized estuaries of Long Island Sound and New York City, USA: a preliminary evaluation of ocean farmed Gracilaria for alternative fish feeds. Algae 2014; 29(3): 227-35.
[http://dx.doi.org/10.4490/algae.2014.29.3.227]

[50] Schiener P, Black KD, Stanley MS, Green DH. The seasonal variation in the chemical composition of the kelp species Laminaria digitata, Laminaria hyperborea, Saccharina latissima and Alaria esculenta. J Appl Phycol 2015; 27(1): 363-73.
[http://dx.doi.org/10.1007/s10811-014-0327-1]

[51] Butler MR. Seasonal variations in *Chondrus crispus*. Biochem J 1936; 30(8): 1338-44.
[http://dx.doi.org/10.1042/bj0301338] [PMID: 16746163]

[52] Black WAP. The seasonal variation in weight and chemical composition of the common British

Laminariaceae. J Mar Biol Assoc U K 1950; 29(1): 45-72.
[http://dx.doi.org/10.1017/S0025315400056186]

[53] Turner NJ. The ethnobotany of edible seaweed (*Porphyra abbottae* and related species; Rhodophyta: Bangiales) and its use by First Nations on the Pacific Coast of Canada. Can J Bot 2003; 81(4): 283-93.
[http://dx.doi.org/10.1139/b03-029]

[54] Shuuluka D, Bolton JJ, Anderson RJ. Protein content, amino acid composition and nitrogen-to-protein conversion factors of *Ulva* rigida and *Ulva* capensis from natural populations and *Ulva* lactuca from an aquaculture system, in South Africa. J Appl Phycol 2013; 25(2): 677-85.
[http://dx.doi.org/10.1007/s10811-012-9902-5]

[55] Lourenço SO, Barbarino E, De-Paula JC, Pereira LOS, Marquez UML. Amino acid composition, protein content and calculation of nitrogen-to-protein conversion factors for 19 tropical seaweeds. Phycol Res 2002; 50(3): 233-41.
[http://dx.doi.org/10.1111/j.1440-1835.2002.tb00156.x]

[56] Safi C, Charton M, Pignolet O, Silvestre F, Vaca-Garcia C, Pontalier PY. Influence of microalgae cell wall characteristics on protein extractability and determination of nitrogen-to-protein conversion factors. J Appl Phycol 2013; 25(2): 523-9.
[http://dx.doi.org/10.1007/s10811-012-9886-1]

[57] Zhou G, Sun Y, Xin H, Zhang Y, Li Z, Xu Z. *In vivo* antitumor and immunomodulation activities of different molecular weight lambda-carrageenans from Chondrus ocellatus. Pharmacol Res 2004; 50(1): 47-53.
[http://dx.doi.org/10.1016/j.phrs.2003.12.002] [PMID: 15082028]

[58] Ninomiya K. Umami: a universal taste. Food Rev Int 2002; 18(1): 23-38.
[http://dx.doi.org/10.1081/FRI-120003415]

[59] Mouritsen OG. Seaweeds: edible, available, and sustainable sustainable. University of Chicago Press, 2013.
[http://dx.doi.org/10.7208/chicago/9780226044538.001.0001]

[60] Niwa K, Furuita H, Yamamoto T. Changes of growth characteristics and free amino acid content of cultivated Porphyra yezoensis Ueda (Bangiales Rhodophyta) blades with the progression of the number of harvests in a nori farm. J Appl Phycol 2008; 20(5): 687-93.
[http://dx.doi.org/10.1007/s10811-007-9273-5]

[61] Wildman REC, Medeiros DM. Advanced human nutrition. CRC Press. 2000; p. 604.

[62] Fujiwara-Arasaki T, Mino N, and Kuroda M. The protein value in human nutrition of edible marine algae in Japan. In Eleventh international seaweed symposium Springer Dordrecht, 1984; pp. 513-6.
[http://dx.doi.org/10.1007/978-94-009-6560-7_104]

[63] Urbano MG, Goñi I. Bioavailability of nutrients in rats fed on edible seaweeds, Nori (*Porphyra tenera*) and Wakame (*Undaria pinnatifida*), as a source of dietary fibre. Food Chem 2002; 76(3): 281-6.
[http://dx.doi.org/10.1016/S0308-8146(01)00273-4]

[64] Marrion O, Schwertz A, Fleurence J, Guéant JL, Villaume C. Improvement of the digestibility of the proteins of the red alga *Palmaria palmata* by physical processes and fermentation. Food/Nahrung 2003; 47: 5-344.: 339-44.

[65] Marrion O, Fleurence J, Schwertz A, *et al.* Evaluation of protein *in vitro* digestibility of *Palmaria palmata* and *Gracilaria verrucosa.* J Appl Phycol 2005; 17(2): 99-102.
[http://dx.doi.org/10.1007/s10811-005-5154-y]

[66] Wong KH, Cheung PCK. Effect of fiber-rich brown seaweeds on protein bioavailability of casein in growing rats. Int J Food Sci Nutr 2003; 54(4): 269-79.
[http://dx.doi.org/10.1080/09637480120092099] [PMID: 12850888]

[67] Rodríguez De Marco E, Steffolani ME, Martínez CS, León AE. Effects of spirulina biomass on the

technological and nutritional quality of bread wheat pasta. Lebensm Wiss Technol 2014; 58(1): 102-8. [http://dx.doi.org/10.1016/j.lwt.2014.02.054]

[68] Harnedy PA, FitzGerald RJ. Extraction of protein from the macroalga *Palmaria palmata*. Lebensm Wiss Technol 2013; 51(1): 375-82. [http://dx.doi.org/10.1016/j.lwt.2012.09.023]

[69] Ursu AV, Marcati A, Sayd T, Sante-Lhoutellier V, Djelveh G, Michaud P. Extraction, fractionation and functional properties of proteins from the microalgae *Chlorella* vulgaris. Bioresour Technol 2014; 157: 134-9. [http://dx.doi.org/10.1016/j.biortech.2014.01.071] [PMID: 24534795]

[70] Maehre HK, Edvinsen GK, Eilertsen KE, Elvevoll EO. Heat treatment increases the protein bioaccessibility in the red seaweed dulse (*Palmaria palmata*), but not in the brown seaweed winged kelp (Alaria esculenta). J Appl Phycol 2016; 28(1): 581-90. [http://dx.doi.org/10.1007/s10811-015-0587-4]

[71] Eyster KM. The membrane and lipids as integral participants in signal transduction: lipid signal transduction for the non-lipid biochemist. Adv Physiol Educ 2007; 31(1): 5-16. [http://dx.doi.org/10.1152/advan.00088.2006] [PMID: 17327576]

[72] Georgianna DR, Mayfield SP. Exploiting diversity and synthetic biology for the production of algal biofuels. Nature 2012; 488(7411): 329-35. [http://dx.doi.org/10.1038/nature11479] [PMID: 22895338]

[73] Bleakley S, Hayes M. Algal proteins: extraction, application, and challenges concerning production. Foods 2017; 6(5): 33. [http://dx.doi.org/10.3390/foods6050033] [PMID: 28445408]

[74] Cottin SC, Sanders TA, Hall WL. The differential effects of EPA and DHA on cardiovascular risk factors. Proc Nutr Soc 2011; 70(2): 215-31. [http://dx.doi.org/10.1017/S0029665111000061] [PMID: 21349231]

[75] Li MH, Robinson EH, Tucker CS, Manning BB, Khoo L. Effects of dried algae *Schizochytrium sp.*., a rich source of docosahexaenoic acid, on growth, fatty acid composition, and sensory quality of channel catfish *Ictalurus punctatus*. Aquaculture 2009; 292(3-4): 232-6. [http://dx.doi.org/10.1016/j.aquaculture.2009.04.033]

[76] Conquer JA, Holub BJ. Supplementation with an algae source of docosahexaenoic acid increases (n-3) fatty acid status and alters selected risk factors for heart disease in vegetarian subjects. J Nutr 1996; 126(12): 3032-9. [http://dx.doi.org/10.1093/jn/126.12.3032] [PMID: 9001371]

[77] Holub BJ. Docosahexaenoic acid (DHA) and cardiovascular disease risk factors. Prostaglandins Leukot Essent Fatty Acids 2009; 81(2-3): 199-204. [http://dx.doi.org/10.1016/j.plefa.2009.05.016] [PMID: 19545988]

[78] Haug IJ, Sagmo LB, Zeiss D, Olsen IC, Draget KI, Seternes T. Bioavailability of EPA and DHA delivered by gelled emulsions and soft gel capsules. Eur J Lipid Sci Technol 2011; 113(2): 137-45. [http://dx.doi.org/10.1002/ejlt.201000450]

[79] Schuchardt J, Schneider I, Meyer H, Neubronner J, von Schacky C, Hahn A. Incorporation of EPA and DHA into plasma phospholipids in response to different omega-3 fatty acid formulations - a comparative bioavailability study of fish oil vs. krill oil. Lipids Health Dis 2011; 10(1): 145. [http://dx.doi.org/10.1186/1476-511X-10-145] [PMID: 21854650]

[80] Mendes A, Reis A, Vasconcelos R, Guerra P, Lopes da Silva T. *Crypthecodinium cohnii* with emphasis on DHA production: a review. J Appl Phycol 2009; 21(2): 199-214. [http://dx.doi.org/10.1007/s10811-008-9351-3]

[81] Barclay W, Weaver C, Metz J, Hansen J. Development of a docosahexaenoic acid production technology using *Schizochytrium*: historical perspective and update. Single cell oils 2010; 75-96.

[http://dx.doi.org/10.1016/B978-1-893997-73-8.50008-6]

[82] Jensen CL, Voigt RG, Prager TC, *et al.* Effects of maternal docosahexaenoic acid intake on visual function and neurodevelopment in breastfed term infants. Am J Clin Nutr 2005; 82(1): 125-32. [http://dx.doi.org/10.1093/ajcn/82.1.125] [PMID: 16002810]

[83] Jensen CL, Voigt RG, Llorente AM, *et al.* Effects of early maternal docosahexaenoic acid intake on neuropsychological status and visual acuity at five years of age of breastfed term infants. J Pediatr 2010; 157(6): 900-5. [http://dx.doi.org/10.1016/j.jpeds.2010.06.006] [PMID: 20655543]

[84] Imhoff-Kunsch B, Stein AD, Villalpando S, Martorell R, Ramakrishnan U. Docosahexaenoic acid supplementation from mid-pregnancy to parturition influenced breast milk fatty acid concentrations at 1 month postpartum in Mexican women. J Nutr 2011; 141(2): 321-6. [http://dx.doi.org/10.3945/jn.110.126870] [PMID: 21178076]

[85] Arterburn LM, Oken HA, Bailey Hall E, Hamersley J, Kuratko CN, Hoffman JP. Algal-oil capsules and cooked salmon: nutritionally equivalent sources of docosahexaenoic acid. J Am Diet Assoc 2008; 108(7): 1204-9. [http://dx.doi.org/10.1016/j.jada.2008.04.020] [PMID: 18589030]

[86] Neff LM, Culiner J, Cunningham-Rundles S, *et al.* Algal docosahexaenoic acid affects plasma lipoprotein particle size distribution in overweight and obese adults. J Nutr 2011; 141(2): 207-13. [http://dx.doi.org/10.3945/jn.110.130021] [PMID: 21178084]

[87] Wang C, Harris WS, Chung M, *et al.* n−3 Fatty acids from fish or fish-oil supplements, but not α-linolenic acid, benefit cardiovascular disease outcomes in primary- and secondary-prevention studies: a systematic review. Am J Clin Nutr 2006; 84(1): 5-17. [http://dx.doi.org/10.1093/ajcn/84.1.5] [PMID: 16825676]

[88] Bernstein AM, Ding EL, Willett WC, Rimm EB. A meta-analysis shows that docosahexaenoic acid from algal oil reduces serum triglycerides and increases HDL-cholesterol and LDL-cholesterol in persons without coronary heart disease. J Nutr 2012; 142(1): 99-104. [http://dx.doi.org/10.3945/jn.111.148973] [PMID: 22113870]

[89] Valera B, Suhas E, Counil E, Poirier P, Dewailly E. Influence of polyunsaturated fatty acids on blood pressure, resting heart rate and heart rate variability among French Polynesians. J Am Coll Nutr 2014; 33(4): 288-96. [http://dx.doi.org/10.1080/07315724.2013.874913] [PMID: 24971659]

[90] Geppert J, Kraft V, Demmelmair H, Koletzko B. Docosahexaenoic acid supplementation in vegetarians effectively increases omega-3 index: A randomized trial. Lipids 2005; 40(8): 807-14. [http://dx.doi.org/10.1007/s11745-005-1442-9] [PMID: 16296399]

[91] Nauroth JM, Liu YC, Van Elswyk M, *et al.* Docosahexaenoic acid (DHA) and docosapentaenoic acid (DPAn-6) algal oils reduce inflammatory mediators in human peripheral mononuclear cells *in vitro* and paw edema *In vivo*. Lipids 2010; 45(5): 375-84. [http://dx.doi.org/10.1007/s11745-010-3406-3] [PMID: 20364438]

[92] Van Biervliet S, Devos M, Delhaye T, Van Biervliet JP, Robberecht E, Christophe A. Oral DHA supplementation in ΔF508 homozygous cystic fibrosis patients. Prostaglandins Leukot Essent Fatty Acids 2008; 78(2): 109-15. [http://dx.doi.org/10.1016/j.plefa.2007.12.005] [PMID: 18276127]

[93] Viso AC, Marty JC. Fatty acids from 28 marine microalgae. Phytochemistry 1993; 34(6): 1521-33. [http://dx.doi.org/10.1016/S0031-9422(00)90839-2]

[94] Legeżyńska J, Kędra M, Walkusz W. Identifying trophic relationships within the high Arctic benthic community: how much can fatty acids tell? Mar Biol 2014; 161(4): 821-36. [http://dx.doi.org/10.1007/s00227-013-2380-8]

[95] Rossoll D, Bermúdez R, Hauss H, *et al.* Ocean acidification-induced food quality deterioration

constrains trophic transfer. PLoS One 2012; 7(4): e34737.
[http://dx.doi.org/10.1371/journal.pone.0034737] [PMID: 22509351]

[96] Chrismadha T, Borowitzka MA. Effect of cell density and irradiance on growth, proximate composition and eicosapentaenoic acid production of *Phaeodactylum tricornutum* grown in a tubular photobioreactor. J Appl Phycol 1994; 6(1): 67-74.
[http://dx.doi.org/10.1007/BF02185906]

[97] Pasquet V, Ulmann L, Mimouni V, *et al.* Fatty acids profile and temperature in the cultured marine diatom Odontella aurita. J Appl Phycol 2014; 26(6): 2265-71.
[http://dx.doi.org/10.1007/s10811-014-0252-3]

[98] Branger B, Cadudal JL, Delobel M, *et al.* Spiruline as a food supplement in case of infant malnutrition in Burkina-Faso. Archives de pediatrie: organe officiel de la Societe francaise de pediatrie 2003; 10(5): 424-31.

[99] Graeve M, Kattner G, Wiencke C, Karsten U. Fatty acid composition of Arctic and Antarctic macroalgae: indicator of phylogenetic and trophic relationships. Mar Ecol Prog Ser 2002; 231: 67-74.
[http://dx.doi.org/10.3354/meps231067]

[100] Schmid M, Guihéneuf F, Stengel DB. Fatty acid contents and profiles of 16 macroalgae collected from the Irish Coast at two seasons. J Appl Phycol 2014; 26(1): 451-63.
[http://dx.doi.org/10.1007/s10811-013-0132-2]

[101] McCauley JI, Meyer BJ, Winberg PC, Ranson M, Skropeta D. Selecting Australian marine macroalgae based on the fatty acid composition and anti-inflammatory activity. J Appl Phycol 2015; 27(5): 2111-21.
[http://dx.doi.org/10.1007/s10811-014-0465-5]

[102] Kumari P, Kumar M, Gupta V, Reddy CRK, Jha B. Tropical marine macroalgae as potential sources of nutritionally important PUFAs. Food Chem 2010; 120(3): 749-57.
[http://dx.doi.org/10.1016/j.foodchem.2009.11.006]

[103] Kumar M, Kumari P, Trivedi N, *et al.* Minerals, PUFAs and antioxidant properties of some tropical seaweeds from Saurashtra coast of India. J Appl Phycol 2011; 23(5): 797-810.
[http://dx.doi.org/10.1007/s10811-010-9578-7]

[104] Jiang Y, Chen F. Effects of temperature and temperature shift on docosahexaenoic acid production by the marine microalge *Crypthecodinium cohnii.* J Am Oil Chem Soc 2000; 77(6): 613-7.
[http://dx.doi.org/10.1007/s11746-000-0099-0]

[105] Jiang H, Gao K. Effects of lowering temperature during culture on the production of polyunsaturated fatty acids in the marine diatom *Phaeodactylum tricornutum* (bacillariophyceae) J Phycol 2004; 40(4): 651-4.
[http://dx.doi.org/10.1111/j.1529-8817.2004.03112.x]

[106] Blouin N, Calder BL, Perkins B, Brawley SH. Sensory and fatty acid analyses of two Atlantic species of Porphyra (Rhodophyta). J Appl Phycol 2006; 18(1): 79-85.
[http://dx.doi.org/10.1007/s10811-005-9017-3]

[107] Mishra VK, Temelli F, Ooraikul B, Shacklock PF, Craigie JS. Lipids of the red alga, *Palmaria palmate* Botanica Marina 1993; 36: 169-74.

[108] Gubelit YI, Makhutova ON, Sushchik NN, Kolmakova AA, Kalachova GS, Gladyshev MI. Fatty acid and elemental composition of littoral "green tide" algae from the Gulf of Finland, the Baltic Sea. J Appl Phycol 2015; 27(1): 375-86.
[http://dx.doi.org/10.1007/s10811-014-0349-8]

[109] Sayanova O, Haslam RP, Calerón MV, *et al.* Identification and functional characterisation of genes encoding the omega-3 polyunsaturated fatty acid biosynthetic pathway from the coccolithophore Emiliania huxleyi. Phytochemistry 2011; 72(7): 594-600.
[http://dx.doi.org/10.1016/j.phytochem.2011.01.022] [PMID: 21316718]

[110] Read BA, Kegel J, Klute MJ, *et al.* Pan genome of the phytoplankton Emiliania underpins its global distribution. Nature 2013; 499(7457): 209-13.
[http://dx.doi.org/10.1038/nature12221] [PMID: 23760476]

[111] Saiki T. The digestibility and utilization of some polysaccharide carbohydrates derived from lichens and marine algae. J Biol Chem 1906; 2(3): 251-65.
[http://dx.doi.org/10.1016/S0021-9258(17)46046-7]

[112] Terada A, Hara H, Mitsuoka T. Effect of dietary alginate on the faecal microbiota and faecal metabolic activity in humans. Microb Ecol Health Dis 1995; 8(6): 259-66.

[113] Michel C, Macfarlane GT. Digestive fates of soluble polysaccharides from marine macroalgae: involvement of the colonic microflora and physiological consequences for the host. J Appl Bacteriol 1996; 80(4): 349-69.
[http://dx.doi.org/10.1111/j.1365-2672.1996.tb03230.x] [PMID: 8849638]

[114] Hehemann JH, Correc G, Barbeyron T, Helbert W, Czjzek M, Michel G. Transfer of carbohydrate-active enzymes from marine bacteria to Japanese gut microbiota. Nature 2010; 464(7290): 908-12.
[http://dx.doi.org/10.1038/nature08937] [PMID: 20376150]

[115] Paulsen BS, Barsett H. Bioactive pectic polysaccharides. Polysaccharides 2005; I: 69-101.

[116] Pal A, Kamthania MC, Kumar A. Bioactive compounds and properties of seaweeds—a review. OAlib 2014; 1(4): 1-17.
[http://dx.doi.org/10.4236/oalib.1100752]

[117] Rioux LE, Turgeon SL. Seaweed carbohydrates. Seaweed sustainability 2015; 141-92.
[http://dx.doi.org/10.1016/B978-0-12-418697-2.00007-6]

[118] Stiger-Pouvreau V, Bourgougnon N, Deslandes E. Carbohydrates from seaweeds. Seaweed in health and disease prevention 2016; 223-74.
[http://dx.doi.org/10.1016/B978-0-12-802772-1.00008-7]

[119] Cofrades S, López-López I, Solas MT, Bravo L, Jiménez-Colmenero F. Influence of different types and proportions of added edible seaweeds on characteristics of low-salt gel/emulsion meat systems. Meat Sci 2008; 79(4): 767-76.
[http://dx.doi.org/10.1016/j.meatsci.2007.11.010] [PMID: 22063041]

[120] Gupta S, Abu-Ghannam N. Bioactive potential and possible health effects of edible brown seaweeds. Trends Food Sci Technol 2011; 22(6): 315-26. a
[http://dx.doi.org/10.1016/j.tifs.2011.03.011]

[121] Gupta S, Abu-Ghannam N. Recent developments in the application of seaweeds or seaweed extracts as a means for enhancing the safety and quality attributes of foods. Innov Food Sci Emerg Technol 2011; 12(4): 600-9. b
[http://dx.doi.org/10.1016/j.ifset.2011.07.004]

[122] Griffin JA. An investigative study into the beneficial use of seaweed in bread and the broader food industry. Dissertation presented to Dublin Institute of Technology, School of Culinary Arts and Food Technology in partial fulfilment of the requirements for the Batchelor's Degree BSc (Hons.) Baking and Pastry Arts Management, 2015.

[123] Mabeau S, Fleurence J. Seaweed in food products: biochemical and nutritional aspects. Trends Food Sci Technol 1993; 4(4): 103-7.
[http://dx.doi.org/10.1016/0924-2244(93)90091-N]

[124] Watson DB. Public health and carrageenan regulation: a review and analysis. J App Phyc 2007; 20(5): 505-13.

[125] Barlow S, Pascal G, Cressey P, Folmer D, Smith J. Carrageenan (addendum). *Safety evaluation of certain food additives*. WHO Food Addit Ser 2015; 70.

[126] Fleurence J, Levine I, Eds. Seaweed in health and disease prevention. Academic Press 2016; p. 476.

[127] Rupérez P, Saura-Calixto F. Dietary fibre and physicochemical properties of edible Spanish seaweeds. Eur Food Res Technol 2001; 212(3): 349-54. [http://dx.doi.org/10.1007/s002170000264]

[128] McDermid KJ, Stuercke B, Haleakala OJ. Total dietary fiber content in Hawaiian marine algae. Botanica Marina. 2005; 48: pp. 437-80. [http://dx.doi.org/10.1515/bot.2005.057]

[129] Benjama O, Masniyom P. Biochemical composition and physicochemical properties of two red seaweeds (Gracilaria fisheri and G. tenuistipitata) from the Pattani Bay in Southern Thailand. Songklanakarin J Sci Technol 2012; 34(2): 223.

[130] Lahaye M. Marine algae as sources of fibres: Determination of soluble and insoluble dietary fibre contents in some 'sea vegetables'. J Sci Food Agric 1991; 54(4): 587-94. [http://dx.doi.org/10.1002/jsfa.2740540410]

[131] Cantarel BL, Lombard V, Henrissat B. Complex carbohydrate utilization by the healthy human microbiome. PLoS One 2012; 7(6): e28742. [http://dx.doi.org/10.1371/journal.pone.0028742] [PMID: 22719820]

[132] Nicholson JK, Holmes E, Kinross J, *et al.* Host-gut microbiota metabolic interactions. Science 2012; 336(6086): 1262-7. [http://dx.doi.org/10.1126/science.1223813] [PMID: 22674330]

[133] Bäckhed F, Ley RE, Sonnenburg JL, Peterson DA, Gordon JI. Host-bacterial mutualism in the human intestine. Science 2005; 307(5717): 1915-20. [http://dx.doi.org/10.1126/science.1104816] [PMID: 15790844]

[134] Fernando WMADB, Ranaweera KKDS, Bamunuarachchi A, Brennan CS. The influence of rice fibre fractions on the *in vitro* fermentation production of short chain fatty acids using human faecal micro flora. Int J Food Sci Technol 2008; 43(12): 2237-44. [http://dx.doi.org/10.1111/j.1365-2621.2008.01861.x]

[135] O'Sullivan L, Murphy B, McLoughlin P, *et al.* Prebiotics from marine macroalgae for human and animal health applications. Mar Drugs 2010; 8(7): 2038-64. [http://dx.doi.org/10.3390/md8072038] [PMID: 20714423]

[136] de Jesus Raposo M, de Morais A, de Morais R. Marine polysaccharides from algae with potential biomedical applications. Mar Drugs 2015; 13(5): 2967-3028. [http://dx.doi.org/10.3390/md13052967] [PMID: 25988519]

[137] Dhargalkar V. Uses of seaweeds in the Indian diet for sustenance and well-being. Sci Cult 2015; 80: 192-202.

[138] Pulz O, Gross W. Valuable products from biotechnology of microalgae. Appl Microbiol Biotechnol 2004; 65(6): 635-48. [http://dx.doi.org/10.1007/s00253-004-1647-x] [PMID: 15300417]

[139] Plaza M, Herrero M, Cifuentes A, Ibáñez E. Innovative natural functional ingredients from microalgae. J Agric Food Chem 2009; 57(16): 7159-70. [http://dx.doi.org/10.1021/jf901070g] [PMID: 19650628]

[140] Chacón-Lee TL, González-Mariño GE. Microalgae for "healthy" foods—possibilities and challenges. Compr Rev Food Sci Food Saf 2010; 9(6): 655-75. [http://dx.doi.org/10.1111/j.1541-4337.2010.00132.x] [PMID: 33467820]

[141] Lordan S, Ross RP, Stanton C. Marine bioactives as functional food ingredients: potential to reduce the incidence of chronic diseases. Mar Drugs 2011; 9(6): 1056-100. [http://dx.doi.org/10.3390/md9061056] [PMID: 21747748]

[142] Vo TS, Ngo DH, Ta QV, Kim SK. Marine organisms as a therapeutic source against herpes simplex virus infection. Eur J Pharm Sci 2011; 44(1-2): 11-20.

[http://dx.doi.org/10.1016/j.ejps.2011.07.005] [PMID: 21782018]

[143] Hasegawa T, Kimura Y, Hiromatsu K, *et al.* Effect of hot water extract of *Chlorella vulgaris* on cytokine expression patterns in mice with murine acquired immunodeficiency syndrome after infection with Listeria monocytogenes. Immunopharmacology 1997; 35(3): 273-82. [http://dx.doi.org/10.1016/S0162-3109(96)00150-6] [PMID: 9043941]

[144] Cheng FC, Lin A, Feng JJ, *et al.* Effects of *Chlorella* on activities of protein tyrosine phosphatases, matrix metalloproteinases, caspases, cytokine release, B and T cell proliferations, and phorbol ester receptor binding. J Med Food 2004; 7(2): 146-52. [http://dx.doi.org/10.1089/1096620041224076] [PMID: 15298760]

[145] Kralovec JA, Power MR, Liu F, *et al.* An aqueous *Chlorella* extract inhibits IL-5 production by mast cells *in vitro* and reduces ovalbumin-induced eosinophil infiltration in the airway in mice *In vivo*. Int Immunopharmacol 2005; 5(4): 689-98. [http://dx.doi.org/10.1016/j.intimp.2004.11.016] [PMID: 15710338]

[146] Sheng J, Yu F, Xin Z, Zhao L, Zhu X, Hu Q. Preparation, identification and their antitumor activities *in vitro* of polysaccharides from *Chlorella pyrenoidosa*. Food Chem 2007; 105(2): 533-9. [http://dx.doi.org/10.1016/j.foodchem.2007.04.018] [PMID: 26059131]

[147] Hsu HY, Jeyashoke N, Yeh CH, Song YJ, Hua KF, Chao LK. Immunostimulatory bioactivity of algal polysaccharides from *Chlorella pyrenoidosa* activates macrophages via Toll-like receptor 4. J Agric Food Chem 2010; 58(2): 927-36. [http://dx.doi.org/10.1021/jf902952z] [PMID: 19916503]

[148] Watanabe S, Seto A. US Patent No 4,831,020 Washington, DC: US Patent and Trademark Office 1989.

[149] Pasco D, Pugh N. US Patent No 7,846,452 Washington, DC: US Patent and Trademark Office 2010.

[150] Suárez ER, Kralovec JA, Bruce Grindley T. Isolation of phosphorylated polysaccharides from algae: the immunostimulatory principle of *Chlorella pyrenoidosa*. Carbohydr Res 2010; 345(9): 1190-204. [http://dx.doi.org/10.1016/j.carres.2010.04.004] [PMID: 20466355]

[151] Kwak JH, Baek SH, Woo Y, *et al.* Beneficial immunostimulatory effect of short-term *Chlorella* supplementation: enhancement of Natural Killer cell activity and early inflammatory response (Randomized, double-blinded, placebo-controlled trial). Nutr J 2012; 11(1): 53. [http://dx.doi.org/10.1186/1475-2891-11-53] [PMID: 22849818]

[152] Umezawa I, Komiyama K. US Patent No 4,533,548 Washington, DC: US Patent and Trademark Office 1985.

[153] Umezawa IWAO, Komiyama K, Shinbukawa N, Mori M, Kojima Y. An acidic polysaccharide, chlon A, from *Chlorella pyrenoidosa*. Chemotherapy 1982; 30(10): 1041-5.

[154] Kralovec JA. U.S. Patent No. 6,974,576. Washington, DC: U.S. Patent and Trademark Office 2005.

[155] Admassu H, Zhao W, Yang R, Gasmalla MA, Alsir E. Development of Functional Foods: Sea Weeds (Algae) Untouched Potential and Alternative Resource–A Review. Int J Sci Technol Res 2015; 4(1).

[156] Aquino RS, Landeira-Fernandez AM, Valente AP, Andrade LR, Mourão PA. Occurrence of sulfated galactans in marine angiosperms: evolutionary implications. Glycobiology 2004; 15(1): 11-20. [http://dx.doi.org/10.1093/glycob/cwh138] [PMID: 15317737]

[157] Popper ZA, Michel G, Hervé C, *et al.* Evolution and diversity of plant cell walls: from algae to flowering plants. Annu Rev Plant Biol 2011; 62(1): 567-90. [http://dx.doi.org/10.1146/annurev-arplant-042110-103809] [PMID: 21351878]

[158] Bourgougnon N, Stiger-Pouvreau V. Red and Brown Macroalgae along the French Coasts, Metropole and Overseas Departements and Territories. Handbook of marine macroalgae: biotechnology and applied phycology 2012; 58-105.

[159] Mak W, Hamid N, Liu T, Lu J, White WL. Fucoidan from New Zealand *Undaria pinnatifida*: Monthly

variations and determination of antioxidant activities. Carbohydr Polym 2013; 95(1): 606-14. [http://dx.doi.org/10.1016/j.carbpol.2013.02.047] [PMID: 23618312]

[160] Stabler SP, Allen RH. Vitamin B12 deficiency as a worldwide problem. Annu Rev Nutr 2004; 24(1): 299-326. [http://dx.doi.org/10.1146/annurev.nutr.24.012003.132440] [PMID: 15189123]

[161] Martin C, Butelli E, Petroni K, Tonelli C. How can research on plants contribute to promoting human health? Plant Cell 2011; 23(5): 1685-99. [http://dx.doi.org/10.1105/tpc.111.083279] [PMID: 21586682]

[162] Norziah MH, Ching CY. Nutritional composition of edible seaweed *Gracilaria changgi*. Food Chem 2000; 68(1): 69-76. [http://dx.doi.org/10.1016/S0308-8146(99)00161-2]

[163] Ferraces-Casais P, Lage-Yusty MA, Rodríguez-Bernaldo de Quirós A, López-Hernández J. Evaluation of bioactive compounds in fresh edible seaweeds. Food Anal Methods 2012; 5(4): 828-34. [http://dx.doi.org/10.1007/s12161-011-9321-2]

[164] Hernández-Carmona G, Carrillo-Domínguez S, Arvizu-Higuera DL, *et al.* Monthly variation in the chemical composition of *Eisenia arborea* J.E. Areschoug. J Appl Phycol 2009; 21(5): 607-16. [http://dx.doi.org/10.1007/s10811-009-9454-5]

[165] Ortiz J, Uquiche E, Robert P, Romero N, Quitral V, Llantén C. Functional and nutritional value of the Chilean seaweeds ***Codium fragile, Gracilaria chilensis*** and ***Macrocystis pyrifera***. Eur J Lipid Sci Technol 2009; 111(4): 320-7. [http://dx.doi.org/10.1002/ejlt.200800140]

[166] Chan JCC, Cheung PCK, Ang PO. Comparative studies on the effect of three drying methods on the nutritional composition of seaweed *Sargassum hemiphyllum* (turn.) C. Ag. J Agric Food Chem 1997; 45(8): 3056-9. [http://dx.doi.org/10.1021/jf9701749]

[167] Ling ALM, Yasir S, Matanjun P, Abu Bakar MF. Effect of different drying techniques on the phytochemical content and antioxidant activity of *Kappaphycus alvarezii*. J Appl Phycol 2015; 27(4): 1717-23. [http://dx.doi.org/10.1007/s10811-014-0467-3]

[168] Croft MT, Lawrence AD, Raux-Deery E, Warren MJ, Smith AG. Algae acquire vitamin B12 through a symbiotic relationship with bacteria. Nature 2005; 438(7064): 90-3. [http://dx.doi.org/10.1038/nature04056] [PMID: 16267554]

[169] Haddad EH, Berk LS, Kettering JD, Hubbard RW, Peters WR. Dietary intake and biochemical, hematologic, and immune status of vegans compared with nonvegetarians. Am J Clin Nutr 1999; 70(3) (Suppl.): 586s-93s. [http://dx.doi.org/10.1093/ajcn/70.3.586s] [PMID: 10479236]

[170] Waldmann A, Koschizke JW, Leitzmann C, Hahn A. Homocysteine and cobalamin status in German vegans. Public Health Nutr 2004; 7(3): 467-72. [http://dx.doi.org/10.1079/PHN2003540] [PMID: 15153278]

[171] Allen LH. Causes of vitamin B12 and folate deficiency. Food Nutr Bull 2008; 29(2_suppl1) (Suppl.): S20-34. [http://dx.doi.org/10.1177/15648265080292S105] [PMID: 18709879]

[172] Helliwell KE, Wheeler GL, Leptos KC, Goldstein RE, Smith AG. Insights into the evolution of vitamin B12 auxotrophy from sequenced algal genomes. Mol Biol Evol 2011; 28(10): 2921-33. [http://dx.doi.org/10.1093/molbev/msr124] [PMID: 21551270]

[173] Warren MJ, Raux E, Schubert HL, Escalante-Semerena JC. The biosynthesis of adenosylcobalamin (vitamin B12). Nat Prod Rep 2002; 19(4): 390-412. [http://dx.doi.org/10.1039/b108967f] [PMID: 12195810]

[174] Watanabe F, Takenaka S, Katsura H, *et al.* Dried green and purple lavers (Nori) contain substantial amounts of biologically active vitamin B(12) but less of dietary iodine relative to other edible seaweeds. J Agric Food Chem 1999; 47(6): 2341-3.
[http://dx.doi.org/10.1021/jf981065c] [PMID: 10794633]

[175] Miyamoto E, Yabuta Y, Kwak CS, Enomoto T, Watanabe F. Characterization of vitamin B12 compounds from Korean purple laver (Porphyra sp.) products. J Agric Food Chem 2009; 57(7): 2793-6.
[http://dx.doi.org/10.1021/jf803755s] [PMID: 19256490]

[176] Takenaka S, Sugiyama S, Ebara S, *et al.* Feeding dried purple laver (nori) to vitamin B_{12}-deficient rats significantly improves vitamin B_{12} status. Br J Nutr 2001; 85(6): 699-703.
[http://dx.doi.org/10.1079/BJN2001352] [PMID: 11430774]

[177] Rauma AL, Törrönen R, Hänninen O, Mykkänen H. Vitamin B-12 status of long-term adherents of a strict uncooked vegan diet ("living food diet") is compromised. J Nutr 1995; 125(10): 2511-5.
[PMID: 7562085]

[178] Dagnelie PC, van Staveren WA, van den Berg H. Vitamin B-12 from algae appears not to be bioavailable. Am J Clin Nutr 1991; 53(3): 695-7.
[http://dx.doi.org/10.1093/ajcn/53.3.695] [PMID: 2000824]

[179] Russell-Jones GJ, Arthur L, Walker H. Vitamin B12-mediated transport of nanoparticles across Caco-2 cells. Int J Pharm 1999; 179(2): 247-55.
[PMID: 10053217]

[180] Stupperich E, Kräutler B. Pseudo vitamin B 12 or 5-hydroxybenzimidazolyl-cobamide are the corrinoids found in methanogenic bacteria. Arch Microbiol 1988; 149(3): 268-71.
[http://dx.doi.org/10.1007/BF00422016]

[181] Stupperich E, Nexø E. Effect of the cobalt-N coordination on the cobamide recognition by the human vitamin B12 binding proteins intrinsic factor, transcobalamin and haptocorrin. Eur J Biochem 1991; 199(2): 299-303.
[http://dx.doi.org/10.1111/j.1432-1033.1991.tb16124.x] [PMID: 2070790]

[182] Watanabe F. Vitamin B12 from edible algae-from food science to molecular biology. Vitamins (Japanese) 2007; 81: 49-55.

[183] Yamada K, Yamada Y, Fukuda M, Yamada S. Bioavailability of dried asakusanori (*Porphyra tenera*) as a source of Cobalamin (Vitamin B12). Int J Vitam Nutr Res 1999; 69(6): 412-8.
[http://dx.doi.org/10.1024/0300-9831.69.6.412] [PMID: 10642899]

[184] Brown MR. Effects of storage and processing on the ascorbic acid content of concentrates prepared from *Chaetoceros calcitrans*. J Appl Phycol 1995; 7(5): 495-500.
[http://dx.doi.org/10.1007/BF00003934]

[185] Jiménez-Escrig A, Jiménez-Jiménez I, Pulido R, Saura-Calixto F. Antioxidant activity of fresh and processed edible seaweeds. J Sci Food Agric 2001; 81(5): 530-4.
[http://dx.doi.org/10.1002/jsfa.842]

[186] Lage-Yusty MA, Alvarado G, Ferraces-Casais P, López-Hernández J. Modification of bioactive compounds in dried edible seaweeds. Int J Food Sci Technol 2014; 49(1): 298-304.
[http://dx.doi.org/10.1111/ijfs.12277]

[187] Matanjun P, Mohamed S, Mustapha NM, Muhammad K. Nutrient content of tropical edible seaweeds, *Eucheuma cottonii, Caulerpa lentillifera* and *Sargassum polycystum*. J Appl Phycol 2009; 21(1): 75-80.
[http://dx.doi.org/10.1007/s10811-008-9326-4]

[188] Helliwell KE, Wheeler GL, Smith AG. Widespread decay of vitamin-related pathways: coincidence or consequence? Trends Genet 2013; 29(8): 469-78.
[http://dx.doi.org/10.1016/j.tig.2013.03.003] [PMID: 23623319]

[189] Al-Thawadi S. Public perception of algal consumption as an alternative food in the Kingdom of Bahrain. Arab Journal of Basic and Applied Sciences 2018; 25(1): 1-12. [http://dx.doi.org/10.1080/25765299.2018.1449344]

[190] Almela C, Algora S, Benito V, *et al.* Heavy metal, total arsenic, and inorganic arsenic contents of algae food products. J Agric Food Chem 2002; 50(4): 918-23. [http://dx.doi.org/10.1021/jf0110250] [PMID: 11829668]

[191] Besada V, Andrade JM, Schultze F, González JJ. Heavy metals in edible seaweeds commercialised for human consumption. J Mar Syst 2009; 75(1-2): 305-13. [http://dx.doi.org/10.1016/j.jmarsys.2008.10.010]

[192] van der Spiegel M, Noordam MY, van der Fels-Klerx HJ. Safety of novel protein sources (insects, microalgae, seaweed, duckweed, and rapeseed) and legislative aspects for their application in food and feed production. Compr Rev Food Sci Food Saf 2013; 12(6): 662-78. [http://dx.doi.org/10.1111/1541-4337.12032] [PMID: 33412718]

[193] Grobbelaar JU. Quality Control and Assurance: crucial for the sustainability of the applied phycology industry. J Appl Phycol 2003; 15(2/3): 209-15. [http://dx.doi.org/10.1023/A:1023820711706]

[194] Görs M, Schumann R, Hepperle D, Karsten U. Quality analysis of commercial *Chlorella* products used as dietary supplement in human nutrition. J Appl Phycol 2010; 22(3): 265-76. [http://dx.doi.org/10.1007/s10811-009-9455-4]

[195] Harnedy PA, FitzGerald RJ. Extraction and enrichment of protein from red and green macroalgae. Natural Products From Marine Algae 2015; 103-8. [http://dx.doi.org/10.1007/978-1-4939-2684-8_4]

[196] Machmudah S, Shotipruk A, Goto M, Sasaki M, Hirose T. Extraction of astaxanthin from *Haematococcus pluvialis* using supercritical CO_2 and ethanol as entrainer. Ind Eng Chem Res 2006; 45(10): 3652-7. [http://dx.doi.org/10.1021/ie051357k]

[197] Cuellar-Bermudez SP, Aguilar-Hernandez I, Cardenas-Chavez DL, Ornelas-Soto N, Romero-Ogawa MA, Parra-Saldivar R. Extraction and purification of high-value metabolites from microalgae: essential lipids, astaxanthin and phycobiliproteins. Microb Biotechnol 2015; 8(2): 190-209. [http://dx.doi.org/10.1111/1751-7915.12167] [PMID: 25223877]

[198] Marella C, Muthukumarappan K, Metzger LE. Application of membrane separation technology for developing novel dairy food ingredients. J Food Process Technol 2013; 4(9)

[199] Mundy CJ, Barber DG, Michel C. Variability of snow and ice thermal, physical and optical properties pertinent to sea ice algae biomass during spring. J Mar Syst 2005; 58(3-4): 107-20. [http://dx.doi.org/10.1016/j.jmarsys.2005.07.003]

[200] Mock T, Thomas DN. Recent advances in sea-ice microbiology. Environ Microbiol 2005; 7(5): 605-19. [http://dx.doi.org/10.1111/j.1462-2920.2005.00781.x] [PMID: 15819843]

[201] Pandey R, Sahu A, K K V, M P. Studies on light intensity distribution inside an open pond photo-bioreactor. Bioprocess Biosyst Eng 2015; 38(8): 1547-57. [http://dx.doi.org/10.1007/s00449-015-1398-3] [PMID: 25868715]

[202] Skerratt JH, Nichols PD, McMeekin TA, Burton H. Seasonal and inter-annual changes in planktonic biomass and community structure in eastern Antarctica using signature lipids. Mar Chem 1995; 51(2): 93-113. [http://dx.doi.org/10.1016/0304-4203(95)00047-U]

[203] N R, L P. Evaluation of fatty acids as biomarkers for a natural plankton community. A field study of a spring bloom and a post-bloom period off West Greenland. Mar Biol 2002; 141(3): 423-34. [http://dx.doi.org/10.1007/s00227-002-0841-6]

[204] Leu E, Falk-Petersen S, Kwaśniewski S, Wulff A, Edvardsen K, Hessen DO. Fatty acid dynamics during the spring bloom in a High Arctic fjord: importance of abiotic factors versus community changes. Can J Fish Aquat Sci 2006; 63(12): 2760-79.
[http://dx.doi.org/10.1139/f06-159]

[205] Leu E, Wiktor J, Søreide JE, Berge J, Falk-Petersen S. Increased irradiance reduces food quality of sea ice algae. Mar Ecol Prog Ser 2010; 411: 49-60.
[http://dx.doi.org/10.3354/meps08647]

[206] Smith REH, Clement P, Head E. Biosynthesis and photosynthate allocation patterns of arctic ice algae. Limnol Oceanogr 1989; 34(3): 591-605.
[http://dx.doi.org/10.4319/lo.1989.34.3.0591]

[207] Smith REH, Cavaletto JR, Eadie BJ, Gardner WS. Growth and lipid composition of high Arctic ice algae during the spring bloom at Resolute, Northwest Territories, Canada. Mar Ecol Prog Ser 1993; 97: 19-29.
[http://dx.doi.org/10.3354/meps097019]

[208] Lee SH, Whitledge TE, Kang SH. Spring time production of bottom ice algae in the landfast sea ice zone at Barrow, Alaska. J Exp Mar Biol Ecol 2008; 367(2): 204-12.
[http://dx.doi.org/10.1016/j.jembe.2008.09.018]

[209] Mock T, Kroon BMA. Photosynthetic energy conversion under extreme conditions—I: important role of lipids as structural modulators and energy sink under N-limited growth in Antarctic sea ice diatoms. Phytochemistry 2002; 61(1): 41-51.
[http://dx.doi.org/10.1016/S0031-9422(02)00216-9] [PMID: 12165301]

[210] Shimamatsu H. A pond for edible Spirulina production and its hydraulic studies. In: Twelfth international seaweed symposium. Springer Dordrecht 1987; 83-9.
[http://dx.doi.org/10.1007/978-94-009-4057-4_11]

[211] Seshadri CV, Jeeji Bai N. Spirulina ETTA national symposium MCRC 1992.

[212] Richmond A. Spirulina. Micro-algal biotechnology 1988.

[213] Richmond A, Lichtenberg E, Stahl B, Vonshak A. Quantitative assessment of the major limitations on productivity of Spirulina platensis in open raceways. J Appl Phycol 1990; 2(3): 195-206.
[http://dx.doi.org/10.1007/BF02179776]

[214] Belay A. A. Production of high-quality Spirulina at Earthrise farms. Algal biotechnology, in the Asia-Pacific region 1994; 92-102.

[215] Bizzi A, Chiesara E, Clementi F. Trattamenti prolungati nel ratto con diete conntenenti proteine di Spirulina. Aspetti biochimici, morfologici e tossicologici. Prospettive della coltura di Spirulina in Italia 1980; 205.

[216] Yoshino Y, Hirai Y, Takahashi H, Yamamoto N, Yamazaki N. The chronic intoxication test on Spirulina product fed to Wistar-strain rats. The Japanese Journal of Nutrition and Dietetics 1980; 38(4): 221-5.
[http://dx.doi.org/10.5264/eiyogakuzashi.38.221]

[217] Chamorro G, Salazar M. Teratogenic study of Spirulina in mice. Arch Latinoam Nutr 1990; 40(1): 86-94.
[PMID: 2129477]

[218] Mitchell GV, Grundel E, Jenkins M, Blakely SR. Effects of graded dietary levels of *Spirulina maxima* on vitamins A and E in male rats. J Nutr 1990; 120(10): 1235-40.
[http://dx.doi.org/10.1093/jn/120.10.1235] [PMID: 2213251]

[219] Kapoor R, Mehta U. Effect of supplementation of blue green alga (Spirulina) on outcome of pregnancy in rats. Plant Foods Hum Nutr 1993; 43(1): 29-35.
[http://dx.doi.org/10.1007/BF01088093] [PMID: 8464842]

[220] Schubert LE. The use of *Spirulina* (Cyanophycaea) and *Chlorella* (Chlorophyceae) as food resource for animals and humans. In: Round FE, Chapman DJ, Eds. Progressing Physiological Research 1988; 237.

[221] Kawaguchi K. Microalgae production systems in Asia, in algae biomass 1980.

[222] Running JA, Huss RJ, Olson PT. Heterotrophic production of ascorbic acid by microalgae. J Appl Phycol 1994; 6(2): 99-104.
[http://dx.doi.org/10.1007/BF02186063]

[223] Becker EW, Venkataraman LV. Production and processing of algae in pilot plant scale experiences of the Indo-German project. Algae biomass: production and use/sponsored by the National Council for Research and Development, Israel and the Gesellschaft fur Strahlen-und Umweltforschung (GSF), Munich, Germany; editors, Gedaliah Shelef, Carl J Soeder 1980.

[224] Becker EW, Venkataraman LV. Biotechnology and exploitation of algae: the Indian approach 1982.

[225] Borowitzka MA. Microalgae as sources of essential fatty acids. Aust J Biotechnol 1988; 1: 58-62. a

[226] Bajpai P, Bajpai PK. Arachidonic acid production by microorganisms. Biotechnol Appl Biochem 1992; 15(1): 1-10.
[http://dx.doi.org/10.1111/j.1470-8744.1992.tb00194.x] [PMID: 1550658]

[227] Bajpai P, Bajpai PK. Eicosapentaenoic acid (EPA) production from microorganisms: a review. J Biotechnol 1993; 30(2): 161-83.
[http://dx.doi.org/10.1016/0168-1656(93)90111-Y] [PMID: 7764033]

[228] Chen F, Johns MR. Effect of C/N ratio and aeration on the fatty acid composition of heterotrophic*Chlorella* sorokiniana. J Appl Phycol 1991; 3(3): 203-9.
[http://dx.doi.org/10.1007/BF00003578]

[229] Takeuchi D, Uehara K. Docosahexaenoic acid production eliminating generation of fish malodour. World Patent 1993; (9320225):

[230] Barclay WR, Meager KM, Abril JR. Heterotrophic production of long chain omega-3 fatty acids utilizing algae and algae-like microorganisms. J Appl Phycol 1994; 6(2): 123-9.
[http://dx.doi.org/10.1007/BF02186066]

[231] Bajpai P, Bajpai P, Ward O. Production of docosahexaenoic acid by *Thraustochytrium aureum*. Appl Microbiol Biotechnol 1991; 35(6): 706-10.
[http://dx.doi.org/10.1007/BF00169881]

[232] Bajpai PK, Bajpai P, Ward OP. Optimization of production of docosahexaenoic acid (DHA) by *Thraustochytrium aureum* ATCC 34304. J Am Oil Chem Soc 1991; 68(7): 509-14.
[http://dx.doi.org/10.1007/BF02663823]

[233] Vasumathi KK, Nithiya EM, Pandey R, Premalatha M. Studies on the effect of wind speed on loss of carbon dioxide during bio sequestration. Int J Hydrogen Energy 2017; 42(42): 26555-62.
[http://dx.doi.org/10.1016/j.ijhydene.2017.07.083]

[234] Vasumathi KK, Premalatha M, Subramanian P. Parameters influencing the design of photobioreactor for the growth of microalgae. Renew Sustain Energy Rev 2012; 16(7): 5443-50.
[http://dx.doi.org/10.1016/j.rser.2012.06.013]

[235] Borowitzka MA. Algae as food. Microbiology of fermented foods 1998; 585-602.
[http://dx.doi.org/10.1007/978-1-4613-0309-1_18]

[236] Tredici MR, Carlozzi P, Chini Zittelli G, Materassi R. A vertical alveolar panel (VAP) for outdoor mass cultivation of microalgae and cyanobacteria. Bioresour Technol 1991; 38(2-3): 153-9.
[http://dx.doi.org/10.1016/0960-8524(91)90147-C]

[237] Torzillo G, Carlozzi P, Pushparaj B, Montaini E, Materassi R. A two-plane tubular photobioreactor for outdoor culture of Spirulina. Biotechnol Bioeng 1993; 42(7): 891-8.

[http://dx.doi.org/10.1002/bit.260420714] [PMID: 18613137]

[238] Silva HJ, Cortiñas TI. Vertical thin-layer photoreactor for controlled cultivation of cyanobacteria. World J Microbiol Biotechnol 1994; 10(2): 145-8. [http://dx.doi.org/10.1007/BF00360875] [PMID: 24420935]

[239] Pohl P, Kohlhase M, Martin M. Photobioreactors for the axenic mass cultivation of microalgae. Algal biotechnology, London: Elsevier Applied Science 1988.

[240] Wohlgeschaffen GD, Subba Rao DV, Mann KH. Vat incubator with immersion core illumination — a new, inexpensive setup for mass phytoplankton culture. J Appl Phycol 1992; 4(1): 25-9. [http://dx.doi.org/10.1007/BF00003957]

[241] Hilaly AK, Karim MN, Guyre D. Optimization of an industrial microalgae fermentation. Biotechnol Bioeng 1994; 43(4): 314-20. [http://dx.doi.org/10.1002/bit.260430408] [PMID: 18615695]

[242] Mori K, Ohya H, Matsumoto K, Furune H. Sunlight supply and gas exchange systems in microalgal bioreactor bioreactor Adv Space Res 1987; 7(4): 47-52. [http://dx.doi.org/10.1016/0273-1177(87)90031-7]

[243] Grant Burgess J, Iwamoto K, Miura Y, Takano H, Matsunaga T. An optical fibre photobioreactor for enhanced production of the marine unicellular alga Isochrysis aff. galbana T-Iso (UTEX LB 2307) rich in docosahexaenoic acid. Appl Microbiol Biotechnol 1993; 39(4-5): 456-9. [http://dx.doi.org/10.1007/BF00205032]

[244] Pirt SJ, Lee YK, Walach MR, Pirt MW, Balyuzi HHM, Bazin MJ. A tubular bioreactor for photosynthetic production of biomass from carbon dioxide: Design and performance. J Chem Technol Biotechnol Biotechnol 1983; 33(1): 35-58. [http://dx.doi.org/10.1002/jctb.280330105]

[245] Richmond A, Boussiba S, Vonshak A, Kopel R. A new tubular reactor for mass production of microalgae outdoors. J Appl Phycol 1993; 5(3): 327-32. [http://dx.doi.org/10.1007/BF02186235]

[246] Vasumathi KK, Premalatha M, Subramanian P. Experimental studies on the effect of harvesting interval on yield of *Scenedesmus arcuatus* var. capitatus. Ecol Eng 2013; 58: 13-6. [http://dx.doi.org/10.1016/j.ecoleng.2013.06.012]

CHAPTER 4

Habitat and Resources of Edible Seaweeds

Monica G. Kavale[1,2,*] and **Tejal K. Gajaria**[1,2]

[1] *Academy of Scientific and Innovative Research, Ghaziabad, Uttar Pradesh, India*

[2] *CSIR-Central Salt and Marine Chemicals Research Institute, Bhavnagar, Gujarat, India*

Abstract: Seaweeds are primary producers of the oceanic ecosystem and essential for the survival of marine life forms. As they proliferate and restructure niches, they support the development, reproduction, and growth of a plethora of marine biodiversity habitats. However, the unprecedented climate change and human interventions have severely compromised coastal ecosystems. The physiological parameters such as temperature, pH, and carbonate chemistry have affected the abundance, nutritional dynamics, and, most critically, the geospatial distributions of seaweed communities. These factors play a pivotal role in the design and establishment of artificial cultivation strategies for economically valuable seaweeds. So far, the meager explorations of geochemical interactions among seaweeds have made it difficult to understand the possible environmental interactions with seaweed communities, thus sometimes severely affecting commercial cultivation practices. Furthermore, modeling environmental parameters with coastal geochemistry may effectively address the management and conservation of seaweed biodiversity.

Keywords: Cultivation, Edible Seaweeds, Habitat Diversity, Land Cultivation, Seaweed Farming.

INTRODUCTION

Seaweeds are ecologically valuable primary producers as well as ecosystem engineers that play a central role in structuring coastal habitats ranging from kelp forests to coral reefs [1]. Habitats are crucial to the primary development of seaweed communities wherein changes in the environmental variables viz., temperature, desiccation, salinity, wave height, nutrition supply, and carbon dioxide levels, being the direct factors, affect the survival, reproduction, and growth of seaweed communities [2]. However, other factors also significant in designing coastal habitats like imposing ocean acidification, global warming, sea-level rise, along with the variations in wave heights and terrigenous nutrition run-

[*] **Corresponding author Monica G. Kavale:** CSIR-Central Salt and Marine Chemicals Research Institute, Bhavnagar, Gujarat, India; Tel: +91-278-2567760 (Ext. 6130); Fax: +91-278-2565762; E-mail: mgkavale@csmcri.res.in

offs yet remain to be fully explored [3]. Substantial environmental changes are known to directly affect seaweeds in terms of their eco-physiology, growth, reproduction, and survival, wherein they are considered to get acclimatized to these changes via adaptation or migration [4, 5]. The extensive reshuffling of seaweed communities is also known to occur as a result of ecological balance between competing species change and varying herbivorous activities [3]. There are several key areas in which ecological insights into global habitat shifting are lacking, however, by strategically prioritizing research in the area of complex environmental variations, including geo-chemical changes, evolutionary adaptations, and multiple stressor effects, we can concomitantly advance our current understanding of seaweed biology and climate change ecology to effectively address the management and conservation of coastal and marine ecosystems [3].

Like other photoautotrophic biospheres, seaweeds are key marine habitat structuring agents that harbour incredible oceanic biodiversity. Seaweeds are among the fundamental basis of marine productive food webs and include industrially valuable species intimately linked to our economic systems via food and medicine [6]. There are numerous threats to these biologically important and economically valuable seaweed habitats, predominantly, human interventions have led to the rising CO_2 levels and associated global warming. Variations in global thermal profiles and ocean chemistry associated with increasing CO_2 levels have led to widespread shifts in biological systems. There are reports of a steady decline of kelps and fucoid species in terms of their abundance or reduction attributed to ocean warming together with other anthropogenic stressors [7 - 9]. In addition, climate change is known to facilitate the introduction and proliferation of invasive species, as evidenced by the introduction of immigrant algae in the North-East Atlantic [10]. However, in spite of the socio-economic and commercial importance of various seaweed communities, less attention has been given to their conservation and biodiversity [11]. The present chapter encompasses a detailed discussion of various seaweed habitats and their native seaweed communities, followed by their current cultivation practices and future prospects.

HABITAT DIVERSITY

The surface of the earth comprises about 71% of water reservoirs in various forms viz., oceans, seas, gulfs, bays, estuaries, lakes, rivers, waterfalls, *etc.* Of this, about 97% is from oceans, and the remaining 3% encompasses mostly freshwater in the form of lakes, rivers, and waterfalls. The world's oceans are divided into 5 geographic regions, namely, the Pacific, Atlantic, Indian, Arctic, and the Southern Oceans. The Pacific and Atlantic oceans are the largest basins occupying

321.1×10^6 km^2 area of the earth. Based on light penetration levels, oceans are differentiated vertically into two zones, *i.e.*, photic and aphotic zone. The photic zone is also known as the epipelagic or euphotic zone, where visible light can penetrate up to a depth of 200meters, whereas the aphotic zone comprises areas that are impervious to visible light. The euphotic zone is further differentiated into three zones viz. the supralittoral or sprays zone, the littoral or sublittoral zone (experiences periodic exposure to solar radiation), and the sublittoral zone (the area which is never exposed to solar radiation) [12]. The supralittoral zone, located above the high tide watermark, regularly experiences varying degrees of splash from the oceanic water. The seaweeds growing in this zone show better tolerance to desiccation and extreme climatic variation. The genera *Porphyra*, *Bangia*, *Ulva/Enteromorpha*, and *Fucus* are the common inhabitants of the supralittoral zone. The littoral zone ranges from high tide watermark to low tide watermark, which is regularly covered and uncovered by the oceanic waters during high and low tide, respectively. This zone is subdivided into three sub-zones, namely the upper littoral zone, mid littoral zone, and lower littoral zone. The mid-littoral zone is periodically exposed or submerged during normal tides, whereas the lower littoral zone is a borderline of the sub-littoral zone generally exposed during low spring tide. The species growing in such zones can withstand short periods of solar exposure. A wide variety of green, brown, and red seaweeds can grow in the littoral zone, for example, *Ulva, Padina, Sargassum, Turbinaria, Hypnea,* and *Acanthophora*. The sub-littoral zone is always sheltered by the oceanic waters and ranges from the lowest water tide mark up to a depth of 50 meters, where solar radiation can penetrate. This zone is comparatively stable and experiences subtle fluctuations in temperature and solar radiation. A dense growth of seaweeds in this zone forms huge beds or forests and provides a habitat for thousands of invertebrates and life forms. The dominant seaweed species in this zone are *Laminaria, Undaria, Macrocystis, Alaria, Ecklonia, Sargassum,etc.*

The secondary water reservoirs are the seas, the part of oceans that are partially surrounded by land. The Arabian Sea, Caspian Sea, Caribbean Sea, Mediterranean Sea, South China Sea, and the Sea of Japan, are some of the largest seas in the world. The gulfs are water reservoirs with a narrow mouth and are almost completely surrounded by land, whereas bays are small water reservoirs with a broad inlet separated from the oceans when the land curves inwards. The Gulf of Mexico is a relatively shallow continental shelf and is the largest gulf in the world, with sandy and limestone substrata. It comprises rich seaweed biodiversity, especially that of green algae, which signifies an affinity towards tropical seaweeds flora, for example, *Caulerpa, Codium, Enteromorpha/Ulva, Monostroma, Eucheuma, Kappaphycus, Gelidiella, Gelidium, Gracilaria, Halymenia, Laurencia, Porphyra, Sargassum,* and *Calpomenia* are the commonly reported edible seaweeds, while the seaweed genera namely *Laminaria, Undaria,*

Eklonia, and *Macrocystis* are absent [7, 13]. The Bay of Bengal is one of the largest bays in the world covering approximately 1230,000 km^2 area bounded by India, Bangladesh, Myanmar, Sri Lanka, Sumatra, and Indonesia. The seaweed flora shows similarities to the flora of the Gulf of Mexico [14, 15].

The abiotic factors that influence the diversity and distribution of seaweeds include the geological (substratum), the physical (temperature, light, humidity, rain), the chemical (salinity, nitrate, phosphate oxygen, carbon dioxide, pH, *etc.*), and the biotic factors (grazing and competition). Based on the substratum, the coastline is differentiated into consolidated coasts and unconsolidated coasts. The consolidated coasts are rocky landforms with high wave energy and successive erosion. Rocky coasts are biologically productive habitats composed of solid igneous basaltic, granite, and soft porous laterite and lime material. The laterite rocks are rich in iron and aluminum ore, specifically present in hot and wet tropical countries like India [16]. The rocky substratum is attributed to high species diversity which supports a wide range of animals and invertebrates at various tropic levels. The rocky shores are present across the world. The wave action, tide, and climate are accountable for the creation of cliffs, wave-cut platforms, boulders, and rock pools. Cliffs are vertical rock faces positioned at the edge of the sea, shaped due to erosion and weathering of rocks, and comprise about 80% of the world's oceanic coasts [17]. In the polar region, frost weathering plays a vital role in the development of cliffs. The vegetation is observed on top of the cliff, especially grasses. The area of cliffs under the surf zone shows the growth of some seaweeds and cyanobacteria. The seaweed species like *Porphyra, Enteromorpha,* and *Ulva* are commonly found in this area. Generally, wave-cut platforms are exposed during low tide and submerged during high tide. This area becomes a rich source of seaweed biodiversity. The seaweed species grow in a vertical zonation. The tidal regime, wave action, and water spray directly affect the presence of seaweed species that varies along the coast. The species like *Laminaria, Chondrus, Dictyota,* and *Porphyra*, are typically found on boulders, pebbles, and gravels in the littoral zone, although not in equal abundance as in the case of a stable bedrock. The tidal pools, also known as rock pools, are the isolated pockets formed due to erosion of hard rock along the littoral zone. They can be shallow or deep, always filled with seawater, and provide shelter to the seaweeds from the strong waves. The seaweed species like *Gracilaria dura, Sarconema, Halimeda, Jania,* and *Caulerpa* are some of the common species that are inhabitants of such rock pools.

Unconsolidated coasts are the beaches found on barrier islands and sandy shores. Beaches can be divided into backshore (a part of the shore above high tide), foreshore (a part of the shore exposed during high tide and submerged during low tide), and nearshore zones (the zone where waves break and then re-form). The barrier islands, also called sand islands, are narrow stretches of sand formed parallel to the coastline due to the deposition of sediments. These islands are not stable and are frequently displaced by wind and wave action. Barriers islands are found throughout the world except for Antarctica and are predominant around the Gulf of Mexico. The barrier islands are vulnerable to storms, sea-level rise, and tsunamis, but also protect the coasts from the same. Beaches are long stretches of loose deposits of sand that include gravel and shells and act as buffer zones protecting coastlines and cliffs from direct waves. Sandy beaches are enclosed between two rocky headlands forming a bay. Waves, winds, and tides are the most influential factors that move the sand towards the shore. Sandy beaches have very low seaweed diversity. However, seaweed species like *Enteromorpha, Ulva, Boergesenia forbesii, Gracilaria dura, Gracilaria verrucosa,* and *Padina* are the few species found in the surf zone (region of crashing waves) of the beaches. In addition to rocky and sandy beaches, estuaries, mudflats, and lagoons (Fig. **1**) are also important coastal environments and are briefly described below.

Estuaries

The estuary is an inlet of the sea that is a partially enclosed body of brackish water and muddy habitat that is associated with one or more rivers and streams. The estuarine areas comprise a wide variety of habitats viz. seagrass meadows, mangroves, salt marshes, oyster beds, and mudflats. The estuaries are biologically the most productive areas with unique seaweed biota and less species richness as they are adapted for extreme fluctuations in temperature and salinities. The morphology and composition of the estuaries are variable across the world depending upon climatic change, seasonality, wave action, freshwater run-offs, sediment availability, and tides. The salinity of estuaries varies from 1-32 psu, with a tidal range of 2-4 meters comprising high amounts of inorganic nitrogen and phosphorous. The estuarine areas are vulnerable to nutrient enrichment which is evident from the growth of opportunistic ephemerals, and the bloom-forming green foliose seaweeds. Due to the upsurge of nutrients in Helford Estuary, the seagrass *Zostera marina* was replaced by *Enteromorpha* and *Ulva,* besides the growth of *Ascophyllum nodosum and Fucus serratus* has been drastically reduced. Massive growth was also observed for *Chaetomorpha, Enteromorpha, and Ulva* in Ythan Estuary Scotland [18, 19]. In India, the genus *Monostroma* is associated with *Ulva, Enteromorpha, and Cladophora*. Chilika Lake, one of the largest brackish water lagoon areas in Asia, about 1100 km^2 long and located along the coast of Odisha, India, showed a significant abundance of agarophyte *Gracilaria verrucosa* [20]. Generally, seaweeds like *Ulva/Enteromorpha, Monostroma,* and *Gayralia* thrive in estuarine areas associated with mangroves [21 - 24].

Fig. (1). Various habitats for marine macroalgal species; **a.** Supralittoral, littoral, and sublittoral zones; **b.***Ulva* sp. growing on laterite rock **c.** Shallow water pools **d.** cliff with the wave-cut platform; **e.***Gracilaria* growing on sandy substrata at Veraval; **f.** Basaltic rock boulders at Karwar; **g.** Estuarine area during low tide; **f.** Lagoon area of Lakshadweep.

Mudflats

Mudflats are also known as tidal flats found in sediment-rich areas composed of mud and sand from river runoffs and the inflow from tides. It lies amid an area of marine, fresh, and land environments with generally a less inclined slope that appears flat and frequently flooded, and influenced by wave activity [25]. *Gracilaria vermiculophylla* is gaining attention as it grows on the mudflats of Georgia and South Carolina, where no other native species of seaweed is seen growing.

Seagrass Meadows

Seagrass beds grow in shallow, salty, and brackish water right across the world, from the Tropics to the Arctic region except for Antarctica. They provide food and habitat to a variety of animals, invertebrates, and algae. Seagrass meadows are prone to anthropogenic activities, strong waves, and storms.

Lagoons

Lagoons are the water bodies separated from the larger water bodies by barrier islands, coral reefs, and other natural barriers. Atolls and coastal lagoons are the two types of lagoons. The atolls are the coral reefs that encircle a shallow lagoon. For example, the Union Territory of Lakshadweep is the only coral reef island in India and the largest atoll in the world (Lakshadweep action plan on climate change 2012). The lagoons harbor substantial biodiversity, including corals, fishes, seagrasses, and seaweeds. The seaweed species belonging to the Chlorophyceae, Phaeophyceae, and Rhodophyceae grow abundantly in the lagoon area. About 62 genera with 114 species of seaweeds have been reported from Lakshadweep islands among which *Ulva/Enteromorpha, Caulerpa, Codium, Sargassum, Turbinaria, Gracilaria*, *Acanthophora,* and *Gelidiella* are the common edible seaweed species [26].

HABITAT PREFERENCES OF EDIBLE SEAWEEDS

Green Seaweeds

The green seaweeds grow in marine as well as in estuarine waters. They extend right from the infralittoral fringe to the sublittoral zone, including tidal pools attached by rhizoids and basal disc to a variety of substrata, exposed rocks, Mollusk shells, corals, sand, and silt over stones, mud, and epiphytic on other algae and mangroves such as *Sargassum, Kappaphycus* and, *Gelidiella* or free-floating forms [27, 28]. Amongst the green seaweeds, the species of *Caulerpa* grow from the littoral to shallow sublittoral zone of marine waters on various substrata like sandy and silt-laden lagoons associated with seagrasses, exposed rocky substrata, and even found in mud associated with mangroves forming a dense mat. They also grow in shaded rock pools. However, the species of *Codium, Ulva/Enteromorpha, Gayralia*, and *Monostroma* inhabit both estuarine as well as marine waters. These species are tolerant to high and low temperatures, salinities, and dehydration. The species abundantly are found growing along the entire littoral zone besides shallow subtidal zone and tidal pools.

Generally, the species of *Monostroma* and *Gayralia* are found to be associated with pneumatophores of mangroves [22, 28, 29]. It is noteworthy to mention that both the genera *Ulva* and *Monostroma* were observed growing in freshwater habitats as well [30 - 32].

Brown Seaweeds

The term "Kelp" collectively refers to the largest and structurally complex forms of seaweeds in the order Laminariales comprising of 30 genera that form huge beds or forests, providing food and habitat to the organisms of all at the Tropic level [33, 34]. Among the "Kelp", the genera, namely *Saccharina*, *Laminaria*, *Undaria*, *Macrocystis*, *Alaria*, *Pterygophora*, *Ecklonia*, *Eckloniopsis*, *Egregia*, *Eisenia*, and *Costaria* are sourced as food. These kelp species are perennial, annual, and biennial in occurrence. They are a rich source of minerals and vitamins especially Iodine and vitamin C and B12 and provide for an array of health benefits against hypothyroidism, obesity, diabetes, blood clots, prostate cancer, inflammation, *etc.* The "Kelps" particularly *Saccharina japonica* (Japanese Kombu), *Alaria esculenta* (winged kelp) *Undaria pinnatifida* (Japanese wakame), are considered to be the most important food resources for Asian and European countries. All the species of "Kelps" are found growing in the sublittoral zone and go deep down to a depth of 30meters depending upon the turbidity of the water, whereas the rest of the genera are found in the littoral to the sublittoral zone; also found in intertidal pools on an exposed coast growing on a hard rocky reef or any hard substrata. Furthermore, in the sheltered sites, they grow on gravel and small stones [34, 35]. *Laminaria* exceptionally does not grow in Antarctica. Further, it never grows in the muddy and sandy bottom or below 20 psu of salinity. The maximum growth of the *Laminaria* spp. occurs between the salinity ranges of 25 to 35 psu. Lengthy plants often reach the surface during low tide but hold fast and always remain under the water. *Laminaria* is popularly named 'haidai' in China, 'kombu' in Japan and 'Dasima' in The Republic of Korea. The genus *Alaria esculenta,* commonly called the "winged kelp" or "Atlantic wakame", can grow up to 4 meters in length. It grows on the strong wave-exposed rocky substrata. It cannot grow in places like the South-North Sea and the English Channel where the water temperature exceeds more than 16°C [36]. *Sargassum fusiforme* (Japanese Hiziki), *S. fulvellum* are edible species, especially in China and the Republic of Korea, where growth is abundantly along the littoral to sublittoral zones of rocky coastline.

Red Seaweeds

The division Rhodophyta embraces a huge number of edible species that are available commercially in the market, such as *Porphyra/Pyropia/Phycocallidia* spp. (Purple laver), *Palmaria palmata* (Dulse), *Chondrus crispus* (Irish moss), *Kappaphycus/Eucheuma*, and *Gracilaria* spp. (Ogo). The genera *Porphyra* and *Gracilaria* are cosmopolitan in occurrence and grow widely along with diverse ecological niches from the splash zone of littoral to sublittoral zones, including tidal pools attached to rocky reefs, sand, and pebbles. Some species of *Porphyra* are often found as an epiphyte on several other seaweeds like *Ulva, Grateloupia, etc*. The genera *Chondrus crispus, Kappaphycus*, and *Palmaria* grow on the sheltered to semi-exposed rocky substrates in moderate to strong water currents along shallow littoral to sublittoral zones deep down up to a depth of 24 meters. In European waters, *Chondrus* is often found growing as an epiphyte on *Laminaria* and *Fucus*.

BIOGEOGRAPHY OF EDIBLE SEAWEEDS

A certain number of seaweeds enjoy a wide geographical distribution while some would appear to be almost cosmopolitan, but a vast majority are restricted to definite areas. Temperature and light are undoubtedly the two principal factors determining the latitudinal distribution of seaweeds. The tropical seaweed flora is distinguished especially by the abundance of Siphonales and the dominance of Florideae. The most important Phaeophyceae are Dictyotales and species of *Sargassum* and *Turbinaria*, whereas Laminriales are completely lacking. The algal flora of the Tropics shows a greater resemblance to that of the Southern than the Northern hemisphere. A considerable number of seaweeds are restricted to the Southern hemisphere. The sub-Antarctic and Antarctic floras include several species with a circumpolar distribution. Generally, *Caulerpa* and *Codium* are warm water tropical genera but are also found in cold temperate regions as invasive species. The species of *Ulva* and *Monostroma* are cosmopolitan in occurrence. The brown seaweeds, namely *Laminaria, Undaria, Alaria, Ecklonia,* and *Lessonia* are restricted in occurrence, commonly growing in the cold temperate (The Sea of Japan, Sea of Okhotsk, Tartar Channel, and the North Pacific Ocean) and the subpolar regions (The United States, Canada, New England, Denmark, Norway, Finland, Sweden, Iceland, and Russia) of the world, whereas the genera like Durvillaea, *Hizikia* and *Sargassum*, *Padina*, *Dictyota*, *Turbinaria*, *Colpomenia*, *Hydroclatharus*, and *Rosenvingea* are found in temperate to tropical regions including Japan, China, Republic of Korea, Russia, Philippines, Indonesia, India, Sri Lanka, *etc*. The genera *Porphyra* and *Gracilaria* are found in polar, subpolar, cold and warm temperate, tropical and subtropical regions of the globe. *Chondrus crispus* and *Palmaria palmata* are cold temperate species distributed in North Atlantic (Greenland to Portugal), and Pacific oceans (Canada to Russia and Japan) [37]. The species of *Alaria* are widely distributed in the Northern hemisphere, especially the North Pacific and Atlantic oceans [38].

METHODS OF CULTIVATION EMPLOYED FOR EDIBLE SEAWEEDS ACROSS THE WORLD

Worldwide, seaweed cultivation provides a livelihood option to coastal settings with a market value of USD 6 billion. During 2015, about 29.1 million tons of seaweeds were produced, an encompassing 85% share for food and food products [43]. Generally, seaweeds are commercially cultivated in three different ways i. vegetative propagation, ii. spore/zygote-derived asexual and sexual reproduction in two phases, and iii. land-based cultivation by both vegetative and spore/zygote-derived seedlings. The three groups of seaweeds (Chlorophyta, Phaeophyta, and Rhodophyta) exhibit a multiphasic pattern of the life cycle. Regulating the entire life cycle of algae has endowed the commercial farming of many seaweed species since ancient times, especially in Asian countries. Recently, advancement has taken place to develop species-specific cultivation technology [39].

Farming through Vegetative Propagation

The red seaweeds species of *Gracilaria*, *Kappaphycus/Eucheuma*, and, *Palmaria palmata* are being cultivated by means of vegetative propagation. The individual fragment has the ability to reproduce through meristematic cells. Various methods are used worldwide based on geographical specificity. Generally, off bottom longlines, bamboo raft methods are employed to calm shallow protected areas. Floating monolines and tubular net methods are used where cultivation is carried out in the open seas or the bay area [14, 40]. The algae fragments are inserted into polypropylene ropes and allowed to grow for 30-45 days. The cultivator needs to clean the longlines and rafts and other contaminants on a daily basis. The genus *Eucheuma* is the world's most mariculture seaweed and yielded about 10.2 million tons during the year 2015, whereas *Kappaphycus* and *Gracilaria* produced 1.8 million tons and 3.9 million tons, respectively [43].

The species of *Caulerpa* are widely sold in the local public markets of the Philippines and Vietnam, harvested from natural beds, and internationally exported to Japan and Taiwan; although its cultivation in the Philippines is nascent [41, 42]. The wild harvest is an alternative livelihood for the Philippines coastal population. It costs around USD 0.30-0.79 per kg of fresh weight [41]. *Caulerpa lentillifera* is cultivated in ponds and open lagoon areas of the Philippines, and Thailand [43]. The branches of the thallus are simply buried in the muddy substratum for a specific interval. Harvesting is done when a uniform bed of *Caulerpa* is formed by keeping sufficient seed stock for the future crop. The plants are then uprooted from the mud. The average production of *Caulerpa* in the Philippines was 1219 tons during the year 2015 [43].

Spore/Zygote Derived Seedlings

The green seaweeds like *Ulva/Enteromorpha,* and *Monostroma* the brown seaweeds like *Laminaria/Saccharina, Undaria, Alaria*, and *Ecknonia, Sargassum/Hizikia* the red seaweeds *Porphyra/Pyropia/Phycocalidia* are being cultivated by asexual spores-zoospores, zygotes, and chonchospores respectively.

It involves two-stage cultivation that requires a nursery for rearing seedlings derived from zoospores, zygotes and chonchospores, and out-planting of juvenile seedlings to cultivation fields in the open seas.

Cultivation of Ulva, Monostroma and *Codium*

The genera *Ulva* and *Monostroma* both are cultivated through the nursery rearing of seedlings by zoospores [23, 24, 44 - 47]. In the nursery, plants are facilitated to release the zoospores and collected on nylon or cotton strings and maintained at a specific temperature and light intensity. After attaining certain growth, the seedlings are transplanted to the fields for further growth. The cultivation of *Ulva* and *Monostroma* through nursery rearing of the seedlings is traditionally done in Japan, China, and South Korea and has currently been adopted by various countries like Brazil, the USA, and Australia [23, 45]. The annual production of *Monostroma* in Japan is approximately 2500 tons dry weight, whereas 6159 metric tons fresh weight for *Ulva* spp. in South Korea [46, 48]. About 100 metric tons of dry weight of *Ulva* is produced through cultivation in China [49].

Codium fragile cultivation is popular in South Korea as it has high market value than *Porphyra* and *Undaria* [48]. Generally, zygote-based natural seeding on the cultivation ropes was practiced traditionally since 1987. The artificial seeding and tank culture method was developed by Hwang *et al.* [48] by seeding utricles and medullary filaments of *Codium fragile*. The vegetative thalli are chopped with a hand blender into pieces less than 5 mm in length, followed by isolating utricles and medullary filaments seeded on a seed frame coiled with a 100meter seed fiber. The seeded frames were cultured in tanks for 60 days till the plants grew up to 4 cm and are then out-planted on the floating longlines on the open seas [50]. The annual production of *Codium* in Korea was 3980 metric tons during the year 2017, with a market value of USD $2048 [48].

Cultivation of Laminaria/Saccharina, Undaria, and *Ecklonia*

These seaweeds have a biphasic lifecycle, *i.e.*, the large sporophytes phase alternate with the microscopic gametophytic phase. For cultivation, the large concrete tanks filled with seawater are kept under shaded and controlled environmental conditions. These tanks are maintained with reproductive sporophytic blades, which will release zoospores. The zoospores are then recruited on seed strings tied to a frame of PVC and placed in the artificial tanks in the nursery with 17-25°C temperature and light intensity up to 10 µmoles photons $m^{-2}s^{-1}$. At the commencement of autumn (September-October) when the temperature falls below 23°C, the seeded frames are then maintained in the intermediate culture fields for acclimatization of young sporophytic plants to open seawater conditions. Once the sporophytic plants attain certain growth up to 1 cm in length, the seeded string fibers are removed from the frame and tied to culture ropes. Recently sewing machines have been used to wind the seeded ropes to the main culture ropes. About 8.0 million tons of *Laminaria/Saccharina* and 2.3 million tons of *Undaria* were produced during the year 2015 worldwide.

Sargassum Cultivation

The *Sargassum* cultivation is based on sexual reproduction; asexual reproduction is absent. The plant body is gametophytic, consisting of antheridia (male) and oogonia (female) reproductive organs. The zygote (fertilization product) is used as seed stock for *Sargassum* cultivation. The mature fertile thalli are cultured (two months) in the outdoor concrete tanks for the release of zygotes. The zygote suspension is then collected by rubbing the thalli followed by sieving them through a net (300-500 μm mesh size). The zygotes are then seeded on the cotton and nylon mixed fiber strings (100 meters in length) tied to a PVC frame with the help of a paintbrush. The seeded PVC frames are reared in an outdoor tank for 60 days until the thalli reach 10-15 mm in length. Fresh water and aeration are continuously supplied to the tanks in addition to a light of 60-80 μmol photons $m^{-2}s^{-1}$. The seedlings were then out-planted to an open sea farm and cultivated using a long line method in which the seeded strings are coiled around the polypropylene rope [50]. At present, *Sargassum* is being cultivated in the Republic of Korea with an annual production of 54925 metric tons with a market value of USD 0.029 billion.

Porphyra/Pyropia/Phycocalidia Cultivation

These genera show a heteromorphic lifecycle consisting of filamentous sporophyte *i.e.*, conchocelis and bladed gametophyte. The modern technique of *Porphyra* cultivation comprises 4 steps viz. culture of conchocelis, recruitment of the conchospores on the culture nets, nursery rearing of seedlings, and out-planting. In Asian countries, the conchocelis culture commences in mid-spring (March-April) through zygospores. The conchocelis phase is grown on oysters, scallop shells, or artificial substrata made up of vinyl films and calcium carbonate in shallow tanks (0.25-0.5 meters in depth) containing seawater and zygotospores. The zygotospores suspension can be produced in various ways, for example, drying the fertile thalli overnight followed by immersion into the seawater, chopping the fertile thalli, and putting them into tanks containing oysters or artificially prepared shells, grinding the fertile thalli and separating the zygotospores by filtration. In advanced cultivation techniques, the conchocelis are directly seeded on an oyster or artificially prepared calcium carbonate substrata. The substrata are placed either hanging or scattered over the bottom of the tanks. The conchocelis phase is allowed to grow during summer 16:8L: D photoperiod,

23°C, 35psu salinity, 45μmole photons $m^{-2}s^{-1}$. The conchocelis phase is also propagated in bioreactors under controlled environmental conditions in order to get a huge quantity of conchospores. The conchocelis phase matures from late September to October into stars. The seeding of nets is carried out in autumn, preferably when the water temperature ranges from 23-24°C. The nets are 1.8 to 2 meters wide and 18-45 meters long and made up of nylon or polyethylene or polypropylene fiber with a mesh size varying from 15-18 cm or 30-35 cm. One of the most important characteristics of these nets is they absorb water, swell speedily and dry instantaneously. At the time of cultivation (autumn season), the seeding of nets is done by placing nets in the conchospores tank mechanically by fixing the nets on a roller or a belt that is rotated in the seeding tanks containing mature conchospores. The seeded nets are then transferred to the sea nursery. In the nursery, the seeded nets are then placed on a floating system in such a way that they should be exposed periodically to improve growth and reduce epiphytes. After attaining a thallus length of 1-3 cm, the nets are transferred to the farm site [39, 51]. As per FAO 2018, approximately 1.2 million tons of *Porphyra* were produced through mariculture across the world [43].

Land-Based Cultivation

This is one of the most significant methods that includes pond or tank culture of seaweed species. Generally, the species of *Ulva* and *Gracilaria* are cultivated in the pond and integrated into a multi-trophic system co-cultured with marine animals. Besides, *Ulva* and *Gracilaria*, *Porphyra* is also a potential candidate for outdoor tank cultivation through conchospores and monospores [52, 53]. It is an eco-friendly practice and has the advantage to regulate the growth conditions for the desired valuable products like proteins, sulphated polysaccharides, fatty acids and amino acids, and ensures quality and safety [53].

FORM OF CONSUMPTION AND OTHER CONSUMABLE PRODUCTS

Green Seaweeds

The *Ulva* is commonly called sea-lettuce. Various species of *Ulva* namely, *U. lactuca, U. ohoni U. conglobata, U. fasciata, U. compressa, U. reticulata, U. tubulosa, U. prolifera, U. linza, U. rigida, U. intestinalis* (gut weed) are edible in various Asian and European countries; used as a condiment in tea, vegetables, soups and cooked with meat or fish [47, 54]. Dried sheets of *U. fasciata* roasted in oil are used as snacks in Sri Lanka [55]. Aonori, is a trading name for *Ulva* and *Monostroma* and its powder is often sprinkled on cooked rice. The thalli of *Monostroma* are processed as a preserved food, boiled with soya sauce and sugar. In China *Monostroma nitidum* is used as a condiment; the dried thalli are cooked with vegetables and fish. In coastal China and Turkey, sweet and salty bread slices are baked with cornmeal and *M. angicava, U. intestinalis*. Besides this, egg noodles are also prepared from *M. nitidum* [47, 54, 56].

Brown Seaweeds

The *Laminaria, Undaria and Hizikia* are consumed as raw, boiled, or dried and used as an ingredient in vegetables, and soups. In Vietnam, these are mixed with green beans, coconut milk, jellies, *etc*. The quality i.e. substantiality value of *Laminaria* for human consumption is generally assessed by the thickness of the blade corresponding to its duration of cultivation and evaluated by the formula as wet weight per area of the blade. The highest quality *Laminaria* is produced after two years of cultivation [57]. The price and taste are the key factors for "Kelp" consumption in Asia, but in Europe and America, it is selected on the basis of nutritional value. It is interesting to note that the "Kelp" value in Europe and America is USD 944 tons[1] while in Asia it is USD177 tons[1] of wet weight [58]. The *Sargassum* spp. is popularly known as horsetail seaweed; it is consumed as a soup or along with soya sauce in the Republic of Korea, Hawaii, and the Philippines [39, 59]. *Fucus vesiculosus* is used to make tea. It is not much utilized as a food for human consumption however, at some places it is used in cooked foods, soups, or sprinkled on salads.

Methods of Processing of Undaria

The plants are cut at the bottom and top during harvesting and dried above 65°C temperature. The fronds turn green after drying. The green color of the frond is dependent upon the time and temperature during drying. One has to note that too high a temperature leads to degradation of the pigment chlorophyll to phaeophytin which gives the fronds the brown color. Incorrect boiling and drying lead to a change in the color of fronds from green to brown during storage. It was suggested that a short period of boiling at 80-95°C for 30-60 seconds is best for getting the desirable color i.e. bright green. After drying the frond of *Undaria*, it is cooled and mixed with a saltwater solution. The salted fronds are then dehydrated for several days, after which the midribs are removed and later packed into polyvinyl packages and stored at -20°C. Secondly, the *Undaria* fronds are boiled, chopped, desalted, and re-dried. Drying using this process is carried out by mechanical rolling dryers. A major disadvantage of this process is quality control. Frequently foreign material adheres to the harvested fronds. Roughly 50 extraneous materials per ton of dried wakame are accepted by the manufacturer [39, 60].

Red Seaweeds

Among the red seaweeds, *Porphyra* is one of the most nutritious and esteemed seaweed in the world. It is either eaten in a dried or processed form. The dried and roasted *Porphyra* sheets, commonly called Nori sheets, are available in the retail market [51]. 'Shushi' is one of the most delicious and popular cuisines all over the world. It is also used in the preparation of soups, salads, jams, and jellies [61] The *Palmaria* has a special place in the Irish population; its consumption was reported way back in the 5th century. It is eaten raw or dried, cooked with milk, used as a condiment, sprinkled on salads and vegetables, sandwiched with buttery bread, with fish, used in stew and soup; toasted *Palmaria* is consumed as a snack. Due to its, salty, sweet, nut-like taste, *Palmaria* is often garnished with corn and root vegetables. In Briton, it is also used as animal fodder and popularly named based on the animal for which it has been used, for example, cow-weed in England, and Horseweed in Norway. Along with the above-mentioned countries, it is also consumed as a dietary supplement in North America and Europe [62]. Another important edible seaweed is *Chondrus,* commonly called Irish moss. In Ireland, it is traditionally used in making milk-based puddings [63]. The other seaweeds like *Gracilaria, Kappaphycus, Eucheuma, Grateloupia, Acanthophora, Hypnea, Gelidium, Laurencia,* and *Halymenia*, are commonly used as vegetable salads or eaten as raw or in the boiled form [64, 65].

FUTURE PROSPECTS

Seaweeds show enormous biodiversity in terms of habitat, physiology, and biotic as well as abiotic stress resilience. However, changing climatic conditions have forced these communities towards cumulative migrations and adaptations at various levels. In addition, due to the lack of poorly understood habitat, biology, and their geo-climatic interactions, resolutions to their loss in abundance or extensive invasiveness remain a great matter of concern among phycologists. Efforts for conserving marine macroalgal diversity could be extended to modeling artificially created native habitats to conserve rare or low abundance species that may demand a controlled biosphere with optimized levels of biotic as well as abiotic factors necessary for physiological development and growth. However, seaweeds might be able to quickly adapt to changing climates, but their specific abundance in response to the changing environments remains a matter of great concern for human welfare, such as in the case of "green tides" from the yellow sea and "golden tides" from the Caribbean Sea.

CONSENT FOR PUBLICATION

Not applicable.

CONFLICT OF INTEREST

The author(s) declare no conflict of interest, financial or otherwise.

ACKNOWLEDGEMENTS

The author(s) would like to thank CSIR, New Delhi, for financial support, Director, CSIR-CSMCRI, for support and facilities. TKG acknowledges CSIR for the SRF grant. CSIR-CSMCRI PRIS registration number 94/2021.

REFERENCES

[1] Graham MH. Effects of local deforestation on the diversity and structure of southern California giant kelp forest food webs. Ecosystems (N Y) 2004; 7(4): 341-57. [http://dx.doi.org/10.1007/s10021-003-0245-6]

[2] Brodie J, Ash LV, Tittley I, Yesson C. A comparison of multispectral aerial and satellite imagery for mapping intertidal seaweed communities. Aquat Conserv 2018; 28(4): 872-81. [http://dx.doi.org/10.1002/aqc.2905]

[3] Harley CDG, Anderson KM, Demes KW, *et al.* Effects Of Climate Change On Global Seaweed communities J Phycol 2012; 48(5): 1064-78. [http://dx.doi.org/10.1111/j.1529-8817.2012.01224.x]

[4] Flukes EB, Wright JT, Johnson CR. Phenotypic plasticity and biogeographic variation in physiology of habitat-forming seaweed: response to temperature and nitrate. J Phycol 2015; 51(5): 896-909. [http://dx.doi.org/10.1111/jpy.12330] [PMID: 26986886]

[5] Eggert A. Seaweed Responses to Temperature. In: Wiencke C, Bischof K, Eds, Seaweed Biology. Ecological Studies. Springer, Berlin, Heidelberg 2012; vol 219: pp. 47-66. [http://dx.doi.org/10.1007/978-3-642-28451-9_3]

[6] Rönnbäck P, Kautsky N, Pihl L, Troell M, Söderqvist T, Wennhage H. Ecosystem goods and services from Swedish coastal habitats: identification, valuation, and implications of ecosystem shifts. Ambio 2007; 36(7): 534-44. [http://dx.doi.org/10.1579/0044-7447(2007)36[534:EGASFS]2.0.CO;2] [PMID: 18074889]

[7] Felder DL, Camp DK, Eds. Gulf of Mexico Origin, Waters, and Biota: Biodiversity - Google Books. Vol. 1. Texas A&M University Press, 2009; 1393.

[8] Yesson C, Bush LE, Davies AJ, Maggs CA, Brodie J. Large brown seaweeds of the British Isles: Evidence of changes in abundance over four decades. Estuar Coast Shelf Sci 2015; 155: 167-75. [http://dx.doi.org/10.1016/j.ecss.2015.01.008]

[9] Brodie J, Wilbraham J, Pottas J, Guiry MD, Brodie J. A revised check-list of the seaweeds of Britain. J Mar Biol Assoc U K 2016; 96(5): 1005-29. [http://dx.doi.org/10.1017/S0025315415001484]

[10] Sorte CJB, Williams SL, Zerebecki RA. Ocean warming increases threat of invasive species in a marine fouling community. Ecology 2010; 91(8): 2198-204. [http://dx.doi.org/10.1890/10-0238.1] [PMID: 20836440]

[11] Brodie J, Andersen RA, Kawachi M, Millar AJK. Endangered algal species and how to protect them. Phycologia 2009; 48(5): 423-38. [http://dx.doi.org/10.2216/09-21.1]

[12] Lüning K. Seaweeds: Their Environment, Biogeography, and Ecophysiology, Wiley 1990; 544.

[13] Vázquez-Delfín E, Freile-Pelegrín Y, Pliego-Cortés H, Robledo D. Seaweed resources of Mexico: Current knowledge and future perspectives 2019; Vol. 62: 275-89.

[14] Mantri VA, Kavale MG, Kazi MA. Seaweed biodiversity of India: Reviewing current knowledge to identify gaps, challenges, and opportunities. Diversity 2020; 12(1): 13.

[15] Abdullah Al M, Akhtar A, Rahman MF, Kamal AHM, Karim NU, Hassan ML. Habitat structure and diversity patterns of seaweeds in the coastal waters of Saint Martin's Island, Bay of Bengal, Bangladesh. Reg Stud Mar Sci 2020; 33: 100959. [http://dx.doi.org/10.1016/j.rsma.2019.100959]

[16] Hoshino M, Sanematsu K, Watanabe Y. REE Mineralogy and Resources. Handbook on the Physics and Chemistry of Rare Earths 2016; 129-291.

[17] Trenhaile A. The Geomorphology of rock coasts. Oxford University Press 1987; 384.

[18] Bermejo R, MacMonagail M, Heesch S, *et al.* The arrival of a red invasive seaweed to a nutrient over-enriched estuary increases the spatial extent of macroalgal blooms. Mar Environ Res 2020; 158: 104944. [http://dx.doi.org/10.1016/j.marenvres.2020.104944] [PMID: 32250838]

[19] Raven JA, Taylor R. Macroalgal growth in nutrient-enriched estuaries: A biogeochemical and evolutionary perspective. Water Air Soil Pollut Focus 2003; 3(1): 7-26. [http://dx.doi.org/10.1023/A:1022167722654]

[20] Sahoo D, Sahu N, Sahoo D. A Critical Survey of Seaweed Diversity of Chilika Lake, India. Algae 2003; 18(1): 1-12. [http://dx.doi.org/10.4490/ALGAE.2003.18.1.001]

[21] Braga MRA. Recruitment of two species of monostromatic blade-like chlorophytes, Monostroma sp. and Ulvaria oxysperma (Ulvales, Chlorophyta), in Sao Paulo State, Brazil. Phycol Res 1997; 45(3): 153-61. [http://dx.doi.org/10.1111/j.1440-1835.1997.tb00088.x]

[22] Kavale MG, Italiya B, Veeragurunathan V. Scaling the production of Monostroma sp. by optimizing culture conditions. J Appl Phycol 2020; 32(1): 451-7. [http://dx.doi.org/10.1007/s10811-019-01922-6]

[23] Pellizzari FM, Absher T, Yokoya NS, Oliveira EC. Cultivation of the edible green seaweed Gayralia (Chlorophyta) in southern Brazil. J Appl Phycol 2007; 19(1): 63-9. [http://dx.doi.org/10.1007/s10811-006-9111-1]

[24] Ohno M. Cultivation ofMonostroma nitidum (Chlorophyta) in a river estuary, southern Japan. J Appl Phycol 1995; 7(2): 207-13. [http://dx.doi.org/10.1007/BF00693069]

[25] Jackson NL. Estuaries. Vol. 10, Treatise on Geomorphology. Elsevier Ltd. 2013; 308-27.

[26] N. Kaliaperumal PK and SK. Seaweed and Seagrass Resources. CMFRI bulletin 43: Marine Living Resources of the Union Territory of Lakshadweep-An Indicative Survey With Suggestions For Development James PSBR, Suseelan C, Eds. 1989; 162-75.

[27] Yabe T, Ishii Y, Amano Y, *et al.* Green tide formed by free-floating Ulva spp. at Yatsu tidal flat, Japan. Limnology 2009; 10(3): 239-45. [http://dx.doi.org/10.1007/s10201-009-0278-4]

[28] Krishnamurthy V. Algae of India and Neighbouring Countries I. Chlorophycota. New Delhi: Oxford and IBH Publishing Co. Pvt. Ltd 2000; 210.

[29] Masakiyo Y, Shimada S. Species Diversity of the Genus Ulva (Ulvophyceae, Chlorophyta) in Japanese Waters, with Special Reference to Ulva tepida Masakiyo et S. Shimada sp. nov. Bull Natl Mus Nat Sci Ser B Bot 2014; 40(1): 1-13.

[30] Singh KP, Bhargava KS. A note on the occurence of Monostroma Quaternarium (Kutz.) Desm. from Nainital, India. Curr Sci 1959; 28(5): 205.

[31] Rybak A, Messyasz B. New locality of greenalga *Viva prolifera* (Ulvaceae) in the Wielkopolska Region. Fragm Florist Geobot Pol 2010; 17(1): 199-203.

[32] Miladi R, Manghisi A, Armeli Minicante S, Genovese G, Abdelkafi S, Morabito M. A DNA Barcoding Survey of *Ulva* (Chlorophyta) in Tunisia and Italy Reveals the Presence of the Overlooked Alien *U. ohnoi*. Cryptogam, Algol 2018; 39(1): 85-107. [http://dx.doi.org/10.7872/crya/v39.iss1.2018.85]

[33] Bolton JJ. The biogeography of kelps (Laminariales, Phaeophyceae): A global analysis with new insights from recent advances in molecular phylogenetics 2010; Vol. 64: 263-79.

[34] Santelices B. The discovery of kelp forests in deep-water habitats of tropical regions. Vol. 104, Proceedings of the National Academy of Sciences of the United States of America. National Academy of Sciences 2007; 19163-4. [http://dx.doi.org/10.1073/pnas.0708963104]

[35] Garbary DJ, Deckert RJ. Three Part Harmony — Ascophyllum and Its Symbionts. Symbiosis 2006; 309-21.

[36] Kraan S, Verges Tramullas A, Guiry MD. The edible brown seaweed Alaria esculenta (Phaeophyceae, Laminariales): Hybridization, growth and genetic comparisons of six Irish populations. J Appl Phycol 2000; 12(6): 577-83. [http://dx.doi.org/10.1023/A:1026519030398]

[37] Pina AL, Costa AR, Lage-Yusty MA, López-Hernández J. An evaluation of edible red seaweed (Chondrus crispus) components and their modification during the cooking process. Lebensm Wiss Technol 2014; 56(1): 175-80. [http://dx.doi.org/10.1016/j.lwt.2013.08.006]

[38] Kraan S, Rueness J, Guiry M. Are North Atlantic *Alaria esculenta* and *A. grandifolia* (Alariaceae, Phaeophyceae) conspecific? Eur J Phycol 2001; 36(1): 35-42. [http://dx.doi.org/10.1080/09670260110001735178]

[39] Hwang EK, Choi HG, Kim JK. Seaweed resources of Korea 2020; Vol. 63: 395-405.

[40] Martínez B, Viejo RM, Rico JM, *et al.* Open sea cultivation of *Palmaria palmata* (Rhodophyta) on the northern Spanish coast. Aquaculture 2006; 254(1-4): 376-87. [http://dx.doi.org/10.1016/j.aquaculture.2005.10.025]

[41] Estrada JL, Dionisio-Sese ML. Sea Grapes (Caulerpa Spp.) (Chlorophyta: Caulerpaceae) from Coron, Northern Palawan, the Philippines with Notes on Their Harvest And Production. Trop Nat Hist 2020; 20(3): 255-64.

[42] Nagappan T, Vairappan CS. Nutritional and bioactive properties of three edible species of green algae, genus Caulerpa (Caulerpaceae). J Appl Phycol 2014; 26(2): 1019-27. https://link.springer.com/ article/10.1007/ s10811-013-0147-8 [http://dx.doi.org/10.1007/s10811-013-0147-8]

[43] FAO. The Global Status of Seaweed Production 2018.

[44] Hong DD, Hien HM, Son PN. Seaweeds from Vietnam used for functional food, medicine and biofertilizer. Journal of Applied Phycology 2007; 817-26. [http://dx.doi.org/10.1007/s10811-007-9228-x]

[45] Li Y, Song W, Xiao J, *et al.* Tempo-spatial distribution and species diversity of green algae micro-propagules in the Yellow Sea during the large-scale green tide development. Harmful Algae 2014; 39: 40-7. [http://dx.doi.org/10.1016/j.hal.2014.05.013]

[46] Ohno M, Triet VD. Artificial seeding of the green seaweed Monostroma for cultivation. J Appl Phycol 1997; 417-23.

[47] Kılınç B, Cirik S, Turan G, Tekogul H, Koru E. Seaweeds for Food and Industrial Applications In:

Muzzalupo I Food industry London: IntechOpen 2013.
[http://dx.doi.org/10.5772/53172]

[48] Hwang EK, Park CS. Seaweed cultivation and utilization of Korea. Vol. 35, Algae. Korean Society of Phycology 2020; 107-21.
[http://dx.doi.org/10.4490/algae.2020.35.5.15]

[49] Zhang QC, Yu RC, Chen ZF, *et al.* Genetic evidence in tracking the origin of Ulva prolifera blooms in the Yellow Sea, China. Vol. 78, Harmful Algae. Elsevier B.V 2018; 86-94.

[50] Hwang EK, Park CS, Baek JM. Artificial seed production and cultivation of the edible brown alga, Sargassum fulvellum (Turner) C. Agardh: Developing a new species for seaweed cultivation in Korea. Eighteenth International Seaweed Symposium 2008; 25-31.

[51] Levine IA, Sahoo D. Porphyra: Harvesting Gold from the Sea. I.K. International Publishing House Pvt. Ltd., New Delhi, 2010.

[52] Neori A, Guttman L. Thoughts on Algae Cultivation toward an Expansion of Aquaculture to the Scale of Agriculture. 7th International Conference on Innovation in Chemical, Agricultural, Biological and Environmental Sciences 2017.

[53] Israel A, Levy I, Friedlander M. Experimental tank cultivation of Porphyra in Israel. Eighteenth International Seaweed Symposium 2008; 9-14.

[54] Abbott IA. Ethnobotany of seaweeds: clues to uses of seaweeds. Fifteenth International Seaweed Symposium 1996; 15-20.
[http://dx.doi.org/10.1007/978-94-009-1659-3_2]

[55] Jayakody MM, Vanniarachchy MPG, Wijesekara WLI. Development and characterization of a seaweed snack using *Ulva fasciata.* J Food Sci Technol 2021; 58(4): 1617-22.
[http://dx.doi.org/10.1007/s13197-020-04880-x] [PMID: 33746289]

[56] Chang HC, Wu LC. Texture and quality properties of Chinese fresh egg noodles formulated with green seaweed (Monostroma nitidum) powder. J Food Sci 2008; 73(8): S398-404.
[http://dx.doi.org/10.1111/j.1750-3841.2008.00912.x] [PMID: 19019127]

[57] Peteiro C, Salinas J, Freire O, Fuertes C. Cultivation of the autochtonous seaweed "Laminaria saccharina" off the Galician Coast (NW Spain): production and features of the sporophytes for an Annual and Biennial Harvest. Thalass An Int J Mar Sci 2006; 22(1): 45-53.

[58] Grebe GS, Byron CJ, Gelais AS, Kotowicz DM, Olson TK. An ecosystem approach to kelp aquaculture in the Americas and Europe 2019; Vol. 15: 100215.

[59] Novaczek I, Athy A. Sea Vegetable Recipes for the Pacific Islands. USP Marine Studies Programme / Secretariat of the Pacific Community 2001.

[60] Yamanaka R, Akiyama K. Cultivation and utilization of Undaria pinnatifida (wakame) as food. J Appl Phycol 1993; 5(2): 249-53.
[http://dx.doi.org/10.1007/BF00004026]

[61] Ohno M, Largo DB. The Seaweed Resources of Japan. In: Critchley AT, Ohno M, Eds. Seaweed Resources of the World 1st ed Japan International Cooperation Agency 1998; 1-14.

[62] Mouritsen OG, Dawczynski C, Duelund L, Jahreis G, Vetter W, Schröder M. On the human consumption of the red seaweed dulse (Palmaria palmata (L.) Weber & Mohr). J Appl Phycol 2013; 25(6): 1777-91.
[http://dx.doi.org/10.1007/s10811-013-0014-7]

[63] Imeson AP. Carrageenan and furcellaran. Handbook of Hydrocolloids: Second Edition Elsevier Inc 2009; 164-85.
[http://dx.doi.org/10.1533/9781845695873.164]

[64] Lindsey Zemke-White W, Ohno M. World seaweed utilisation: An end-of-century summary. J Appl Phycol 1999; 11(4): 369-76.

[http://dx.doi.org/10.1023/A:1008197610793]

[65] Trono GC. The Seaweed Resources of Philippines. In: Critchley, Alan T, Ohno M, editors. Seaweed Resources of the World. Yokusuka: Japan International Cooperation Agency 1998; 47-61.

CHAPTER 5

Algal-Based Nutraceuticals: Application and Potential

Jayani Samarathunga[1], **Isuru Wijesekara**[1,*] and **Madhura Jayasinghe**[1]

[1] *Department of Food Science and Technology, Faculty of Applied Sciences, University of Sri Jayewardenepura, Gangodawila, Nugegoda, Sri Lanka*

Abstract: Today, consumers pay more attention to natural therapies rather than pharmaceutical treatments. Contentedly, algal nutraceuticals provide a massive range of medicinal properties. Both macro and microalgae have shown health beneficial bioactivities like antioxidant activity, antidiabetic, anti-inflammatory, anticancer, antitumor activities, *etc*. When evaluating algal nutraceuticals in the global market, several weaknesses can be identified with regard to their food safety and quality. The lack of standards and legislation on handling seaweeds in the food industry is a hindrance to enjoy the actual health benefits of algal nutraceuticals. Trends in microalgal nutraceuticals have increased since they can be cultivated easily in the natural environment. The capacity of biological activities of algae in the human body as nutraceuticals and pharmaceuticals is higher than some terrestrial plants and animal sources. Therefore, bioactive compounds of algae will definitely help to prevent infectious pandemic diseases like COVID-19. Algal proteins also have emerged as a meat alternative. Research studies on the bioavailability and digestibility of these bioactive constituents are found to be still limited. Therefore, consumers show less tendency to consume algal nutraceuticals. Anti-nutritional factors like fiber, tannin and certain cations reduce the bioavailability and digestibility of algal nutrients. However, new technologies are now being developed to reduce these limiting factors. Today, with the influence of the green technology concept, algae will perform a great role in human health as a promising source of nutraceuticals.

Keywords: Algal nutraceuticals, Anticancer, Antidiabetic, Anti-Inflammatory, Anti-nutritional, Antioxidant, Antitumor Activities, Bioactivities, Bioavailability, Digestibility, Food safety, Green technology, Legislations, Meat alternative, Micro algae, Natural therapies, Pandemic diseases, Pharmaceutical treatments, Quality.

* **Corresponding author Isuru Wijesekara:** Department of Food Science and Technology, Faculty of Applied Sciences, University of Sri Jayewardenepura, Gangodawila, Nugegoda, Sri Lanka; Tel: +94768237800; Fax:+94112758000; E-mail: isuruw@sci.sjp.ac.lk

Avinash Mishra (Ed.)

INTRODUCTION

What are Nutraceuticals?

The demand for seaweed as nutraceuticals is increasing in both Asia and Europe. Nutraceuticals are "broad spectrum biological therapies which basically promote protection and prevention" [1]. The term "nutraceutical," which combines "nutrition" and "pharmaceutical", was first introduced in 1989 by Stephen DeFelice, founder and chairman of the Foundation for Innovation in Medicine [2]. According to DeFelice, a nutraceutical can be defined as "a food (or part of a food) that provides medical or health benefits, including the prevention and/or treatment of a disease" [3].

Seaweeds have been consumed as food from ancient times. It is an excellent source of bioactive compounds [4]. Meanwhile, its pharmaceutical properties have been identified according to various scientific research studies [5]. Seaweeds, as a food together with their medicinal properties, finally have been developed into nutraceuticals. Moreover, Asian and European communities have evaluated safety regulations associated with seaweed consumption in human diets [6]. Similar to plant and animal food sources, algae are rich in protein, carbohydrates, vitamins and minerals. The biological activities of algal proteins and polysaccharides provide evidences for the emergence of an algal nutraceutical as a potential source of functional food [7 - 9]. In this chapter, research gaps, trends and future prospects of algal nutraceuticals have been discussed.

Primary and secondary metabolites of algae have shown antioxidant, anticancer, antiulcer, anti-inflammatory, and anti-diabetic properties in in-vitro studies [10 - 12]. It is important to know whether ingestion of algal-based foods is nontoxic or non-allergic. Cytotoxicity of algal glycoprotein extracts of *Ulva* sp. tested on Vero cells had shown no toxicity [13]. This provides evidence that algal metabolites are safe for consumption. Hence, more such research studies have to be encouraged to increase the utilization of seaweeds as a promising source of nutraceuticals. Nutraceuticals have much attention due to the prevention of chronic diseases [14]. There is always an inflated demand for natural bioresources with potential medicinal benefits as food ingredients proving the famous quote "Let food be your medicine and medicine be your food," by the father of medicine, Hippocrates.

Importance of Nutraceuticals in Human Health

Today, the food habits of humans have been corrupted mainly by unhealthy foods like junk foods and fast foods. However, these foods ultimately result in causing various health problems like diabetes, cancer, osteoporosis, obesity, cardiovascular diseases, cholesterol, arthritis, *etc*. Medication practices for these diseases have badly affected patients since they may face various side effects [2]. Health-conscious people always satiate hunger by consuming natural foods. Hence, natural therapies give more health benefits than drug therapy. There is a positive correlation between foods and drugs with the introduction of natural disease-preventing agents called 'nutraceuticals'. The current market for nutraceuticals is growing up with the increase in population. People fed up with high expensive medicines and disease treatments pay much attention to nutraceuticals since these are almost natural. Nutraceuticals have emerged with the concept 'Prevention is better than cure' and can be mainly categorized as dietary fibre, antioxidants, polyunsaturated fatty acids, probiotics, prebiotics, amino acids, vitamins, minerals and polyphenols [15]. These important metabolites in natural plant and animal sources increase the quality of the food due to antioxidant, anticancer, antidiabetic, and anti-inflammatory properties. Seaweed polysaccharides, proteins, vitamins, minerals and lipids (Fig. **1**) have elicited versatile potentials as nutraceuticals in promoting human health [9].

	Polysaccharides	Protein	Lipids	Vitamins & Minerals	Pigments
Red algae -	Agar Carrageenan Floridian starch	Glycoprotein Lectins Phycobiliproteins (Pigment-Red algae) Micosporin like amino acids	Phytosterols (Fucosterols, cholesterol and β-sitosterol) polyunsaturated fatty acids		Carotenoids Chlorophyll Xanthophyll
Brown algae	Alginate Fucoidan Laminarin				
Green algae -	Ulvan				

Fig. (1). Different types of bioactive compounds in algae which are found as nutraceuticals.

ALGAL BASED NUTRACEUTICALS

Seaweed Polysaccharides as Nutraceuticals

The cell wall of seaweeds consists of polysaccharides like ulvan, alginates, fucoidan, agar and carrageenan, which approximately account for 40-50% of the dry weight [16, 17]. The biochemical composition of polysaccharides varies according to species, growth environment, and harvesting period [18]. Hence, these factors affect the quality of nutraceutical products. For example, in algae located in wavy and rough water, the amount of polysaccharides produced is larger than algae grown in calm water since more polysaccharides are produced to provide protection from harsh environmental conditions [16]. Also, the biochemical structure of these polysaccharides can affect bioactive properties like antioxidant activity, antihypertensive activity, *etc* [19].

Alginates, agar and carrageenan are the most common seaweed polysaccharides utilized in the current market. These are sulphated polysaccharides that have shown biological activities of potential interest in the nutraceutical industry [20]. In 2016, European Union exported 2,462 tons of seaweeds for agar; and 40,802 tons for carrageenan [6]. Comparably, in 2016, European Union imported 4,171 tons of seaweeds for agar and 70,627 tons for carrageenan [6].

Applications

All these hydrocolloids are used as gelling agents, stabilizers, thickening agents, emulsifying agents and food coatings [21]. Apart from the functional applications in the food industry, their therapeutic properties have become prominent with the influence of global pandemics. According to the study evaluated by Eccles *et al.* [22], "Iota-Carrageenan," which is one of the three copolymers of Carrageenan, has shown antiviral activity against common cold in humans. However, these findings had made an attempt "Iota-Carrageenan nasal spray" to be regarded as a safe treatment for the common cold [22]. Besides, carrageenan has elicited antiviral activity against influenza virus H1N1, hepatitis A, herpes and dengue [20, 23]. "Kappa carrageenan" extracted from the red algae *Kappaphycus striatum* had shown antitumor and immunostimulating activities [24]. Moreover, carrageenan is used in the meat and dairy industry as a protein binder [25]. For example, carrageenan extracted from red algae *Eucheuma and Kappaphycus* genera is added to meat as a fat replacer [26].

Agar which is mostly extracted from *Gelidium* sp. and *Gracilaria* sp., is used as an alternative to gelatine [16]. It is interesting that agar has shown antioxidant, antitumor activities [2]. In food applications, agar provides these medicinal properties in addition to its major roles in foods as a gelling agent, adhesive, *etc.* Agar has low digestibility and is therefore added to food in low amounts [27]. For this reason, it is used in diet plans for diabetic patients where formulations made with sugar substitutes mixed with agar as "low-calorie food products" [27]. Agar, carrageenan and alginate are good sources of dietary fiber [28]. Pretreatment of polysaccharide extracted from *Ulva lactuca* had exhibited anti-peroxidative and anti-hyperlipidemic properties in rats intoxicated with d-Galactosamine [29]. Antimicrobial and immunomodulation activities have been identified from the *in vitro* analysis of the stimulation of the immune response mediators of the farm animals' gut treated with sulphated polysaccharide extracted from *Ulva armoricana* [30]. Accordingly, the above-stated sulphated polysaccharide extract has the ability to protect farm animals from infectious diseases [30].

Alginates also have a good potential as nutraceuticals. Toxic materials originating from food have been absorbed by the dietary alginates inside the gastrointestinal tract [31]. Brownlee *et al.* [31] reported that alginates reduce plasma cholesterol more effectively than agar and fucoidan. Certain drinks and snack bars containing alginates have reduced "postprandial peak glucose concentrations" and total glucose uptake compared to those which contain guar gum [31, 32].

Algal Proteins as Nutraceuticals

Generally, green algae can contain 9-26% (dry weight) of protein, while brown algae contain 3-15% (dry weight), and red algae contain up to 47% (dry weight) [33]. Dietary protein we consume with food is absorbed by our body in the form of amino acids. It is important to consume "essential amino acid", amino acids that are not produced by our body for the function of our body and cell growth. Seaweeds contain nearly all essential amino acids [33]. Hence, algae can be considered a good nutraceutical. Amino acid digestibility is a prime factor that increases the quality of protein [34]. Several studies have reported proteins extracted from *Porphyra tenera* have been digested by pronase and presence of both pepsin and pancreatin with *in vitro* digestibility of 70% and 56%, respectively [35]. Different *Sargassum* sp. protein concentrates have shown approximately 80% *in vitro* digestibility [34]. Generally, seaweeds contain a low amount of sulphur-containing amino acids (Methionine and cysteine). But research studies have revealed that seaweeds contain higher amounts of sulphur-containing amino acids than terrestrial plants [2, 36]. These facts provide evidences that humans can fulfill their protein requirement with algal proteins. Most of the seaweeds contain a higher amount of aspartic and glutamic amino acids [33]. Interestingly, "aspartic acid gives a unique flavour while glutamic acid is considered as the major element which produces umami flavour" [2]. *Laminaria japonica* is a seaweed used in Japan that contains constituents of the characteristic umami flavour [37].

Applications

Apart from essential amino acids in seaweeds, lectins are "carbohydrate-binding proteins", which have exhibited antifungal activity against human pathogenic yeasts *Candida guilliermondii*, anti-HIV activity and anticancer properties in humans [2, 38]. Algal lectins are a new group of inhibitors that may be useful as "novel platelet aggregation inhibitors" [39]. Glycoproteins are a common type of protein found in the cell wall of seaweeds [40]. Glycoproteins isolated from the green algae *Codium decorticatum* have shown anticancer properties [41]. Phycobiliproteins also play an important role as a nutraceutical in human diets. These include phycoerythrin, phycocyanin, allophycocyanin, and phycoerythrocyanin. These are mainly extracted from cyanobacteria called *Spirulina* and using red seaweeds [42]. Usually, it is used as a natural pigment in the food industry [42]. Application of marine bioactive peptides has done great work in promoting human health. "Sea weed-derived bioactive peptides" could be formed by enzyme hydrolysis of algal proteins. These peptides have shown antioxidant, antihypertensive and antidiabetic properties [43]. Several studies have revealed about ultraviolet (UV) protective compound called "Mycosporine like amino acids (MAA)". These act like UV filters and act as anticancer agents [44]. The extracts of the red alga, *Palmaria palmata* (Dulse), have elicited anti-proliferative activities due to the presence of MAA [45].

Algal Lipids as Nutraceuticals

Lipids are important to the human body as a good energy source and provide insulation and protection. Although seaweeds contain less than 4% of lipid content [46], these lipids are rich in many bioactive compounds. Studies have reported lipids in brown seaweeds contain bioactive compounds like omega-3 polyunsaturated fatty acid, Fucoxanthin and polyphenol [47]. Algal lipids have shown excellent fatty acid profiles which contain essential Polyunsaturated fatty acids (PUFA) like alpha-linoleic acid (ALA), gamma linoleic acid (GLA), linoleic acid (LA), eicosapentaenoic acid (EPA), docosahexaenoic acid (DHA) and arachidonic acid (AA) [46]. Furthermore, algae have exhibited higher or equal amounts of PUFA compared to terrestrial vegetables [46]. When referring to the total fatty acid composition of seaweeds, Palmitic acid is the most abundant fatty acid [46]. According to the study evaluated by Kumari *et al.* [48], the lipid content of marine algae has been increased among the phyla as Rhodophyta, Chlorophyta and Phaeophyta. Phaeophyta and Rhodophyta species were rich in AA and EPA, while Ulvales were rich in DHA [48]. DHA is essential in brain functioning and development. Both EPA and DHA reduce the risk factors for heart diseases. These fatty acids cannot be synthesised by our body and could be obtained through diets. Hence, the consumption of marine algae acts as nutraceuticals that provide these fatty acids.

Applications

Bioactive constituents in algal lipids have shown various health benefits such as antioxidant, anti-obesity, antidiabetic and anti-proliferative activities in the human body [47]. Incorporation of edible seaweeds Spaghetti (*Himanthalia elongata*),

Wakame (*Undaria pinnatifida*), and Nori (*Porphyra umbilicalis*) to "low salt meat emulsion model system" increased the PUFAs and antioxidant capacity [49]. PUFAs prevent many heart diseases by lowering cholesterol and triacylglycerol in plasma. Phytosterols have elicited many therapeutic properties like anti-obesity, anticancer, anti-diabetes, hepatoprotection, anti-Alzheimer's, antioxidant, anti-inflammatory and antiaging activities [50]. Fucosterol, cholesterol and β-sitosterol were the most common phytosterols available in brown algae, red algae and in green algae, respectively. Phytosterols extracted from *Sargassum angustifolium* have shown anticancer properties on breast and colon cancer cells [51].

Vitamins and Minerals of Algae as Nutraceuticals

Marine algae contain a higher amount of minerals than terrestrial plant sources. The most abundant minerals in seaweeds are Iron (Fe) and Iodine (I) [52]. However, the mineral composition varies with factors like the growth environment. Minerals perform a wide range of vital functions in the human body, including nerve transmission, regulation of metabolic activities, structural components of essential body parts (teeth, bones), *etc*. These minerals are mainly categorized into two classes "Major" and "Trace" minerals. Major minerals, including calcium (Ca), magnesium (Mg), potassium (K), sodium (Na), chloride (Cl), phosphorus (P) and sulfur (S), are required in our body in large amounts while trace minerals iodine (I), zinc (Zn), selenium (Se), iron (Fe), manganese (Mn), copper (Cu), cobalt (Co), molybdenum (Mo), fluoride (F), chromium (Cr) and boron (B) are required in small quantities [53]. It is important to maintain these minerals in a balanced ratio. For example, the Na/K ratio is essential for the healthy function of the nervous system. Twenty-two tropical seaweeds from the Saurashtra coast of India have been evaluated for their mineral composition [54]. Accordingly, all seaweeds have shown "Na/K and C/N ratio and high amounts of macroelements (Na, K, Ca, and Mg) as compared to the terrestrial vegetables" [54]. Studies revealed that the participation of seaweeds in diets to fulfil the Reference Dietary Intake (RDI) of minerals is quite satisfactory [52]. The mineral composition of brown seaweeds was relatively higher than red and green seaweeds [52].

Algae contain both water-soluble vitamins (B and C) and fat-soluble (A, D, E and K) vitamins. Interestingly, seaweeds like Asakusanori (*Porphyra tenera*), maruba-amanori (*Porphyra suborbiculata*) and akaba-gin-nansou (*Rhodo giossum pulcherum*) have elicited a higher amount of vitamin B_{12}, the vitamin which was generally found in animal sources [55]. It helps to prevent anaemia and keep red blood cells and nerve cells healthy. Seaweeds also provide remarkable amounts of provitamin A (carotenoids), vitamin B_1, B_2 and C [8]. Vitamins protect our body from diseases and prevent various deficiency diseases. Studies have concluded seaweeds generally exhibit low bioavailability of vitamins [8]. This is one major limiting factor that has reduced the potential of seaweeds as a source of vitamins.

Algal Pigments as a Potential Source of Nutraceuticals

Pigments in marine algae, such as chlorophylls, carotenoids, phycobiliproteins and xanthophylls *etc*., have been shown to have beneficial health bioactivities like antioxidant, anti-inflammatory, antidiabetic and antiangiogenic properties [56]. These pigments are used as natural food colourings in the food industry [56]. For example, these have been commonly used in jelly products and candies Previous studies have concluded phycobiliproteins could be used as a colourant in puddings [56, 57].

ALGAL NUTRACEUTICALS IN THE GLOBAL MARKET

In the past few decades, various seaweed-based food products using whole seaweed, parts or extracts of seaweeds in either cooked or raw form have been found on the market. Among them, "Sushi" is the most popular Japanese food, which is a roll wrapped in red algae called "Nori" (*Porphyra purpurea*). The study of Taboada *et al.* [58] revealed that nori is an excellent source of nutrients and is well accepted by experimental animals. According to this study, both nori and "Wakame" (*Undaria pinnatifida*), brown algae had reduced the level of LDL cholesterol [58]. Wakame is often served with salads and soups.

Today, there is an increasing demand for micro algae-based nutraceuticals (Table **1**). Micro algae production is one of the most profitable industries since they have a rapid growth rate. DIC Corporation (Japan), Algaetech International Sdn Bhd (Malaysia), AlgaeCan Biotech Ltd. (Canada) and Qponics Limited (Australia) are a few companies that produce algal-based products worldwide. The annual market for phycocyanin is around 5–10 million US dollars [57]. Phycocyanin from *Spirulina* spp. and phycoerythrin from *Porphyridium* spp. are two of the most commercially produced phycobiliproteins [42]. Today there is a good demand for natural colouring agents among health-conscious consumers. According to an exclusive report by Meticulous Research®, the algal pigments market will expect a worth of $452.4 Million by 2025 [62].

Table 1. Algal-based nutraceutical products and their beneficial properties.

Product/Application	Algae	Description	References
Hana Tsunomata	*Chondrus crispus*	Used in mixed salads. Gluten-free and low-calorie product rich in Fe, Mg and fiber.	[2]
Spirulina pacifica®	*Spirulina*	Single cell proteins.	[2]

(Table 1) cont.....

Product/Application	Algae	Description	References
Astaxanthin-60	*Haematococcus pluvialis*	Contains carotenoids with antioxidant properties.	[2]
Micro algae as a food ingredient	*Arthrospira platensis* (Spirulina)	PUFAs (n-3) fatty acids and phycobiliproteins reduce the risk of heart diseases and show antioxidant properties.	[59]
Micro algae as a food ingredient	*Chlorella* spp.	Rich in carotenoids, PUFAs (n-3), fatty acids and sterols. Exhibit antiviral, antitumour, anti hyperlipidaemia, and anticoagulant activities.	[59]
Micro algae as a food ingredient (Vegetarian food gels, pasta)	*Arthrospira maxima*	Rich in PUFAs (n-3) fatty acids. Provides antioxidant properties.	[60]
Aphanizomenon	*Aphanizomenon* cyanobacterium	Antibacterial properties, stimulate liver function and increase bile secretion. Reduce the risk of heart diseases.	[61]

POTENTIAL OF ALGAL NUTRACEUTICALS AS A SAFE FOOD

Algal nutraceuticals have already captured the global market with the introduction of many commercially available food products. However, proper extraction methods are needed to provide quality and safe products to the customers. Previous studies have reported that seaweeds have shown nontoxic behaviour in in-vitro analysis. The crude dichloromethane: chloroform extract of *Stypopodium zonale* have been exhibited good cytotoxic activity against human melanoma cells (skin cancer cells) [63].

According to the United States Food and Drug Administration, carrageenan has been labelled as "Generally Recognized as Safe" [22]. Daily dose level of carrageenan has been limited to up to 2% (~30 mg/person) for topical formulations in the pharmaceutical industry [22]. Similarly, agar also universally falls under the category "Generally Recognized as Safe" [27]. A survey carried out on the edible seaweed market in the United Kingdom (UK) has revealed that only 10% of total products had exhibited iodine content in the product label, and 26 products from the total 226 products surveyed had led to an iodine intake above the tolerable adult upper level of 600 µg/day [64]. This study implies that it is essential to ensure the nutritional quality of algal-based foods to fulfill the actual nutritional needs of customers. Previous studies have reported a great possibility of contaminating seaweeds during harvesting, cultivation and processing [6]. The knowledge of food safety hazards in the seaweed industry is at the bottom level. When incorporating seaweeds into food products, they should be checked for safe levels of heavy metals like Arsenic (As), Lead (Pb), Cadmium

(Cd) and Mercury (Hg). Accumulation of heavy metals and contaminants is high when algae are grown near industrial and agricultural sites [6].

CURRENT POTENTIAL FOR USING ALGAL-BASED NUTRACEUTICALS.

Up to date, many studies have evaluated algal-based nutraceuticals related to their biological activities. Interestingly, the application of nanotechnology to produce algal-based nutraceuticals has emerged. Due to the high expenditure on producing nutraceuticals, researchers are looking for novel methods with high yields. Therefore, nanotechnology has been mediated to produce algal-based nutraceuticals. Metallic nanoparticles like gold and silver nanoparticles can be produced by reducing metal salts of algal biomolecules to metallic nanoparticles easily [65]. Algal polysaccharides have exhibited cancer-fighting properties and reduced toxicities of chemotherapy [66]. These act as nanocarriers where algal polysaccharides combine with nanoparticles and drugs, which may reduce the side effects of chemotherapy [66]. A new potential has been studied by Ansorena and Astiasarán [67] to produce a nutraceutical containing algae oil. These contain essential fatty acids like omega-3 (docosahexaenoic acid). Nutraceutical products are made of algae oil from *Schizochytrium* sp., *Ulkenia* sp. and *Crypthecodinium cohnii* [67]. For example, DHA-rich oil obtained from *Schizochytrium sp* has been commercialized as a 200 mg DHA capsule and later, this formulation has been incorporated into foods as nutraceuticals [67]. These nutraceutical bioactive compounds of algae can be determined through chromatography techniques and NMR (Nuclear magnetic resonance). It is advantageous that most of the biological compounds can be extracted by green extraction methods like Supercritical fluid extraction and subcritical water extraction, which were considered economical [68]. There is a potential to use algal nutraceuticals in the process of eliminating outbreaks like COVID-19. Several studies have reported that bioactive compounds of the micro algae *Spirulina* can perform an antioxidant activity in the human body [69]. Cold water extracts of *Spirulina* exhibited antiviral properties against the Influenza virus [70]. The study of Ratha *et al.* [69], suggests that *spirulina*-derived nutraceuticals can act as an immune booster to overcome bad health issues caused due to COVID-19. It contains Angiotensin-converting enzyme inhibitory peptides. Peptides of *Spirulina* regulated the respiratory cycle and reduced the damage caused to the lungs [69]. There are several studies that have reported the therapeutic effects of *Spirulina* supplements [69, 71].

The structure of Ulvan, which is a polysaccharide of Ulva, has exhibited healthy aging effects [72]. The study of Liu *et al.* [72] has suggested that these Ulvan has a nutraceutical potential and improves the shelf life and quality of food. They had performed desirable physiochemical and gelling properties [72]. Shah *et al.* [73]

have reported the nutraceutical potential of North Bornean Seaweeds. This study revealed that the presence of nutraceutical compounds like phthalic anhydride, 3,4-ethylenedioxythiophene, 2-pentylthiophene, furoic acid in *Kappaphycus alvarezii*, eicosapentaenoic acid, palmitoleic acid, β-carotene *Eucheuma Denticulatum etc*. was responsible for their therapeutic properties like antioxidant, antimicrobial, anticancer, cardiovascular protective and neuroprotective activities [73]. Three strains of the Chlorophyta, *Tetraselmis Chui,* have been evaluated for their nutraceutical potential. Those three strains of *Tetraselmis chui* have produced higher content of polyunsaturated fatty acids [74]. The reports of Panggabean *et al*. [75] also reveal that sulphated polysaccharides of algae can be an effective tool that can show antiviral properties against different COVID 19 viruses (Alpha, beta, gamma, delta and omicron). Carrageenan, agaran, fucoidan, porphyrin, ulvan and exopolysaccharides are some of the sulphated polysaccharides found in algae. Different forms of carrageenan like λ-carrageenan, κ- carrageenan and ι- carrageenan have shown antiviral activities [75]. The study of Morokutti-Kurz *et al*. [76] revealed that ι-carrageenan had inhibited the cell entry of SARS-CoV-2 pseudo-typed lentivirus in a dose-dependent manner. All these imply that algal polysaccharides, peptides, fatty acids and other important bioactive compounds found in algae can be utilized as nutraceuticals with therapeutic values.

CONCLUSION

Algae contain a large number of bioactive compounds, which have shown many health benefits. Extraction and purification of these biomolecules should be developed; hence it affects the price and quality of nutraceutical products. Although many risk factors which affect the quality of these products have been identified, food safety regulations have not yet been introduced. Therefore, limitation levels of heavy metal contents, other contaminants and standards for algal nutraceuticals have to be legislated. However, new trends have arisen to utilize algal-based nutraceuticals as a solution to cure chronic diseases since these are more economical and natural.

CONSENT FOR PUBLICATION

Not applicable.

CONFLICT OF INTEREST

The author(s) declare no conflict of interest, financial or otherwise.

ACKNOWLEDGEMENT

The authors acknowledge the funding from the National Research Council Sri Lanka [Grant number 19-095].

REFERENCES

[1] Aronson JK. Defining 'nutraceuticals': neither nutritious nor pharmaceutical. Br J Clin Pharmacol 2017; 83(1): 8-19.
[http://dx.doi.org/10.1111/bcp.12935] [PMID: 26991455]

[2] Tanna B, Mishra A. Metabolites Unravel Nutraceutical Potential of Edible Seaweeds: An Emerging Source of Functional Food. Compr Rev Food Sci Food Saf 2018; 17(6): 1613-24.
[http://dx.doi.org/10.1111/1541-4337.12396] [PMID: 33350143]

[3] DeFelice SL. The nutraceutical revolution: its impact on food industry R&D. Trends Food Sci Technol 1995; 6(2): 59-61.
[http://dx.doi.org/10.1016/S0924-2244(00)88944-X]

[4] Holdt SL, Kraan S. Bioactive compounds in seaweed: functional food applications and legislation. J Appl Phycol 2011; 23(3): 543-97.
[http://dx.doi.org/10.1007/s10811-010-9632-5]

[5] Rengasamy KRR, Mahomoodally MF, Aumeeruddy MZ, Zengin G, Xiao J, Kim DH. Bioactive compounds in seaweeds: An overview of their biological properties and safety. Food Chem Toxicol 2020; 135: 111013.
[http://dx.doi.org/10.1016/j.fct.2019.111013] [PMID: 31794803]

[6] Banach JL, Hoek-van den Hil EF, Fels-Klerx HJ. Food safety hazards in the European seaweed chain. Compr Rev Food Sci Food Saf 2020; 19(2): 332-64.
[http://dx.doi.org/10.1111/1541-4337.12523] [PMID: 33325177]

[7] Wijesekara I, Karunarathna WKDS. Usage of Seaweed Polysaccharides as Nutraceuticals. Seaweed Polysaccharides: Isolation, Biological and Biomedical Applications. Elsevier Inc. 2017; pp. 341-8.

[8] Škrovánková S. Seaweed vitamins as nutraceuticals. Adv Food Nutr Res 2011; 64: 357-69.
[http://dx.doi.org/10.1016/B978-0-12-387669-0.00028-4] [PMID: 22054961]

[9] Shannon E, Abu-Ghannam N. Seaweeds as nutraceuticals for health and nutrition. Phycologia 2019; 58(5): 563-77.
[http://dx.doi.org/10.1080/00318884.2019.1640533]

[10] Meenakshi S, Umayaparvathi S, Arumugam M, Balasubramanian T. *in vitro* antioxidant properties and FTIR analysis of two seaweeds of Gulf of Mannar. Asian Pac J Trop Biomed 2011; 1(1): S66-70.
[http://dx.doi.org/10.1016/S2221-1691(11)60126-3]

[11] Rocha D, Seca A, Pinto D. Seaweed secondary metabolites *in vitro* and *in vivo* anticancer activity. Mar Drugs 2018; 16(11): 410.
[http://dx.doi.org/10.3390/md16110410] [PMID: 30373208]

[12] Ismail GA, Gheda SF, Abo-Shady AM, Abdel-Karim OH. *in vitro* potential activity of some seaweeds as antioxidants and inhibitors of diabetic enzymes. Food Sci Technol (Campinas) 2020; 40(3): 681-91.
[http://dx.doi.org/10.1590/fst.15619]

[13] Wijesekara I, Lang M, Marty C, *et al.* Different extraction procedures and analysis of protein from Ulva sp. in Brittany, France. J Appl Phycol 2017; 29(5): 2503-11.
[http://dx.doi.org/10.1007/s10811-017-1239-7]

[14] Ganesan AR, Tiwari U, Rajauria G. Seaweed nutraceuticals and their therapeutic role in disease prevention. Food Sci Hum Wellness 2019; 8(3): 252-63.
[http://dx.doi.org/10.1016/j.fshw.2019.08.001]

[15] Das L, Bhaumik E, Raychaudhuri U, Chakraborty R. Role of nutraceuticals in human health. J Food Sci Technol 2012; 49(2): 173-83.
[http://dx.doi.org/10.1007/s13197-011-0269-4] [PMID: 23572839]

[16] Cotas J, Leandro A, Pacheco D, Gonçalves AMM, Pereira L. A comprehensive review of the nutraceutical and therapeutic applications of red seaweeds (Rhodophyta). Life (Basel) 2020; 10(3): 19.
[http://dx.doi.org/10.3390/life10030019] [PMID: 32110890]

[17] Torres MD, Flórez-Fernández N, Domínguez H. Integral utilization of red seaweed for bioactive production. Mar Drugs 2019; 17(6): 314.
[http://dx.doi.org/10.3390/md17060314] [PMID: 31142051]

[18] Britton D, Schmid M, Revill AT, *et al.* Seasonal and site-specific variation in the nutritional quality of temperate seaweed assemblages: implications for grazing invertebrates and the commercial exploitation of seaweeds. J Appl Phycol 2020; 1-14.

[19] Makkar F, Chakraborty K. Antioxidative sulphated polygalactans from marine macroalgae as angiotensin-I converting enzyme inhibitors. Nat Prod Res 2018; 32(17): 2100-6.
[http://dx.doi.org/10.1080/14786419.2017.1363756] [PMID: 28814101]

[20] Cunha L, Grenha A. Sulfated seaweed polysaccharides as multifunctional materials in drug delivery applications. Mar Drugs 2016; 14(3): 42.
[http://dx.doi.org/10.3390/md14030042] [PMID: 26927134]

[21] Qin Y. Seaweed hydrocolloids as thickening, gelling, and emulsifying agents in functional food products.Bioactive Seaweeds for Food Applications. Natural Ingredients for Healthy Diets 2018; pp. 135-52.
[http://dx.doi.org/10.1016/B978-0-12-813312-5.00007-8]

[22] Eccles R, Meier C, Jawad M, Weinmüllner R, Grassauer A, Prieschl-Grassauer E. Efficacy and safety of an antiviral Iota-Carrageenan nasal spray: a randomized, double-blind, placebo-controlled exploratory study in volunteers with early symptoms of the common cold. Respir Res 2010; 11(1): 108.
[http://dx.doi.org/10.1186/1465-9921-11-108] [PMID: 20696083]

[23] Wang W, Zhang P, Hao C, Zhang XE, Cui ZQ, Guan HS. *in vitro* inhibitory effect of carrageenan oligosaccharide on influenza A H1N1 virus. Antiviral Res 2011; 92(2): 237-46.
[http://dx.doi.org/10.1016/j.antiviral.2011.08.010] [PMID: 21867732]

[24] Yuan H, Song J, Li X, Li N, Dai J. Immunomodulation and antitumor activity of κ-carrageenan oligosaccharides. Cancer Lett 2006; 243(2): 228-34.
[http://dx.doi.org/10.1016/j.canlet.2005.11.032] [PMID: 16410037]

[25] Tanna B, Mishra A. Nutraceutical Potential of Seaweed Polysaccharides: Structure, Bioactivity, Safety, and Toxicity. Compr Rev Food Sci Food Saf 2019; 18(3): 817-31.
[http://dx.doi.org/10.1111/1541-4337.12441] [PMID: 33336929]

[26] Peñalver R, Lorenzo JM, Ros G, Amarowicz R, Pateiro M, Nieto G. Seaweeds as a functional ingredient for a healthy diet. Mar Drugs 2020; 18(6): 301.
[http://dx.doi.org/10.3390/md18060301] [PMID: 32517092]

[27] Armisén R, Galatas F. Agar. In: Zheng L. Handbook of Hydrocolloids: Second Edition. CRC Press, New Delhi 2009; pp. 82-107.

[28] Rupérez P, Saura-Calixto F. Dietary fibre and physicochemical properties of edible Spanish seaweeds. Eur Food Res Technol 2001; 212(3): 349-54.
[http://dx.doi.org/10.1007/s002170000264]

[29] Sathivel A, Raghavendran HRB, Srinivasan P, Devaki T. Anti-peroxidative and anti-hyperlipidemic nature of *Ulva lactuca* crude polysaccharide on d-Galactosamine induced hepatitis in rats. Food Chem Toxicol 2008; 46(10): 3262-7.
[http://dx.doi.org/10.1016/j.fct.2008.07.016] [PMID: 18706469]

[30] Berri M, Slugocki C, Olivier M, *et al.* Marine-sulfated polysaccharides extract of Ulva armoricana green algae exhibits an antimicrobial activity and stimulates cytokine expression by intestinal epithelial cells. J Appl Phycol 2016; 28(5): 2999-3008. [http://dx.doi.org/10.1007/s10811-016-0822-7]

[31] Brownlee IA, Seal CJ, Wilcox M, Dettmar PW, Pearson JP. Applications of Alginates in Food. In: Rehm B, Eds. Alginates: Biology and Applications. Microbiology Monographs. Springer, Berlin, Heidelberg 2009; vol. 13: pp. 211-8. [http://dx.doi.org/10.1007/978-3-540-92679-5_9]

[32] Williams JA, Lai CS, Corwin H, *et al.* Inclusion of guar gum and alginate into a crispy bar improves postprandial glycemia in humans. J Nutr 2004; 134(4): 886-9. [http://dx.doi.org/10.1093/jn/134.4.886] [PMID: 15051842]

[33] Fleurence J, Morançais M, Dumay J. Seaweed proteins. 2nd ed. Proteins in Food Processing 2018; pp. 245-62.

[34] Kazir M, Abuhassira Y, Robin A, *et al.* Extraction of proteins from two marine macroalgae, Ulva sp. and Gracilaria sp., for food application, and evaluating digestibility, amino acid composition and antioxidant properties of the protein concentrates. Food Hydrocoll 2019; 87(87): 194-203. [http://dx.doi.org/10.1016/j.foodhyd.2018.07.047]

[35] Fujiwara-Arasaki T, Mino N, Kuroda M. The protein value in human nutrition of edible marine algae in Japan. Hydrobiologia 1984; 116-117(1): 513-6. [http://dx.doi.org/10.1007/BF00027735]

[36] Marinho GS, Holdt SL, Angelidaki I. Seasonal variations in the amino acid profile and protein nutritional value of Saccharina latissima cultivated in a commercial IMTA system. J Appl Phycol 2015; 27(5): 1991-2000. [http://dx.doi.org/10.1007/s10811-015-0546-0]

[37] Milinovic J, Campos B, Mata P, Diniz M, Noronha JP. Umami free amino acids in edible green, red, and brown seaweeds from the Portuguese seashore. J Appl Phycol 2020; 32(5): 3331-9. [http://dx.doi.org/10.1007/s10811-020-02169-2]

[38] Cordeiro RA, Gomes VM, Carvalho AFU, Melo VMM. Effect of proteins from the red seaweed Hypnea musciformis (Wulfen) Lamouroux on the growth of human pathogen yeasts. Braz Arch Biol Technol 2006; 49(6): 915-21. [http://dx.doi.org/10.1590/S1516-89132006000700008]

[39] Matsubara K, Sumi H, Hori K. Platelet aggregation is inhibited by phycolectins. Experientia 1996; 52(6): 540-3. [http://dx.doi.org/10.1007/BF01969724] [PMID: 8698085]

[40] Domozych DS, Ciancia M, Fangel JU, Mikkelsen MD, Ulvskov P, Willats WGT. The cell walls of green algae: A journey through evolution and diversity. Front Plant Sci 2012; 3: 82. [http://dx.doi.org/10.3389/fpls.2012.00082] [PMID: 22639667]

[41] Senthilkumar D, Jayanthi S. Partial characterization and anticancer activities of purified glycoprotein extracted from green seaweed Codium decorticatum. J Funct Foods 2016; 25: 323-32. [http://dx.doi.org/10.1016/j.jff.2016.06.010]

[42] Osório C, Machado S, Peixoto J, *et al.* Pigments content (Chlorophylls, fucoxanthin and phycobiliproteins) of different commercial dried algae. Separations 2020; 7(2): 33. [http://dx.doi.org/10.3390/separations7020033]

[43] Admassu H, Gasmalla MAA, Yang R, Zhao W. bioactive peptides derived from seaweed protein and their health benefits: antihypertensive, antioxidant, and antidiabetic properties. J Food Sci 2018; 83(1): 6-16. [http://dx.doi.org/10.1111/1750-3841.14011] [PMID: 29227526]

[44] Chrapusta E, Kaminski A, Duchnik K, Bober B, Adamski M, Bialczyk J. mycosporine-like amino

acids: potential health and beauty ingredients. Mar Drugs 2017; 15(10): 326. [http://dx.doi.org/10.3390/md15100326] [PMID: 29065484]

[45] Yuan YV, Westcott ND, Hu C, Kitts DD. Mycosporine-like amino acid composition of the edible red alga, Palmaria palmata (dulse) harvested from the west and east coasts of Grand Manan Island, New Brunswick. Food Chem 2009; 112(2): 321-8. [http://dx.doi.org/10.1016/j.foodchem.2008.05.066]

[46] Sánchez-Machado DI, López-Cervantes J, López-Hernández J, Paseiro-Losada P. Fatty acids, total lipid, protein and ash contents of processed edible seaweeds. Food Chem 2004; 85(3): 439-44. [http://dx.doi.org/10.1016/j.foodchem.2003.08.001]

[47] Susanto E, Fahmi AS, Abe M, Hosokawa M, Miyashita K. lipids, fatty acids, and fucoxanthin content from temperate and tropical brown seaweeds. Aquat Procedia 2016; 7: 66-75. [http://dx.doi.org/10.1016/j.aqpro.2016.07.009]

[48] Kumari P, Kumar M, Gupta V, Reddy CRK, Jha B. Tropical marine macroalgae as potential sources of nutritionally important PUFAs. Food Chem 2010; 120(3): 749-57. [http://dx.doi.org/10.1016/j.foodchem.2009.11.006]

[49] López-López I, Bastida S, Ruiz-Capillas C, *et al.* Composition and antioxidant capacity of low-salt meat emulsion model systems containing edible seaweeds. Meat Sci 2009; 83(3): 492-8. [http://dx.doi.org/10.1016/j.meatsci.2009.06.031] [PMID: 20416671]

[50] Hannan MA, Sohag AAM, Dash R, *et al.* Phytosterols of marine algae: Insights into the potential health benefits and molecular pharmacology. Phytomedicine 2020; 69: 153201. [http://dx.doi.org/10.1016/j.phymed.2020.153201] [PMID: 32276177]

[51] Ostad SN, Khanavi M, Gheidarloo R, *et al.* Cytotoxicity of fucosterol containing fraction of marine algae against breast and colon carcinoma cell line. Pharmacogn Mag 2012; 8(29): 60-4. [http://dx.doi.org/10.4103/0973-1296.93327] [PMID: 22438665]

[52] Mišurcová L, Machů L, Orsavová J. Seaweed minerals as nutraceuticals. Adv Food Nutr Res 2011; 64: 371-90. [http://dx.doi.org/10.1016/B978-0-12-387669-0.00029-6] [PMID: 22054962]

[53] Gharibzahedi SMT, Jafari SM. The importance of minerals in human nutrition: Bioavailability, food fortification, processing effects and nanoencapsulation. Trends Food Sci Technol 2017; 62: 119-32. [http://dx.doi.org/10.1016/j.tifs.2017.02.017]

[54] Kumar M, Kumari P, Trivedi N, *et al.* Minerals, PUFAs and antioxidant properties of some tropical seaweeds from Saurashtra coast of India. J Appl Phycol 2011; 23(5): 797-810. [http://dx.doi.org/10.1007/s10811-010-9578-7]

[55] Yamada S, Shibata Y, Takayama M, Narita Y, Sugawara K, Fukuda M. Content and characteristics of vitamin B12 in some seaweeds. J Nutr Sci Vitaminol (Tokyo) 1996; 42(6): 497-505. [http://dx.doi.org/10.3177/jnsv.42.497] [PMID: 9089476]

[56] Aryee ANA, Agyei D, Akanbi TO. Recovery and utilization of seaweed pigments in food processing. Curr Opin Food Sci 2018; 19: 113-9. [http://dx.doi.org/10.1016/j.cofs.2018.03.013]

[57] Sudhakar MP, Jagatheesan A, Perumal K, Arunkumar K. Methods of phycobiliprotein extraction from gracilaria crassa and its applications in food colourants. Algal Res 2015; 8: 115-20.

[58] Taboada MC, Millán R, Miguez MI. Nutritional value of the marine algae wakame (Undaria pinnatifida) and nori (Porphyra purpurea) as food supplements. J Appl Phycol 2013; 25(5): 1271-6. [http://dx.doi.org/10.1007/s10811-012-9951-9]

[59] Van der Voort MP, Vulsteke E, de Visser CL. Macro-economics of algae products: Output WP2A7 02. EnAlgae Swansea University 2015.

[60] Piwowar A, Harasym J. The importance and prospects of the use of algae in agribusiness.

Sustainability (Basel) 2020; 12(14): 5669.
[http://dx.doi.org/10.3390/su12145669]

[61] M. Zubeck H. Evaluation of Microalgae for use as Nutraceuticals and Nutritional Supplements. J Nutr Food Sci 2012; 2(5): 1-6.

[62] Algal Pigments Market by Type (Beta Carotene, Astaxanthin, Fucoxanthin, Phycocyanin, Phycoerythrin), Application (Food & Beverages, Nutraceuticals, Aquaculture, Cosmetics, Pharmaceuticals), and Geography - Global Forecast to 2025, Meticulous Research®, London 2019.https://www.meticulousresearch.com/download-sample-report/cp_id=5059

[63] Rocha FD, Soares AR, Houghton PJ, Pereira RC, Kaplan MAC, Teixeira VL. Potential cytotoxic activity of some Brazilian seaweeds on human melanoma cells. Phytother Res 2007; 21(2): 170-5.
[http://dx.doi.org/10.1002/ptr.2038] [PMID: 17133653]

[64] Bouga M, Combet E. Emergence of seaweed and seaweed-containing foods in the uk: Focus on labeling, iodine content, toxicity and nutrition. Foods 2015; 4(4): 240-53.
[http://dx.doi.org/10.3390/foods4020240] [PMID: 28231201]

[65] Koyande AK, Chew KW, Manickam S, Chang JS, Show PL. Emerging algal nanotechnology for high-value compounds: A direction to future food production. Trends Food Sci Technol 2021; 116: 290-302.
[http://dx.doi.org/10.1016/j.tifs.2021.07.026]

[66] Ouyang Y, Qiu Y, Liu Y, *et al.* Cancer-fighting potentials of algal polysaccharides as nutraceuticals. Food Res Int 2021; 147: 110522.
[http://dx.doi.org/10.1016/j.foodres.2021.110522] [PMID: 34399500]

[67] Ansorena D, Astiasarán I. Development of nutraceuticals containing marine algae oils. Funct Ingredients from Algae Foods Nutraceuticals 1999; 2013: 634-57.

[68] Sarkar S, Gayen K, Bhowmick TK. Green extraction of biomolecules from algae using subcritical and supercritical fluids. Biomass Convers Biorefin 2022; 14: 1-23.

[69] Ratha SK, Renuka N, Rawat I, Bux F. Prospective options of algae-derived nutraceuticals as supplements to combat COVID-19 and human coronavirus diseases. Nutrition 2021; 83: 111089. https://linkinghub.elsevier.com/retrieve/pii/S0899900720303725
[http://dx.doi.org/10.1016/j.nut.2020.111089] [PMID: 33412367]

[70] Chen YH, Chang GK, Kuo SM, *et al.* Well-tolerated Spirulina extract inhibits influenza virus replication and reduces virus-induced mortality. Sci Rep 2016; 6(1): 24253.
[http://dx.doi.org/10.1038/srep24253] [PMID: 27067133]

[71] Selmi C, Leung PSC, Fischer L, *et al.* The effects of Spirulina on anemia and immune function in senior citizens. Cell Mol Immunol 2011; 8(3): 248-54.
[http://dx.doi.org/10.1038/cmi.2010.76] [PMID: 21278762]

[72] Liu D, Ouyang Y, Chen R, *et al.* Nutraceutical potentials of algal ulvan for healthy aging. Int J Biol Macromol 2022; 194: 422-34.
[http://dx.doi.org/10.1016/j.ijbiomac.2021.11.084] [PMID: 34826453]

[73] Shah MD, Venmathi Maran BA, Shaleh SRM, Zuldin WH, Gnanaraj C, Yong YS. therapeutic potential and nutraceutical profiling of north bornean seaweeds: A review. Mar Drugs 2022; 20(2): 101.
[http://dx.doi.org/10.3390/md20020101] [PMID: 35200631]

[74] Moser GAO, Barrera-Alba JJ, Ortega MJ, Alves-de-Souza C, Bartual A. Comparative characterization of three Tetraselmis chui (Chlorophyta) strains as sources of nutraceuticals. J Appl Phycol 2022; 34(2): 821-35.
[http://dx.doi.org/10.1007/s10811-021-02675-x]

[75] Panggabean JA, Adiguna SP, Rahmawati SI, *et al.* antiviral activities of algal-based sulfated polysaccharides. Molecules 2022; 27(4): 1178.

[http://dx.doi.org/10.3390/molecules27041178] [PMID: 35208968]

[76] Morokutti-Kurz M, Fröba M, Graf P, *et al.* Iota-carrageenan neutralizes SARS-CoV-2 and inhibits viral replication *in vitro*. PLoS One. 2021; 17;16(2): e0237480.

CHAPTER 6

Algal Based Nutraceuticals: Trends and Prospects

Antony V. Samrot[1,*], **R. Emilin Renitta**[2] and **S. Saigeetha**[3]

[1] *School of Bioscience, Faculty of Medicine, Bioscience and Nursing, MAHSA University, Jalan SP2, Bandar Saujana Putra, Jenjarom, Selangor, Malaysia*

[2] *Department of Food Processing, Karunya Institute of Technology and Science, Coimbatore, Tamil Nadu, India*

[3] *Department of Biotechnology, School of Bio and Chemical Engineering, Sathyabama Institute of Science and Technology, Chennai, Tamil Nadu, India*

Abstract: Among various emerging food sectors, the nutraceutical sector has gained the attraction of most industries and researchers. Nutraceuticals are generally considered to be the alternative to the pharmaceutical dosage which provides physiological benefits. They are usually functional ingredient present which supplements nutrition to the body. These nutraceuticals are also derived from algae, where the most commonly found products of algae are derived from *Spirulina* sp, *Chlorella* sp and *Klamath* sp. It is reported that microalgae tend to strengthen the immune system by increasing nutritional constituents present in the body. In this review, detailed elaboration of various nutraceutical compounds from algal sources, their recent trends and future prospects are discussed.

Keywords: Nutraceuticals, Pigments, Polysaccharides, Microalgae, Macroalgae.

INTRODUCTION

'Let food be thy medicine and medicine be thy food' was stated by the Greek physician Hippocrates, father of Medicine (460–377 B.C). A rather impressive and suitable word for this statement, "Nutraceutical," was forwarded by Stephen DeFelice in 1989 [1 - 3]. The combinatorial approach of infusing nutrition and pharmaceutical sector in medicine resulted in the term 'Nutraceutical'. According to DeFelice, [4] 'A nutraceutical can be defined as any substance that may be considered a food or part of a food and provides medical or health benefits including the prevention and treatment of disease'.

* **Corresponding author Antony V. Samrot:** School of Bioscience, Faculty of Medicine, Bioscience and Nursing, MAHSA University, Jalan SP2, Bandar Saujana Putra, Jenjarom, Selangor, Malaysia; Tel: +603 5102 2200; E-mail: antonysamrot@gmail.com

Avinash Mishra (Ed.)

Technically, nutraceuticals are bioactive chemical compounds enriched with medicinal properties that aid in disease prevention and are worthwhile promoting health [5]. 'Beyond the diet, before the drugs' is the new definition of nutraceuticals proposed by Santini *et al.* [6]. Nutraceuticals offer potential nutritional and multiple therapeutic effects being a better alternative to medicine and nutrition [7]. In the current era of a sedentary lifestyle, extensive health issues among humans and the encroachment of fast foods on the human diet, several health regulatory measures to encourage the consumption of health-promoting foods are introduced [8]. In such a situation, nutraceuticals could stand as a better alternative as they offer null side effects with increased health value and are also economically affordable [9]. Nutraceuticals are found in natural foods with key ingredients such as anti-oxidants, phytochemicals, minerals, nutritious vitamins, minerals and lipids [10]. It could be genetically engineered (designer) foods, herbal products or certain processed foods like oats, cereals, soups and beverages [11]. Both the microalgal and macroalgal products have anticancer, and antidiabetic activity [12, 13]. Nutraceuticals are classified into traditional and non-traditional nutraceuticals [14] and are represented in Fig. (**1**) below.

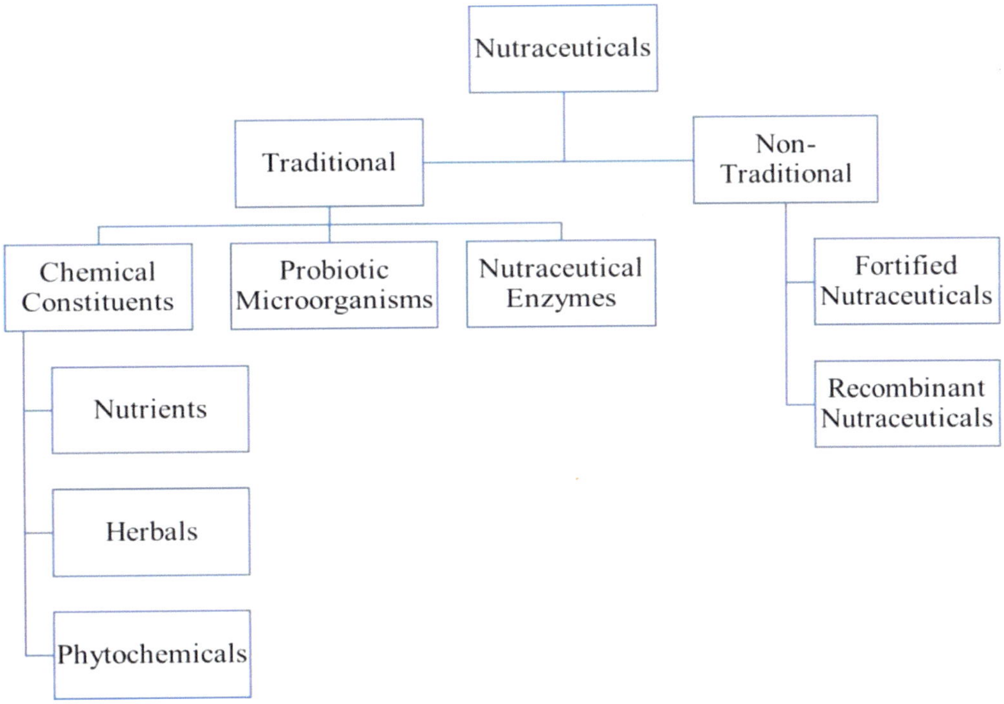

Fig. (1). Classification of nutraceuticals.

NUTRACEUTICALS FROM ALGAL SOURCE

Nutraceuticals can be extracted from any microbe, plant or animal source that is rich in dietary fibre, probiotics, prebiotics, polyunsaturated fatty acids, anti-oxidants, vitamins, polyphenols and spices [9]. One such source is the micro and macro algae. They are the diversified group of organisms enriched with nutrients. Microalgae are regarded as bio-factories for pharmaceutical and therapeutic applications. Various nutraceuticals were extracted, developed and formulated for commercial use. Macroalgae are consumed as whole food to serve as nutraceutical while it is extracted and formulated from microalgae [15].

MICROALGAL NUTRACEUTICALS

All ecosystems are richly embodied with diverse species of Microalgae [16, 17]. The nutraceutical composition of microalgae is of high importance in the food and pharmaceutical industries. Microalgae such as *Arthrospira platensis* (*Spirulina*), and *Chlorella vulgaris* are excessively consumed among humans [18, 19]. They are consumed for their nutritive value and were also regarded as safe (GRAS) by US FDA (Food and Drug Administration) [20]. Nutraceuticals such as sterols, polyunsaturated fatty acids (PUFAs), pigments, proteins, phenolic compounds, anti-oxidants, polysaccharides, vitamins and minerals can be effectively extracted from microalgae [15, 21, 22] and formulated to be sold in the form of tablets and powders [20]. Spirulina, Klamath, and the chloroficean *Chlorella* spp. are the dominant microalga in the nutraceutical market [23]. Apart from these, many other microalgae enriched with nutraceuticals such as Vitamin B12, E, β-carotene, xanthins, phycobiliproteins, chlorophyll, EPA (Eicosapentaenoic acid), DHA (Docosahexaenoic acid), sulphated polysaccharides and many phenolic and volatile compounds are mentioned in Garcia *et al.* [24]. Nutraceuticals, especially from microalgae, promise to lessen the occurrence of malnourishment [25]. Bishop and Zubeck [26] reported that microalgal nutraceuticals aid in the prevention and treatment of viral infections, heart diseases and cancers. The commercialization of microalgal products has gained global attention owing to their inherent medicinal properties and nutraceutical value.

Polysaccharides

Microalgal polysaccharides are of commercial importance [20] and are valued for their therapeutic properties such as anti-viral, anti-tumor and anti-oxidant properties. They are preferred for their immune-boosting ability [27]. Guzman *et al.* [28] reported that the crude polysaccharide extracts of the microalgae *Chlorella stigmatophora* and *Phaeodactylum tricornutum* possessed immunosuppressant and immunostimulatory effects, respectively. The red-tide microalga *Gyrodinium impudicum* is enriched with sulphated exopolysaccharide.

An investigation by Yim *et al.* [29] stated that the high-sulfated exopolysaccharide of this microalga has immunostimulatory activity and hence can be incorporated into pharmaceutical products. Moreover, microalgal exopolysaccharides are reported to hold antihyperlipidemia, antibacterial, anti-oxidant, antitumor and anticoagulant properties [24]. They also retain antiviral, antilipidemic, antiglycemic, antiadhesive, anti-oxidant, free radical scavenging, and anti-tumor properties [30, 31]. All these properties of polysaccharides of microalgae prove that they can be a promising source of nutraceuticals.

Proteins

Arthrospira spp. and *Chlorella* spp. have a great demand in the market owing to their proteinaceous nutraceuticals that aid in the prevention of cell damage and disease occurrence. The amino acids and peptides found in microalgae are indirectly involved in enhancing immunogenic activity. Enzymes produced by microalgae are also involved in enhancing the metabolic activity in human tissues [32]. *Chaetoceros* spp., *Isochrysi* spp., *Nannochloropsis* spp., *Pavlova* spp., *Phaeodactylum* spp., *Skeletonema* spp., *Thalassiosira* spp. and *Tetraselmis* spp. are the microalgal species mainly utilized as animal nutrition owing to their protein content [30]. *Spirulina* act as a single cell protein, a rich reservoir of protein that eventually is a better ingredient to be added in the human diet to prevent and control obesity [33].

Pigments

Spirulina contains β carotene, a carotenoid pigment that possesses provitamin A which enlightens vision and aids in immune function. It also prevents cancer, heart diseases, atherosclerosis, oxidative stress, aging and tissue degeneration [32 - 36]. Astaxanthin is a powerful anti-oxidant that is involved in reducing protein degradation, muscular degeneration, cardiovascular diseases, rheumatoid arthritis and several neurodegenerative diseases. *Haematococcus pluvialis* is the major producer of astaxanthin and has a rich commercial value in the food and animal feed industry [32]. Its anti-oxidant effect is stronger than β carotene [33]. Ikeuchi *et al.* [37] reported that astaxanthin suppresses obesity. It is also available in many commercial formulations, including oils, syrups, soft gels, tablets and is an effective nutritional supplement [38]. Similarly, other pigments such as leutin, zeaxanthin, violaxanthin, antheraxanthin, neoxanthin and a few carotenoids were reported to possess anti-cancerous effects [24, 27]. The photosynthetic pigment chlorophyll is an effective anti-oxidant that aids in reducing oxidative stress. It also aids liver function, bile secretion, and increases hemoglobin levels in the blood [26].

Lipids

Marine microalgae are highly enriched with omega-3 fatty acids, Eicosapentaenoic acid (EPA) and Docosahexaenoic acid (DHA) which are generally categorized as PUFAs (Poly Unsaturated Fatty Acids). *Ulkenia* sp. and *Schizochytrium* sp., play a major role in contributing lipid nutraceuticals. The lipid content of these microalgae benefits lactating women and heart patients [20]. A high amount of PUFAs can be efficiently extracted from marine microalgae for nutritional and therapeutic purposes. Apart from these, arachidonic acid (ARA), gamma-linolenic acid (or γ-linolenic acid; GLA) were also reported in *Arthrospira platensis, Isochrysis galbana, Odontella* and *Porphyridium cruentum* [32, 27]. Microalgal lipids also prevent adiposity [33]. DHA and EPA are used as adult dietary supplements in cheeses, yogurts, spreads and dressings, and breakfast cereals [36]. The commercial demand for PUFAs and the emerging microalgal nutraceuticals paves way for microalgal intake in the human diet.

Sterols

A diversity of sterols can be observed in microalgae. Brassicasterol, sitosterol, and stigmasterol, phytosterol were found in *Pavlova* sp., *Tetraselmis* sp. and *Nannochloropsis* sp. Sterol content in microalgae varies based on the nature of its habitat. Phytosterols can be utilized as a dietary supplement as they could act like a secondary messenger and reduce cholesterol levels [27]. Margarine and vegetable oils are supplemented with microalgal polysterols to reduce cholesterol levels. They also possess anti-oxidant, anticancer, antibacterial, anti-inflammatory and antidiabetic effects [39]. Microalgal sterols can effectively function as nutraceuticals as an alternative to pharmaceutical sterols derived from other sources.

Vitamins

Vitamin E is reported to be present in *Dunaliella tertiolecta* and *Nannochloropsis oculate* [40, 41]; thus, it can be used as vitamin supplement. Even more, these microalgae contain more vitamin E than the plants [42]. Vitamin C is profoundly found in microalgae, whereas *Skeletonema marinoi* is reported to have more vitamin C [43, 44].

MACROALGAL NUTRACEUTICALS

Macroalgae (seaweed) are commonly consumed as food in east Asian countries. The rich polysaccharide content of seaweed is endowed with enormous medicinal and nutritional value. Moreover, they also provide essential nutrients for human nutrition. Polysaccharides such as cellulose, agar, agarose, carrageenan, laminarian, and alginate are excellent nutraceuticals [45]. Sterols, lipids, polyphenols, carotenoids and bioactive peptides in seaweeds are found to possess nutraceutical value [46]. Several researchers have mentioned the beneficiary effect encountered in humans through the consumption of various nutraceuticals present in seaweeds [47 - 49]. Glucose metabolism is enriched by dieckol, fucoidan, polyphenol, fucoxanthin, sterols, phlorotannin, and anthocyanin present in macroalgae. Nutraceuticals such as fucophlorethol, phlorotannins, monoglycosyldiacylglycerol, polyphenols, amarouciaxanthin A, octaphlorethol A (OPA), ascorbic acid, octadecadienoate, fucosterol, and flavonoids found in seaweeds aid in lipid metabolism. Mineral deficiency in humans could also be treated through seaweed consumption [50], while seaweed lipid is a better alternative for PUFAs from fish as they emanate a fishy odour [51]. All these rich health benefits of seaweed offer them a good market value. Many commercialized seaweed products were developed by ICAR (Indian Council of Agricultural Research) - Central Marine Fisheries Research Institute (CMFRI) to treat diabetes, arthritis, heart diseases, bacterial infections, dyslipidemia and hypothyroid disorder [52]. All these properties of seaweed and their abundant growth make them economically affordable food and medicinal commodity.

Polysaccharides

Polysaccharides are abundant in seaweeds and have diversified applications in the food, cosmetic and pharmaceutical industries. Polysaccharides such as agar, alginate, carrageenan, fucoidan, ulvan and laminarian are of commercial importance and are primarily used as food hydrocolloids [53, 54]. Agar is generally extracted from the red seaweeds *Gelidium* sp. and *Gracilaria* sp. It is made up of α (1-4)-3, 6-anhydro-L-galactose and β 9(1-3)-D-galactose residues possess cell apoptosis effect, thus acting as an anti-cancer agent. Carrageenan from red seaweeds possesses anti-coagulant activity, while fucoidan and alginate in *Sargassum* sp. reduced cholesterol and triglyceride level, increasing high-density lipoprotein [55, 56]. Algal polysaccharides offer a low glycemic index as most of them are non-starchy fibre made up of various monosaccharides that are branched together. Regarding this, laminarians from brown seaweed can act as dietary fibre. Hence, algal polysaccharides are reported as compatible food for diabetic patients [57, 58]. Algal polysaccharides can also be employed as prebiotics [59].

Proteins

The protein content is relatively high in red and green algae than in brown algae, but it is also said that the protein composition in macroalgae varies during seasonal variation [60]. Seaweed protein could be a better alternative for animal protein as it is richly enriched with essential amino acids required for metabolic activity in human body. Therefore, seaweeds are highly recommended for a vegan diet and are regarded as a complete protein source as they contain all the nine essential amino acids [57, 61, 62]. Phycobiliproteins are the pigment-protein compounds present in seaweeds. Phycoerythrins, phycocyanins, phycoerythrocyanins, and allophycocyanins are the four main groups of phycobiliproteins. They exhibit a combinatorial effect as the pigment offer anti-oxidant activity, natural colorant and fluorescent marker while the protein offers therapeutic properties such as anti-inflammatory and anti-tumor activity. These phycobiliproteins can be added to food as functional ingredients owing to their hepatoprotective nature [47, 48].

Pigments

Carotenoids, xanthophylls and chlorophylls are the major pigments in seaweed. Xanthophyll derivatives- leutin and zeaxanthin, are effective in preventing muscular degeneration and are also employed as a dietary nutraceutical. Besides, they also exhibit anti-oxidant, anticancerous, antihypercholesterolemic and neurodegenerative properties. Ganesan *et al.* [49] reported that the marine macroalgal pigments siphonaxanthin from the green seaweed *Codium fragile* and fucoxanthin from the brown seaweed *Undaria pinnatifida* demonstrated cytotoxic effect against human leukemia. These pigments were isolated from the edible seaweeds. This highlights the possibility of utilizing these seaweeds as nutraceuticals [47]. α and β carotenoids in edible red macroalgae possess antioxidative, anti-inflammatory and antitumor activity. It also lessens oxidative stress and reduces the risk of ophthalmological diseases in humans [48]. Anti-oxidant, antitumor, anti-degranulation and chemo-protective activity are rendered by chlorophyll pigments in edible seaweeds. Chlorophyll is an effective anti-oxidant that is primarily known for its free radical scavenging activity. This property is also retained by chlorophyll pigments in seaweeds and therefore is an effective nutraceutical Susanto *et al.* [63].

Lipids

The lipid content in macroalgae is preferably low but is enriched with the essential PUFAs which are highly recommended for the prevention of cardiovascular diseases. These lipids also possess antitumor, antidiabetic, and antimicrobial properties and can be used to treat osteoarthritis [64]. Some seaweeds also possess MUFAs (Monounsaturated Fatty Acids). Lipids in seaweeds are biologically active and are associated with the treatment and prevention of chronic diseases. Moreover, the presence of lipid-soluble pigments such as chlorophylls and carotenoids ease the prophylactic approach of edible seaweeds [63]. Essential fatty acid and PUFAs are richly constituted in brown seaweeds compared with green and red macroalgae. This makes brown seaweeds commercially important and demands their rich nutritional and therapeutic profile [47]. PUFAs, especially EPA and DHA, are involved in cell signalling, regulation of factors engaged in transcription, aid in brain development and act as immunomodulators. They also forbid the occurrence of cardiovascular, neurodegenerative, ocular and autoimmune diseases [65, 48].

Sterols

Sterols are the precursors of hormones with anti-oxidant, antiviral, anti-fungal, and anti-bacterial properties. Cholesterol is one such sterol that plays a vital role in the human circulatory system. It reduces total cholesterol and LDL cholesterol in blood [47]. Red macroalgae are richly bestowed with cholesterol and its derivatives [48]. These phytosterols also act as precursors of vitamins [66]. Apart from cholesterol, sterols such as fucosterol, ergosterol, desmosterol, ergosterol, campesterol, and stigmasterol were identified in macroalgae [67]. Anticancer, antidiabetic, anti-oxidant, hepatoprotective, antihyperlipidemic, antifungal, antihistaminic, anticholinergic, antiadipogenic, antiphotodamaging, anti-osteoporotic, blood cholesterol reducing, blood vessel thrombosis preventive and butyrylcholinesterase inhibitory activities are exhibited by fucosterols present in seaweeds [68]. Phytosterols from seaweeds also act as neuroprotective compounds [69].

EXTRACTION OF ALGAL NUTRACEUTICALS

Phong *et al.* [70] extracted proteins from Chlamydomonas sp. by solvent extraction methods. Four different solvents were used. After harvesting, the *Chlamydomonas* sp. was lyophilised and further grounded to a fine powder using mortar and pestle and 0.1g of powder was dispersed in 10ml of solvents like ethanol, methanol, water and 1- propanol and was subjected to centrifugation. After centrifuging, supernatant was collected to identify protein presence. Algal nutraceuticals are extracted by means of various methods like alkaline, acidic, and aqueous extraction, which is further followed by ultrasonication, centrifugation, ultrafiltration and chromatography (Fig. **2**) [71]. Species involved in the production of algal nutraceuticals, along with their mode of extraction have been listed in Table **1**.

Table 1. Extraction of algal nutraceuticals and their percentage yield.

Species	Extraction name	Extraction method	Protein yield	References
Palmaria palmata	Polysaccharide degradation	Enzymatic hydrolysis	67%	[72]

(Table 1) cont.....

Species	Extraction name	Extraction method	Protein yield	References
Palmaria palmata	High shear force	Physical process	40%	[72]
Ascophyllum nodosum	Acid alkaline treatment	Chemical extraction	59.76%	[71]
Palmaria palmata	Alkaline and aqueous treatment	Chemical extraction	24%	[72]

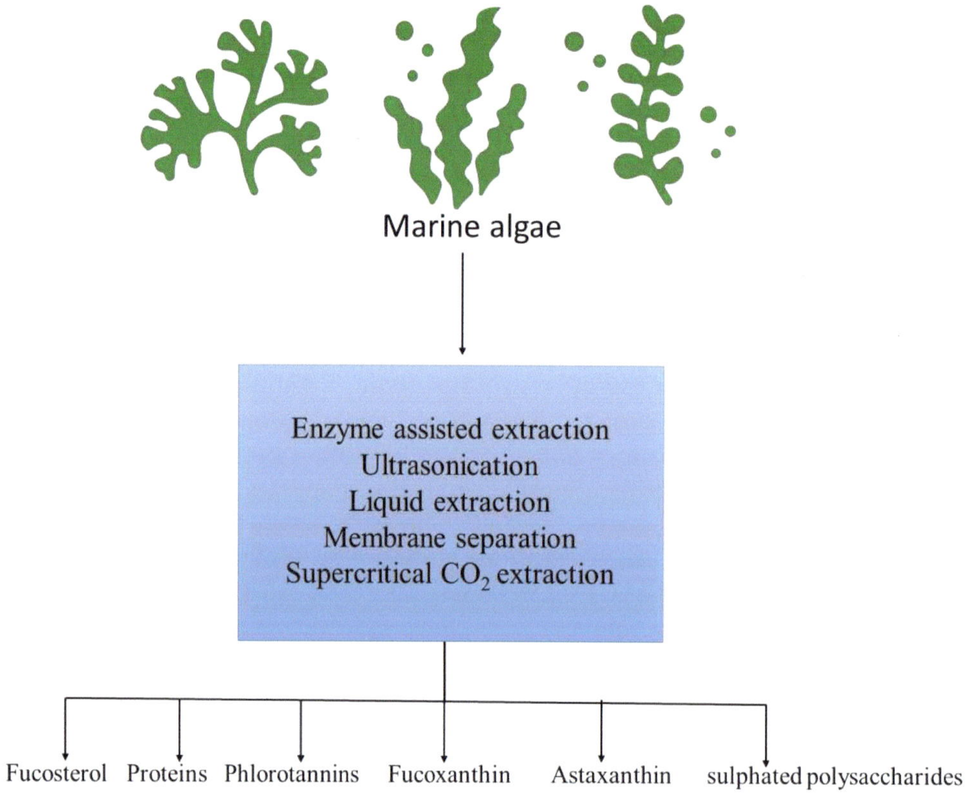

Fig. (2). Extraction of compounds from algae.

Seaweed is considered to be rich in various types of polysaccharides, namely galactans, xylans, alginates, carrageenans, laminarins, fucoidans and floridean starch. These polysaccharides will reduce the availability of proteins and decrease the extraction efficiency. In order to avoid this, the polysaccharides must be subjected to cell disruption techniques [19].

Carotenoids and chlorophyll pigments can be extracted from thylakoid membrane using solvents like dimethyl sulfoxide, ethanol and methanol whereas phycoerythrin and phycocyanin can be extracted from algal tissues as they are soluble in water [73]. Wiltshire *et al.* [74] extracted fatty acids and pigments from the algal species *Scenedesmus obliquus* using the ultrasound method. He also reported that this ultrasound method is also suitable for extraction of pigments and fatty acids from other algal species like *Cryptomonas erosa, Cyclotella meneghiniana, Microcystis aeruginosa* and *Staurastrum paradoxum* and is thus applicable to a wide spectrum of algae. There are many methods available for the extraction of nutraceutical compounds from algal species. Some of the methods were under study, and still more studies must be carried forward to understand other suitable methods for extracting compounds.

TRENDS

Nowadays, the microalgae sector is growing rapidly and is the best alternative and renewable source for producing pharmaceutical as well as nutraceutical compounds [75, 76]. Microalgae are generally recognized as safe (GRAS), making them usable in both the pharmaceutical and food industries [77]. In the 21st-century, a major problem faced is the increasing human population with limited resources. By utilizing microalgae as a food supplement, essential food supply needs can be satisfied in an eco-friendly way [78]. Due to their easy cultivation and composition, microalgae have gained application as a great solution to feeding nourishing food to the human population [79]. Microalgae can serve as a functional agents such as nutraceuticals under abiotic stress conditions. Microalgae are the predominant producers of omega 3 and 6 polyunsaturated fatty acids (PUFAs), pigments, carotenoids, carbohydrates, proteins and lipids. These compounds are used as additives in cosmetic applications. (PUFAs) and its derivatives like Ɣ-linolenic acid, and docosapentaenoic acid are especially used in the treatment of cardiovascular disease, asthma and type 2 diabetes [80, 81]. Microalgae have an enormous amount of phosphorous ad nitrogen, which are the products of wastewater which helps algal biomass rich in biomolecules [82]. On the other hand, genetically modified microalgae play a major role in enhancing nutrition and are safer for the health of human population and the environment. According to the current market portfolio, the microalgae-based products are smaller but are increasing steadily. Awareness and safety clearance for microalgae consumption, regulatory issues prevailing to genetically engineered microalgae, sustainable production, and a growing market are likely to help the commercialization of microalgae-based products [83, 84].

PUFAs and omega-fatty acids can be used as food supplements and can prevent cardiovascular disease. Chlorophylls, carotenoids and phycobiliproteins are the most important pigments produced from microalgae and used for imparting colors in pharmaceutical and other food industries [85]. By the end of 2021, nutraceuticals market would cross about $278.96 billion. Similarly, polyunsaturated fatty acids (PUFAs) produced from microalgae, such as arachidonic acid, docosahexaenoic acid, etc., are also an important commercial product, and their global market is around USD 9.0 billion/Year [86]. Macroalgae, also called seaweed, are a source of great nutrition in the human diet as it contains very low-calorie content and are rich in nutrients like minerals, vitamins, proteins etc [87]. Macroalgae are sources of many natural products and metabolites with unique therapeutic and nutritional properties [88]. These metabolites include polyunsaturated fatty acids, proteins, L-α kainic acid, furanone, pigments, phycocolloids and phlorotannins. Red and brown seaweeds are alternative sources of proteins, vitamins and minerals. They have been used to improve protein digestibility and prepare bioactive peptides. Recently, antihypertensive bioactive peptides have been isolated that may act as angiotensin-converting enzyme (ACE) inhibitors. Due to their biological activity, the alginates from brown seaweeds are used as hydrocolloids. Fucans from brown seaweeds are used in Food and cosmetics industries [89 - 92].

Algin, agar and carrageenan are obtained from algae [93]. Alginates are obtained from the cell wall of brown algae [94]. They are used in the pharmaceutical and food industries in the form of stabilizers for emulsions and suspension [95]. Agar is obtained from red algae such as *Gracilaria* spp., *Gelidium* spp., *Gelidiella* spp. and *Pterocladia* spp. [96]. Agar is used as a substrate for bacteriologic culture and tissue culture of eukaryotic cells in research laboratories because of its unique physical properties as it melts at 100 °C and remains liquid until cooled to 40°C [97]. Algin is obtained from brown algae like *Lallinaria* spp., *Ascophyllum* spp., *Macrocystis* spp., *Cystoseira* spp., *Sargassum* spp. and *Turbinaria* spp. [98]. These seaweeds obtained from algae are used as gelling, thickening, and stabilizing agents in food, textile, pharmaceutical, paint industries etc. Carrageenan are not only used in the food but also in textile, cosmetics and medicines. Water soluble Pigments like Phycobillins especially blue phycobilin from *Arthrospira* are used as colorants for food and cosmetic products. Xanthophyll pigment also has a large application in the coloration of cosmetic and drugs. AX is a xanthophyll carotenoid found in micro-algae. It is responsible for the red or pink colour of seafood such as lobster, salmon, crab, trout fish eggs and shrimp [95, 99].

NUTRACEUTICALS FROM MARINE ALGAE- HEALTH BENEFITS

Brown sea vegetable namely *E. cava* is responsible for the production of phlorotannin which has the potential of reducing the growth of human breast cancer cells by inducing apoptosis. Sulphated polysaccharide purified from the E. cava induces apoptosis at *in vitro* condition and have efficient antiproliferative effect on human leukemic monocyte lymphoma cell line (U-937) [100]. The chemoprevention against hepatic cancer can be assisted by fucoxanthin, carotenoid. The antiproliferative activity of dietary fucoidan was found to be associated with the induction of apoptosis and with autophagy in human gastric cancer (AGS) cells. Fucoxanthin has potential of reducing the viability of HepG2 cells along with the induction of cell cycle arrest at G0/G1 phase and also induces apoptosis in human leukemia HL-60 [101, 102].

Thrombin time, prothrombin time and activated partial thromboplastin time are the methods used for determining the anticoagulant activity of bio actives from marine algae [103, 104]. Novel anticoagulant nutraceuticals are manufactured by the usage of sulphate polysaccharides derived from the algal seaweed and also from fucoidans from brown seaweeds [105] and carrageenan from red sea weeds in the pharmaceutical industry phloroglucinol can be developed as a novel anticoagulant agent [106].

On free radicals phlorotannin produced from brown sea vegetables exhibit strong anti-oxidant activity [107]. The useful effects of fucoxanthin as an anti-oxidant is based on the properties of singlet oxygen quenching and its ability of trapping free radicals depending upon the number of conjugated double bonds of the molecule. It also has the capacity of inhibiting apoptosis induced by H_2O_2, intracellular ROS formation and DNA damage. Fucosterol increases the activity of free radical scavenging enzymes like catalase, superoxide dismutase and glutathione peroxidase [108]. 6,6′ -bieckol produced from *E. cava* at IC50 of 1.07 μM inhibits the activity of HIV-1 reverse transcriptase enzyme and also prevents the entry of HIV1 and also it completely inhibited HIV-1 replication and exhibited no cytotoxicity at any concentration [109]. Brown sea vegetables produces fucoidans which was reported to have antiviral activity against HIV, HSV-1 and HSV-2 [110]. Fucosterol derived from saringosterol found in brown sea vegetable shows inhibitory action against *Mycobacterium tuberculosis* [111].

FUTURE PROSPECTS

Marine resources are well recognized for their bioactive substance, which can be used as nutraceuticals. Algae and Cyanobacteria contain rich sources of various compounds like lectins, pigments, anti-oxidants, fibers, steroids, vitamins, polysaccharides and essential lipids. They are used widely in different countries due to their applications in pharmaceuticals as well as in nutraceuticals. Marine farming serves as a main source of all nutraceutical compounds extracted from the algal source. It is the major producer of various types of polysaccharides, which can be further utilized for practical applications. Microalgae produce various compounds like polyunsaturated fatty acids, carotenoids and other secondary metabolites, which can be substituted in food due to their efficient medicinal value but for the extraction of these compounds photobioreactor must be available with defined growth conditions. There is an urge to find how the process can be converted into green technology by avoiding the usage of other compounds that causes pollution to the environment [112, 113]. The algal-based products must increase from small scaled to large-scale, and their necessity must be explained to the mass crowd. Further research must be carried out to find regarding the nutritional value of algal-derived compounds.

CONCLUSION

Nutraceuticals from micro and macroalgae exhibit rich nutritious and therapeutic properties. The abundant occurrence of algae has paved the way for widespread utilization and culturing of algae for commercialization. The nutraceutical properties bestowed in them are preferred over those available in other sources as they are cost-effective and easily reproducible through algal farming.

CONSENT FOR PUBLICATION

Not applicable.

CONFLICT OF INTEREST

The author(s) declare no conflict of interest, financial or otherwise.

ACKNOWLEDGEMENT

Declared none.

REFERENCES

[1] Pandey M, Verma RK, Saraf SA. Nutraceuticals: new era of medicine and health. Asian J Pharm Clin Res 2010; 3(1): 11-5.

[2] Verma G, Mishra MK. A review on nutraceuticals: classification and its role in various diseases. Int J Pharm Ther 2016; 7(4): 152-260.

[3] Nwosu OK, Ubaoji KI. Nutraceuticals: History, classification and market demand InFunctional foods and nutraceuticals. Cham: Springer 2020; pp. 13-22.

[4] DeFelice SL. The nutraceutical initiative: a recommendation for US Economic and regulatory reforms. Genetic engineering news (USA). 1992.

[5] Majaz Q. MolviKhurshid I, Nazim S, Asir Q, Shoeb Q. Nutraceuticals: importance and advances in medicine and health. Int Res J Pharm 2012; 4: 71-3.

[6] Santini, A., Cammarata, S. M., Capone, G., Ianaro, A., Tenore, G. C., Pani, L., & Novellino, E. (2018). Nutraceuticals: Opening the debate for a regulatory framework. British journal of clinical pharmacology 2017; 84(4): 659-72.

[7] Palthur MP, Palthur SS, Chitta SK. Nutraceuticals: Concept and regulatory scenario. Int J Pharm Pharm Sci 2010; 2(2): 14-20.

[8] Wadolowska L, Hamulka J, Kowalkowska J, *et al.* Prudent-active and fast-food-sedentary dietary-lifestyle patterns: The association with adiposity, nutrition knowledge and sociodemographic factors in Polish teenagers—The ABC of healthy eating project. Nutrients 2018; 10(12): 1988. [http://dx.doi.org/10.3390/nu10121988] [PMID: 30558296]

[9] Ansari SH, Chauhan B, Kalam N, Kumar G. Current concepts and prospects of herbal nutraceutical: A review. J Adv Pharm Technol Res 2013; 4(1): 4-8. [http://dx.doi.org/10.4103/2231-4040.107494] [PMID: 23662276]

[10] Shrestha R, Shrestha S, K C B, Shrestha S. Evaluation of nutritional supplements prescribed, its associated cost and patients knowledge, attitude and practice towards nutraceuticals: A hospital based cross-sectional study in Kavrepalanchok, Nepal. PLoS One 2021; 16(6): e0252538. [http://dx.doi.org/10.1371/journal.pone.0252538] [PMID: 34086755]

[11] G S, Srinath D. Nutraceuticals and their health benefits. International Journal of Pure & Applied Bioscience 2017; 5(4): 1151-5. [http://dx.doi.org/10.18782/2320-7051.5407]

[12] Samrot AV, Thirunalasundari T. Effect of Oscillatoria willei–a marine cyanobacterium on hydrazine induced toxicity. Malays J Microbiol 2012; 8(4): 236-41.

[13] Siloto RMP, Weselake RJ. Site saturation mutagenesis: Methods and applications in protein engineering. Biocatal Agric Biotechnol 2012; 1(3): 181-9. [http://dx.doi.org/10.1016/j.bcab.2012.03.010]

[14] Singh J, Sinha S. Classification, regulatory acts and applications of nutraceuticals for health. Int J Pharma Bio Sci 2012; 2(1): 177-87.

[15] Alsenani F, Ahmed F, Schenk PM. Nutraceuticals from microalgae. In: Bagchi D, Preuss HG, Swaroop A. Nutraceuticals and functional foods in human health and disease prevention. Boca Raton, FL, United States: CRC Press. 673-84. [http://dx.doi.org/10.1201/b19308-59]

[16] Van Thang Duong YL, Nowak E, Schenk PM. Microalgae isolation and selection for prospective biodiesel production. Advances Biofuel Production: Algae Aquatic Plants 2013; 5(6): 1835-49.

[17] Alam MM, Mumtaz AS, Russell M, Grogger M, Veverka D, Hallenbeck PC. Isolation and characterization of microalgae from diverse Pakistani habitats: Exploring third-generation biofuel potential. Energies 2019; 12(14): 2660. [http://dx.doi.org/10.3390/en12142660]

[18] Liu J, Chen F. Biology and industrial applications of Chlorella: advances and prospects. Adv Biochem Eng Biotechnol 2016; 153: 1-35. [PMID: 25537445]

[19] Mosquera-Murillo Z, Peña-Salamanca EJ. Effect of salinity on growth of the green alga Caulerpa sertularioides (Bryopsidales, Chlorophyta) under laboratory conditions. Hidrobiologica 2016; 26(2): 277-82. [http://dx.doi.org/10.24275/uam/izt/dcbs/hidro/2016v26n2/Mosquera]

[20] Molino A, Iovine A, Casella P, *et al.* Microalgae characterization for consolidated and new application in human food, animal feed and nutraceuticals. Int J Environ Res Public Health 2018; 15(11): 2436. [http://dx.doi.org/10.3390/ijerph15112436] [PMID: 30388801]

[21] Islam MN, Alsenani F, Schenk PM. Chapter 1. Microalgae as a sustainable source of nutraceuticals. In: Gupta VK, Treichel H, Shapaval VO, de Oliveira LA, Tuohy MG. Eds. Microbial Functional Foods and Nutraceuticals; John Wiley & Sons, Inc.: Hoboken, NJ, USA 2017; 1-19. [http://dx.doi.org/10.1002/9781119048961.ch1]

[22] Subudhi S. Bioprospecting for algal based nutraceuticals and high value added compounds. J Pharm Pharm 2017; 4(2): 145-50. [http://dx.doi.org/10.15436/2377-1313.17.1651]

[23] Nicoletti M. Microalgae Nutraceuticals. Foods 2016; 5(4): 54. [http://dx.doi.org/10.3390/foods5030054] [PMID: 28231149]

[24] García JL, de Vicente M, Galán B. Microalgae, old sustainable food and fashion nutraceuticals. Microb Biotechnol 2017; 10(5): 1017-24. [http://dx.doi.org/10.1111/1751-7915.12800] [PMID: 28809450]

[25] Koyande AK, Chew KW, Rambabu K, Tao Y, Chu DT, Show PL. Microalgae: A potential alternative to health supplementation for humans. Food Sci Hum Wellness 2019; 8(1): 16-24. [http://dx.doi.org/10.1016/j.fshw.2019.03.001]

[26] Bishop WM, Zubeck HM. Evaluation of microalgae for use as nutraceuticals and nutritional supplements. J Nutr Food Sci 2012; 2(5): 1-6. [http://dx.doi.org/10.4172/2155-9600.1000147]

[27] Galasso C, Gentile A, Orefice I, *et al.* Microalgal derivatives as potential nutraceutical and food supplements for human health: A focus on cancer prevention and interception. Nutrients 2019; 11(6): 1226. [http://dx.doi.org/10.3390/nu11061226] [PMID: 31146462]

[28] Guzmán S, Gato A, Lamela M, Freire-Garabal M, Calleja JM. Anti-inflammatory and immunomodulatory activities of polysaccharide from *Chlorella stigmatophora* and *Phaeodactylum tricornutum*. Phytother Res 2003; 17(6): 665-70. [http://dx.doi.org/10.1002/ptr.1227] [PMID: 12820237]

[29] Yim, J. H., Son, E., Pyo, S., & Lee, H. K. (2005). Novel sulfated polysaccharide derived from red-tide microalga Gyrodinium impudicum strain KG03 with immunostimulating activity in vivo. Marine Biotechnology, 7(4): 331-8.

[30] Raposo M, De Morais R, Bernardo de Morais A. Bioactivity and applications of sulphated polysaccharides from marine microalgae. Mar Drugs 2013; 11(1): 233-52. [http://dx.doi.org/10.3390/md11010233] [PMID: 23344113]

[31] Park GT, Go RE, Lee HM, *et al.* Potential anti-proliferative and immunomodulatory effects of marine microalgal exopolysaccharide on various human cancer cells and lymphocytes *in vitro*. Mar Biotechnol (NY) 2017; 19(2): 136-46. [http://dx.doi.org/10.1007/s10126-017-9735-y] [PMID: 28161850]

[32] de Jesus Raposo MF, de Morais RMSC, de Morais AMMB. Health applications of bioactive compounds from marine microalgae. Life Sci 2013; 93(15): 479-86. [http://dx.doi.org/10.1016/j.lfs.2013.08.002] [PMID: 23994664]

[33] Patias LD, Maroneze MM, Siqueira SF, de Menezes CR, Zepka LQ, Jacob-Lopes E. Chapter 11: Single-cell protein as a source of biologically active ingredients for the formulation of antiobesity foods. In: Holban AM, Grumezescu AM. Alternative and Replacement Foods, Handbook of Food Bioengineering 2018; 317-53. [http://dx.doi.org/10.1016/B978-0-12-811446-9.00011-3]

[34] Mobin S, Alam F. Some promising microalgal species for commercial applications: a review. Energy Procedia 2017; 110: 510-7. [http://dx.doi.org/10.1016/j.egypro.2017.03.177]

[35] Hemantkumar JN, Rahimbhai MI. Microalgae and its use in nutraceuticals and food supplements. In: Vítová M. Microalgae: from Physiology to Application. London: IntechOpen 2019; 10.

[36] Sathasivam R, Radhakrishnan R, Hashem A, Abd Allah EF. Microalgae metabolites: A rich source for food and medicine. Saudi J Biol Sci 2019; 26(4): 709-22. [http://dx.doi.org/10.1016/j.sjbs.2017.11.003] [PMID: 31048995]

[37] Ikeuchi M, Koyama T, Takahashi J, Yazawa K. Effects of astaxanthin in obese mice fed a high-fat diet. Biosci Biotechnol Biochem 2007; 71(4): 893-9. [http://dx.doi.org/10.1271/bbb.60521] [PMID: 17420580]

[38] Raja R, Coelho A, Hemaiswarya S, Kumar P, Carvalho IS, Alagarsamy A. Applications of microalgal paste and powder as food and feed: An update using text mining tool. Beni Suef Univ J Basic Appl Sci 2018; 7(4): 740-7.
[http://dx.doi.org/10.1016/j.bjbas.2018.10.004]

[39] Luo X, Su P, Zhang W. Advances in microalgae-derived phytosterols for functional food and pharmaceutical applications. Mar Drugs 2015; 13(7): 4231-54.
[http://dx.doi.org/10.3390/md13074231] [PMID: 26184233]

[40] Carballo-Cárdenas EC, Tuan PM, Janssen M, Wijffels RH. Vitamin E (α-tocopherol) production by the marine microalgae Dunaliella tertiolecta and Tetraselmis suecica in batch cultivation. Biomol Eng 2003; 20(4-6): 139-47.
[http://dx.doi.org/10.1016/S1389-0344(03)00040-6] [PMID: 12919791]

[41] Durmaz Y. Vitamin E (α-tocopherol) production by the marine microalgae Nannochloropsis oculata (Eustigmatophyceae) in nitrogen limitation. Aquaculture 2007; 272(1-4): 717-22.
[http://dx.doi.org/10.1016/j.aquaculture.2007.07.213]

[42] Mokrosnop VM, Zolotareva EK. Microalgae as tocopherol producers Biotechnologia Acta 2014; 7, 26-33.
[http://dx.doi.org/10.15407/biotech7.02.026]

[43] De Roeck-Holtzhauer Y, Quere I, Claire C. Vitamin analysis of five planktonic microalgae and one macroalga. J Appl Phycol 1991; 3(3): 259-64.
[http://dx.doi.org/10.1007/BF00003584]

[44] Smerilli A, Orefice I, Corato F, Gavalás Olea A, Ruban AV, Brunet C. Photoprotective and antioxidant responses to light spectrum and intensity variations in the coastal diatom *S keletonema marinoi*. Environ Microbiol 2017; 19(2): 611-27.
[http://dx.doi.org/10.1111/1462-2920.13545] [PMID: 27712003]

[45] Suleria H, Osborne S, Masci P, Gobe G. Marine-based nutraceuticals: An innovative trend in the food and supplement industries. Mar Drugs 2015; 13(10): 6336-51.
[http://dx.doi.org/10.3390/md13106336] [PMID: 26473889]

[46] Liu D, Keesing JK, He P, Wang Z, Shi Y, Wang Y. The world's largest macroalgal bloom in the Yellow Sea, China: Formation and implications. Estuar Coast Shelf Sci 2013; 129: 2-10.
[http://dx.doi.org/10.1016/j.ecss.2013.05.021]

[47] Ganesan AR, Tiwari U, Rajauria G. Seaweed nutraceuticals and their therapeutic role in disease prevention. Food Sci Hum Wellness 2019; 8(3): 252-63.
[http://dx.doi.org/10.1016/j.fshw.2019.08.001]

[48] Cotas J, Leandro A, Pacheco D, Gonçalves AMM, Pereira L. A comprehensive review of the nutraceutical and therapeutic applications of red seaweeds (Rhodophyta). Life (Basel) 2020; 10(3): 19.
[http://dx.doi.org/10.3390/life10030019] [PMID: 32110890]

[49] Ganesan P, Noda K, Manabe Y, *et al.* Siphonaxanthin, a marine carotenoid from green algae, effectively induces apoptosis in human leukemia (HL-60) cells. Biochim Biophys Acta, Gen Subj 2011; 1810(5): 497-503.
[http://dx.doi.org/10.1016/j.bbagen.2011.02.008] [PMID: 21371530]

[50] Mišurcová L, Ambrožová J, Samek D. Seaweed lipids as nutraceuticals. Adv Food Nutr Res 2011; 64: 339-55.
[http://dx.doi.org/10.1016/B978-0-12-387669-0.00027-2] [PMID: 22054960]

[51] Mišurcová L, Machů L, Orsavová J. Seaweed minerals as nutraceuticals. Adv Food Nutr Res 2011; 64: 371-90.
[http://dx.doi.org/10.1016/B978-0-12-387669-0.00029-6] [PMID: 22054962]

[52] Chakraborty K, Vijayagopal P, Gopalakrishnan A. Nutraceutical products from seaweeds-wonder herbs of the oceans. Marine Fisheries Information Service. Technical and Extension Series 2018;

(237): 7-12.

[53] Williams PA, Phillips GO. 11. The use of hydrocolloids to improve food texture. In: McKenna BM Ed. Texture in Food Volume 1: Semi-Solid Foods. Woodhead Publishing Limited 2003; 251-74. [http://dx.doi.org/10.1533/9781855737082.2.251]

[54] Venkatesan J, Lowe B, Anil S, *et al.* Seaweed polysaccharides and their potential biomedical applications. Stärke 2015; 67(5-6): 381-90. [http://dx.doi.org/10.1002/star.201400127]

[55] Chen SL, Ma HI, Han JM, *et al.* Antinociceptive effects of morphine and naloxone in mu-opioid receptor knockout mice transfected with the MORS196A gene. J Biomed Sci 2010; 17(1): 28. [http://dx.doi.org/10.1186/1423-0127-17-28] [PMID: 20403204]

[56] Ahmed ABA, Adel M, Karimi P, Peidayesh M. Pharmaceutical, cosmeceutical, and traditional applications of marine carbohydrates. Adv Food Nutr Res 2014; 73: 197-220. [http://dx.doi.org/10.1016/B978-0-12-800268-1.00010-X] [PMID: 25300548]

[57] Shannon E, Abu-Ghannam N. Seaweeds as nutraceuticals for health and nutrition. Phycologia 2019; 58(5): 563-77. [http://dx.doi.org/10.1080/00318884.2019.1640533]

[58] Jönsson M, Allahgholi L, Sardari RRR, Hreggviðsson GO, Nordberg Karlsson E. Extraction and modification of macroalgal polysaccharides for current and next-generation applications. Molecules 2020; 25(4): 930. [http://dx.doi.org/10.3390/molecules25040930] [PMID: 32093097]

[59] Gomez-Zavaglia A, Prieto Lage MA, Jimenez-Lopez C, Mejuto JC, Simal-Gandara J. The potential of seaweeds as a source of functional ingredients of prebiotic and anti-oxidant value. Antioxidants 2019; 8(9): 406. [http://dx.doi.org/10.3390/antiox8090406] [PMID: 31533320]

[60] Wells ML, Potin P, Craigie JS, *et al.* Algae as nutritional and functional food sources: revisiting our understanding. J Appl Phycol 2017; 29(2): 949-82. [http://dx.doi.org/10.1007/s10811-016-0974-5] [PMID: 28458464]

[61] Leandro A, Pacheco D, Cotas J, Marques JC, Pereira L, Gonçalves AMM. Seaweed's bioactive candidate compounds to food industry and global food security. Life (Basel) 2020; 10(8): 140. [http://dx.doi.org/10.3390/life10080140] [PMID: 32781632]

[62] Biris-Dorhoi ES, Michiu D, Pop CR, *et al.* Macroalgae—A sustainable source of chemical compounds with biological activities. Nutrients 2020; 12(10): 3085. [http://dx.doi.org/10.3390/nu12103085] [PMID: 33050561]

[63] Susanto E, Fahmi AS, Hosokawa M, Miyashita K. Variation in lipid components from 15 species of tropical and temperate seaweeds. Mar Drugs 2019; 17(11): 630. [http://dx.doi.org/10.3390/md17110630] [PMID: 31698797]

[64] Kendel M, Wielgosz-Collin G, Bertrand S, Roussakis C, Bourgougnon N, Bedoux G. Lipid composition, fatty acids and sterols in the seaweeds *Ulva armoricana*, and *Solieria chordalis* from Brittany (France): An analysis from nutritional, chemotaxonomic, and antiproliferative activity perspectives. Mar Drugs 2015; 13(9): 5606-28. [http://dx.doi.org/10.3390/md13095606] [PMID: 26404323]

[65] Gonçalves A, Marques JC, Gonçalves F. Fatty acids' profiles of aquatic organisms: Revealing the impacts of environmental and anthropogenic stressors. Fatty Acids 2017; 89-118.

[66] Ibañez E, Herrero M, Mendiola JA, Castro-Puyana M. Extraction and characterization of bioactive compounds with health benefits from marine resources: macro and micro algae, cyanobacteria, and invertebrates InMarine bioactive compounds. Boston, MA: Springer 2012; pp. 55-98.

[67] Lopes G, Sousa C, Valentao P, Andrade PB. Sterols in Algae and Health.Bioactive Compounds from Marine Foods: Plant and Animal Sources. John Wiley & Sons, Ltd. 2013.

[http://dx.doi.org/10.1002/9781118412893.ch9]

[68] Abdul QA, Choi RJ, Jung HA, Choi JS. Health benefit of fucosterol from marine algae: a review. J Sci Food Agric 2016; 96(6): 1856-66. [http://dx.doi.org/10.1002/jsfa.7489] [PMID: 26455344]

[69] Vanmierlo T, Schepers M, Martens N, *et al.* Edible seaweed-derived constituents: an undisclosed source of neuroprotective compounds. Neural Regen Res 2020; 15(5): 790-5. [http://dx.doi.org/10.4103/1673-5374.268894] [PMID: 31719238]

[70] Phong WN, Le CF, Show PL, Lam HL, Ling TC. Evaluation of different solvent types on the extraction of proteins from microalgae. Chem Eng Trans 2016; 52: 1063-8.

[71] Kadam SU, Álvarez C, Tiwari BK, O'Donnell CP. Extraction and characterization of protein from Irish brown seaweed *Ascophyllum nodosum*. Food Res Int 2017; 99(Pt 3): 1021-7. [http://dx.doi.org/10.1016/j.foodres.2016.07.018] [PMID: 28865612]

[72] Harnedy PA, FitzGerald RJ. Extraction of protein from the macroalga *Palmaria palmata*. Lebensm Wiss Technol 2013; 51(1): 375-82. [http://dx.doi.org/10.1016/j.lwt.2012.09.023]

[73] Alam T, Najam L, Al-Harrasi A. Extraction of natural pigments from marine algae J Agr Mar Sci 2018; 23(1): 81-91.

[74] Wiltshire KH, Boersma M, Möller A, Buhtz H. Extraction of pigments and fatty acids from the green alga Scenedesmus obliquus (Chlorophyceae). Aquat Ecol 2000; 34(2): 119-26. [http://dx.doi.org/10.1023/A:1009911418606]

[75] Saha S, Murray P. Exploitation of microalgae species for nutraceutical purposes: cultivation aspects. Fermentation (Basel) 2018; 4(2): 46. [http://dx.doi.org/10.3390/fermentation4020046]

[76] Udayan A, Arumugam M, Pandey A. Nutraceuticals from algae and cyanobacteria. In: Rastogi R, Madamwar D, Pandey A (Eds.) Algal Green Chemistry: Recent Progress in Biotechnology. Elsevier 2017; 65-89. [http://dx.doi.org/10.1016/B978-0-444-63784-0.00004-7]

[77] Jha D, Jain V, Sharma B, Kant A, Garlapati VK. Microalgae-based Pharmaceuticals and Nutraceuticals: An Emerging Field with Immense Market Potential. ChemBioEng Rev 2017; 4(4): 257-72. [http://dx.doi.org/10.1002/cben.201600023]

[78] Tang DYY, Khoo KS, Chew KW, Tao Y, Ho SH, Show PL. Potential utilization of bioproducts from microalgae for the quality enhancement of natural products. Bioresour Technol 2020; 304: 122997. [http://dx.doi.org/10.1016/j.biortech.2020.122997] [PMID: 32094007]

[79] Levasseur W, Perré P, Pozzobon V. A review of high value-added molecules production by microalgae in light of the classification. Biotechnol Adv 2020; 41: 107545. [http://dx.doi.org/10.1016/j.biotechadv.2020.107545] [PMID: 32272160]

[80] Leu S, Boussiba S. Advances in the production of high-value products by microalgae. Ind Biotechnol (New Rochelle NY) 2014; 10(3): 169-83. [http://dx.doi.org/10.1089/ind.2013.0039]

[81] Kusmayadi A, Leong YK, Yen HW, Huang CY, Chang JS. Microalgae as sustainable food and feed sources for animals and humans – Biotechnological and environmental aspects. Chemosphere 2021; 271: 129800. [http://dx.doi.org/10.1016/j.chemosphere.2021.129800] [PMID: 33736224]

[82] Balamurugan S, Sathishkumar R, Li HY. Biotechnological perspectives to augment the synthesis of valuable biomolecules from microalgae by employing wastewater. J Water Process Eng 2020; 101713.

[83] M U N, Mehar JG, Mudliar SN, Shekh AY. Mehar JG, Mudliar SN, Shekh AY. Recent advances in

microalgal bioactives for food, feed, and healthcare products: commercial potential, market space, and sustainability. Compr Rev Food Sci Food Saf 2019; 18(6): 1882-97. [http://dx.doi.org/10.1111/1541-4337.12500] [PMID: 33336956]

[84] Stirk WA, van Staden J. Potential of phytohormones as a strategy to improve microalgae productivity for biotechnological applications. Biotechnol Adv 2020; 44: 107612. [http://dx.doi.org/10.1016/j.biotechadv.2020.107612] [PMID: 32810563]

[85] Bhatia L, Bachheti RK, Garlapati VK, Chandel AK. Third-generation biorefineries: a sustainable platform for food, clean energy, and nutraceuticals production. Biomass Convers Biorefin 2020; 1-6. [http://dx.doi.org/10.1007/s13399-020-00843-6]

[86] Dhandayuthapani K, Malathy S, Mulla SI, Gupta SK. An Insight into the Potential Application of Microalgae in Pharmaceutical and Nutraceutical Production InAlgae. Singapore: Springer 2021; pp. 135-79.

[87] Rodrigues D, Freitas AC, Pereira L, *et al.* Chemical composition of red, brown and green macroalgae from Buarcos bay in Central West Coast of Portugal. Food Chem 2015; 183: 197-207. [http://dx.doi.org/10.1016/j.foodchem.2015.03.057] [PMID: 25863629]

[88] Barbosa M, Valentão P, Andrade P. Bioactive compounds from macroalgae in the new millennium: implications for neurodegenerative diseases. Mar Drugs 2014; 12(9): 4934-72. [http://dx.doi.org/10.3390/md12094934] [PMID: 25257784]

[89] Kim EY, Kim DG, Kim YR, Hwang HJ, Nam TJ, Kong IS. An improved method of protein isolation and proteome analysis with *Saccharina japonica* (Laminariales) incubated under different pH conditions. J Appl Phycol 2011; 23(1): 123-30. [http://dx.doi.org/10.1007/s10811-010-9550-6]

[90] Plaza M, Cifuentes A, Ibáñez E. In the search of new functional food ingredients from algae. Trends Food Sci Technol 2008; 19(1): 31-9. [http://dx.doi.org/10.1016/j.tifs.2007.07.012]

[91] MacArtain P, Gill CIR, Brooks M, Campbell R, Rowland IR. Nutritional value of edible seaweeds. Nutr Rev 2007; 65(12): 535-43. [http://dx.doi.org/10.1111/j.1753-4887.2007.tb00278.x] [PMID: 18236692]

[92] Fleurence J. Seaweed proteins. Trends Food Sci Technol 1999; 10(1): 25-8. [http://dx.doi.org/10.1016/S0924-2244(99)00015-1]

[93] Sridhar S, Rengasamy R. Sridhar S, Rengasamy R. Potential of seaweed liquid fertilizers (SLFS) on some agricultural crop with special reference to protein profile of seedlings. Int J Dev Res 2011; 1(7): 055-7.

[94] Helmiyati, Aprilliza M. Characterization and properties of sodium alginate from brown algae used as an ecofriendly superabsorbent. In: Materials Science and Engineering Conference Series 2017; 188(1): 012019. [http://dx.doi.org/10.1088/1757-899X/188/1/012019]

[95] Anis M, Ahmed S, Hasan MM. Algae as nutrition, medicine and cosmetic: the forgotten history, present status and future trends. World J Pharm Pharm Sci 2017; 6(6): 1934-59.

[96] Armisen R, Gaiatas F. Agar. Handbook of hydrocolloids 2009; 82-107.

[97] Todar K. Todar's online textbook of bacteriology. University of Wisconsin-Madison Dept. of Bacteriology, Madison, WI.

[98] Chennubhotla VS, Rao MU, Rao KS. Commercial importance of marine macro algae. Seaweed Research and Utilization 2013; 35(1 & 2): 118-28.

[99] Williamson EM, Liu X, Izzo AA. Trends in use, pharmacology, and clinical applications of emerging herbal nutraceuticals. Br J Pharmacol 2020; 177(6): 1227-40. [http://dx.doi.org/10.1111/bph.14943] [PMID: 31799702]

[100] Park HS, Kim GY, Nam TJ, Deuk Kim N, Hyun Choi Y. Antiproliferative activity of fucoidan was associated with the induction of apoptosis and autophagy in AGS human gastric cancer cells. J Food Sci 2011; 76(3): T77-83.
[http://dx.doi.org/10.1111/j.1750-3841.2011.02099.x] [PMID: 21535865]

[101] Das SK, Hashimoto T, Kanazawa K. Growth inhibition of human hepatic carcinoma HepG2 cells by fucoxanthin is associated with down-regulation of cyclin D. Biochim Biophys Acta, Gen Subj 2008; 1780(4): 743-9.
[http://dx.doi.org/10.1016/j.bbagen.2008.01.003] [PMID: 18230364]

[102] Kim KN, Heo SJ, Kang SM, Ahn G, Jeon YJ. Fucoxanthin induces apoptosis in human leukemia HL-60 cells through a ROS-mediated Bcl-xL pathway. Toxicol in vitro 2010; 24(6): 1648-54.
[http://dx.doi.org/10.1016/j.tiv.2010.05.023] [PMID: 20594983]

[103] McLellan DS, Jurd KM. McLellan DS, Jurd KM. Anticoagulants from marine algae. Blood Coagul Fibrinolysis 1992; 3(1): 69-77.
[http://dx.doi.org/10.1097/00001721-199202000-00011]

[104] Church FC, Meade JB, Treanor RE, Whinna HC. Antithrombin activity of fucoidan. The interaction of fucoidan with heparin cofactor II, antithrombin III, and thrombin. J Biol Chem 1989; 264(6): 3618-23.
[http://dx.doi.org/10.1016/S0021-9258(18)94111-6] [PMID: 2914965]

[105] Colliec S, Fischer AM, Tapon-Bretaudiere J, Boisson C, Durand P, Jozefonvicz J. Anticoagulant properties of a fucoidan fraction. Thromb Res 1991; 64(2): 143-54.
[http://dx.doi.org/10.1016/0049-3848(91)90114-C] [PMID: 1811335]

[106] Bae JS. Antithrombotic and profibrinolytic activities of phloroglucinol. Food Chem Toxicol 2011; 49(7): 1572-7.
[http://dx.doi.org/10.1016/j.fct.2011.04.003] [PMID: 21501649]

[107] Li YX, Wijesekara I, Li Y, Kim SK. Phlorotannins as bioactive agents from brown algae. Process Biochem 2011; 46(12): 2219-24.
[http://dx.doi.org/10.1016/j.procbio.2011.09.015]

[108] Heo SJ, Ko SC, Kang SM, *et al.* Cytoprotective effect of fucoxanthin isolated from brown algae Sargassum siliquastrum against H2O2-induced cell damage. Eur Food Res Technol 2008; 228(1): 145-51.
[http://dx.doi.org/10.1007/s00217-008-0918-7]

[109] Artan M, Li Y, Karadeniz F, Lee SH, Kim MM, Kim SK. Anti-HIV-1 activity of phloroglucinol derivative, 6,6′-bieckol, from Ecklonia cava. Bioorg Med Chem 2008; 16(17): 7921-6.
[http://dx.doi.org/10.1016/j.bmc.2008.07.078] [PMID: 18693022]

[110] Witvrouw M, De Clercq E. Sulfated polysaccharides extracted from sea algae as potential antiviral drugs. Gen Pharmacol 1997; 29(4): 497-511.
[http://dx.doi.org/10.1016/S0306-3623(96)00563-0] [PMID: 9352294]

[111] Wächter GA, Franzblau SG, Montenegro G, Hoffmann JJ, Maiese WM, Timmermann BN. Inhibition of *Mycobacterium t uberculosis* Growth by Saringosterol from *Lessonia n igrescens*. J Nat Prod 2001; 64(11): 1463-4.
[http://dx.doi.org/10.1021/np010101q] [PMID: 11720535]

[112] Remya RR, Julius A, Suman TY, *et al.* Role of Nanoparticles in Biodegradation and Their Importance in Environmental and Biomedical Applications. Journal of Nanomaterials 2022, 6090846, 15 pages..

[113] Samrot AV, Purohit K, Saigeetha S, Shobana N, Stalin Dhas T, Jane Cypriyana PJ. Citrus sinensis cellulose fibres incorporated with SPIONs for effective removal of crystal violet dye. Biocatal Agric Biotechnol 2022; 39: 102211.
[http://dx.doi.org/10.1016/j.bcab.2021.102211]

Algal Functional Foods and Nutraceuticals, 2022, 131-183

CHAPTER 7

Extraction of Microalgal Bioactive Compounds Towards Functional Ingredients: A Biorefinery Approach with Prospects And Challenges

Joana Assunção[1,2], Helena M. Amaro[1], F. Xavier Malcata[2,3,4] and A. Catarina Guedes[1,*]

[1] *CIIMAR/CIMAR-LA – Interdisciplinary Centre of Marine and Environmental Research, University of Porto, Novo Edifício do Terminal de Cruzeiros do Porto de Leixões, Av. General Norton de Matos, s/n, Matosinhos, Portugal*

[2] *LEPABE – Laboratory for Process Engineering, Environment, Biotechnology and Energy, University of Porto, Rua Dr. Roberto Frias, s/n, Porto, Portugal*

[3] *FEUP – Department of Chemical Engineering, Faculty of Engineering, University of Porto, Rua Dr. Roberto Frias, s/n, Porto, Portugal*

[4] *ALICE - Associate Laboratory in Chemical Engineering, Faculty of Engineering, University of Porto, Rua Dr. Roberto Frias,Porto, Portugal*

Abstract: Algal bioactive compounds, including pigments, polyunsaturated fatty acids, vitamins, proteins and phenolic compounds, have a wide range of increasing applications in the pharmaceutical, nutraceutical, feed, food, and cosmetics sectors. Extraction of such compounds with the purpose of obtaining nutraceutical and food ingredients at large has been under scrutiny over the last years towards greener and more sustainable bio-based processes. The step of the extraction is indeed considered the most expensive in algal-based industrial bioprocess, mainly owing to the high processing costs and environmental impacts. Therefore, the application of a biorefinery strategy (for biomass reutilization, via choice of specific green and generally recognized as safe solvent), will certainly help attain a more feasible and competitive solution. Furthermore, emerging extraction technologies are potentially suitable for application at a large scale, as part of a multi-integrated extraction strategy. This chapter accordingly addresses the most important (micro)algal-compounds possessing bioactivities, and already or likely to be used as functional ingredients; in addition, extraction procedures thereof are discussed, bearing in mind a biorefinery strategy – with major prospects and associated challenges.

* **Corresponding author A. Catarina Guedes:** CIIMAR/CIMAR-LA - Interdisciplinary Centre of Marine and Environmental Research, Matosinhos, Portugal; Tel: +351-22-340-1800; E-mail: acatarinaguedes@gmail.com

Avinash Mishra (Ed.)

Keywords: Algae, Assisted Extraction, Bioactive Compounds, Biomass Reuse, Bioprocess, Biorefinery, Circular Economy, Cyanobacteria, Enzyme-Assisted Extraction, Functional Ingredients, Generally Recognized As Safe Solvents, Microalga Farms, Microalgae, Microwave-Assisted Extraction, Pressurized Liquid Extraction, Pulse-Electric Field, Subcritical Water Extraction, Supercritical Fluid Extraction, Ultrasound-Assisted Extraction.

INTRODUCTION

Given the current economic and environmental context, arising from an ever-increasing population, with a longer life expectancy and a stressful urban lifestyle, a higher incidence of diseases such as obesity, diabetes, cancer, pulmonary, cardiovascular, neurological or infectious conditions by multi-resistant bacteria [1 - 5] is anticipated. Hence, searching for new sources of bioactive natural products is of paramount importance to find new drug candidates [6], as well as new products and alternatives to enrich human or/and animal diets aimed at fighting or at least preventing certain diseases (*i.e.*, nutraceuticals).

The global expectations concerning the production of new natural products bearing health-related advantages upon consumption are increasing, along with environmental awareness encompassing sustainable products. In this context, several companies all over the world are starting to readapt their business model towards a more eco-friendly process – in which the production of daily products is based on renewable raw materials and upgrade of their by-products, thus minimizing waste and energy consumption.

The search/production for alternative compounds of interest for the nutraceutical and food industries have been gaining strength from natural non-conventional sources such as macroalgae, microalgae and cyanobacteria. Focusing particularly in microalgae and cyanobacteria; they are a particular group of biotechnological significance – their number is estimated to have more than 70,000 species but are yet a largely untapped resource [7]. Such microorganisms are able to perform photosynthesis, by converting chemical energy from sunlight and use inorganic carbon sources to release oxygen, and produce biomass and a wide range of organic compounds and unique metabolites with economical interest [8 - 11]. They have been receiving increased attention in the past years owing to their rich-composition in bioactive active compounds, with nutraceutical, food, feed or pharmaceutical applications [12 - 14]. Such compounds include pigments (*i.e.*, chlorophylls (Chls), carotenoids, phycobiliproteins), fatty acids (polyunsaturated fatty acids – PUFAs), phenolic compounds, exopolysaccharides, vitamins and proteins [15]. Most of these molecules display potent bioactive properties – antioxidant, anti-tumoral, immuno-modulatory, anti-bacterial, and anti-inflammatory, among others, widely sought by the industry of natural and functional products for nutraceutical, dietary supplements, food additives, food colorants or feed applications [16]. Compounds such as phycocyanin, carotenoids or PUFAs have been explored for their remarkable antioxidant and nutraceutical properties [15, 17, 18], yet the only way to gain benefits is by consuming dried algal powder via incorporation in algal-based products. In fact, most of the algal-based market is directed to single-cell protein (whole biomass) for use as a health food, dietary supplements or feed [19]. At present, all marketed algae-based products are intracellular, and therefore biomass production is the main asset for the sustainability of the industrial route [19].

With an increasing turn out of the market to natural and bioactive compounds, with strong perspectives in food and nutraceutical applications, the fractionation of biomass into more specialized compounds (*e.g.*, phycocyanin, phycoerythrin, phenolic compounds, exopolysaccharides) is a priority to reduce the overall cost of algal-based bioprocessing. Besides the cost of microalga cultivation (*i.e.* photobioreactors and facilities, energy consumption), downstream processing steps as fractioning and purification of the compounds of interest account for 2 to 90%, the total cost of manufacturing microalgae-based bioproducts – depending on the purity of the biochemical and purity specifications [19, 20]. This is considered the major bottleneck to obtaining a viable process, with competitive prices at the commercial level [21]. To face the expected demands for algae-based natural and healthier ingredients, optimization of extraction and fractionation of biomass into rich-extracts of value-added compounds have been under scrutiny in the last years, towards a greener and sustainable bio-based process. By resorting to a biorefinery strategy – *e.g.*, biomass extraction maximization and reutilization, coupled with a choice of certain green and cheaper solvents, it is expected to increase the economic feasibility of the process and thus, achieve more competitive algal-based processes towards the nutraceutical and food markets. Improvement of biorefinery techniques will indeed prove essential to exploit all products produced via algae cultivation [21].

This chapter will accordingly provide a perspective on the most common extraction procedures of bioactive compounds, namely food and nutraceutical purposes, in light of the biorefinery concept – with a discussion of major prospects and associated challenges.

ALGAL-DERIVED BIOACTIVE COMPOUNDS AS FUNCTIONAL INGREDIENTS

Microalgae and cyanobacteria and their extracts represent a diverse and remarkable natural source of diverse bioactive compounds with nutritional and therapeutical value – including carotenoids (*i.e.*, β-carotene, zeaxanthin, lutein, astaxanthin), phycobiliproteins or PBPs (phycocyanin, phycoerythrin, allophycocyanin), PUFAs, exopolysaccharides, phenolic compounds, vitamins, and proteins, with a potential market. Those are considered value-added compounds, as they possess important pharmacological bioactivities; they may be used as functional ingredients in food or cosmetics, and included in nutraceutical formulae for human and animal welfare (aquaculture, pets) [14, 22 - 26]. Most such compounds are reported to possess important biological activities – *e.g.*, anti-aging, anti-tumoral, anti-viral, anti-inflammatory, antimicrobial, anticoagulant, anti-obesity, anti-diabetes, and immune modulators agents, among others [14, 18, 24, 27]; further to antioxidant features helpful in fighting and preventing action of reactive oxygen species (ROS) – widely responsible several oxidative stress-related diseases (*i.e.*, cholesterol, atherosclerosis, diabetes, *etc.*) [16, 24, 27] (Table **1**).

Table 1. Main microalgal sources of bioactive compounds with potential application as a functional ingredient(s).

Bioactive Compounds		**Algal Sources**	**Bioactive Features**	**Application as Functional Ingredient**	**Refs**
Pigments					
Chlorophylls (Chls)	Chl *a* Chl *b* Pheophytin Other chlorophyll derivatives	*Arthorospira maxima* *Arthorospira platensis* *Chlorella vulgaris* *Chlorella sp.* *Scendesmus obliquus*	Antioxidant Anti-acne Anti-bacterial Anti-mutagenic Anti-obesity Anti-thrombotic Chemoprotective effect Inducer of hemoglobin production Immuno-modulator Stimulator of liver and bile secretion Hyaluronidase activity (anti-ageing) Anti-inflammatory Antimicrobial properties Photosensitizer agent	*Pharmaceutical* *Cosmetics*: dyeing, hydrating or anti-acne agent, anti-ageing (cream formulas, soaps, shampoo, tooth brush). *Nutraceutical*: dietary supplements (capsules and powder forms). *Food*: colorant (soup, milk, beverage, *etc*) *Other applications*: diagnostic tool (*i.e.* cancer)	[28, 30–33, 35, 88–90]

(Table 1) cont.....

Carotenoids	β-carotene	*Dunaliella salina* *Arthorospira platensis* *Arthorospira maxima* *Chlorella zofingiensi*	Provitamin A function Antioxidant Chemoprotective effect Antihypertensive effect Anti-ageing Anti-asthmatic/allergic	*Pharmaceutical* *Cosmetics*: UV-filter, anti-ageing, antioxidant, anti-pigmentation and hydration agents	[33, 40, 53, 54, 91–97, 41–43, 47, 49–52]
	-	-	effect Combat eye-related and cardiovascular diseases Hepatic-protection Immunomodulation Photoprotection	(sunscreen, serum, moisturizers, cream formulas). *Nutraceutical*: food additive, nutritional supplements	
-	Astaxanthin	*Haematococcus pluvialis* *Chlorella zofingiensi*	Antioxidant Anti-ageing Anti-inflammatory Antihypertensive effect Anti-bacterial Anti-diabetic Anti-obesity Combat of eye-related and cardiovascular diseases, fatigue, and bone loss Enhancement of immune function Neuro-protection Photoprotection	fortification of food beverage *Feed*: cattle and fish feed, pet food; improvement of fish and bird coloring	-
	Zeaxanthin	*Chlorella* sp.	Antioxidant Chemoprotective effect Anti-ageing Prevention of eye-related and cardiovascular diseases Photoprotection	-	-
	Lutein	*Muriellopsis sp.* *Scenedesmus almeriensis* *Chlorella vulgaris*	Antioxidant Anti-inflammatory Anti-bacterial Chemoprotective effect Prevention of eye-related and cardiovascular diseases Hepatic-protection Photoprotection	-	-

(Table 1) cont.....

	Cantaxanthin	*Coelastrella striolata* var. *multistriata* *Chlorella vulgaris*	Antioxidant Anti-inflammatory Chemoprotective effect Neuro-protection Prevention of eye-related and cardiovascular diseases Immuno-modulator Photoprotection	-	-
	Fucoxanthin	*Odontella aurita, Phaeodactylum tricornutum Isochrysis galbana*	Antioxidant Anti-tumor Anti-inflammatory Anti-malarial Anti-obesity Antiangiogenic effect	-	-
Phycobiliproteins	Phycocyanin Allophycocyanin	*Aphanizomenon flos-aquae* *Arthorospira platensis* *Arthorospira maxima* *Arthrospira fusiformis* *Oscillatoria tenuis* *Anabaena oryzae* *Anabaena sp.* *Nostoc sp.* *Phormidium valderianum* *Phorphyridium aerugineum*	Antioxidant Anti-allergic Anti-bacterial Anti-cancer Anti-fungal Anti-inflammatory Anti-obesity Anti-viral Hepatic-protection Immunomodulation Neuro-protection Photosentizer agent (phycocyanin)	*Pharmaceutical Cosmetics*: dye and antioxidant agents (cream formulas, make-up). *Nutraceutical*: food additive and supplement. *Food*: colorant, additive (chewing gum, pasta, ice creams, desserts, jellies, candies, dairy products, soft drinks, chips) *Other applications*: research tool (*i.e.* fluorescence, molecular probes, labeled antibodies) Diagnostic and therapeutic tool (*i.e.* cancer) (phycocyanin)	[23,56, 57, 98–112]

(Table 1) cont.....

	Phycoerythrin	*Porphyridium cruentum* *Rhodosorus marinus* *Halomicronema sp.*	Antioxidant Anti-allergic Anti-bacterial Anti-cancer Anti-fungal Anti-inflammatory Anti-obesity Anti-viral Hepatic-protection Immunomodulation Neuro-protection	-	-
PUFAs (Poly unsaturated fatty acids)	Eicosapentaenoic acid (EPA) Docosahexaenoic acid (DHA) Arachidonic acid (ARA) Gamma-Linolenic acid (GLA)	*Arthorospira platensis* (GLA) *Arthorospira flos-aquae* (GLA) *Crypthecodinium cohnii* (EPA) *Isochrysis galbana* (DHA) *Porphyridium cruentum* (DHA) *Parietochloris incisa* (ARA) *Phaeodactylum tricornutum* (DHA, EPA) *Nannochloropsis gaditana* (EPA) *Nannochloropsis oculata* (EPA) *Ulkenia* (DHA) *Pavlova lutheri* (EPA)	Antioxidant Anti-inflammatory Anti-cholesteremic Anti-coagulant Anti-diabetic Anti-tumor Anti-obesity Anti-microbial Anti-arteriosclerotic Neuro-protection (*i.e.* combat of Alzheimer's disease)	*Pharmaceutical Nutraceutical* and *food*: food supplements and nutraceutical formulas, infant enriched-formulas *Feed*: positive influence on the growth, immune response and gut function of fish and poultry	[59,65, 68, 69, 113–120]
Phenolic compounds	Benzoic acid derivatives Hydroxybenzaldehydes Cinnamic acid derivatives	*Arthorospira platensis* *Arthorospira maxima* Nannochloropsis sp.	Antioxidant Anti-inflammatory Anti-allergic Anti-mycotoxigenic Anti-microbial (*i.e.* fungus, bacteria) Anti-tumor Prevention cardiovascular diseases Combat diabetes mellitus Neuro-protection (*i.e.* combat Alzheimer's disease)	*Pharmaceutical Nutraceutical* and *food* – nutritional supplement *Feed*: food enriched-formulas to prevent certain diseases	[14,68, 77, 79, 80, 119, 121]

(Table 1) cont.....

Polysaccharides	EPS (exopolysacharydes) Sulphated polysaccharides	*Arthorospira platensis* *Chlorella pyrenoidosa* *Chlorella* sp. *Botryococcus* sp. *Porphyridium spp.* *Rhodella* sp.	Antioxidant Anticoagulant Anti-adhesive Anti-allergic Anti-inflammatory Anti-tumor Anti-viral Anti-metastasis Anti-microbial Bio-adhesive properties Immunomodulator Radioprotective Stimulate collagen synthesis Protect against enzymatic proteolysis	*Pharmaceutical Cosmetics*: antiviral, anti-inflammatory, antioxidant, emulsifying and hydrating agents, skin protection (*i.e.* cream and cosmetic enriched-formulas, shampoo) *Food*: (food coating, gelling, flocculant and hydrating agents, stabilizers, emulsifiers, shelf-life extenders) *Other applications*: regenerative medicine (medical engineering scaffolds for stem cell cultivation), drug design (*i.e.* encapsulation); oil/wastewater bioremediation	[17, 71–73, 76, 115, 122–125]
Vitamins	Vit B1, B2, B3, B7, B12, vit. C vit. D vit. E Folic acid Pantothenic acid Pyridoxine Riboflavin Nicotinic acid	*Arhrospira* sp. *Chlorella* sp. *Isochrysis galbana* *Nannochloropsis* gaditana *Scenedesmus* sp.	Antioxidant Precursor of some enzyme co-factors Protection of the immune system Help in cell formation and blood coagulation	*Cosmetics*: antioxidant (serum and cream-enriched formulas,) *Nutraceutical*: multicomplex vitamin enriched-products	[14, 19, 81, 82, 126]

(Table 1) cont.....

Proteins	Enzymes Amino acids (glutamate, aspartame) and peptides	*Arthrospira* sp. *Chlorella* sp. *Dunaliella salina* *Phaeodactylum* sp. *Tetraselmis* sp. *Chlamydomonas reinhardtii*	Antioxidant Anti-ageing Anti-appetite Anti-cancer Anti-cholesteremic Anti-hypoglycemic Anti-microbial Anti-obesity Anti-freeze properties Collagen stimulation	*Pharmaceutical Cosmetic*: antioxidant, collagen stimulator, anti-ageing agent, photoprotective properties (UV filters, cream or serum enriched-formulas) *Nutraceutical*: nutritional supplement *Food*: thickening, gelling and emulsifying agent, dietary supplements	[19, 84–87, 126, 127]

Among pigments, chlorophylls, carotenoids and PBPs stand out as non-toxic and non-carcinogenic natural compounds with relevant biological activities [16].

Chl *a* and their derivatives (*i.e.*, chl *b*, pheophytin) are commercially important natural green pigments, which may be extracted from cyanobacteria and algal biomass [24, 28]. Owing to its particularly bright green color, Chl *a* has been used as a natural colorant in the food industry (*i.e.*, soap, cheeses, edible oils, pasta, *etc.*) [29, 30]. Chl and its derivatives also possess relevant antioxidant [30], anti-inflammatory [31], anti-mutagenic and anticancer effects [32], via induction of carcinogen-detoxifying enzymes – thereby decreasing the risk of cancer [33]. In addition, it can be used in nutraceutical and cosmetic industries as inclusion as an ingredient in skin care products, owing to anti-ageing, antioxidant, anti-acne and hydrating properties. Despite plant leaves (*i.e.*, alfafa) being still the major source for industrial applications, extraction of microalgae (*i.e. Chlorella vulgaris*) [34] and cyanobacteria, namely *Arthrospira platensis* [35], have gained some significance.

Carotenoids are bright-colored molecules ranging from yellow to red, and possess unique antioxidant properties (*i.e.*, scavenging of free radicals and ROS) – quite appealing to the nutraceutical, food (*i.e.*, dietary supplements, colorants), animal feed, pharmaceutical and cosmetic industries [22, 33, 36]. For instance, β-carotene is a well-known active form of vitamin A – commonly added to multivitamin formulae in nutraceutical products. It fights and prevents cardiovascular diseases (*i.e.*, atherosclerosis, cholesterol and low-density protein oxidation), and is described to improve the immune system [37]. One of the major sources of this carotene is *Dunalilla salina,* which can accumulate up to 10-14% of its total dry weight (DW) [38]; however, *Chlorellla zofingiensis* [39] and cyanobacterium *Arthrospira platensis* (commercial name *Spirulina*) are also gaining ground in the production of this carotenoid. Astaxanthin is another carotenoid widely applied as a functional ingredient (*i.e.*, cosmetics, nutraceuticals, food and feed industries), proven to have a potent antioxidant capacity – even higher than other carotenoids (*i.e.*, β-carotene) [33, 40, 41]. Its application as a natural additive for aquaculture feed is of particular importance, to enhance growth metabolism and improve the immune system [42]. The freshwater species *Haematococcus pluvialis* is one of the richest sources of this pigment; it can accumulate up 7% of its DW [43], and is already produced at a large scale by several companies [44]. Other carotenoids, including zeaxanthin, lutein, canthaxanthin or fucoxanthin, also exhibit potent free radical scavenging features, and cancer prevention/suppression and immunomodulatory properties – thus unfolding promising pharmaceutical and nutraceutical applications [40, 45 - 48]. Most of them are already employed as nutritional supplements (*i.e.*, animal and fish feed formulation) or natural food colorant, besides having the potential use as functional ingredients in cosmetic formulas for their antioxidant properties. Most of these carotenoids have been obtained via the cultivation of microalgae, such as *Chlorella* sp. for zeaxanthin [49], *Muriellopsis* sp. and *Scenedesmus almeriensis* for lutein [50], and *Coalastrella* var. *striolata* and *Chlorellla zofingiensis* for canthaxanthin [51, 52]; in the case of fucoxanthin, both diatoms (*Odontella aurita, Phaeodactylum tricornutum*) and haptophytes (*i.e.*, *Isochrysis galbana*), have served as main sources thereof [47, 53]. Brown (macro)algae (*i.e.*, *Laminaria Japonica*) have also been explored for fucoxanthin production [47]. However, the overall production of carotenoids by cyanobacteria, has been limited to just a few species, like *Arthrospira platensis* and *Arthrospira maxima* [54].

The PBPs (*i.e.*, phycocyanin, allophycocyanin, phycoerythrin) are water-soluble colorant pigments (red and blue), chiefly present in the cyanobacterial photosynthetic apparatus, but also in red algae and cryptophytes [41]. They are generally described as strong antioxidant, anti-inflammatory, anti-diabetes, neuroprotective, immunomodulatory, antitumor and hepatoprotective agents, as well as anti-ageing agents for being able to combat free-radicals on the epidermis skin, and reducing cell damage, wrinkles and skin marks [23, 25, 55, 56]. They are used as natural colorants in human food (*i.e.*, chewing gum, dairy products and jellies) [57], dyes and cosmetics; due to their fluorescence at a particular wavelength (*i.e.*, phycoerythrin), they can be used as chemical probes in research and disease diagnosis [23, 57]. The major applications of PBPs are listed in Table **1** to further detail. The principal PBR sources exploited for functional ingredients are cyanobacterial cultures (*i.e. Arthrospira plantensis*) for phycocyanin and allophycocyanin, and red algae (Rodophyta) such as *Porphyridium cruentum* for phycoerythrin production [57, 58].

Other bioactive compounds of interest include fatty acids, in particular PUFAs, such as docosahexaenoic acid (DHA) (22:6 omega-3), eicosapentaenoic acid (EPA) (20:5 omega-3), gamma-linolenic acid (GLA) 18:3 (omega-6) and arachidonic acid (ARA) (20:4 omega-6) [59]; they play an important role on human and animal health, by regulating their cellular and tissue metabolism [27, 60]. They exhibit anti-inflammatory and antioxidant capacities [61, 62], and are extremely important to maintain brain and cardiovascular health (*i.e.* hypotriglyceridemia), besides helping fight other diseases like depression, arthritis or cancer [63]. Such compounds are produced to integrate nutritional supplements, in infant enriched-formulas and feed in aquaculture [64, 65]. Despite the existing major sources of those compounds being vegetable oils or fish oil, algae have been gaining some market share. DHA oil, for instance, is already commercially available [14] from dinoflagellate *Crypthecodinium cohnii* [66], genera *Schizochytrium* and *Ulkenia* (found in the kingdom Chromista) [67]; it may account for up to more than 40% of the total fatty acids content produced by these species [66]. Other PUFAs, such as GLA, ARA or EPA, are generally found in cyanobacteria *Arthrospira platensis*, *Aphanizomenon flos-aquae* [68 - 70], but also in other algae species as *Porphyridium cruentum*, *Nannochloropsis* sp., *Isochrysis galbana* or *Phaeodactylum tricornutum* [59, 65, 69].

Algal polysaccharides, in particular extracellular polymeric substances or exopolysaccharides (EPS), are of increasing interest as functional ingredients for their potential applications in the food, cosmetic and pharmaceutical industries [19, 71]. They are complex and heterogeneous macromolecules, constituted by different monosaccharides – and, in some cases, glucuronic acid and sulfate groups, usually bound externally to the algal cell-wall or that are excreted to the medium [72, 73]. With unique biochemical features, their main biological capacities include anti-tumoral, anti-viral, anti-inflammatory, antioxidant, anticoagulant, antimicrobial or bio-adhesive properties [71, 74]. In addition, they were described as having important rheological features – being useful as thickening agents and for their gelling properties within the food field [14, 74]. They also have biosurfactant features, potentially useful for water and oil bioremediation [73, 75]. Table **1** presents their application as functional ingredient in more detail. *Arthrospira platensis* cyanobacterium, and other microalgae such as *Botryococcus* sp., *Chlorella* sp., *Porphyridium* spp., and *Rhodella* sp. are known to produce and excrete polysaccharides to relatively high extents [17, 72, 76].

Phenolic compounds and phenolic-derived compounds (*i.e.*, benzoic acid derivatives, hydroxybenzaldehydes, and cinnamic acid derivatives) are another class of bioactive metabolites produced by microalga and cyanobacteria, claimed to be potent active antioxidants, and to protect cells against harmful oxidative stress while preventing cellular lipid oxidation [77]. They act as protective agents against fungus, inflammation, degenerative diseases (*i.e.*, Alzeimer's disease) and cancer [77 - 79]. Diets rich in phenolic substances benefit human and animal health by preventing cardiovascular diseases and cancer; they have also been associated with the prevention of neurodegenerative diseases and diabetes mellitus [77]. Therefore, their potential as a functional ingredient in pharmaceuticals, food supplements or cosmetics (*i.e.*, enriched-cream formulas) would be a plus. *Arthrospira* spp. and *Nannochloropsis* are pointed out as potential sources of such compounds [80].

Algal cells are rich in several kinds of vitamins, besides pro-vitamin A (α- and β-carotene) – namely several vitamins of the B group, *e.g.* B1 (thiamine), B2 (riboflavin), B3 (niacin), B12 (cobalamin) and B7 (biotin), vitamin C (ascorbic acid), vitamin D, vitamin E (tocopherols and tocotrienols) and folic acid, pantothenic acid, pyridoxine, riboflavin or nicotinic acid [14, 19, 81]. They are widely known for their antioxidant features toward protection of the immune system, cell formation and blood coagulation. They are also precursors of some important enzyme cofactors with application in nutraceutics. Moreover, they are relevant within the context of feed in aquaculture [14]. *Isochrysis galbana* sp [81], as well as other microalgae (*Chlorella* sp., *Nannochloropsis gaditana*) and cyanobacteria (*i.e. Arhrospira platensis)*, are excellent sources of such compounds [82].

Algal proteins (*i.e.* enzymes, peptides and amino acids such as aspartate or glutamate) represent a highly beneficial nutritional source toward human and animal health. Important biological activities, including anti-cholesteremic, anti-hypoglycemic, anti-obesity, anti-cancer, anti-bacterial or anti-ageing, have been reported [83] – so algal based-proteins are directly added to diet supplements and function as nutraceuticals. Some algal can produce some enzymes (*i.e.* metalloenzymes), which possess antioxidant activity and other specific types of proteins, called antifreeze proteins – reported to prevent recrystallization thus, avoiding damage to other proteins [84]. This has interesting cryopreservation applications in food and pharmaceutical fields (*i.e.* treatment of gout). Algal proteins have also shown emulsifying and thickening properties, quite useful in the food industry [85]. In cosmetics, they improve skin appearance by collagen stimulation with antioxidant and photoprotective effects [86, 87] – properties useful for functional ingredients in cream-enriched formulas. *Arthrospira* sp. and *Chlorella* sp. available in the market represent excellent sources of proteins; *Dunaliella salina*, *Tetraselmis* sp. or *Phaeodactylum* sp. also appear promising sources of such compounds [69].

ALGAL-BASED PROCESS AND BIOREFINERY – SOME CONSIDERATIONS

Given the importance of algal-derived bioactive compounds and expected demand for natural and healthier bioactive ingredients, novel and innovative cost-effective industrial technologies are urged. The R&D necessary to obtain algal-derived bioactive compounds and the development of win-win technologies – towards sustainable, economical, greener and safer approaches, have been increasing; a biorefinery strategy appears useful here, to reach the broader context of a circular bioeconomy. The latter, is an industrial model focused on the positive society-wide benefits entailed by economic activities, with growth based on the consumption of renewable resources, by taking advantage of the raw-materials used in the process in the first place, and exploring alternative energy sources [128].

Overall algal-based bioprocesses are still struggling to achieve large-scale and commercial feasibility, in part because biomass production generates low yields, and requires intensive downstream processing [19, 20].

Currently, algal bioprocessing is based on the production of whole dried biomass, to be included in all marketed algae-based products (*e.g.*, *Spirulina* or *Chlorella* powder as a nutraceutical, dietary supplements or in feed) [129].

Furthermore, microalgal-based processes are usually developed, considering one-specific applications [19]. However, market price increases depending on application and quantity/purification degree at stake. For instance, production focused only in more refined products, such as PBPs or carotenoids for food/pharmaceutical industry, justifies a premium price increment in the market (*e.g.* astaxanthin attains 1600-6500€ per kg, or PBPs up to 12 300 €) [19, 130] compared to whole freeze-dried biomass used by food industry (up to 15€ per kg for *Chlorella*) [19] (Fig. **1**). However, focus on value-added compounds itself does not convert the process necessarily into a more economically feasible (or sustainable) approach, in part owing to the low algal yields and costly downstream processing. Hence, integrated multi-component extraction of the bioactive compounds, and reuse of biomass by resorting to the biorefinary concept should capitalize the process, and consolidate the market of functional ingredients. In addition, integration of equipment with inexpensive and low energy consumption as part of greener and safer approaches should be adopted, towards an overall sustainable and feasible algal-based process [21, 131].

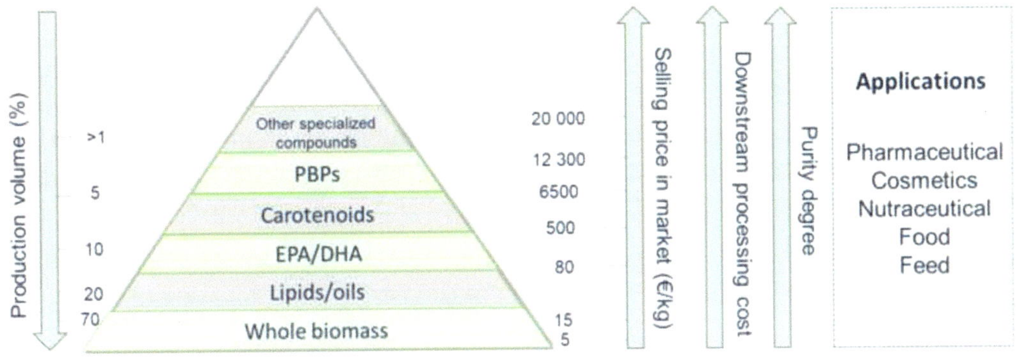

Fig. (1). Trends of production volume (%), selling price in the market (€/kg), downstream processing cost and purity degree with corresponding applications for different microalgae-based products and compounds. Adapted from Jacob-Lopes *et al.* [19]. Data references [17, 19, 129, 132, 133]:

In the algal based-biorefinery towards bioactive compound production, there are several steps to take into account in attempts to improve overall algal bioprocessing: i) strain selection ii) microalgal cultivation (optimization of cultivation conditions), iiii) optimization of algal biomass harvesting, iv) optimization of biomass disruption and compound extraction, and iv) fractionation of biomass and purification of bioactive compounds for application as functional ingredients (Fig. **2**). In the first case, it is important to select the best and more productive algal strains, besides making a prospective screening to preselect bioactive compounds that may exhibit a potential market value. Only a few cyanobacteria and microalgae (*i.e. Chlorella* sp., *Dunaliella* sp., *Botryococcus braunii, Tetraselmis* sp., *Haematoccocus* sp., *Scenedesmus* sp., *Phaeodactylum tricornutum*, *Crypthecodinium cohnii*, *Isochrysis* sp., *Nannochloropsis* sp., *Porphyridium* sp., *Arthrospira platensis*, *Aphanizomenon* sp.) have been effectively explored to date as biotechnological resources [14, 25, 67, 134]. However, countless diversity of (micro)algal found in nature unfolds a great potential to screen and diversify the genetic pool, while unfolding species and strains highly suitable for the production of bioactive metabolites to a larger scale [18, 135]. Such strains can also be improved through targeted genetic modification, by using new emerging technologies like CRISPR-Cas9 [136]. This technology is a transformation technique that could be used to improve biomass production and metabolites by overexpression/knock down/knock out regulatory genes of interest [136].

Therefore, in the step of microalgal cultivation (*i.e.*, upstream processing), it is important to anticipate the best cultivation system (open raceway ponds or photobioreators, PBRs) that can meet the growth requirements of selected strains. Each of the cultivation systems presents advantages and constraints; for instance, open raceways have lower building costs and algae can use sunlight to grow, yet they are more prone to contamination and undergo continuous water evaporation while experiencing poor CO_2 dissolution; concomitant with many diluted cultures and lower productivities. On the other hand, PBRs allow better control of environmental conditions, are less prone to contamination, can achieve higher cell densities and require less land area than open raceways. However, they are more expensive in terms of construction, maintenance and energy consumption [137].

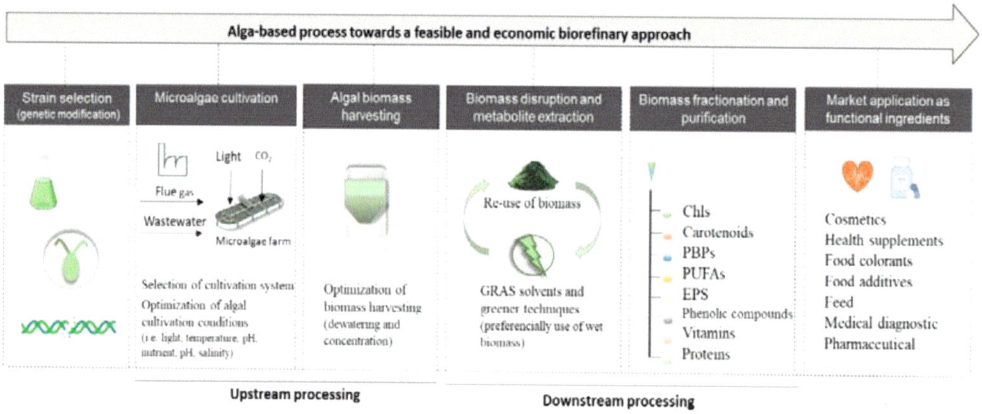

Fig. (2). Microalga-based biorefinery bioprocess towards economic feasibility of producing functional ingredients.

In addition, culture conditions (*e.g.*, light, temperature, nutrient media, salinity, pH) should be optimized, because they influence both growth rates and accumulation rates of bioactive compounds [17, 135]. In certain cases, the best conditions of growth are not the ones optimal for the accumulation of certain bioactive compounds – with the latter, being triggered by environmental stress conditions [17]; hence, a compromise between achieving high biomass yields and high bioactive compound contents should be considered. This step encompasses the so-called microalga farms, in which microalgae can obtain their nutrients in a more sustainable way. Microalga farms include premises for sustainable cultivation of microalga biomass by resorting to sunlight, atmospheric CO_2 or CO_2 streams (*i.e.*, flue gases) from other industries (*i.e.*, power plants, cement industry), in tandem with use of wastewaters or even marine/salt waters [65] that would be available at a minimal cost. This strategy envisages processes of bioremediation, in particular mitigation of greenhouse gas and fossil fuel emissions, thus helping fight global warming. It also adds a component of effluents treatment (when using wastewaters), by taking advantage of the consumption of organic nutrients (*e.g.*, phosphorous, nitrogen) required by microalgae to grow. Moreover, microalgae have the advantage of high photosynthetic efficiencies; when coupled with (bio)energy production techniques (*i.e.*, pyrolysis, fermentation, liquefaction), they can support sustained production of renewable energy sources (*e.g.* biodiesel, bioethanol, biomethane) [138, 139].

The latter steps of the overall bioprocess refer to downstream-processing – which urges considerable improvements in harvesting, biomass disruption/separation, and bioactive compound recovery. On the one hand, the high percentage of water in the culture (*i.e.*, low yields) requires a process of dewatering and concentration of biomass that is highly demanding of energy [140]; microfiltration, centrifugation and gravity sedimentation are commonly employed methods, yet optimization is still required. Once this process is complete, microalgal biomass may follow two alternative routes for target-product extraction – wet or dried. The process of dewatering is cheaper than the subsequent steps of drying; hence, a huge effort to develop methods of extraction directly from wet biomass to reduce the cost of energy consumed has been on the way [131, 140]. On the other hand, biomass disruption is an essential step, which also requires optimization depending on the selected strain; for example, some microalgae are filamentous and others possess mucilaginous layers, which can make it more difficult to extract their intracellular compounds (and thus more costly) [21]. Finally, the fractionation and purification of the bioactive compounds need to be optimized, considering their different chemical features and accessibility inside the cell. Several literature studies, chiefly at the laboratory scale, have focused on optimization of extraction enhancement, but of one compound at a time. Therefore, and under the true meaning of Biorefinery, the idea is to fractionate the biomass into more specific bioactive compounds, by taking advantage of the biomass in one single batch; this is indeed one of the chief bottlenecks towards a competitive commercial process – the success of which depends largely on the selected algae strain [69]. The major constraint relates to extraction of the different fractions without damaging the others. Hence, there is an urgent need for a strategy to maximize extraction content of certain algal bioactive compounds without negative effects upon the other fractions – by taking into account their accessibility and polarity properties. An optimized fractionation approach would be expected to recover both lipophilic of hydrophilic bioactive fractions, by taking advantage of different solvent polarities. Furthermore, a sustainable, greener and safe approach will require use of green solvents (preferentially cheap) and generally recognized as safe (GRAS), complemented by re-use of biomass when necessary. GRAS solvents are set and regulated by the U.S. Food and Drug Administration (FDA) – and are relevant within this context, because algal-derived GRAS extracts are more likely applied into food applications.

In light of these considerations, the development of efficient extraction techniques within the biorefinery concept is relevant towards a vast ingredient functional market – with high quality and enough quantity, to keep up with the globally increasing demands for algal-based products [21]. The next sections will address in more detail the most common strategies used to extract bioactive metabolites, and those potentially useful to integrate a commercial scale approach within the biorefinery framework.

BIOMASS DISRUPTION AND ASSISTED EXTRACTION TECHNIQUES OF ALGAL-BIOACTIVE COMPOUNDS

Efficient algal-downstream processes are essential to ensure the high-quality of biomass; however, they represent challenging steps, namely quite expensive and energy-intensive [131, 141]. In order to reach the bioactive metabolites in the algal-biomass, it is necessary to disrupt algal cells and break their cell wall. Depending on the particular algal cell wall features and target compound, different approaches have been employed. The release of bioactive metabolites accumulated intracellularly can be achieved via mechanical action (*e.g.* mechanical homogenization, bead milling, high shear homogenization) or non-mechanical action (*e.g.* thermal action, solvent/chemical or enzyme application). Oftentimes, algal wall breakage and metabolite extraction are associated processes. In conventional alga metabolite recovery, mechanical technologies are frequently used for disintegration of the cell wall, in combination with chemical/solvent methods for permeabilization of the algal membrane [142].

Such disruption should occur in a mild way, so as to keep some cellular components intact and/or prevent metabolite degradation; however, most works entailing algal-bioactive extraction have been focused on specific target metabolite, and used disruption techniques that ultimately lead to cell wall breakage yet at the expense of also degrading other important cell fractions [13, 21].

Some of the most conventional extraction and separation techniques – widely reported to extract bioactive compounds in plants and algae [13, 143], including bead milling/high shear homogenization, high pressure, heating or osmotic shock, freezing-thawing cycles, or use of chemicals/solvents that are highly toxic (*i.e.* Soxhlet extraction chloroform:methanol) [21, 144].

The extraction of bioactive algal fatty acids and pigments (*i.e.* chl, carotenoids) has resorted to some of the aforementioned techniques [144] – but in most cases, classical extraction tends to focus on the extraction of a single, particular bioactive chemical, which in the end lead to the degradation of other algal cellular fractions. For instance, fatty acid extraction based on Soxhlet technique (one of the harsh and commonly used ones) – which employs consecutive passage of high amounts of solvent at high temperatures, until the sample is fully exhausted [145], harms the protein fraction via thermal denaturation. In most cases, the use of classical techniques implies high amounts of solvents; this raises an issue concerning the cost of extraction process, and serious concerns to the environment as per solvent disposal. In addition, employment of high temperatures holds a positive effect upon extraction yields by increasing the rates of mass transfer and helping solubilization of active compounds – thus reducing the time of extraction; nevertheless, important nutritional and bioactive compounds, such as vitamins, proteins or phenolic compounds, are likely to undergo decay [144, 146].

Since classical techniques have been widely applied to extract bioactive compounds from plant tissues, their efficacy with microalgal biomass is lower, because of the rigid cell wall composition of the latter, that hampers constant solvent dissolution rates [147]. In addition, the power and nature of the solvent used, and biomass-to-solvent ratio strongly impact the extraction yields and efficiencies [143, 148]. Finally, such efficiencies are also determined by the selected strain and the final target-product and application [149]. For instance, several algal strains (*i.e.*, frustule diatoms, armored dinoflagellates, or algae/cyanobacteria with thick exopolysaccharide envelopes) hold cell wall coverage more resistant to shear stress or solvent permeabilization, and breakage thereof becomes extremely difficult to attain [144]. In other cases, *e.g.*, algae that are widely sensitive to shear stress forces as unarmored dinoflagellates, any mechanical processing can theoretically rupture their cell wall [150]. In both cases, it is essential to assess whether the yields and cost of extraction balances the selling price of the metabolite in question, and still leave a profit margin. The low yields of cultivation also play a role in cost; if the alga exhibits low growth rates or yields, the energy effort to retrieve a certain compound will not compensate the elevated processing costs.

Bearing in mind the above statement, new promising alternatives exist to conventional extraction techniques – known to be time-consuming and have heavy consumption of energy. Moreover, purity of the solvents, low extraction yields and thermal decomposition of some of the bioactive ingredients dictate the major drawbacks in the conventional extraction techniques employed (*i.e.*, Soxhlet technique) [148] (see Table **2**). Hence, some options have been proposed as potential new extraction techniques for algal bioactive compounds towards industrial scale, such as ultrasound-assisted extraction (UAE), microwave-assisted extraction (MAE), pulsed electric field (PEF), sub- or supercritical fluid extraction (SFE), subcritical water extraction (SWE), supersonic fluid processing, pressurized liquid extraction (PLE), and enzyme-assisted extraction (EAE) [21, 144]. Some of the main techniques, will be approached with more detail in the next section; their advantages and disadvantages are listed in Table **2**, *via-à-vis* with Soxhlet technique.

Table 2. Comparison of soxhlet conventional technique with new emerging extraction technologies, with corresponding force/driving concept, advantages and disadvantages.

Extraction Techniques	Driving Force/ Concept	Advantages	Disadvantages	Refs.
Conventional				
Soxhlet	Heat + solvent recirculation (exhaust of samples)	• Allows recirculation of solvent at small scale units • Does not use complex equipment	• Use high temperatures • Loss of several bioactive fractions of interest • Time-consuming • Use of high amounts of toxic solvents • Solvent disposal issues • High energy consumption • Poor application at large scale	[13, 142, 145]
New emerging technologies				
Ultrasound assisted-extraction (UAE)	Acoustic cavitation	• Minimized loss of bioactive compounds • High efficient extraction • Enhanced mass transfer and facilitated solvent access to cell content • Low extraction temperatures • Use of low amount of solvent • Application of GRAS solvents • Environmentally friendly • Short periods of time *per* cycle of extractions • Easy to handle/reproducible • Easy to scale-up	• Low selectivity • Efficiency dependent on biomass quantity and container shape/unit design • Moderate-high energy consumption (longer treatment adds cost to the process)	[142, 151, 152]

(Table 2) cont.....

Microwave assisted-extraction (MAE)	Electromagnetic field/Thermal energy (disruption of weak hydrogen bonds of algal matrix)	• Increased yields and purity of extract • Prevention of degradation of thermosensitive compounds • Homogeneous thermoregulation of medium • Use of low amount of solvent (easily recycled) • Application of GRAS solvents • Environmentally friendly • Short periods of time *per* cycle of extractions • Easy to handle Easy to scale-up (already employed at commercial scale for other purposes)	• Possible generation of heat (depending on solvent used) • Risk of explosion • High energy consumption • Cooling requirements High maintenance cost	[151–153]
Pulsed-electric-field (PEF)	Electric pulses - Cell wall/membrane-electroporation	• Gentle disruption technique • Selective extraction • Isolation of compounds without previous biomass dewatering • Application of GRAS solvents • Environmentally friendly • Low energy requirements • Easy to scale-up	• High capital cost • Biomass may need previous treatment with deionized solutions • Not suitable for biomass grown in salt water medium without washing • High capital cost	[13, 144, 154]

(Table 2) cont.....

Pressurized liquid extraction (PLE)	High pressure and temperature	• Rapid extraction • Less solvent consumption • Application of GRAS solvents • Environmentally friendly • Allows recirculation of solvent (saving costs) • High temperature gives an opportunity to explore potentially newly formed bioactives of interest	• Low compound selectivity • Degradation of thermo-labile compounds • Possible problems of saturation/obstruction • High energy requirements (*i.e.* use of pumps) • High capital cost	[144, 145, 155]
Supercritical-fluid extraction (SFE)	Temperature and pressure (above critical point) – use of CO_2 that becomes a supercritical fluid	• Potential high extraction yields • High compound selectivity • Low deterioration compound • No use of solvents (just CO_2) • Allows recirculation of CO_2 (saving cost) • Possible to combine with GRAS solvents, to increase efficiency • Environmentally friendly • Easy to scale-up/already implemented at large scale for other purposes	• CO_2 solubility limited to apolar compounds • Longer extraction times compared to other non-conventional techniques • High processing costs • High capital cost • High maintenance cost	[144, 156]

(Table 2) cont.....

Subcritical-water extraction (SWE)	Temperature and pressure (above critical point)	• Rapid extraction • High extract quality and selectivity • Low cost of extracting agent (water) • Environmentally friendly • Potential newly formed compounds to be explored with antioxidant capacity	• Possible high capital cost • High maintenance cost • Not available at large scale	[144]
Enzymatic-assisted extraction (EAE)	Cell wall/membrane hydrolysis by enzyme catalytic process	• Mild disruption technique • Highly selective approach • Fast and efficient approach • Possible combination with GRAS solvents for more efficient extraction	• High-cost commercial enzymes • Recovery enzymes from algal cell debris can make the process more expensive • After some cycles of extraction, enzymes may become deactivated	[13]

Ultrasound-Assisted Extraction (UAE)

UAE is a strategic technology employing ultrasound waves (high frequencies within 20khz-100 MHz), which create cavitation bubbles in a solvent material (or liquid environment where algal biomass lies). The growth and collapse of the bubbles generate macroturbulence, high-velocity interparticle collisions and liquid jets that cause cell wall perturbation by surface breaking, solvent internal diffusion, and release of intracellular contents into the solvent [146, 157]. Therefore, ultrasound exerts a mechanical effect, allowing greater penetration of solvent into the sample matrix, and thus, increasing the contact surface area between solid and liquid phase [158, 159].

Several studies have shown their assisted efficiency in extracting bioactive compounds, when combined with solvents [34, 160 - 162] (Table **3**). For instance, when using an ultrasound-assisted process, regular extraction yields of carotenoids (60-75%) achieved a maximum of 70-80%, compared to plain exhaust extraction. Yet, the same process showed no influence on lipid extraction, even with different intensities and different solvent concentrations [160]. The level of efficiency of this technique, is highly dependent on extraction time, size of sample and volume used (volume to biomass ratio), frequency and number of cycles used, and electrical acoustic intensity (Table **3**). The latter ones, greatly influence the exposure to acoustic cavitation, helping in cell disruption and release of intracellular compounds [146].

This technique is demonstrated to improve extraction yields and the associated economics, holding a greener impact. This method is not time-consuming, and is able to complete a reproducible and complete set of extractions in a short period of time. In addition, the amount of solvent used is considerably lower than conventional extraction techniques (*e.g.*, Soxhlet, percolation) [158]; it can also be used in combination with GRAs solvents, such as ethanol or ethyl acetate, to process algal biomass for foodstuffs or food ingredients. This is a common technique used in laboratory research, but it can be easily scaled-up to an industrial level. Furthermore, this process significantly increased the rate effects, in several processes of the chemical and food industries, with improved levels of purity of final products and reduced energy consumption, compared to classical methods [158].

Table 3. Comparison of bioactive compound yields/contents of extraction from algae, of different new emergent extraction techniques.

Extraction Technique	Algae Source	Bioactive Compound	Processing Conditions and Factors to be Optimized	Best processing condition	Maximum Extraction Yield/ Bioactive Compound Content	Refs.
Ultrasound-assisted extraction (UAE)	*Arthospira platensis*	Functional proteins	**S:** dH_2O **T**: 4±1ºC **pH**: 3,5,7,9,10 **S/B**: 0.015 $mL.g^{-1}$ **ET**: 60 min	**pH**: 7, 9, 10	84% (total proteins extract) + 20% than thermal extraction	[202]
	Chlorella vulgaris	Lutein	**Pre-treatment**: enzymatic (2h, 1.23% (*w/v*), pH=4.5, T: 50 ºC) **S:** EtOH (90 % (*v/v*)) **T**: 20, 30, 40ºC **S/B**: 10, 30, 50 $mL.g^{-1}$ **ET**: 2h,5h,8h **UP**: 35 kHz **UF**: 56.6 $W.cm^{-2}$	**Pre-treatment**: enzymatic **T**: 37.7ºC **S/B**: 31 $mL.g^{-1}$ **ET**: 162 min	3.16 $mg.g_{fresh\ biomass}^{-1}$	[203]
	Chlorella vulgaris	Chl *a*	**S**: EtOH 70, 80, 90% (*v/v*) **T**: 50,60,70ºC **ET**: 30, 60, 90 min **UP**: 200 W	**S**: EtOH 79.4% (*v/v*) **T**: 61.4 °C **ET**: 78.7 min	88.9% (extract) 31.1 $mg.g_{DW}^{-1}$	[34]
	Heterochlorella luteoviridis	Car	**S**: EtOH 50-75% (*v/v*) **T**: 30ºC **ET**: 50 min **UF**: 0-100% (435 $kJ.kg^{-1}$)	**S**: EtOH 60–75% (*v/v*) **UP**: 40-80%	77-80% (extract)/1.26-1.31 $mg.mL^{-1}$	[160]

(Table 3) cont.....

	Oscillatoria sp. BTA170	PBPs	**S**: Britton–Robinson buffer (0.05M) **T**: room temperature **pH**: 5, 7, 9 **B/S**: 0.1, 0.2, 0.3 $mg.mL^{-1}$ **ET**: 20 min **UF**: 50, 70, 100% **UP**: 12.73, 16.99, 21.23 $W.cm^{-2}$	**pH**: 7 **B/S**: 0.2 $mg.mL^{-1}$ **ET**: 20 min **UF**: 75% **UP**: 16.99 $W.cm^{-2}$	PC – 92.42% (extract)/131 $mg.g_{DW}^{-1}$ APC – 87.01% (extract)/108.24 $mg.g_{DW}^{-1}$ PE – 91.89% (extract)/30.10 $mg.\ g_{DW}^{-1}$)	[204]
	Spirulina sp.	β-Car	**Pre-treatment**: MeOH (0, 2, 4, 6, 10min) **S**: hexane, n-heptane,diethyl ether, n-heptane **B/S**: 0.01-0.05 $mg.ml^{-1}$ **T**: 10-50 ºC **ET**: 4-8 min **UP**: 64, 115, 149, 167, 185 and 210 $W.cm^{-1}$	**Pre-treatment:** MeOH (2 min) **S**: n-heptane **B/S**: 0.03 $mg.mL^{-1}$ **T**: 30ºC **ET**: 8 min **UP**: 167 $W.cm^{-2}$	47.10% (extract)	[161]
	Spirulina platensis	PC	**S**: EtOH **T**: 30,45,60 ºC **ET**: 20,35,50 min **UF**: 28, 35, 42 kHz **UP**: 200 W	**T**: 52.2ºC **ET**: 42 min **UF**: 42 kHz	$14.94\%_{DW}$	[205]
	Tetraselmis (KCTC 12236BP)	Phenolic compounds	**S**: EtOH - 21.8,40.0,58.3,76.5,94.8% (*v/v*) **T**: 50,60,70,80,90ºC **ET**: 15,30, 45,60,75 min **UP**: 300,400,500,600,700 W	**S**: EtOH – 55.4% (*v/v*) **T**: 74.4ºC **ET**: 59.6 min **UP**: 700 W	9.8 $mg_{gallic\ acid\ equivalents}$ (GAE).g_{DW}^{-1}	[146]
Microwave-assisted extraction (MAE)	*Cylindrotheca closterium*	Chl *a* Fuc	**S:** Acetone **T:** 56 ºC **IT:** 2,5,10,15 min **IP**: 25,50,75,100 W **Other observations**: Magnetic stirring	**IT:** 5 min **IP**: 50 W	Chl *a*: 8.65 ± 0.29 $\mu g.mg_{DW}^{-1}$ Fuc: 4.24 ± 0.09 $\mu g.mg_{DW}^{-1}$	[206]
	Arthrospira platensis	Polysaccharides	**Pre-treatment:** Soxhlet (EtOH, 80ºC, 1/30 B/S, 20 min **S:** Water **B/S**: 1:15, 1:25, 1:30, 1:35, 1:45 $ml.mg^{-1}$ **IT:** 0,20,40,60 min **IP**: 186, 310, 434, 558, 620 W **Other observations**: Magnetic stirring	**B/S**: 1:30 $ml.mg^{-1}$ **IT:** 20 min **IP**: 434 W	127 $mg.g_{DW}^{-1}$	[170]
	Dunaliella bardawil	Chl *a* β-Car	**S:** pentane, cyclopentane, n-hexane, cyclohexane, benzene, and toluene **SV**: 60-85% (*v/v*) **T**: 40-60ºC **IT**: 7.5-12.5 min **IP**: 15-25%	**S:** n-hexane **SV**: 85 % (*v/v*) **T**: 40ºC **IT**: 12.5 min **IP**: 25%	46.27 $mg.g_{DW}^{-1}$	[169]

(Table 3) cont.....

-	*Nannochloropsis* sp.	PUFAs (EPA)	**S:** dH_2O + NaCl (ranging concentration 1-35% (*w/v*) **S/B:** 0.1 $ml.mg^{-1}$ **T**: 60,70,80,90,100ºC **IT**: 5,10,17.5,25,30 min **IP**: 800 W **IF**: 2.45 GHz	**S:** dH_2O + NaCl (10% (*w/v*) **T**: 100ºC **IT**: 5 min	10.42 $mg.g_{DW}^{-1}$ (EPA - 8.15 $mg.g_{DW}^{-1}$)	[207]
	Spirulina sp.	PC	**S:** Acetate buffer (pH=5), phosphate buffer pH=6,7),Tris-Cl buffer (pH=8) **S/B**: 0.08 $ml.mg^{-1}$ **IT**: 0-50 s (5 s intervals) **IP**: 90 W **IF**: 2.45 GHz	**S**: Tris-Cl buffer pH=8) **IT**: 22.5 s	0.022 $mg.mL^{-1}$ (IC_{50})	[168]
	Porphyridium purpureum	PBPs	**S**: Deonized water **S/B**: 0.35 $ml.mg^{-1}$. **T**: 40-120ºC **IT:** 1/6,1/2, 1, 2, 5 min **IP**: n/a **IF**: 2.45 GHz **Other observations**: Magnetic stirring	PC – **T**: 100ºC, **IT**:10 sec APC – **T**: 100ºC, **IT**: 1 min PE – **T**: 40ºC, **IT**:10 sec	PC - 7.37$\%_{DW}$ APC - 3.51$\%_{DW}$ PE - 3.48$\%_{DW}$	[167]
Pulsed-electric field (PEF)-Assisted extraction	*Chlorella vulgaris*	Phenolic compounds	**S**: n/a **T**: 20±5ºC **ED**: n/a; **PA**: n/a **PL**: 5 µs (1h treatment) **EF**: 27, 35 $kV.cm^{-1}$ W_T: 50, 100, 150 $kJ.kg_{susp}^{-1}$ **PF**: 1-6 Hz **Other observations**: PEF as pre-treatment for the extraction of bioactive compounds with solvent stirred agitation (140 rpm) for 1h (25ºC)	**EF**: 27 $kV.cm^{-1}$ W_T: 100 $kJ.kg_{susp}^{-1}$	1.21 $mg.L^{-1}$	[177]
	Chlorella vulgaris	Lutein	**S**: EtOH 96% (*v/v*) **T**: 10, 25, 40ºC **ED**: 0.25 cm; **PA**: n/a **PL**: 6,15,75,150 µs **EF**: 10,15, 20, 25 $kV.cm^{-1}$ W_T: 0.009, 0.021, 0.038,0.059 $kJ.kg_{DW}^{-1}$ **PF**: 0.5 Hz **Other observations**: PEF as pre-treatment for the extraction of lutein and mixed with solvent by vortexing sample	**T**: 25-30Cº **PL**: 100 µs **EF**: 25 $kV.cm^{-1}$ W_T: n/a	524 $\mu g.g_{DW}^{-1}$	[176]

(Table 3) cont.....

	Chlorella vulgaris	Car	**S**: EtOH 96% (*v/v*) **T**: n/a **ED**: 0.25 cm; **PA**: 1.76 cm^2 **PL**: 6,15,75,150 µs **EF**: 10,15, 20, 25 $kV.cm^{-1}$ $\mathbf{W_T}$: 0.009, 0.021, 0.038,0.059 $kJ.kg_{DW}^{-1}$ **PF**: 0.5 Hz **Other observations**: PEF as pre-treatment for the extraction of Car with solvent for 1h (25ºC)	**PL**: 75 µs **EF**: 20 $kV.cm^{-1}$ $\mathbf{W_T}$: n/a	790 $\mu g.g_{DW}^{-1}$ (104% increase compared to no PEF treatment)	[175]
-	*Phaeodactylum tricornutum*	Phenolic compounds	**S**: dH_2O and DMSO 50% (*v/v*) **T**: n/a **ED**: 10 cm; **PA**: n/a **PL**: 100 µs **EF**: 1 $kV.cm^{-1}$/400 pulses; 3 $kV.cm^{-1}$/45 pulses $\mathbf{W_T}$: 100 $kJ.kg^{-1}_{DW}$ **PF**: 2 Hz **Other observations**: PEF as pre-treatment for the extraction of phenolic compounds with solvent stirred agitation (400 rpm) for 4h or 24h	$\mathbf{EF}_{dH2O}$: 1 $kV.cm^{-1}$/400 pulses (4h subsequent extraction) $\mathbf{EF}_{DMSO\ 50\%\ (v/v)}$: 3 $kV.cm^{-1}$/45 pulses (24h subsequent extraction)	dH_2O – 4.3 $mg_{GAE}.g_{DW}^{-1}$ DMSO – 8 $mg_{GAE}.g_{DW}^{-1}$	[178]
	Tetraselmis chuii	Phenolic compounds	**S**: dH_2O and DMSO 50% (*v/v*) **T**: n/a **ED**: 10 cm; **PA**: n/a **PL**: 100 µs **EF**: 1 $kV.cm^{-1}$/400 pulses; 3 $kV.cm^{-1}$/45 pulses $\mathbf{W_T}$: 100 $kJ.kg^{-1}_{DW}$ **PF**: 2 Hz **Other observations**: PEF was used as pre-treatment for the extraction of bioactive compounds with solvent stirred agitation for 4h or 24h	**S**: DMSO 50% (*v/v*) and dH_2O **EF**: 3 $kV.cm^{-1}$/45 pulses (4h)	dH_2O – 6.42 $mg_{GAE}.g_{DW}^{-1}$ DMSO – 6.70 $mg_{GAE}.g_{DW}^{-1}$	[178]

(Table 3) cont.....

Supercritical-fluid extraction (SFE)	*Arthrospira platensis*	GLA	**SF**: CO_2 **CO**: EtOH – 0, 10 % (*v/v*) **T**: 50, 60 ºC **P**: 25, 35 MPa **ET**: 2,4,6 h	**CO**: 10% (*v/v*) **T**: 60ºC **P**: 35 MPa **ET**: 4h	0.44%$_{DW}$	[185]
	Chlorella pyrenoidosa	Lutein	**SF**: CO_2 **CO**: EtOH 0,10,20, 30, 50, 70% (*v/v*) **T**: 40, 50, 60ºC **P**: 20,25,30 MPa **ET**: 2,4,6 h	**T**: 50ºC **P**: 25 MPa **ET**: 4h	84% (extraction degree)	[187]
	Synechococcus sp.	Chl *a* Car	**SF**: CO_2 **CO**: n/a **T**: 40, 50, 60ºC **P**: 100, 200, 300, 400, 500 bar **ET**: 180 min	Chl *a* – **T**: 60ºC; **P**: 500 bar Car – **T**:50ºC; **P**: 300 bar	Chl *a* - 1.511 $mg.g_{DW}^{-1}$ Car - 0.715 $mg.g_{DW}^{-1}$	[186]
	Spirulina pacifica	β-Car Zea	**SF**: CO_2 **CO**: Ethanol 5, 10, 15% (*v/v*) **T**: 40, 60, 80ºC **P**: 150,250, 350 bar **ET**: 40,70,100 min	β-car - **T**: 60ºC; **P**: 350 bar; **ET**: 70 min; **CO**: 15% (*v/v*) Zea - **T**: 80ºC; **P**: 350 bar; **ET**: 100 min; **CO**: 15% (*v/v*)	118 mg per 100 g_{DW}^{-1} 48 mg per 100 $g_{\cdot DW}^{-1}$	[208]
	Schizochytrium limacinum	DHA	**SF**: CO_2 **CO**: Ethanol 30,40,50,70,100 % (*v/v*) **T**: 30,40,50, 60 ºC **P**: 15, 20,25,30,35,40 MPa **ET**: 2 h	**CO**: Ethanol 100 % (*v/v*) **T**: 40 ºC **P**: 35 MPa **ET**:2 h	27.5%$_{\text{Total fatty acids}}$	[209]
Subcritical-water extraction (SWE)	*Chlorella vulgaris*	Phenolic compounds	**SF**: Water **BC**: 5,12.5,20 wt.% **T**: 100,175,200 ºC **P**: n/a **ET**: 5,12.5,20 min	**BC**: 20 wt.% **T**: 165 ºC **ET**: 5 min	58.73 $mg_{GAE}\cdot g_{DW}^{-1}$	[192]
	Chlorella vulgaris	Phenolic compounds	**SF**: Ultrapure water **BC**: 1 g_{DW} **T**: 100, 200ºC **P**: 6.89 MPa **ET**: 20 min	**T**: 200 ºC	32.91 $mg_{GAE}\cdot g_{Extract}^{-1}$ (increase of 2-fold total phenolic content compared to T=100ºC)	[188]
	Haematococcus pluvialis	Phenolic compounds and vitamin E	**SF**: Ultrapure water **BC**: 1 g_{DW} **T**: 50, 100, 150, 200ºC **P**: 6.89 MPa **ET**: 20 min	**T**: 200 ºC	33%$_{DW}$	[191]

(Table 3) cont.....

Pressurized liquid extraction (PLE)	*Gloeothece sp.*	β-Car PUFAs	**S**: EtOH 100%, *v/v* **BC**: 50, 100,150 mg **T**: 30,40,50,60,70 ºC **P**: 72, 142, 210, 260 bar **ET**:4 ,8, 12, 16, 20 min (1-5 cycles o solvent recirculation)	**BC**: 50 mg **T**: 70 ºC **P**: 80 bar **ET**: 12-20 min (3-5 cycles)	β-Car – 2.9 $mg_{PI} \cdot mg_{Extract}^{-1}$ (3-fold higher than ultrasound) PUFAs – 56.8 $\mu g_{FA} \cdot mg_{Extract}^{-1}$ (1.7-fold increase compared to ultrasound)	[195]
	Neochloris oleoabundans	Car	**S**: EtOH 100%, *v/v* **BC**: 600 mg **T**: 100ºC **P**: 103 bar **ET**: 20 min **Other observations**: pre-treated samples with cryogenic milling	-	35.5%/76.7 $mg \cdot mg_{Extract}^{-1}$ (samples of algae subjected to low nitrogen conditions)	[210]
	Nannochloropsis oculata	EPA	**S**: n-hexane, n-hexane/--PrOH (2:1 vol.%) and EtOH (96% *v/v*) **BC**: 3-6 g **T**: 60ºC **P**: 10-12 MPa **ET**: 50, 150 min (10 min per cycle) **Other observations**: pre-treated samples manually grounded with mortar/pestle or homogenized using a centrifugal mill	**S**: EtOH (96%, *v/v*) **ET**: 50 min (5 cycles 1 min)	$3.7\%_{mass}$	[196]
	Dunaliella Salina	Lutein	**S**: n-hexane, EtOH, acetone, chloroform:MeOH (1:1), EtOH **BC**: n/a **T**: 20, 40, 60, 80 ºC **P**: 50, 100 bar **ET**: 5, 10 min (2 cycles)	**S**: Chloroform:MeOH (1:1) **T**: 20 ºC **P**: 100 bar **ET**: 10 min (2 cycles)	92% recovery	[211]
Enzyme-assisted extraction (EAE)	*Haematococcus pluvialis*	Astaxanthin	**E**: β-1,3-glucanase **BC**: 41 mg **EI**: 0.2, 0.4 and 0.6 $U.mL^{-1}$ **pH medium**: 4.5 6.5, 8.5 **T**: 35, 45, 55ºC **ET**: 30, 60, 90 min **Other observations**: ultrasound and freezing-assisted extraction (dichloromethane as solvent)	**EI**: 0.6 $U.mL^{-1}$ **pH medium reaction**: 4.5 **T**: 55ºC **ET**: 30 min	83.90%/1235.89 $\mu g.g_{DW}^{-1}$	[212]
	Haematococcus pluvialis	Astaxanthin	**E**: protease-K **BC**: n/a **EI**: n/a **pH medium**: 5.8 **T**: 30 ºC **ET**: 60 min	-	4.9 $mg.g_{DW}^{-1}$	[213]

(Table 3) cont.....

-	*Arthrospira platensis*	PUFAs (oil extract)	**E**: Alcalase®, Flavourzyme®, Ultraflo®, Vinoflow® (commercial enzymes) **BC**: 0.2 g.mL^{-1} **EL**: 0.5, 1, 2, 3, 4 wt % **pH medium**: 5, 5.5, 6, 6.5, 7, 8 **T**: 25, 30, 40, 50 ºC **ET**: 4, 24h	**E**: Vinoflow® **EL**: 2 wt % **pH medium**: 6.5 **T**: 40ºC **ET**: 24h	38.10 % (oil phase)	[214]

β–car–beta-carotene; APC–allophycocyanin; B/S–biomass to solvent ratio; BC–biomassconcentration; Car–carotenoids; Chl a–chlorophyll a; CO–co-solvent; dH2O–distilled water; DHA–docosahexaenoic acid; DMSO–dimethyl sulfoxide; E–enzyme used; ED–electrodedistance; EF–electric field strength; EI–enzyme initial activity; EL–enzyme loading; EPA–eicosapentaenoic acid; ET–extraction time; EtOH–Ethanol; Fuc–fucoxanthin; GLA–gamma-linolenic acid; IF-irradiation frequency ;IP-irradiation power; IT–irradiation time; MeOH–methanol; P–pressure; PA–plate area; PBPs–phycobiliproteins; PC–phycocyanin; PF–Pulsefrequency; PE–phycoerythrin; PL–pulse length; PrOH–isopropyl alcohol; PUFAS–polyunsaturated fatty acids; UF–ultrasonic frequency; UP–ultrasonic power/ultrasonic acousticintensity; S–solvent; S/B–solvent to biomass ratio; SC–solvent concentration; SF–supercriticalfluid; T–temperature; WT-total specific energy input; Zea–zeaxanthin; n/a-not available.

Microwave-Assisted Extraction (MAE)

MAE approach employs microwave dielectric heating or electromagnetic (microwave) radiation, with a frequency near 2.45 GHz – which causes dielectric heating, primarily absorbed by polar molecules available in the wet biomass. This results in the dispersion of electromagnetic waves that increase the vibration of water molecules and, thus, increase temperature; this, subsequently generates high cell internal pressure (by water evaporation), which results in cell disruption [163]. This technique also destroys the hydrogen bonds present, induces electroporation [164], and consequently, leads to breakage of the cell wall and facilitates penetration of the solvent inside the cell, while allowing the release of target compounds into the cell matrix [131, 143, 144].

MAE is a fast method that couples extraction with reaction – in which microwave temperature is not restrained by thermal or conductive currents; the maximum temperature of the material at stake depends on the power supplied and rate of heat loss [165]. This strategy is widely used in industry for the extraction of value-added compounds to be used in functional foods, such as phytonutrients and phenolic compounds from plants, fruits or vegetables [145, 166]; but until now, it has not yet been applied to algal biomass at large scale. However, several studies at the laboratory scale have demonstrated effective extraction of microalgal bioactive compounds [153, 167 - 170] (Table **3**), including from microalgal cells with high mechanical resistance due to the presence of frustules (*e.g.*, diatoms) or extracellular polysaccharides [153].

Pasquet and colleagues have studied two different microalgae – frustulated

Cylindrotheca closterium diatom and non-frustulated *Dunaliella tertiolecta*. They reported that microwaves helped to overcome the mechanical barrier present in the frustulated microalgae and extract pigments (chl *a* and fucoxanthin) at higher dissolution rates (within a few minutes) compared to conventional techniques (*i.e.*, solvent soaking takes hours); however, no differences in extraction yields regarding the non-frustulated *Dunaliella tertiolecta,* were observed [153].

As UAE uses low amounts of solvent, it is an environmental friendly and green technique – with possible application of GRAS solvents [153]; overall temperature and energy consumption are low, thus preventing thermo-labile compound degradation. In fact, this unique approach generates high yields in terms of quality extracts and value-added biological compounds with a market interest [144]. Several key factors influence the efficiency of MAE, namely the used species, the dielectric properties of solvent, temperature, volume solvent/biomass ratio, extraction time, and power of applied microwaves [171].

This technique can also be performed at low temperatures under vacuum (VMAE) [153], or without the addition of external solvent; the molecules are then, solvated in the water sample, and collected afterwards in a cold trap [165, 167]. The latter technology, called microwave hydrodistillation, displays low energetic consumption, as it does not need secondary processes of distillation or evaporation; it has not been applied to microalgal biomass to large scale – only to essential oils and other bioactive fractions in plants, but it could reveal a promising technology towards bioactive ingredients [163].

Pulse-Electric Field (PEF)

PEF is an approach investigated for extraction of algal-derived compounds, which takes advantage of the strength of electric pulses, to permeabilize algal membrane cells – thus, leading to the release of intracellular components by electroporation. The level of efficiency is highly dependent on the conditions of the electrical field (intensity and pulse treatment time) [172].

This technique has already been applied at the industrial level for pasteurizing food at low temperatures; and increasing the shelf-life of certain food products [173, 174]. Owing to their application in industry, this technology has received some attention in microalga research, and proved efficient in improving the extraction of intracellular bioactive compounds (*e.g.*, pigments, phenolic compounds), from different species of microalgae in combination with organic solvents [175 - 178] (see Table **3**). In addition, its use was studied with fresh algal biomass toward recovery of lipids for biodiesel production purposes, with highly effective results [131, 179]. Hence, the recovery of algal bioactive compounds with PEF, applied to wet biomass, creates an opportunity for efficient isolation of compounds from the culture, without the need of prior dewatering or drying, and corresponding costs [154].

All in all, PEF seems to be a promising strategy to maximize the recovery of microalgal bioactive metabolites, at the expense of low energy requirements when compared to conventional techniques [180] – because it can be applied directly to wet biomass, with low energetic requirements and thus, at low cost. However, more research is required to ascertain the behavior of different cell species, cultures and densities. In addition, the equipment design, as well as the establishment of optimal processing parameters [21] also have to be improved. Since it resorts to electric pulses, some samples are required to be added with deionized solutions [13].

Supercritical-Fluid Extraction (SFE)

SFE approach uses supercritical fluids that exhibit solvent-like properties, to help in the extraction of metabolites of interest from a microalgal matrix [143]. Above certain temperatures and pressure (the so-called critical point), those fluids present physico-chemical features such as high solvation power and low surface tension – with advantages in the recovery of bioactive compounds for many applications [181, 182]. This approach has demonstrated high selectivity (by varying temperature and pressure), low deterioration of metabolites of interest, and short processing times [183,184). CO_2 is one of the preferred supercritical fluids, owing to its low critical point (T=31°C and P=73 bar). The lower critical point of CO_2 allows easy alteration of its physico-chemical features (*i.e.*, intermediate viscosities between liquid and gas, intermediate solubility and diffusivity), thus, improving penetration of the biological materials and extracting thermolabile compounds [143, 184, 185]. Furthermore, CO_2 possesses the advantage of being chemically inert, not inflammable, and easy available; it indeed represents the choice to promote green and environmental savings [143]. Toxic solvents can be avoided (dependent on the target molecule), and CO_2 can be recirculated in the system, thus allowing a decrease in the overall cost of the process [184]. In addition, SFE with CO_2 allows combining the extraction with other solvents (*e.g.* ethanol). CO_2 has in fact a non-polar nature, which, when combined with more polar solvents (*i.e.* ethanol) magnifies its potential of extracting (bioactive) compounds with different polarities and to high extraction yields [184].

The study of SFE towards extraction of lipids and bioactive compounds from algal matrices for health and food applications, has gained relevance in the last years [144, 182]. Such studies have unfolded optimum extraction conditions in SFE – highly influenced by temperature, pressure and use of co-solvents [143] – at laboratory level, for improved extraction yields of specific algal-target metabolites (*e.g.*, vitamins, PUFAs, and pigments) [182, 185 - 187] (see Table **3**). For instance, lutein extraction yield from *Chlorella pyrenoidosa* cultivated under heterotrophic conditions reached 87% after 4h extraction, with 50ºC and 25 MPa of CO_2 combined with 50% (*w/v*) ethanol for optimal conditions in SFE [187]. In other study, *Synechococcus* sp. underwent an increase in extraction yield of total carotenoids, under an operating temperature of 50ºC and pressure of 300 bar; but the highest car/chl *a* ratio, was obtained under 60ºC at 200 bar [186]. On the other hand, Mendes *et al.* [186] studied the influence of SFE with CO_2, and its combination with ethanol in *Arthospira maxima*, for improved GLA extraction and compared to traditional organic solvent extraction. They observed a maximum extraction yield of 0.44% GLA per DW, with 10% ethanol, at 60ºC and 350 bar, a 4.4 fold increase, when compared with plain hexane extraction.

The SFE technology is not yet available for large-scale application, in microalga metabolite extraction. However, its application in industrial extraction is already available for many other purposes (*i.e.*, coffee decaffeination, food ingredients such as aromas, colorants, vitamin-rich extracts, specific lipids and nutraceuticals from plants and vegetables) [144, 156, 184] – which anticipates good perspectives toward economic feasibility of this method within a biorefinery framework.

Subcritical-Water Extraction (SWE)

SWE, also known as "hot water extraction under pressure" [155], is a very promising technique that resorts to water as solvent at high temperatures and pressures to produce extraction in vegetable matrices, including plants and algae [188, 189]. During the process, water is maintained in a liquid state by applying temperatures (from 100 to 374ºC) and pressure (10-60 bar) below supercritical conditions [190]; this allows an increase in solubility of compounds, which permits higher yields of extraction. By increasing temperature, the mass transfer and surface tension are increased, and viscosity is reduced, which facilitates the penetration of solvent into algal matrices. The temperature is an important factor that allows modulation of both the dielectric constant of water (*i.e.* polarity) and the corresponding solubility, while assuring a higher selectivity for target compounds. For instance, at room temperature (25ºC), the dielectric constant of water is 80, which drastically decreases at 200ºC; this value is quite similar to those solvents such as ethanol or methanol, and compounds with more non-polar nature can be better accessed [155]. Other parameters, such as pressure and time are extremely important during the extraction process, as well as the solid-to-solvent ratio, particle size, and initial matrix composition [143, 189].

SWE has been widely employed in plants (*i.e.*, rosemary, soybean) for extraction of antioxidant compounds with nutraceutical applications; in recent years, its application to microalgal matrices towards extraction of metabolites with bioactive properties has attracted some interest, with promising high yields [189, 191] (Table **3**). When Zakaria *et al.* [190] employed SWE to obtain *Chlorella vulgaris* extracts, they obtained an improved extraction yield of phenolic compounds; a gradual increase of temperature (up to 175ºC) led to an increase of total phenolic compounds content, reflected in higher antioxidant capacity. SWE was described to enhance the antioxidant capacity of microalgal natural extracts (*i.e.* in *Chlorella vulgaris)* [188]; the compounds released during extraction can interact with each other, and generate other compounds with bioactive features distinct from the original targets. Plaza *et al.* have demonstrated that extracts of macroalga *Undaria pinnatifida* processed by SWE exhibit a higher antioxidant capacity than the original ones, and confirmed the generation of new antioxidants during the process – via chemical events such as Maillard and caramelization reactions [189]. The interaction between parameters, in particular at elevated temperature and pressure, can lead to modification of the final properties of extracted compounds [188]. Despite this technology still being in its beginning regarding microalgal bioactive compound extraction, it shows a great opportunity to explore new different compounds with potential bioactive properties [189]. This technology is not yet available on a large-scale [190], but an effort is underway toward improved strategies and processing conditions to extract certain algal compounds. This extraction technique also offers an environmentally friendly approach by using non-toxic compounds and low amounts of organic solvents, being fast, and allowing a better selectivity and higher quality of extracts [193]. This would be a strategy to consider in the future to integrate within the biorefinery concept.

Pressurized Liquid Extraction (PLE)

PLE is an advanced technique, also known as enhanced solvent extraction (ESE), high-pressure solvent extraction (HPSE), or accelerated solvent extraction (ASE) [143, 194]; it follows the same principles of SWE, but uses different solvents, including GRAS solvents (*e.g.*, ethanol) suitable to obtain algal food-grade extracts, instead of water. It is performed under high temperatures (50º to 200ºC) and pressures (up to 200 bar); before reaching the critical point, it accelerates the passage of solvent through the matrix that enables higher yields of extraction [144]. As with SWE, it requires less solvent than conventional techniques, and solubility and mass transfer rate of solvent are increased, as well as viscosity decreased – thus, favoring dissolution of certain bioactive metabolites. Several studies have demonstrated its efficiency towards the extraction of algal metabolites, as well as the advantage of using high pressure and temperature via the forced flow of solvent from solid and semi-solid matrices [194 - 196] (Table **3**). For instance, Amaro *et al.* observed an improvement in carotenoids and

PUFAs extraction in a hand-built continuous pressurized solvent extraction system (CPSE) with ethanol. The authors optimized, in a sequential mode, biomass amount, flow-rate/pressure of solvent, temperature and solvent volume – including recirculation cycles. The best conditions obtained for the bioactive compounds at large were 60ºC and 180 bar, when extraction yields were 3-fold for β-carotene and 1.7-fold for PUFAs when compared to ultrasound technique. PLE method holds the advantage of using less solvent and allowing its recirculation in two or more extraction cycles; this minimizes production costs, besides the green chemical principle by using GRAS solvents – quite suitable for extracts that aim at food and nutraceutical applications [194].

This approach is still expensive to develop on a large scale because it requires specific infrastructures and operations that constrain the extraction of natural products [195]. This procedure can be implemented under a static or dynamic mode, yet the former is the most frequently used, based on the availability of commercial equipment [148]. Furthermore, loss of thermolabile compounds and degradation of some compounds due to the high temperatures employed, and pressures up to 200 bar or more [144] have been reported; in photosynthetic matrices, such conditions can also promote alteration of compound properties, *i.e.*, pyropheophytinization (formation of pheophytin), a Chl derivative with antimutagenic and antioxidant features, reported in *Chlorella vulgaris* extracts [194]; which could also be a good chance to explore newly formed compounds, for potential application within a biorefinery approach.

Enzymatic-Assisted Extraction (EAE)

EAE is a non-conventional and selective method that applies at the alga cell wall through enzymatic hydrolysis. This is a catalytic process that causes the breakdown of even the strongest algal cell walls (*i.e.*, mucilaginous sheaths, exopolysaccharides) composed by, *i.e.* (hemi)cellulose with glycoproteins or polysaccharides, materials known to pose a challenge for efficient recovery of intracellular bioactive compounds [197]. This method is strongly influenced by the type of cell wall in algae; hence, the selection of enzyme or cocktail of enzymes to be employed should be addressed a priori. Furthermore, conditions of an enzymatic reaction, including medium pH, temperature, the initial activity of the selected enzyme(s), enzyme loading, and reaction time, are relevant for its efficiency [13]. EAE is a highly selective approach – dependent on the algal matrix composition and the type of enzyme(s) used, which works through the removal of external layers of polysaccharides *via* a biological mechanism, and helps in the induction and release of algal intracellular bioactive compounds of interest. In addition, it offers high overall efficiency, and is an eco-friendly procedure for avoiding the use of harsh/toxic chemicals. It has the advantage of process recyclability and recovery [148]. The process of EAE promotes mass transfer – and, when combined with GRAS solvents or associated to other techniques such as PLE, permits recovery of bioactive compounds (*i.e.* phenolic compounds, carotenoids) in a quite effective manner [13, 198].

Most work with EAE has focused on the recovery of different types of seaweed derived-bioactive compounds (*i.e.*, phenolic compounds) [148, 197, 198]; this is in part due to the discovery of novel enzymes from microbial sources that can degrade alginates, glucuronan, carrageen, laminarin or agarases [199]. These indeed represent physical barriers in such algal sources, known to reduce the extraction efficacy of bioactive compounds at large. Few studies with EAE have focused on bioactive compounds from microalgae (see Table **3**) – and most such studies report EAE towards lipid or carbohydrate extraction [13, 200, 201].

However, enzymes are not commonly used in the algal industry because of the high costs of cell lysing enzymes; care should be taken with such process, owing to natural deactivation or denaturation along the process of recovery [13]. However, this approach has received more and more attention – and is susceptible to becoming quite promising for biotechnological applications in the near future [13, 148, 197]. This is particularly valid for microalgae/cyanobacteria with intricate external sheaths in integrated biorefinery processes.

FUTURE PROSPECTS AND CHALLENGES

Bioactive Compound Fractionation Process Strategy – GRAS Solvents and Re-use of Biomass

As discussed above, valorization of biomass and exploitation of their components to the maximum extent is relevant towards feasible and suitable biorefinery processes. Hence, an integrated multi-component extraction system with re-use of biomass should be considered to contribute to the economic feasibility of the process, and to gain a share in the functional ingredients market. The urgent need for efficient processes of fractionation of biomass to individual constituents; or sequential process assuring the quality of the raw material, to subsequent steps with minimal microalgal damage is essential. The re-use of microalgal biomass (*i.e.*, successive extraction), preferably with extraction methods resorting to GRAS solvents to turn the process safer for food, is an important approach. By selecting different GRAS solvents – *e.g.*, ethanol and water, there is a benefit to playing with different polarities to obtain bioactive/antioxidant compounds with different chemical properties (Fig. **3**). Furthermore, the specificity of the solvent may allow a higher purification level of the compound(s) involved.

This topic is still at an early stage, but some studies at the lab scale have denoted interesting results with this approach [215 - 218]. For instance, Assunção *et al.* [218] focused on the biotechnological potential of *Chroococcidiopsis* sp. by reusing its biomass through a series of successive extractions with different GRAS solvents and distinct polarities – phosphate buffer saline, ethanol, acetone and methanol, towards extraction of several value added-compounds with antioxidant properties (*i.e.*, polysaccharides, PBPs, phenolic compounds, carotenoids and chls). In particular, the series including methanol and phosphate buffer saline led to bioactive extracts with enhanced content in pigments with different chemical features – PBPs, carotenoids and chls, besides polysaccharides or/and phenolic compounds. Hence, a good understanding of microalgal morphology and localization of bioactive compounds inside the cell is relevant to choose the most appropriate technologies to minimize compound loss and damage [13]. Furthermore, the extraction yields, stability and value of the compounds of interest will probably dictate the cascade order [134]. Since re-use of biomass could be a good approach to extract the different compounds, reutilization of GRAS solvents during the process could also be a potential strategy in light of saving costs and turning the process more feasible. The selection of a good GRAS solvent should provide low toxicity, low boiling point, preservative action, fast mass transfer, and prevention of precipitates [13].

Despite several promising approaches at the lab scale, industrial scale-level technology is still inexistent in most cases; there is indeed a huge gap between theoretical extraction studies and experimental application on a large scale – so a detailed exploration of this field is required to improve quality and yield of relevant microalgal compounds.

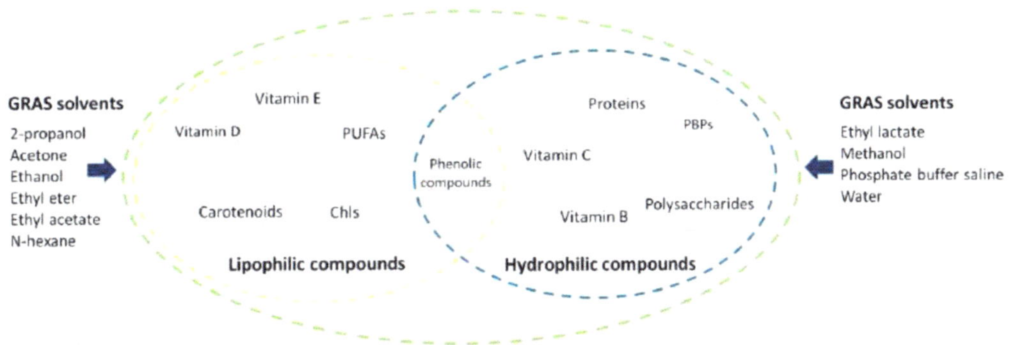

Fig. (3). Hydrophilic and lipophilic bioactive compounds present in microalgal with potential for use in functional foods – with the corresponding GRAS solvents to be used in their extraction, according to polarity.

Biorefinery Safety Issues and Economic Considerations

Concerning the extraction of the algal compounds when biomass valorization is tackled as a whole process – from the choice of organisms (*i.e.*, genetic modified

(GM) microalgae, potentially toxic organisms) through the coupling of the process of cultivation with CO_2 from or with wastewater in microalgae farms, several issues have been raised concerning the safety and quality of the final product. For instance, the development of GM microalgae strain could indeed speed up the production of several bioactives of interest to putatively high yields; however, for now, the costs involved to develop a biorefinery process would be extremely high, and limited by the inexistence of an integrated platform of production. In addition, the use of GM organisms can raise safety issues regarding containment throughout production and downstream processing [134]. The production of algal metabolites, in particular in microalgal open pond systems, and waste produced over metabolite extraction can indeed bear a high risk of release of those organisms, with unwanted repercussions to the environment. Incorporation of bioactive metabolites into functional foods from a GM strain can also be a hard road – because, in general, public opinion does not accept so well those products to a lack of knowledge of the underlying technology, along with a lack of legislation associated with the process. The extraction of bioactive metabolites from some organisms such as cyanobacteria (except *Arthrospira platensis*) with the purpose of integrating functional foods can be an issue; the use of cyanobacteria is not fully regulated for this purpose because some of them may have a potential toxic nature and there still are regulatory gaps. Algal material should be analyzed for the presence of toxins, nucleic acids and heavy metal components prior to commercialization [64]; food safety regulations have become the major constraint regarding application of algal sources towards functional foods [219].

Extraction of compounds from biomass produced in microalgae farms may offer an opportunity for microalgal biomass to be used as part of a biorefinery approach, while decreasing the price of microalgae based-process; however, care should be taken in the extraction of bioactives for food purposes. Typically, biomass is prone to accumulate contaminant compounds that could interfere with target bioactive compounds and thus, compromise their quality as required for the formulation of functional foods. Hence, the integration of specialized and monitoring units, to assess the quality of algal material and final is a must to minimize such risk.

Several challenges regarding bioactive metabolite, in terms of extraction, are still ahead; integrated multi-step approaches based on sound technical and economic analyses, along with safety and wide acceptance of compounds to integrate functional foods. Several new techniques have emerged, such as UAE, MAE, PEF, SFE, and EAE, with more advantages than old classic methods such as Soxhlet. Those techniques appear promising, at least at small scale, for their relatively high efficient yields; they also entail the incorporation of GRAS solvents, are less time-consuming and use less solvent than old methods. Yet, the selection of adequate technologies for algal biorefinery will probably depend on the required energy input, maintenance expenditures and the availability of the current technology. For now, life cycle analyses (LCA) have shown that the cost of such technologies overpass production costs [13]. In this context, the integrated extraction multi-approach within biorefinery still calls for a consistent and efficient proof of concept – to make microalgal industrial units for bioactives economically feasible in the near future, along with minimization of environmental impacts and maintenance of product quality towards functional ingredients [13, 134]. A multidisciplinary approach is indeed needed – not only for the choice of the biological platform needed to succeed but also for the correct integration of engineering principles and infrastructures necessary to make the algal biorefinery a case for success.

CONCLUDING REMARKS

Microalgae have proven a source of several biomolecules with relevant bioactive features, for application as functional ingredients. For now, the costs of overall microalgal-based processes surpass the prices attained by the products in the market, in particular, due to the extraction procedure; this constrains industrial-scale production and distribution of algal products. A biorefinery approach as a conceptual model is expected to become a realistic possibility in the coming future, to be integrated into industrial approaches and thus, definitely contribute to reaching feasible and competitive algal processing.

In recent years, emerging extraction techniques have been studied; despite the difficulty to prove the concept at large scale, integration of GRAS solvents principles and reuse of biomass toward extraction of bioactives within a Biorefinery approach are indeed the route to follow, in attempts to save costs and implement greener processes. This will help in the optimization of the production and maximization its value towards functional ingredients.

CONSENT FOR PUBLICATION

Not applicable.

CONFLICT OF INTEREST

The author(s) declare no conflict of interest, financial or otherwise.

ACKNOWLEDGEMENTS

The authors thank the fund institutions that support this work: FEDER funds through COMPETE2020 - Programa Operacional Competitividade e

Internacionalização (POCI), and national funds through FCT - Fundação para a Ciência e a Tecnologia, within the scope of UIDB/04423/2020 and UIDP/04423/2020 granted to CIIMAR and UIDB/00511/2020 granted to Laboratory for Process Engineering, Environment, Biotechnology and Energy – LEPABE, funded by national funds through FCT/MCTES (PIDDAC). A PhD fellowship (PD/BD/137887/2018) for author J.A. was granted by FCT, under the auspices of Programa Operacional Capital Humano (POCH), supported European Union (UE) by the European Social Fund (ESF) and Portuguese funds (MECTES) and the POCI (Norte 2020).

REFERENCES

[1] Shaw JE, Sicree RA, Zimmet PZ. Global estimates of the prevalence of diabetes for 2010 and 2030. Diabetes Res Clin Pract 2010; 87(1): 4-14. [http://dx.doi.org/10.1016/j.diabres.2009.10.007] [PMID: 19896746]

[2] Kelly T, Yang W, Chen C-S, Reynolds K, He J. Global burden of obesity in 2005 and projections to 2030. Int J Obes 2008; 32(9): 1431-7. [http://dx.doi.org/10.1038/ijo.2008.102] [PMID: 18607383]

[3] Bray F, Jemal A, Grey N, Ferlay J, Forman D. Global cancer transitions according to the Human Development Index (2008–2030): a population-based study. Lancet Oncol 2012; 13(8): 790-801. [http://dx.doi.org/10.1016/S1470-2045(12)70211-5] [PMID: 22658655]

[4] Barnes PJ. Inflammatory mechanisms in patients with chronic obstructive pulmonary disease. J Allergy Clin Immunol 2016; 138(1): 16-27. [http://dx.doi.org/10.1016/j.jaci.2016.05.011] [PMID: 27373322]

[5] Lafferty KD, Mordecai EA. Lafferty KD, Mordecai EA. The rise and fall of infectious disease in a warmer world. F1000Research. 2016.5: F1000 Faculty Rev-2040..

[6] Patterson GML, Larsen LK, Moore RE. Bioactive natural products from blue-green algae. J Appl Phycol 1994; 6(2): 151-7. [http://dx.doi.org/10.1007/BF02186069]

[7] Guiry MD. How many species of algae are there? J Phycol 2012; 48(5): 1057-63. [http://dx.doi.org/10.1111/j.1529-8817.2012.01222.x] [PMID: 27011267]

[8] Tan LT. Bioactive natural products from marine cyanobacteria for drug discovery. Phytochemistry 2007; 68(7): 954-79. [http://dx.doi.org/10.1016/j.phytochem.2007.01.012] [PMID: 17336349]

[9] Singh RK, Tiwari SP, Rai AK, Mohapatra TM. Cyanobacteria: an emerging source for drug discovery. J Antibiot (Tokyo) 2011; 64(6): 401-12. [http://dx.doi.org/10.1038/ja.2011.21] [PMID: 21468079]

[10] Vijayakumar S, Menakha M. Pharmaceutical applications of cyanobacteria—A review. Journal of Acute Medicine 2015; 5(1): 15-23. [http://dx.doi.org/10.1016/j.jacme.2015.02.004]

[11] Urtubia HO, Betanzo LB. Microalgae and Cyanobacteria as Green Molecular Factories: Tools and Perspectives. In: Thajuddin N. and Dhanasekaran D. Eds. Algae - Organisms for Imminent Biotechnology. IntechOpen 2016; 1-28. [http://dx.doi.org/10.5772/63006]

[12] Del Campo JA, Rodríguez H, Moreno J, Vargas MÁ, Rivas J, Guerrero MG. Lutein production by *Muriellopsis* sp. in an outdoor tubular photobioreactor. J Biotechnol 2001; 85(3): 289-95. [http://dx.doi.org/10.1016/S0168-1656(00)00380-1] [PMID: 11173095]

[13] Dixon C, Wilken LR. Green microalgae biomolecule separations and recovery. Bioresour Bioprocess 2018; 5(1): 14.
[http://dx.doi.org/10.1186/s40643-018-0199-3]

[14] Levasseur W, Perré P, Pozzobon V. A review of high value-added molecules production by microalgae in light of the classification. Biotechnol Adv 2020; 41: 107545.
[http://dx.doi.org/10.1016/j.biotechadv.2020.107545] [PMID: 32272160]

[15] Khan MI, Shin JH, Kim JD. The promising future of microalgae: current status, challenges, and optimization of a sustainable and renewable industry for biofuels, feed, and other products. Microb Cell Fact 2018; 17(1): 36.
[http://dx.doi.org/10.1186/s12934-018-0879-x] [PMID: 29506528]

[16] Rizwan M, Mujtaba G, Memon SA, Lee K, Rashid N. Exploring the potential of microalgae for new biotechnology applications and beyond: A review. Renew Sustain Energy Rev 2018; 92: 394-404.
[http://dx.doi.org/10.1016/j.rser.2018.04.034]

[17] Markou G, Nerantzis E. Microalgae for high-value compounds and biofuels production: A review with focus on cultivation under stress conditions. Biotechnol Adv 2013; 31(8): 1532-42.
[http://dx.doi.org/10.1016/j.biotechadv.2013.07.011] [PMID: 23928208]

[18] Stoyneva-Gärtner M, Uzunov B, Gärtner G. Enigmatic Microalgae from Aeroterrestrial and Extreme Habitats in Cosmetics: The Potential of the Untapped Natural Sources. Cosmetics 2020; 7(2): 27.
[http://dx.doi.org/10.3390/cosmetics7020027]

[19] Jacob-Lopes E, Maroneze MM, Deprá MC, Sartori RB, Dias RR, Zepka LQ. Bioactive food compounds from microalgae: an innovative framework on industrial biorefineries. Curr Opin Food Sci 2019; 25: 1-7.
[http://dx.doi.org/10.1016/j.cofs.2018.12.003]

[20] Rammuni MN, Ariyadasa TU, Nimarshana PHV, Attalage RA. Comparative assessment on the extraction of carotenoids from microalgal sources: Astaxanthin from H. pluvialis and β-carotene from D. salina. Vol. 277, Food Chemistry. Elsevier 2019; 1128-34.

[21] Vanthoor-Koopmans M, Wijffels RH, Barbosa MJ, Eppink MHM. Biorefinery of microalgae for food and fuel. Bioresour Technol 2013; 135: 142-9.
[http://dx.doi.org/10.1016/j.biortech.2012.10.135] [PMID: 23186688]

[22] Abed RMM, Dobretsov S, Sudesh K. Applications of cyanobacteria in biotechnology. J Appl Microbiol 2009; 106(1): 1-12.
[http://dx.doi.org/10.1111/j.1365-2672.2008.03918.x] [PMID: 19191979]

[23] Saini DK, Pabbi S, Shukla P. Cyanobacterial pigments: Perspectives and biotechnological approaches. Food Chem Toxicol 2018; 120: 616-24.
[http://dx.doi.org/10.1016/j.fct.2018.08.002] [PMID: 30077705]

[24] Mandal MK, Chanu NK, Chaurasia N. [24] Mandal MK, Chanu NK, Chaurasia N. Cyanobacterial pigments and their fluorescence characteristics: applications in research and industry. In: Singh PK, Kumar A, Singh VK, Shrivastava AK Eds. Advances in Cyanobacterial Biology. Academic Press Inc. 2020; 55-72.
[http://dx.doi.org/10.1016/B978-0-12-819311-2.00005-X]

[25] Pagels F, Guedes AC, Amaro HM, Kijjoa A, Vasconcelos V. Phycobiliproteins from cyanobacteria: Chemistry and biotechnological applications. Biotechnol Adv 2019; 37(3): 422-43.
[http://dx.doi.org/10.1016/j.biotechadv.2019.02.010] [PMID: 30797095]

[26] Lau NS, Matsui M, Abdullah AAA. Cyanobacteria : Photoautotrophic microbial factories for the sustainable synthesis of industrial products. BioMed Res Int 2015; 2015(754934): 1-9.
[http://dx.doi.org/10.1155/2015/754934] [PMID: 26199945]

[27] Sathasivam R, Radhakrishnan R, Hashem A, Abd Allah EF, Allah A. Microalgae metabolites: A rich source for food and medicine. Saudi J Biol Sci 2019; 26(4): 709-22.

[http://dx.doi.org/10.1016/j.sjbs.2017.11.003] [PMID: 31048995]

[28] da Silva Ferreira V, Sant'Anna C. Impact of culture conditions on the chlorophyll content of microalgae for biotechnological applications. World J Microbiol Biotechnol 2017; 33(1): 20. [http://dx.doi.org/10.1007/s11274-016-2181-6] [PMID: 27909993]

[29] Solymosi K, Mysliwa-kurdziel B. Chlorophylls and Their Derivatives Used in Food Industry and Medicine. Mini-reviews. Med Chem 2014; 0(0): 1-29.

[30] Perez-Galvez A, Viera I, Roca M. Chemistry in the bioactivity of chlorophylls: an overview. Curr Med Chem 2018; 24(40): 4515-36. [http://dx.doi.org/10.2174/0929867324666170714102619] [PMID: 28714384]

[31] Subramoniam A, Asha VV, Nair SA, *et al.* Chlorophyll revisited: anti-inflammatory activities of chlorophyll a and inhibition of expression of TNF-α gene by the same. Inflammation 2012; 35(3): 959-66. [http://dx.doi.org/10.1007/s10753-011-9399-0] [PMID: 22038065]

[32] Marková I, Ja J, Subhanová I, *et al.* Chlorophyll-Mediated Changes in the Redox Status of Pancreatic Cancer Cells Are Associated with Its Anticancer Effects. Oxid Med Cell Longev 2018; 4069167: 1-11.

[33] Alam T. Extraction of Natural Colors from Marine Algae. J Agr Marine Sci 2019; 23: 81. [http://dx.doi.org/10.24200/jams.vol23iss0pp81-91]

[34] Kong W, Liu N, Zhang J, *et al.* Optimization of ultrasound-assisted extraction parameters of chlorophyll from *Chlorella vulgaris* residue after lipid separation using response surface methodology. J Food Sci Technol 2014; 51(9): 2006-13. [http://dx.doi.org/10.1007/s13197-012-0706-z] [PMID: 25190857]

[35] Tong Y, Gao L, Xiao G, Pan X. Microwave pretreatment-assisted ethanol extraction of chlorophylls from *Spirulina platensis.* J Food Process Eng 2012; 35(5): 792-9. [http://dx.doi.org/10.1111/j.1745-4530.2010.00629.x]

[36] Demay J, Bernard C, Reinhardt A, Marie B. Natural products from cyanobacteria: Focus on beneficial activities. Mar Drugs 2019; 17(6): 320. [http://dx.doi.org/10.3390/md17060320] [PMID: 31151260]

[37] Hughes DA. Effects of carotenoids on human immune function. Proc Nutr Soc 1999; 58(3): 713-8. [http://dx.doi.org/10.1017/S0029665199000932] [PMID: 10604207]

[38] Blanco AM, Moreno J, Del Campo JA, Rivas J, Guerrero MG. Outdoor cultivation of lutein-rich cells of *Muriellopsis* sp. in open ponds. Appl Microbiol Biotechnol 2007; 73(6): 1259-66. [http://dx.doi.org/10.1007/s00253-006-0598-9] [PMID: 17033775]

[39] Sathasivam R, Juntawong N. Modified medium for enhanced growth of *Dunaliella* strains. Int J Curr Sci 2013; 5: 67-73.

[40] Fakhri S, Abbaszadeh F, Dargahi L, Jorjani M. Astaxanthin: A mechanistic review on its biological activities and health benefits. Pharmacol Res 2018; 136: 1-20. [http://dx.doi.org/10.1016/j.phrs.2018.08.012] [PMID: 30121358]

[41] Eriksen N. Research trends in the dominating microalgal pigments, β-carotene, astaxanthin, and phycocyanin used in feed, in foods, and in health applications. J Nutr Food Sci 2016; 6: 1000507.

[42] Torrissen OJ, Christiansen R. Requirements for carotenoids in fish diets. J Appl Ichthyology 1995; 11(3-4): 225-30. [http://dx.doi.org/10.1111/j.1439-0426.1995.tb00022.x]

[43] Molino A, Iovine A, Casella P, *et al.* Microalgae characterization for consolidated and new application in human food, animal feed and nutraceuticals. Int J Environ Res Public Health 2018; 15(11): 2436. [http://dx.doi.org/10.3390/ijerph15112436] [PMID: 30388801]

[44] Olaizola M. Commercial development of microalgal biotechnology : from the test tube to the marketplace marketplace 2003; 20: 459-66.

[http://dx.doi.org/10.1016/S1389-0344(03)00076-5]

[45] Jiao Y, Reuss L, Wang Y. β-Cryptoxanthin: Chemistry, Occurrence, and Potential Health Benefits. Curr Pharmacol Rep 2019; 5(1): 20-34.
[http://dx.doi.org/10.1007/s40495-019-00168-7]

[46] Tanaka T, Shnimizu M, Moriwaki H. Cancer chemoprevention by carotenoids. Molecules 2012; 17(3): 3202-42.
[http://dx.doi.org/10.3390/molecules17033202] [PMID: 22418926]

[47] Galasso C, Corinaldesi C, Sansone C. Carotenoids from marine organisms: Biological functions and industrial applications. Antioxidants 2017; 6(4): 96.
[http://dx.doi.org/10.3390/antiox6040096] [PMID: 29168774]

[48] Shegokar R, Mitri K. Carotenoid lutein: a promising candidate for pharmaceutical and nutraceutical applications. J Diet Suppl 2012; 9(3): 183-210.
[http://dx.doi.org/10.3109/19390211.2012.708716] [PMID: 22889143]

[49] Christaki E, Bonos E, Giannenas I, Florou-Paneri P. Functional properties of carotenoids originating from algae. J Sci Food Agric 2013; 93(1): 5-11.
[http://dx.doi.org/10.1002/jsfa.5902] [PMID: 23044813]

[50] Del Campo JA, García-González M, Guerrero MG. Outdoor cultivation of microalgae for carotenoid production: current state and perspectives. Appl Microbiol Biotechnol 2007; 74(6): 1163-74.
[http://dx.doi.org/10.1007/s00253-007-0844-9] [PMID: 17277962]

[51] Abe K, Hattori H, Hirano M. Accumulation and antioxidant activity of secondary carotenoids in the aerial microalga *Coelastrella striolata* var. *multistriata.* Food Chem 2007; 100(2): 656-61.
[http://dx.doi.org/10.1016/j.foodchem.2005.10.026]

[52] Pelah D, Sintov A, Cohen E. The effect of salt stress on the production of canthaxanthin and astaxanthin by *Chlorella zofingiensis* grown under limited light intensity. World J Microbiol Biotechnol 2004; 20(5): 483-6.
[http://dx.doi.org/10.1023/B:WIBI.0000040398.93103.21]

[53] Xia S, Wang K, Wan L, Li A, Hu Q, Zhang C. Production, characterization, and antioxidant activity of fucoxanthin from the marine diatom *Odontella aurita.* Mar Drugs 2013; 11(7): 2667-81.
[http://dx.doi.org/10.3390/md11072667] [PMID: 23880936]

[54] El-Baky HHA, El Baz FK, El-Baroty GS. *Spirulina* species as a source of carotenoids and a-tocopherol and its anticarcinoma factors. Biotechnology (Faisalabad) 2003; 2(3): 222-40.
[http://dx.doi.org/10.3923/biotech.2003.222.240]

[55] Tundis R, Loizzo MR, Bonesi M, Menichini F. Potential role of natural compounds against skin aging. Curr Med Chem 2015; 22(12): 1515-38.
[http://dx.doi.org/10.2174/0929867322666150227151809] [PMID: 25723509]

[56] Zheng J, Inoguchi T, Sasaki S, *et al.* Phycocyanin and phycocyanobilin from *Spirulina platensis* protect against diabetic nephropathy by inhibiting oxidative stress. Am J Physiol Regul Integr Comp Physiol 2013; 304(2): R110-20.
[http://dx.doi.org/10.1152/ajpregu.00648.2011] [PMID: 23115122]

[57] Sekar S, Chandramohan M. Phycobiliproteins as a commodity: trends in applied research, patents and commercialization. J Appl Phycol 2008; 20(2): 113-36.
[http://dx.doi.org/10.1007/s10811-007-9188-1]

[58] Bermejo Román R, Alvárez-Pez JM, Acién Fernández FG, Molina Grima E. Recovery of pure B-phycoerythrin from the microalga *Porphyridium cruentum.* J Biotechnol 2002; 93(1): 73-85.
[http://dx.doi.org/10.1016/S0168-1656(01)00385-6] [PMID: 11690696]

[59] Borowitzka MA. High-value products from microalgae—their development and commercialisation. J Appl Phycol 2013; 25(3): 743-56.
[http://dx.doi.org/10.1007/s10811-013-9983-9]

[60] Simopoulos AP. The importance of the omega-6/omega-3 fatty acid ratio in cardiovascular disease and other chronic diseases. Exp Biol Med (Maywood) 2008; 233(6): 674-88. [http://dx.doi.org/10.3181/0711-MR-311] [PMID: 18408140]

[61] Talero E, García-Mauriño S, Ávila-Román J, Rodríguez-Luna A, Alcaide A, Motilva V. Bioactive compounds isolated from microalgae in chronic inflammation and cancer. Mar Drugs 2015; 13(10): 6152-209. [http://dx.doi.org/10.3390/md13106152] [PMID: 26437418]

[62] Montero-Lobato Z, Vázquez M, Navarro F, *et al.* Chemically-induced production of anti-inflammatory molecules in microalgae. Mar Drugs 2018; 16(12): 478. [http://dx.doi.org/10.3390/md16120478] [PMID: 30513601]

[63] Siriwardhana N, Kalupahana NS, Moustaid-Moussa N. Health benefits of n-3 polyunsaturated fatty acids: eicosapentaenoic acid and docosahexaenoic acid. Adv Food Nutr Res 2012; 65: 211-22. [http://dx.doi.org/10.1016/B978-0-12-416003-3.00013-5] [PMID: 22361189]

[64] Spolaore P, Joannis-Cassan C, Duran E, Isambert A, De Génie L, Paris EC. Commercial applications of microalgae. J Biosci Bioeng 2006; 101(2): 87-96. [http://dx.doi.org/10.1263/jbb.101.87] [PMID: 16569602]

[65] Suganya T, Varman M, Masjuki HH, Renganathan S. Macroalgae and microalgae as a potential source for commercial applications along with biofuels production: A biorefinery approach. Renew Sustain Energy Rev 2016; 55: 909-41. [http://dx.doi.org/10.1016/j.rser.2015.11.026]

[66] Mendes A, Reis A, Vasconcelos R, Guerra P, Lopes da Silva T. *Crypthecodinium cohnii* with emphasis on DHA production: a review. J Appl Phycol 2009; 21(2): 199-214. [http://dx.doi.org/10.1007/s10811-008-9351-3]

[67] Hamed I. The Evolution and Versatility of Microalgal Biotechnology: A Review. Compr Rev Food Sci Food Saf 2016; 15(6): 1104-23. [http://dx.doi.org/10.1111/1541-4337.12227] [PMID: 33401835]

[68] Matos J, Cardoso CL, Falé P, Afonso CM, Bandarra NM. Investigation of nutraceutical potential of the microalgae *Chlorella vulgaris* and *Arthrospira platensis*. Int J Food Sci Technol 2020; 55(1): 303-12. [http://dx.doi.org/10.1111/ijfs.14278]

[69] Bhalamurugan GL, Valerie O, Mark L. Valuable bioproducts obtained from microalgal biomass and their commercial applications: A review. Environ Eng Res 2018; 23(3): 229-41. [http://dx.doi.org/10.4491/eer.2017.220]

[70] Choopani A, Poorsoltan M, Fazilati M, Latifi AM, Salavati H. *Spirulina*: A source of gamma-linoleic acid and its applications. J Appl Biotechnol Reports 2016; 3: 483-8.

[71] de Jesus Raposo M, de Morais A, de Morais R. Marine polysaccharides from algae with potential biomedical applications. Mar Drugs 2015; 13(5): 2967-3028. [http://dx.doi.org/10.3390/md13052967] [PMID: 25988519]

[72] Delattre C, Pierre G, Laroche C, Michaud P. Production, extraction and characterization of microalgal and cyanobacterial exopolysaccharides. Biotechnol Adv 2016; 34(7): 1159-79. [http://dx.doi.org/10.1016/j.biotechadv.2016.08.001] [PMID: 27530696]

[73] Cruz D, Vasconcelos V, Pierre G, Michaud P, Delattre C. Exopolysaccharides from Cyanobacteria: Strategies for Bioprocess Development. Appl Sci (Basel) 2020; 10(11): 3763. [http://dx.doi.org/10.3390/app10113763]

[74] Raposo M, De Morais R, Bernardo de Morais A. Bioactivity and applications of sulphated polysaccharides from marine microalgae. Mar Drugs 2013; 11(1): 233-52. [http://dx.doi.org/10.3390/md11010233] [PMID: 23344113]

[75] Singh S, Kant C, Yadav RK, Reddy YP, Abraham G. Cyanobacterial exopolysaccharides: composition, biosynthesis, and biotechnological applications. In: Mishra AK, Tiwari DN, Rai AN. (Eds) Cyanobacteria: From Basic Science to Applications. Academic Press 2019; 347-58. [http://dx.doi.org/10.1016/B978-0-12-814667-5.00017-9]

[76] Plaza M, Herrero M, Cifuentes A, Ibáñez E. Innovative natural functional ingredients from microalgae. J Agric Food Chem 2009; 57(16): 7159-70. [http://dx.doi.org/10.1021/jf901070g] [PMID: 19650628]

[77] Freile-Pelegrín Y, Robledo D. Bioactive phenolic compounds from algae. In: Ledesma-Hernández B, Herrero M, Eds. Bioactive Compounds from Marine Foods: Plant and Animal Sources 2013; 113-29. [http://dx.doi.org/10.1002/9781118412893.ch6]

[78] Abd El-Baky HH, El Baz FK, El-Baroty GS. Production of phenolic compounds from *Spirulina maxima* microalgae and its protective effects *in vitro* toward hepatotoxicity model. Afr J Pharm Pharmacol 2009; 3: 133-9.

[79] Jerez-Martel I, García-Poza S, Rodríguez-Martel G, Rico M, Afonso-Olivares C, Gómez-Pinchetti JL. Phenolic profile and antioxidant activity of crude extracts from microalgae and cyanobacteria strains. J Food Qual. 2017.ID 2924508.

[80] Scaglioni PT, de Oliveira Garcia S, Badiale-Furlong E. Inhibition of *in vitro* trichothecenes production by microalgae phenolic extracts. Food Res Int 2019; 124: 175-80. [http://dx.doi.org/10.1016/j.foodres.2018.07.008] [PMID: 31466637]

[81] Ruiz-Domínguez MC, Cerezal P, Salinas F, Medina E, Renato-Castro G. Application of box-behnken design and desirability function for green prospection of bioactive compounds from isochrysis galbana. Appl Sci (Basel) 2020; 10(8): 2789. [http://dx.doi.org/10.3390/app10082789]

[82] Edelmann M, Aalto S, Chamlagain B, Kariluoto S, Piironen V. Riboflavin, niacin, folate and vitamin B12 in commercial microalgae powders. J Food Compos Anal 2019; 82: 103226. [http://dx.doi.org/10.1016/j.jfca.2019.05.009]

[83] Apone F, Barbulova A, Colucci MG. Plant and microalgae derived peptides are advantageously employed as bioactive compounds in cosmetics. Front Plant Sci 2019; 10: 756. [http://dx.doi.org/10.3389/fpls.2019.00756] [PMID: 31244874]

[84] Brasil BSAF, Siqueira FG, Salum TFC, Zanette CM, Spier MR. Microalgae and cyanobacteria as enzyme biofactories 2017; 76-89.

[85] Bernaerts TMM, Gheysen L, Foubert I, Hendrickx ME, Van Loey AM. The potential of microalgae and their biopolymers as structuring ingredients in food: A review. Biotechnol Adv 2019; 37(8): 107419. [http://dx.doi.org/10.1016/j.biotechadv.2019.107419] [PMID: 31340183]

[86] Mourelle M, Gómez C, Legido J. The potential use of marine microalgae and cyanobacteria in cosmetics and thalassotherapy. Cosmetics 2017; 4(4): 46. [http://dx.doi.org/10.3390/cosmetics4040046]

[87] Yaakob Z, Ali E, Zainal A, Mohamad M, Takriff MS. An overview: biomolecules from microalgae for animal feed and aquaculture. J Biol Res (Thessalon) 2014; 21(1): 6. [http://dx.doi.org/10.1186/2241-5793-21-6] [PMID: 25984489]

[88] Baldea I, Ion RM, Olteanu DE, Nenu I, Tudor D, Filip AG. Photodynamic therapy of melanoma using new, synthetic porphyrins and phthalocyanines as photosensitisers - a comparative study. Clujul Med 2015; 88(2): 175-80. [PMID: 26528068]

[89] Freitas S, Silva NG, Sousa ML, *et al.* Chlorophyll derivatives from marine cyanobacteria with lipid-reducing activities. Mar Drugs 2019; 17(4): 229. [http://dx.doi.org/10.3390/md17040229] [PMID: 30999602]

[90] McCook J, Dorogi P, Vasily D, Cefalo D. *in vitro* inhibition of hyaluronidase by sodium copper chlorophyllin complex and chlorophyllin analogs. Clin Cosmet Investig Dermatol 2015; 8: 443-8. [http://dx.doi.org/10.2147/CCID.S86863] [PMID: 26300653]

[91] Demmig-Adams B, Adams WW III. Antioxidants in photosynthesis and human nutrition. Science 2002; 298(5601): 2149-53. [http://dx.doi.org/10.1126/science.1078002] [PMID: 12481128]

[92] Mayne ST. Beta□carotene, carotenoids, and disease prevention in humans. FASEB J 1996; 10(7): 690-701. [http://dx.doi.org/10.1096/fasebj.10.7.8635686] [PMID: 8635686]

[93] Jomova K, Valko M. Health protective effects of carotenoids and their interactions with other biological antioxidants. Eur J Med Chem 2013; 70: 102-10. [http://dx.doi.org/10.1016/j.ejmech.2013.09.054] [PMID: 24141200]

[94] Lopes G, Clarinha D, Vasconcelos V. Carotenoids from cyanobacteria: A biotechnological approach for the topical treatment of psoriasis. Microorganisms 2020; 8(2): 302. [http://dx.doi.org/10.3390/microorganisms8020302] [PMID: 32098255]

[95] Heydarizadeh P, Poirier I, Loizeau D, *et al.* Plastids of marine phytoplankton produce bioactive pigments and lipids. Mar Drugs 2013; 11(9): 3425-71. [http://dx.doi.org/10.3390/md11093425] [PMID: 24022731]

[96] Peng J, Yuan JP, Wu CF, Wang JH. Fucoxanthin, a marine carotenoid present in brown seaweeds and diatoms: metabolism and bioactivities relevant to human health. Mar Drugs 2011; 9(10): 1806-28. [http://dx.doi.org/10.3390/md9101806] [PMID: 22072997]

[97] Tominaga K, Hongo N, Karato M, Yamashita E. Cosmetic benefits of astaxanthin on humans subjects. Acta Biochim Pol 2012; 59(1): 43-7. [http://dx.doi.org/10.18388/abp.2012_2168] [PMID: 22428137]

[98] Benedetti S, Benvenuti F, Scoglio S, Canestrari F. Oxygen radical absorbance capacity of phycocyanin and phycocyanobilin from the food supplement Aphanizomenon flos-aquae. J Med Food 2010; 13(1): 223-7. [http://dx.doi.org/10.1089/jmf.2008.0257] [PMID: 20136460]

[99] Gantar M, Dhandayuthapani S, Rathinavelu A. Phycocyanin induces apoptosis and enhances the effect of topotecan on prostate cell line LNCaP. J Med Food 2012; 15(12): 1091-5. [http://dx.doi.org/10.1089/jmf.2012.0123] [PMID: 23134462]

[100] Romay C, Armesto J, Remirez D, González R, Ledon N, García I. Antioxidant and anti-inflammatory properties of C-phycocyanin from blue-green algae. Inflamm Res 1998; 47(1): 36-41. [http://dx.doi.org/10.1007/s000110050256] [PMID: 9495584]

[101] González R, Rodríguez S, Romay C, *et al.* Anti-inflammatory activity of phycocyanin extract in acetic acid-induced colitis in rats. Pharmacol Res 1999; 39(1): 55-9. [http://dx.doi.org/10.1006/phrs.1998.0409] [PMID: 10366332]

[102] Liu Q, Huang Y, Zhang R, Cai T, Cai Y. Medical application of Spirulina platensis derived C-Phycocyanin. Evidence-based Complement Altern Med.2016; 2016.ID 7803846.

[103] Ge B, Qin S, Han L, Lin F, Ren Y. Antioxidant properties of recombinant allophycocyanin expressed in *Escherichia coli.* J Photochem Photobiol B Biol 84: 175-80.

[104] Koníčková R, Vaňková K, Vaníková J, *et al.* Anti-cancer effects of blue-green alga *Spirulina platensis*, a natural source of bilirubin-like tetrapyrrolic compounds. Ann Hepatol 2014; 13(2): 273-83. [http://dx.doi.org/10.1016/S1665-2681(19)30891-9] [PMID: 24552870]

[105] Bhat VB, Madyastha KM. Scavenging of peroxynitrite by phycocyanin and phycocyanobilin from Spirulina platensis: protection against oxidative damage to DNA. Biochem Biophys Res Commun 2001; 285(2): 262-6.

[http://dx.doi.org/10.1006/bbrc.2001.5195] [PMID: 11444835]

[106] Patel SN, Sonani RR, Jakharia K, *et al.* Antioxidant activity and associated structural attributes of *Halomicronema* phycoerythrin. Int J Biol Macromol 2018; 111: 359-69. [http://dx.doi.org/10.1016/j.ijbiomac.2017.12.170] [PMID: 29307804]

[107] Thangam R, Suresh V, Asenath Princy W, *et al.* C-Phycocyanin from Oscillatoria tenuis exhibited an antioxidant and *in vitro* antiproliferative activity through induction of apoptosis and G0/G1 cell cycle arrest. Food Chem 2013; 140(1-2): 262-72. [http://dx.doi.org/10.1016/j.foodchem.2013.02.060] [PMID: 23578642]

[108] Liu Q, Huang Y, Zhang R, Cai T, Cai Y. Medical application of *Spirulina platensis* derived C-Phycocyanin. Evid Based Complement Alternat Med 2016; 2016: 7803846. [PMID: 27293463]

[109] Maurya SS, Maurya JN, Pandey VD. Factors regulating phycobiliprotein production in cyanobacteria. Int J Curr Microbiol Appl Sci 2014; 3(5): 764-71.

[110] Prasanna R, Pabby A, Saxena S, Singh PK. Modulation of pigment profiles of *Calothrix elenkenii* in response to environmental changes. J Plant Physiol 2004; 161(10): 1125-32. [http://dx.doi.org/10.1016/j.jplph.2003.09.001] [PMID: 15535122]

[111] Abd El-Baky HH, El-Baroty GS. Characterization and bioactivity of phycocyanin isolated from *Spirulina maxima* grown under salt stress. Food Funct 2012; 3(4): 381-8. [http://dx.doi.org/10.1039/c2fo10194g] [PMID: 22234291]

[112] Gabr GA, El-Sayed SM, Hikal MS. Antioxidant Activities of Phycocyanin: A Bioactive Compound from *Spirulina platensis.* J Pharm Res Int 2020; 32: 73-85. [http://dx.doi.org/10.9734/jpri/2020/v32i230407]

[113] Gacheva G, Gigova L, Ivanova N, *et al.* Suboptimal growth temperatures enhance the biological activity of cultured cyanobacterium *Gloeocapsa* sp. J Appl Phycol 2013; 25(1): 183-94. [http://dx.doi.org/10.1007/s10811-012-9852-y]

[114] Hu H, Li Y, Yin C, Ouyang Y. Isolation and characterization of a mesophilic *Arthrospira maxima* strain capable of producing docosahexaenoic acid. J Microbiol Biotechnol 2011; 21(7): 697-702. [http://dx.doi.org/10.4014/jmb.1101.12040] [PMID: 21791955]

[115] Singh S, Kate BN, Banerjee UC. Bioactive compounds from cyanobacteria and microalgae: an overview. Crit Rev Biotechnol 2005; 25(3): 73-95. [http://dx.doi.org/10.1080/07388550500248498] [PMID: 16294828]

[116] Amaro H, Fernandes F, Valentão P, *et al.* Effect of solvent system on extractability of lipidic components of *Scenedesmus obliquus* (M2-1) and *Gloeothece* sp. on antioxidant scavenging capacity thereof. Mar Drugs 2015; 13(10): 6453-71. [http://dx.doi.org/10.3390/md13106453] [PMID: 26492257]

[117] Guedes AC, Amaro HM, Barbosa CR, Pereira RD, Malcata FX. Fatty acid composition of several wild microalgae and cyanobacteria, with a focus on eicosapentaenoic, docosahexaenoic and α-linolenic acids for eventual dietary uses. Food Res Int 2011; 44(9): 2721-9. [http://dx.doi.org/10.1016/j.foodres.2011.05.020]

[118] Catarina Guedes A, Barbosa CR, Amaro HM, Pereira CI, Xavier Malcata F. Microalgal and cyanobacterial cell extracts for use as natural antibacterial additives against food pathogens. Int J Food Sci Technol 2011; 46(4): 862-70. [http://dx.doi.org/10.1111/j.1365-2621.2011.02567.x]

[119] Plaza M. S. S, Jaime L, García-Blairsy Reina G, Herrero M, Señoráns FJ, Ibanéz E. Screening for bioactive compounds from algae. Adv Biofuels Bioprod 2009; 51: 833-72.

[120] El-Sheekh MM, Osman MEH, Dyab MA, Amer MS. Production and characterization of antimicrobial active substance from the cyanobacterium *Nostoc muscorum.* Environ Toxicol Pharmacol 2006; 21(1): 42-50.

[http://dx.doi.org/10.1016/j.etap.2005.06.006] [PMID: 21783637]

[121] El-Baky HHA, El Baz FK, El-baroty GS. Production of phenolic compounds from *Spirulina maxima* microalgae.pdf. Afr J Biotechnol 2009; 8: 7059-67.

[122] Senni K, Pereira J, Gueniche F, *et al.* Marine polysaccharides: a source of bioactive molecules for cell therapy and tissue engineering. Mar Drugs 2011; 9(9): 1664-81. [http://dx.doi.org/10.3390/md9091664] [PMID: 22131964]

[123] Li P, Harding SE, Liu Z. Cyanobacterial exopolysaccharides: their nature and potential biotechnological applications. Biotechnol Genet Eng Rev 2001; 18(1): 375-404. [http://dx.doi.org/10.1080/02648725.2001.10648020] [PMID: 11530697]

[124] Hayashi K, Hayashi T, Kojima I. A natural sulfated polysaccharide, calcium spirulan, isolated from *Spirulina platensis*: *in vitro* and ex vivo evaluation of anti-herpes simplex virus and anti-human immunodeficiency virus activities. AIDS Res Hum Retroviruses 1996; 12(15): 1463-71. [http://dx.doi.org/10.1089/aid.1996.12.1463] [PMID: 8893054]

[125] Challouf R, Trabelsi L, Ben Dhieb R, *et al.* Evaluation of cytotoxicity and biological activities in extracellular polysaccharides released by cyanobacterium *Arthrospira platensis.* Braz Arch Biol Technol 2011; 54(4): 831-8. [http://dx.doi.org/10.1590/S1516-89132011000400024]

[126] Galasso C, Gentile A, Orefice I, *et al.* Microalgal derivatives as potential nutraceutical and food supplements for human health: A focus on cancer prevention and interception. Nutrients 2019; 11(6): 1226. [http://dx.doi.org/10.3390/nu11061226] [PMID: 31146462]

[127] Becker EW. Micro-algae as a source of protein. Biotechnol Adv 2007; 25(2): 207-10. [http://dx.doi.org/10.1016/j.biotechadv.2006.11.002] [PMID: 17196357]

[128] Rajesh Banu J, Preethi , Kavitha S, Gunasekaran M, Kumar G. Microalgae based biorefinery promoting circular bioeconomy-techno economic and life-cycle analysis. Bioresour Technol 2020; 302: 122822. [http://dx.doi.org/10.1016/j.biortech.2020.122822] [PMID: 32007307]

[129] Khanra S, Mondal M, Halder G, Tiwari ON, Gayen K, Bhowmick TK. Downstream processing of microalgae for pigments, protein and carbohydrate in industrial application: A review. Food Bioprod Process 2018; 110: 60-84. [http://dx.doi.org/10.1016/j.fbp.2018.02.002]

[130] de Morais MG, da Fontoura Prates D, Moreira JB, Duarte JH, Costa JAV. Phycocyanin from microalgae: Properties, extraction and purification, with some recent applications. Ind Biotechnol (New Rochelle NY) 2018; 14(1): 30-7. [http://dx.doi.org/10.1089/ind.2017.0009]

[131] Pôjo V, Tavares T, Malcata FX. Processing Methodologies of Wet Microalga Biomass Toward Oil Separation: An Overview. Molecules 2021; 26(3): 641. [http://dx.doi.org/10.3390/molecules26030641] [PMID: 33530628]

[132] Assunção J, Guedes A, Malcata F. Biotechnological and pharmacological applications of biotoxins and other bioactive molecules from dinoflagellates. Mar Drugs 2017; 15(12): 393. [http://dx.doi.org/10.3390/md15120393] [PMID: 29261163]

[133] Draaisma RB, Wijffels RH, Slegers PM, Brentner LB, Roy A, Barbosa MJ. Food commodities from microalgae. Curr Opin Biotechnol. Elsevier Current Trends 2013; 24: 169-77.

[134] Schiano di Visconte G, Spicer A, Chuck CJ, Allen MJ. The microalgae biorefinery: A perspective on the current status and future opportunities using genetic modification. Appl Sci (Basel) 2019; 9(22): 4793. [http://dx.doi.org/10.3390/app9224793]

[135] Minhas AK, Hodgson P, Barrow CJ, Adholeya A. A review on the assessment of stress conditions for

simultaneous production of microalgal lipids and carotenoids. Front Microbiol 2016; 7: 546. [http://dx.doi.org/10.3389/fmicb.2016.00546] [PMID: 27199903]

[136] Shin SE, Lim JM, Koh HG, *et al.* CRISPR/Cas9-induced knockout and knock-in mutations in *Chlamydomonas reinhardtii.* Sci Rep 2016; 6(1): 27810. [http://dx.doi.org/10.1038/srep27810] [PMID: 27291619]

[137] Assunção J, Malcata FX. Enclosed "non-conventional" photobioreactors for microalga production: A review. Algal Res 2020; 52: 102107. [http://dx.doi.org/10.1016/j.algal.2020.102107]

[138] Chew KW, Yap JY, Show PL, *et al.* Microalgae biorefinery: High value products perspectives. Bioresour Technol 2017; 229: 53-62. [http://dx.doi.org/10.1016/j.biortech.2017.01.006] [PMID: 28107722]

[139] Guedes AC, Amaro HM, Sousa-Pinto I, Malcata FX. Applications of Spent Biomass. 2013; 205-33.

[140] Molina Grima E, Belarbi EH, Acién Fernández FG, Robles Medina A, Chisti Y. Recovery of microalgal biomass and metabolites: process options and economics. Biotechnol Adv 2003; 20(7-8): 491-515. [http://dx.doi.org/10.1016/S0734-9750(02)00050-2] [PMID: 14550018]

[141] Wang B, Li Y, Wu N, Lan CQ. CO_2 bio-mitigation using microalgae. Appl Microbiol Biotechnol 2008; 79(5): 707-18. [http://dx.doi.org/10.1007/s00253-008-1518-y] [PMID: 18483734]

[142] Sajid M, Płotka-Wasylka J. Combined extraction and microextraction techniques: Recent trends and future perspectives. Trends Analyt Chem 2018; 103: 74-86. [http://dx.doi.org/10.1016/j.trac.2018.03.013]

[143] Ibanéz E, Heerero M, Mendiola JA, Castro-Puyana M. Extraction and Characterization of Bioactive Compounds with Health Benefits from Marine Resources: Macro and Micro Algae, Cyanobacteria, and Invertebrates. In: Hayes M, Ed. Marine Bioactive Compounds: Sources, Characterization and Applications 2012; 55-98. [http://dx.doi.org/10.1007/978-1-4614-1247-2_2]

[144] Barba FJ, Grimi N, Vorobiev E. New approaches for the use of non-conventional cell disruption technologies to extract potential food additives and nutraceuticals from microalgae. Food Eng Rev 2015; 7(1): 45-62. [http://dx.doi.org/10.1007/s12393-014-9095-6]

[145] Soquetta MB, Terra LM, Bastos CP. Green technologies for the extraction of bioactive compounds in fruits and vegetables. CYTA J Food 2018; 16(1): 400-12. [http://dx.doi.org/10.1080/19476337.2017.1411978]

[146] Gam DH, Kim SY, Kim JW. Optimization of ultrasound-assisted extraction condition for phenolic compounds, antioxidant activity, and epigallocatechin gallate in lipid-extracted microalgae. Molecules 2020; 25(3): 454. [http://dx.doi.org/10.3390/molecules25030454] [PMID: 31973236]

[147] Wiltshire KH, Boersma M, Möller A, Buhtz H. Extraction of pigments and fatty acids from the green alga *Scenedesmus obliquus* (Chlorophyceae). Aquat Ecol 2000; 34(2): 119-26. [http://dx.doi.org/10.1023/A:1009911418606]

[148] Sosa-Hernández JE, Escobedo-Avellaneda Z, Iqbal HMN, Welti-Chanes J. State-of-the-Art Extraction Methodologies for Bioactive Compounds from Algal Biome to Meet Bio-Economy Challenges and Opportunities. Molecules 2018; 23(11): 2953. [http://dx.doi.org/10.3390/molecules23112953] [PMID: 30424551]

[149] Olaizola M. Commercial development of microalgal biotechnology: from the test tube to the marketplace. Biomol Eng 2003; 20(4-6): 459-66. [http://dx.doi.org/10.1016/S1389-0344(03)00076-5] [PMID: 12919832]

[150] Assunção J, Malcata FX. Dinoflagellates and Toxin Production. In: Malcata FX, Guedes AC, Sousa-Pinto I, Eds. Marine Macro- and Microalgae: An Overview 1st ed. 2018; 216-34.

[151] Cravotto G, Boffa L, Mantegna S, Perego P, Avogadro M, Cintas P. Improved extraction of vegetable oils under high-intensity ultrasound and/or microwaves. Ultrason Sonochem 2008; 15(5): 898-902. [http://dx.doi.org/10.1016/j.ultsonch.2007.10.009] [PMID: 18093864]

[152] Abdel Ghaly MA hattab. Microalgae Oil Extraction Pre-treatment Methods: Critical Review and Comparative Analysis. J Fundam Renew Energy Appl 2015; 5: 1-26.

[153] Pasquet V, Chérouvrier JR, Farhat F, *et al.* Study on the microalgal pigments extraction process: Performance of microwave assisted extraction. Process Biochem 2011; 46(1): 59-67. [http://dx.doi.org/10.1016/j.procbio.2010.07.009]

[154] Eing C, Goettel M, Straessner R, Gusbeth C, Frey W. Pulsed electric field treatment of microalgae - Benefits for microalgae biomass processing. IEEE Trans Plasma Sci 2013; 41(10): 2901-7. [http://dx.doi.org/10.1109/TPS.2013.2274805]

[155] Alam A, Sarker ZI, Ghafoor K, Happy RA. Bioactive Compounds and Extraction Techniques. In: Nguyen VT, editor. Recovering Bioactive Compounds from Agricultural Wastes; John Wiley & Sons Ltd. 2017; 33-53. [http://dx.doi.org/10.1002/9781119168850.ch2]

[156] Perrut M. Supercritical fluid applications: Industrial developments and economic issues. Ind Eng Chem Res 2000; 39(12): 4531-5. [http://dx.doi.org/10.1021/ie000211c]

[157] Munir N, Sharif N, Shagufta N, Saleem F, Manzoor F. Harvesting and processing of microalgae biomass fractions for biodiesel production (a review). Sci Tech Dev 2013; 32: 235-43.

[158] Chemat F, Rombaut N, Sicaire A, Meullemiestre A, Abert-vian M. Ultrasound assisted extraction of food and natural products. Mechanisms, techniques, combinations, protocols and applications. A review. Ultrason - Sonochemistry. 2017; 34: 540-60.

[159] Ravishankar GA, Ambati RR, Alzahrani MAJ, Perera CO. Ultrasound-Assisted Extraction of Bioactive Compounds from Microalgae. Handbook of Algal Technologies and Phytochemicals 2019; 81-8.

[160] Jaeschke DP, Rech R, Marczak LDF, Mercali GD, Domeneghini G. Ultrasound as an alternative technology to extract carotenoids and lipids from *Heterochlorella luteoviridis.* Bioresour Technol 2017; 224: 753-7. [http://dx.doi.org/10.1016/j.biortech.2016.11.107] [PMID: 27955863]

[161] Dey S, Rathod VK. Ultrasound assisted extraction of b-carotene from Spirulina platensis. Ultrason - Sonochemistry 2013; 20: 271-6.

[162] Stramarkou M, Papadaki S, Kyriakopoulou K, Krokida M. Recovery of Functional Pigments from four Different Species of Microalgae. IOSR J Environ Sci Toxicol Food Technol 2016; 10(9): 26-30. [http://dx.doi.org/10.9790/2402-1009022630]

[163] Kapoore R, Butler T, Pandhal J, Vaidyanathan S. Microwave-assisted extraction for microalgae: From biofuels to biorefinery. Biology (Basel) 2018; 7(1): 18. [http://dx.doi.org/10.3390/biology7010018] [PMID: 29462888]

[164] Kadam SU, Tiwari BK, O'Donnell CP. Application of novel extraction technologies for bioactives from marine algae. J Agric Food Chem 2013; 61(20): 4667-75. [http://dx.doi.org/10.1021/jf400819p] [PMID: 23634989]

[165] Chemat F, Rombaut N, Meullemiestre A, *et al.* Review of Green Food Processing techniques. Preservation, transformation, and extraction. Innov Food Sci Emerg Technol 2017; 41: 357-77. [http://dx.doi.org/10.1016/j.ifset.2017.04.016]

[166] Zin MM, Anucha CB, Bánvölgyi S. Recovery of phytochemicals via electromagnetic irradiation

(microwave-mssisted-extraction): betalain and phenolic compounds in perspective. Foods 2020; 9(7): 918.
[http://dx.doi.org/10.3390/foods9070918] [PMID: 32664694]

[167] Juin C, Chérouvrier JR, Thiéry V, *et al.* Microwave-assisted extraction of phycobiliproteins from *Porphyridium purpureum.* Appl Biochem Biotechnol 2015; 175(1): 1-15.
[http://dx.doi.org/10.1007/s12010-014-1250-2] [PMID: 25231233]

[168] Aftari RV, Rezaei K, Bandani AR, Mortazavi A. Antioxidant activity optimisation of *Spirulina platensis* C-phycocyanin obtained by freeze-thaw, microwave-assisted and ultrasound-assisted extraction methods. Qual Assur Saf Crops Foods 2017; 9(1): 1-9.
[http://dx.doi.org/10.3920/QAS2015.0708]

[169] Kunjiappan S, Panneerselvam T, Kannan S. Optimization of microwave assisted extraction of bioactive compounds from *Dunaliella bardawil* using RSM and ANFIS modeling and assessment of anticancer activity of bioactive compou ... Optimization of Microwave-assisted Extraction of Bioactive Compounds f. Curr Microw Chem 2018; 5: 1-16.

[170] Silva AS, de Magalhães WT, Moreira LM, Rocha MVP, Bastos AKP, Rocha P, *et al.* Microwave-assisted extraction of polysaccharides from *Arthrospira* (*Spirulina*) *platensis* using the concept of green chemistry. Algal Res 2018; 35: 178-84.
[http://dx.doi.org/10.1016/j.algal.2018.08.015]

[171] Gong M, Bassi A. Carotenoids from microalgae: A review of recent developments. Biotechnol Adv 2016; 34(8): 1396-412.
[http://dx.doi.org/10.1016/j.biotechadv.2016.10.005] [PMID: 27816618]

[172] Geada P, Rodrigues R, Loureiro L, *et al.* Electrotechnologies applied to microalgal biotechnology – Applications, techniques and future trends. Renew Sustain Energy Rev 2018; 94: 656-68.
[http://dx.doi.org/10.1016/j.rser.2018.06.059]

[173] Min S, Jin ZT, Zhang QH. Commercial scale pulsed electric field processing of tomato juice. J Agric Food Chem 2003; 51(11): 3338-44.
[http://dx.doi.org/10.1021/jf0260444] [PMID: 12744664]

[174] Min S, Jin ZT, Min SK, Yeom H, Zhang QH. Commercial-Scale Pulsed Electric Field Processing of Orange Juice. J Food Sci 2003; 68(4): 1265-71.
[http://dx.doi.org/10.1111/j.1365-2621.2003.tb09637.x]

[175] Luengo E, Condón-Abanto S, Álvarez I, Raso J. Effect of pulsed electric field treatments on permeabilization and extraction of pigments from *Chlorella vulgaris.* J Membr Biol 2014; 247(12): 1269-77.
[http://dx.doi.org/10.1007/s00232-014-9688-2] [PMID: 24880235]

[176] Luengo E, Martínez JM, Bordetas A, Álvarez I, Raso J. Influence of the treatment medium temperature on lutein extraction assisted by pulsed electric fields from *Chlorella vulgaris.* Innov Food Sci Emerg Technol 2015; 29: 15-22.
[http://dx.doi.org/10.1016/j.ifset.2015.02.012]

[177] Pataro G, Goettel M, Straessner R, Gusbeth C, Ferrari G, Frey W. Effect of PEF Treatment on extraction of valuable compounds from microalgae *C. vulgaris.* Chem Eng Trans 2017; 57.

[178] Kokkali M, Martí-Quijal FJ, Taroncher M, Ruiz MJ, Kousoulaki K, Barba FJ. Improved Extraction Efficiency of Antioxidant Bioactive Compounds from *Tetraselmis chuii* and *Phaedoactylum tricornutum* Using Pulsed Electric Fields. Molecules 2020; 25(17): 3921.
[http://dx.doi.org/10.3390/molecules25173921] [PMID: 32867350]

[179] Sheng J, Vannela R, Rittmann BE. Disruption of *Synechocystis* PCC 6803 for lipid extraction. Water Sci Technol 2012; 65(3): 567-73.
[http://dx.doi.org/10.2166/wst.2012.879] [PMID: 22258690]

[180] Golberg A, Sack M, Teissie J, *et al.* Energy-efficient biomass processing with pulsed electric fields for

bioeconomy and sustainable development. Biotechnol Biofuels 2016; 9(1): 94. [http://dx.doi.org/10.1186/s13068-016-0508-z] [PMID: 27127539]

[181] Joelsson E, Erdei B, Galbe M, Wallberg O. Techno-economic evaluation of integrated first- and second-generation ethanol production from grain and straw. Biotechnol Biofuels 2016; 9(1): 1-22. [http://dx.doi.org/10.1186/s13068-015-0423-8]

[182] Núñez GA, del Valle JM, Navia D. Supercritical CO_2 oilseed extraction in multi-vessel plants. 3. Effect of extraction pressure and plant size on production cost. J Supercrit Fluids 2017; 122: 109-18. [http://dx.doi.org/10.1016/j.supflu.2016.11.002]

[183] Michalak I, Dmytryk A, Wieczorek PP, *et al.* Supercritical Algal Extracts: A Source of Biologically Active Compounds from Nature. J Chem 2015; 2015: 1-14. [http://dx.doi.org/10.1155/2015/597140]

[184] Santana A, Jesus S, Larrayoz MA, Filho RM. Supercritical carbon dioxide extraction of algal lipids for the biodiesel production. Procedia Eng 2012; 42: 1755-61. [http://dx.doi.org/10.1016/j.proeng.2012.07.569]

[185] Crampon C, Boutin O, Badens E. Supercritical carbon dioxide extraction of molecules of interest from microalgae and seaweeds. Ind Eng Chem Res 2011; 50(15): 8941-53. [http://dx.doi.org/10.1021/ie102297d]

[186] Mendes RL, Nobre BP, Cardoso MT, Pereira AP, Palavra AF. Supercritical carbon dioxide extraction of compounds with pharmaceutical importance from microalgae. Inorg Chim Acta 2003; 356: 328-34. [http://dx.doi.org/10.1016/S0020-1693(03)00363-3]

[187] Macías-Sánchez MD, Mantell C, Rodríguez M, Martínez de la Ossa E, Lubián LM, Montero O. Supercritical fluid extraction of carotenoids and chlorophyll a from *Synechococcus* sp. J Supercrit Fluids 2007; 39(3): 323-9. [http://dx.doi.org/10.1016/j.supflu.2006.03.008]

[188] Wu Z, Wu S, Shi X. Supercritical fluid extraction and determination of lutein in heterotrophically cultivated Chlorella pyrenoidosa. J Food Process Eng 2007; 30(2): 174-85. [http://dx.doi.org/10.1111/j.1745-4530.2007.00102.x]

[189] Plaza M, Amigo-Benavent M, del Castillo MD, Ibáñez E, Herrero M. Facts about the formation of new antioxidants in natural samples after subcritical water extraction. Food Res Int 2010; 43(10): 2341-8. [http://dx.doi.org/10.1016/j.foodres.2010.07.036]

[190] Zakaria SM, Kamal SMM. Subcritical water extraction of bioactive compounds from plants and algae: applications in pharmaceutical and food ingredients. Food Eng Rev 2016; 8(1): 23-34. [http://dx.doi.org/10.1007/s12393-015-9119-x]

[191] Herrero M, Cifuentes A, Ibañez E. Sub- and supercritical fluid extraction of functional ingredients from different natural sources: Plants, food-by-products, algae and microalgaeA review. Food Chem 2006; 98(1): 136-48. [http://dx.doi.org/10.1016/j.foodchem.2005.05.058]

[192] Rodríguez-Meizoso I, Jaime L, Santoyo S, Señoráns FJ, Cifuentes A, Ibáñez E. Subcritical water extraction and characterization of bioactive compounds from *Haematococcus pluvialis* microalga. J Pharm Biomed Anal 2010; 51(2): 456-63. [http://dx.doi.org/10.1016/j.jpba.2009.03.014] [PMID: 19376663]

[193] Zakaria SM, Kamal SMM, Harun MR, Omar R, Siajam SI. Subcritical water technology for extraction of phenolic compounds from *Chlorella* sp. microalgae. Molecules 2017; 22(7): 1105. [http://dx.doi.org/10.3390/molecules22071105] [PMID: 28671617]

[194] Joana Gil-Chávez G, Villa JA, Fernando Ayala-Zavala J, *et al.* Technologies for extraction and production of bioactive compounds to be used as nutraceuticals and food ingredients: an overview. Compr Rev Food Sci Food Saf 2013; 12(1): 5-23.

[http://dx.doi.org/10.1111/1541-4337.12005]

[195] Herrero M, Castro-Puyana M, Mendiola JA, Ibañez E. Compressed fluids for the extraction of bioactive compounds. Trends Analyt Chem 2013; 43: 67-83. [http://dx.doi.org/10.1016/j.trac.2012.12.008]

[196] Amaro H, Guedes A, Preto M, Sousa-Pinto I, Malcata F. *Gloeothece* sp. as a nutraceutical source-an improved method of extraction of carotenoids and fatty acids. Mar Drugs 2018; 16(9): 327. [http://dx.doi.org/10.3390/md16090327] [PMID: 30208611]

[197] Pieber S, Schober S, Mittelbach M. Pressurized fluid extraction of polyunsaturated fatty acids from the microalga *Nannochloropsis oculata.* Biomass Bioenergy 2012; 47: 474-82. [http://dx.doi.org/10.1016/j.biombioe.2012.10.019]

[198] Sabeena SF, Alagarsamy S, Sattari Z, *et al.* Enzyme-assisted extraction of bioactive compounds from brown seaweeds and characterization. J Appl Phycol 2020; 32: 615-29. [http://dx.doi.org/10.1007/s10811-019-01906-6]

[199] Sánchez-Camargo AP, Montero L, Stiger-Pouvreau V, *et al.* Considerations on the use of enzyme-assisted extraction in combination with pressurized liquids to recover bioactive compounds from algae. Food Chem 2016; 192: 67-74. [http://dx.doi.org/10.1016/j.foodchem.2015.06.098] [PMID: 26304321]

[200] Michel G, Czjzek M. Polysaccharide-degrading enzymes from marine bacteria. In: Trincone A, Ed. Marine Enzymes for Biocatalysis: Sources, Biocatalytic Characteristics and Bioprocesses of Marine Enzymes 2013; 429-64. [http://dx.doi.org/10.1533/9781908818355.3.429]

[201] Zuorro A, Miglietta S, Familiari G, Lavecchia R. Enhanced lipid recovery from *Nannochloropsis* microalgae by treatment with optimized cell wall degrading enzyme mixtures. Bioresour Technol 2016; 212: 35-41. [http://dx.doi.org/10.1016/j.biortech.2016.04.025] [PMID: 27078205]

[202] Rodrigues MA, Da Silva Bon EP. Evaluation of *Chlorella* (Chlorophyta) as source of fermentable sugars via cell wall enzymatic hydrolysis. Enzyme Res 2011; 2011; 405603.

[203] Mahali M, G S. Extraction Methods and Functional Properties of Protein from *Arthrospira platensis* for ioaBbvailability of algal proteins. International Journal of Pharmacy and Chemistry 2019; 5(2): 20-5. [http://dx.doi.org/10.11648/j.ijpc.20190502.12]

[204] Deenu A, Naruenartwongsakul S, Kim SM. Optimization and economic evaluation of ultrasound extraction of lutein from Chlorella vulgaris. Biotechnol Bioprocess Eng; BBE 2013; 18(6): 1151-62. [http://dx.doi.org/10.1007/s12257-013-0213-8]

[205] Sharma R, Bhunia B, Mondal A, *et al.* Statistical optimization of process parameters for improvement of phycobiliproteins (PBPs) yield using ultrasound-assisted extraction and its kinetic study. Ultrason Sonochem 2020; 60: 104762. [http://dx.doi.org/10.1016/j.ultsonch.2019.104762] [PMID: 31546084]

[206] Hadiyanto H, Suttrisnorhadi S. Response surface optimization of ultrasound assisted extraction (UAE) of phycocyanin from microalgae *Spirulina platensis.* Emir J Food Agric 2016; 28(4): 227-34. [http://dx.doi.org/10.9755/ejfa.2015-05-193]

[207] Zghaibi N, Omar R. Kamal SM azlina M, Biak DRA. Microwave-assisted brine extraction for enhancement of the quantity and quality of lipid production from microalgae *Nannochloropsis* sp. Molecules 2019; 24(3581): 1-21.

[208] Careri M, Furlattini L, Mangia A, *et al.* Supercritical fluid extraction for liquid chromatographic determination of carotenoids in *Spirulina Pacifica* algae: a chemometric approach. J Chromatogr A 2001; 912(1): 61-71. [http://dx.doi.org/10.1016/S0021-9673(01)00545-3] [PMID: 11307988]

[209] Tang S, Qin C, Wang H, Li S, Tian S. Study on supercritical extraction of lipids and enrichment of DHA from oil-rich microalgae. J Supercrit Fluids 2011; 57(1): 44-9. [http://dx.doi.org/10.1016/j.supflu.2011.01.010]

[210] Castro-Puyana M, Pérez-Sánchez A, Valdés A, *et al.* Pressurized liquid extraction of *Neochloris oleoabundans* for the recovery of bioactive carotenoids with anti-proliferative activity against human colon cancer cells. Food Res Int 2017; 99(Pt 3): 1048-55. [http://dx.doi.org/10.1016/j.foodres.2016.05.021] [PMID: 28865616]

[211] Sharma N, Mehariya S, Martino M, *et al.* Extraction of Carotenes and Lutein from Dunaliella salina 2019; 112-8.

[212] Machado FRS Jr, Trevisol TC, Boschetto DL, *et al.* Technological process for cell disruption, extraction and encapsulation of astaxanthin from *Haematococcus pluvialis.* J Biotechnol 2016; 218: 108-14. [http://dx.doi.org/10.1016/j.jbiotec.2015.12.004] [PMID: 26685712]

[213] Mendes-Pinto MM, Raposo MFJ, Bowen J, Young AJ, Morais R. Evaluation of different cell disruption processes on encysted cells of Haematococcus pluvialis: Effects on astaxanthin recovery and implications for bio-availability. J Appl Phycol 2001; 13(1): 19-24. [http://dx.doi.org/10.1023/A:1008183429747]

[214] Verdasco-Martín CM, Díaz-Lozano A, Otero C. Advantageous enzyme selective extraction process of essential spirulina oil. Catal Today 2020; 346: 121-31. [http://dx.doi.org/10.1016/j.cattod.2019.02.066]

[215] Ansari FA, Shriwastav A, Gupta SK, Rawat I, Bux F. Exploration of Microalgae Biorefinery by Optimizing Sequential Extraction of Major Metabolites from *Scenedesmus obliquus*. Ind Eng Chem Res 2017; 56(12): 3407-12. [http://dx.doi.org/10.1021/acs.iecr.6b04814]

[216] Pagels F, Salvaterra D, Amaro HM, *et al.* Bioactive potential of Cyanobium sp. pigment-rich extracts. J Appl Phycol 2020; 32(5): 3031-40. [http://dx.doi.org/10.1007/s10811-020-02213-1]

[217] Zhang W, Wang F, Gao B, Huang L, Zhang C. An integrated biorefinery process: Stepwise extraction of fucoxanthin, eicosapentaenoic acid and chrysolaminarin from the same *Phaeodactylum tricornutum* biomass. Algal Res 2018; 32: 193-200. [http://dx.doi.org/10.1016/j.algal.2018.04.002]

[218] Assunção J, Amaro HM, Lopes G, Tavares T, Malcata FX, Guedes AC. Exploration of marine genus Chroococcidiopsis sp.: a valuable source for antioxidant industry? J Appl Phycol 2021; 33(4): 2169-87. [http://dx.doi.org/10.1007/s10811-021-02435-x]

[219] Choudhary B, Chauhan OP, Mishra A. Edible seaweeds: a potential novel source of bioactive metabolites and nutraceuticals with human health benefits. Front Mar Sci 2021; 8: 740054. [http://dx.doi.org/10.3389/fmars.2021.740054]

CHAPTER 8

Bioactive Compounds from Algae: Potential Applications

K.R. Jayappriyan[1,*], **C. Kurinjimalar**[1], **M. Kaviraj**[1], **M. Vijayakumar**[1], **R. Rajkumar**[2] and **Rathinam Raja**[3]

[1] *Centre for Advanced Studies in Botany, University of Madras, Chennai-600 025, India*

[2] *Department of Environmental Sciences, Bharathiar University, Coimbatore-641 046, India*

[3] *Research and Development Wing, Sree Balaji Medical College and Hospital (SBMCH), Bharath Institute of Higher Education and Research (BIHER), Chromepet, Chennai - 600 044, India*

Abstract: Algae are ancient oxygen-producing photosynthetic organism that lives in water bodies all over the world. The algal biodiversity in the aquatic ecosystem represents a new field of product discovery with the potential to manage human ailments. They are untapped resources, and in recent years, algal culture technology has been a business-oriented field owing to its diverse applications. The consumption of algae is traditional in human diets and dates back thousand years ago. The recent increasing algal demand is due to its safety and functional benefits. Algae are an edible source of food with potential therapeutic activities of anti-oxidant, antiobesity, anti-diabetic, anti-inflammatory, anti-cancer, anti-viral, antifungal, and antibacterial properties, which could be employed as medical ailments. Different types of algal-derived bioactive compounds in a biological system are evidenced by *in vitro* and *in vivo* examinations. Raw algae or its compound are under a pipeline of market production due to their cultivable nature, which actually paves the way to generate larger business in the food and nutraceutical industry. This chapter summarizes several algal bioactive compounds with proven activities against metabolic disorders.

Keywords: Bioactive Compounds, Human Health, Macroalgae, Microalgae, Nutraceutical.

INTRODUCTION

Algae are considered to be a vast renewable feedstock for a variety of bioactive compounds. Algae have the potency to thrive in harsh environmental conditions due to the presence of different compounds as a primary defensive mechanism [1]. From the past evidences, bioactive metabolites of natural origin have shown numerous activities and health benefits, such as antibiotics, anti-oxidants,

* **Corresponding author K.R. Jayappriyan:** Centre for Advanced Studies in Botany, University of Madras, Chennai, India; Tel: +91 44-22202752; E-mail: jayapriyan@gmail.com

antidiabetics, anti-cancer drugs, insecticides, herbicides, immunosuppressants, anti-viral, antiparasitics, and pesticides [2, 3]. In general, secondary metabolites come under the bioactive compound groups, which are highly effective in protecting living cells, tissues, and organisms as it does not involve directly in the growth and development of an organism. Bioactive metabolites can act as potential anti-oxidants and neutralize the free radicals present in the cells. The neutralization of reactive oxygen species will prevent the development of human disorders like cancer and infectious diseases.

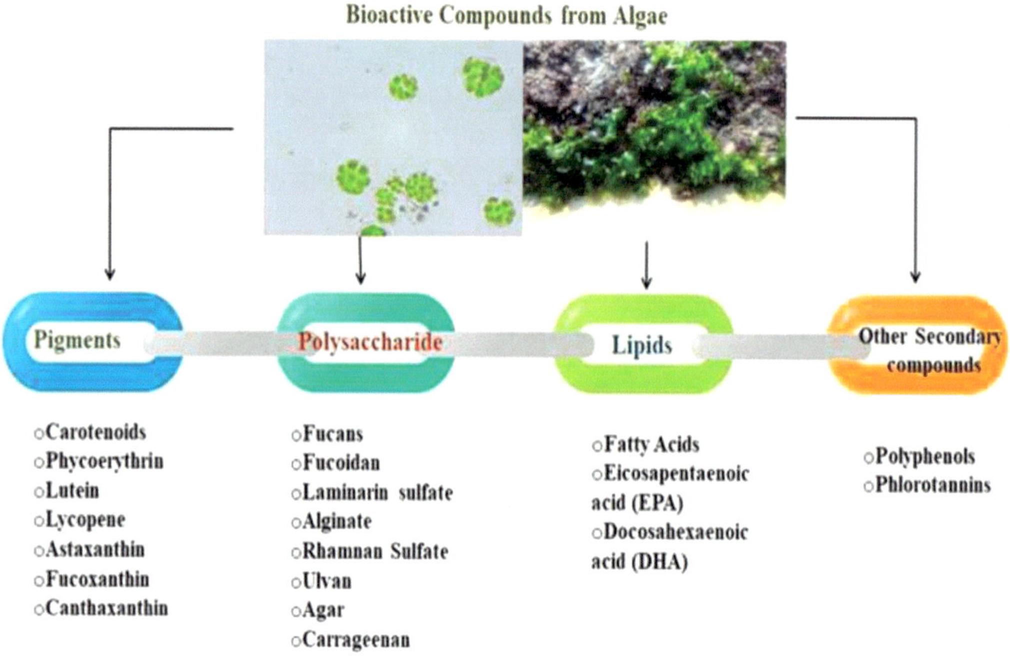

Fig. (1). Types of Bioactive compounds from algae.

The mechanistic tedious principles behind several bioactive compounds are their isolation, purification, and characterization. There will be a huge demand to discover safer, environmental-friendly, renewable, and cost-effective drugs to control the spread of multi-drug resistant microorganisms. Once after the discovery of penicillin, microbes and other organisms are widely studied for creating a new source of bioactive substances. Hence, the naturally available compounds have put forward the preliminary step for the chemical synthesis of bioactive molecules for human utilization. The algae can live in various ecological habitats of mountains, icebergs, freshwater, marine water, marshes, sediments, hot springs, and deserts [4]. They also form a symbiotic association with other organisms for their survival and thus create a particular ecological niche. Algae produce a wide range of natural bioactive compounds, such as alkaloids, phenolics, xanthones, steroids, terpenoids, flavanones, chalcones, carotenoids, flavonoids, flavones, and non-ribosomal peptides. These natural compounds will act as anti-viral, antiprotozoal, antibiotic, antifungal, anticholesterolemic, anti-tumor, immunosuppressor, and antihyperglycemic agents for improving human wellbeing [5] (Figs. **1** & **2**). The current work plans to review the bioactive compounds of algae and their various applications for future production and development.

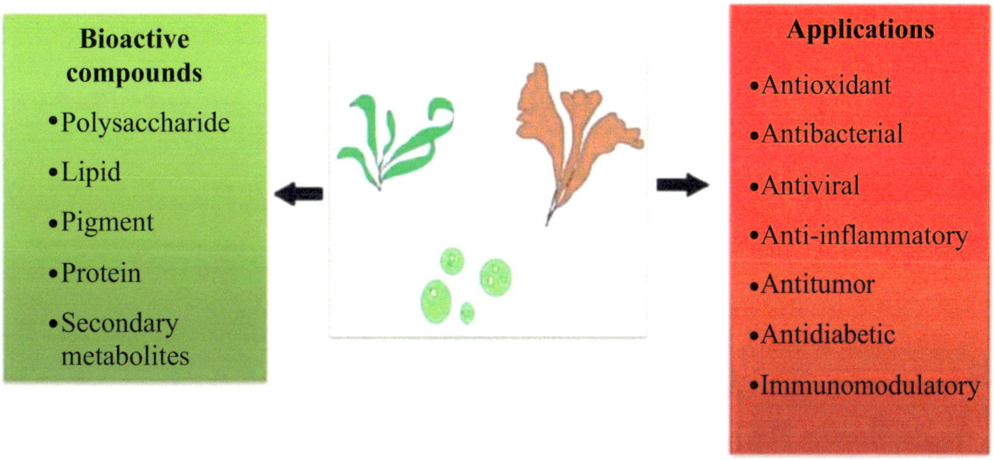

Fig. (2). Applications of Bioactive compounds derived from algae.

ALGAL COMPOUNDS

Lipid-Based Compounds

Fatty Acids (EPA and DHA)

The essential lipids like PUFAs (Poly Unsaturated Fatty Acids) are bioactive molecules, including EPA (Eicosapentaenoic acid) and DHA (docosahexaenoic acid). Fatty acids contain more than 16 carbon atoms with multiple unsaturated double bonds representing PUFAs. PUFAs were classified based on the carbon-carbon double bond and its position in the molecule. The omega-3 fatty acid has a double bond at the third position while omega-6 fatty acids have at the sixth position. Microalgae are a primary rich source of PUFAs like EPA and DHA with biological functions for various aquatic feeders [6]. It is well known that humans need to derive Ω-3 and Ω-6 classes of essential fatty acids that cannot be metabolically synthesized. Different marine microalgae and seaweeds belonging to Chlorophyceae, Dinophyceae, Phaeophyceae, and Rhodophyceae are well explored as rich PUFA content [7]. The different group of algal species, namely

Schizochytrium, Isochrysis, Chaetoceros, Pavlova, Tetraselmis, Nitzschia, Nannochloropsis, Thalassiosira, Aphanizomenon and Haematococcus produces a significant amount of EPA and DHA. Of all the groups of algae, diatom contains EPA and DHA concentrations between 17.5-30.2% and 0.7-6.1%, respectively, with a maximum EPA content of 4.6% of dry weight in diatom species. Guedes [8] reported among the different microalgae studied a high amount of PUFAs in *Pavlova lutheri*. The percentage of PUFAs varies in different algae, with 2.8-4.3% in *Nannochloropsis* sp. and 2.2-3.9% in *Phaeodactylum tricornutum* [9]. Cyanobacteria *Nostoc* is rich in PUFAs, including essential fatty acids, namely alpha-linolenic acid, gamma-linoleic acid, and EPA [10]. In contrast to the cod liver of oil animal origin, a high amount of omega-3 from microalgae reported in *Isochrysis galbana* and *Phaeodactylum tricornutum* with ≥2 folds improved EPA, while the *Crypthecodinium cohnii* yields ≥6 folds improved DHA. Marine fish-derived oil is the major source of PUFAs, which is apparently obtained from the phytoplanktonic diet, which forms the basis of the food chain.

The PUFAs have been recommended as a dietary supplement for proper brain functioning in adults and proper brain development among children. Kris-Etherton *et al.* [11] report that several reports are found in the literature on lowering heart disease when omega-3 fatty acids are administered. The EPA and DHA are given as a dietary nutritional supplement, pharmaceutical, and functional food with anti-oxidant and anti-microbial properties [12]. PUFAs effectively combat non-communicable diseases such as cardiovascular disease; reduce inflammation, lower hypertension, and arthritis with neural development. Therefore, these essential fatty acids are recommended for human consumption to maintain brain and retina fluidity [13]. These two fatty acids protect against neurogenerative diseases like Parkinson and Alzheimer disease [7]. Thus, algae are a demanding new and alternative source of EPA and DHA of natural origin.

The possible cultivation of algal biomass in saline water with high nutritional value and productivity on non-arable land is cost-effective for PUFA production, and can be grown without competing with agricultural land. Commercial production of EPA and DHA are made with the promising algal strain *Nannochloropsis* [14]. The Market in the USA has been a key producer of DHAs, reaching productivity maxima of 240 tons. *Schizochytrium* is a promising candidate for the commercial production of Docosahexaenoic acid due to its capability of producing approximately 20% DHA, which makes them markedly noted for n-3 LC-PUFA production. OmegaTech (USA) develops *Schizochytrium* for low-cost oil production with the brand name DHA Gold. The algal oil has been used as a diet supplement for healthy food and beverages, animal feeds and mariculture products.

Sterols

Sterols are one of the lipid-based compounds usually seen in the majority of Eukaryotic cells. Phytosterols are moreover similar to the characteristics of cholesterol present in animals. The only difference is the presence of an extra carbon atom in the side chain of sterol [15]. Sterol plays a major role as hormone or hormone precursor to maintain the membrane fluidity, permeability, and cell signaling process [16]. Sterols extracted from marine macroalgae like *Himanthalia elongate, Undaria pinnatifida, Porphyra* sp., *Chondrus crispus, Cystoseira* sp., and *Ulva* sp. have shown some promising effects on reducing the total and LDL-C levels [17].

Pigment Based Compounds

Phycocyanin

The phycocyanin is pigmented protein complex increasing oxygen production under low light conditions apart from its usual photosynthetic activity. These water-soluble polypeptide subunits called phycobiliproteins (PBPs) (α and β C-phycocyanin and allophycocyanin) are brilliantly colored and highly non-toxic fluorescent pigment-protein complexes arranged on the surface of the thylakoid membrane. The pigment absorbance wavelength varies from 615 to 620 nm with fluorescence maxima around 640 nm, and the molecular weight is between 70 and 110 kDa. Phycobiliproteins (PBPs) are characterized by Cyanobacteria (Blue-green algae), Rhodophyceae, and Cryptophyceae classes of algae. The protein-rich cyanobacterium *Spirulina* is a major source of inexpensive phycocyanin (blue pigment) with essential amino acids and polyunsaturated fatty acids [18]. The dried algae consist of 55-75% protein, 5-6% lipid, and some vitamins and minerals.

Phycobiliproteins are a group of colored proteins employed in food products as natural pigments. Phycocyanins possess free radical scavenging properties and thus protecting the living cells from oxidative stress. Furthermore, phycocyanin has potent anti-cancer, anti-inflammatory, hepatoprotective, and immuno-modulatory as a stimulation factors and anti-viral properties [19]. *Spirulina* is thus sold in the market as simple dietary supplements for humans and feed for farmed fish and chickens. Therefore, *Spirulina* is grown commercially and harvested, which grows well in warmer climates and alkaline waters. Generally, direct intake of phycocyanin-rich *Spirulina* in non-purified form is proven to be a healthy human food [20]. The purified natural phycocyanin blue color is applied in food applications, including ice cream, candies, cakes, soft drinks, and other milk products. Compared to synthetic dyes, various applications of natural colorants are gaining importance, as they are non-toxic and non-carcinogenic to living cells.

FDA accredited this natural pigment additive in 2013 to meet the demand as natural dyes in industries such as food, textile, pharmaceutical, and cosmetics to evade synthetic dyes' toxic effects.

Consumption of metabolite of cyanobacterium, *Spirulina* sp. possess anti-oxidant and radical scavenging activities are well established [21], from the ancient to recent days since it releases the oxidative stress in the body cell by mitigating excessive ROS found to be one of the natural therapeutic agents. Ou *et al.* [22] demonstrated that alloxan-injured mice treated with oral administration of phycocyanin have reduced blood glucose in serum and controlled cell division expressing p53 gene in the pancreas. This shows the potential of phycocyanin against diabetes. Liu *et al.* [23] reported oxygen-free radical scavenging activity with anti-inflammatory effect of *S. platensis*. The anti-tumor activity of phycocyanin against HeLa cells with inhibitory mechanism changes in the G1 phase, therefore attenuating DNA synthesis in cancer cell proliferation. This shows apoptotic properties such as cell shrinkage, membrane blebbing, nuclear condensation, and DNA fragmentation [24].

Phycoerythrin

The multi-subunit phycoerythrin pigment protein is red fluorescent with light-harvesting property found abundantly in Cyanophyceae and Rhodophyceae. Depending on the source of algae, two categories of phycoerythrin were identified and named as R-PE and B-PE. It is widely used in biomedical applications for labeling the antibodies and receptors to carry out various immunoassays [25]. Several studies on Phycoerythrin revealed that it showed promising applications in human health. Biological properties of natural phycocyanin include non-toxic fluorescent protein pigment with potent anti-oxidant, anti-diabetic, anti-inflammatory, anti-cancer, antimalarial, anti-obese, antiangiogenic, and hepatoprotective activities [10, 26]. Regardless of phycoerythrin origin, it could be considered an effective nourishment with essential pharmaceutical and nutraceutical importance.

Carotenoids

The natural carotenoid is red-orange or yellow-colored pigments of algae and plants are bioactive compounds with six carbon rings. Carotenoids are protected from Ultraviolet rays damage when applied externally via skin or by dietary means [27]. Two major carotenoid groups based on chemical structure include carotenes (consist of carbon and oxygen) and xanthophylls (oxygenated derivatives). The function of carotenoids in photosynthetic organisms is to carry out photosynthetic reactions and protect the skin cells from the exposure of high light intensities. β-carotene is a member of primary and necessary carotenoids in photosynthesis, which is important for good health and is extensively used in pharmaceutical, cosmetics, and food industries [28]. Several types of carotenoids are used in food applications, and most of these carotenoids possess anti-oxidant activity. There are about 400 known carotenoids have been discovered to date, among these, some are used for commercial purposes like β-carotene, astaxanthin, lutein, lycopene, bixin, and zeaxanthin. Carotenoids show promising results in anti-inflammatory, anti-tumor, and chemo-preventive activities [28, 29]. As anti-oxidants, carotenoids neutralize the free radicals therefore preventing or slowing down chronic diseases, cellular damage, and aging. Carotenoids possess antibacterial, anti-viral, antifungal, antioxidative, anti-inflammatory, and anti-tumor properties. It is assumed that carotenoids are thought to protect cells from oxidative stress by quenching singlet oxygen damage with various mechanisms [30]. Consumption of carotenoids can lower the risk of inflammation, heart disease, cancer, type 2 diabetes, obesity, Alzheimer's disease, and Parkinson's disease, and it is also used as a therapeutic agent [31]. In nutraceutical industries, the production profit was grown from $562 million in 2014 to $665 million in 2019 at a 3.4% annual growth rate. In cosmetics, carotenoids are widely used as an ingredient in creams and lotions, and it is also used in cosmetic products for their anti-oxidant properties and nutritional value to the skin and hair.

Beta Carotene

Beta-carotene is one of the carotenoids that belong to a group of more than 600 compounds with biological activities. β-carotene, a polyene compound, is a secondary metabolite with a structure of $C_{40}H_{56}$ possesses synthesized richly in microalgae. *Dunaliella salina* is the best commercial source than any other available source in nature for natural β-carotene, producing up to 14% of its dry weight under high light intensity, increased temperature, high salinity, and nutrient deficiency [32]. Natural β-carotene from *Dunaliella* synthesizes 9 cis isomers, apart from its all-trans isomers, which are stronger anti-oxidants and could not synthesize chemically. Anti-oxidant activity of β-carotene helps to prevent free radical damage in the body cells and stimulation of immune function. The major biological activities of β-carotene include anti-oxidant, anti-angiogenic, and anti-tumor properties [32]. The mode of action is interfering with serum cytokine levels and inhibiting activation and nuclear translocation of transcription factors by the action of β-carotene activity. The anti-diabetic activity of β-carotene against type-2-diabetes Mellitus (T2DM) is evidenced from the previous study [33]. The first industrial production of β-Carotene is obtained from the microalga *D. salina* [34]. Natural β-carotene has many beneficial roles in cataract and in age-related macular degeneration. It also influences the biochemical pathway and shows increased anti-cancer activity than its synthetic alternative [35].

Lutein

Carotenoids may be composed of cyclic groups and oxygen-containing functional groups. Thus, hydrocarbon carotenoids are termed carotenes, while the oxygenated derivatives are identified as xanthophylls (lutein), oxi-groups (cantaxanthin) or a combination of both (astaxanthin) [36] (Table **1**). The primary carotenoids are denoted as structural and functional components of the cellular photosynthetic apparatus (xanthophylls). The microalgae produce secondary carotenoids after exposure to particular environmental stimuli. In general, green algae are intrinsic protein complexes associated with chl a and b and xanthophylls, particularly lutein, and are mentioned as Cab (for chl a/b) or Light-harvesting complex(LHC) proteins. Thus, lutein serves in the absorption of light and has an essential role in secondary photosynthesis. The carotenoid lutein is produced from microalgae, such as *Chlorella protothecoides*, *Chlorella vulgaris*, *Chlorella zofingiensis*, *Scenedesmus almeriensis, Chlorococcum citriforme* and *Muriellopsis* sp [37]. The important medical application of lutein is its anti-oxidant activity. It can reduce the number of free radicals that can interact with other intracellular compounds like DNA, proteins and unsaturated lipids. Adding to this, it has been intended that 6 mg/day is essential to decrease the risk of cataract development and macular degeneration [38]. It also protects from the photo-damage and supports the transmission and visual communication during the development of the eyes. Lutein has also been suggested to show a significant function in the brain's cognitive functions. These functions include mainly neuroprotective effects, *viz.* capability to alter cell membrane fluidity, oxygen diffusion and ion exchange, and its effects on intracellular communication and metabolic pathways in the brain [39]. Though other plants produce lutein, algae are found to be suitable for renewable and feasible production as of now; however, several encapsulation tools are under practice to enhance its bioactivity and industrial applications.

Lycopene

Lycopene is a carotenoid that is responsible for the red to pink colors seen in tomatoes, pink grapefruit, and other foods. However, the major source is tomato which is considered an effective sunscreen agent, and other biological activity includes anti-oxidant, cardioprotective, anti-carcinogenic and anti-atherogenic properties. Algal lycopene is very meagerly reported, nevertheless *Chlorella marina* shows strong anti-oxidant and anti-inflammatory effects in high cholesterol-fed rats [40]. Algal lycopene is one of the promising alternatives for commercial production of lycopene since one of the source materials, such as tomato, is an edible and demanding market product. Moreover, lycopene from edible marine microalgae *Chlorella marina* has shown higher bioactivity than the tomato, particularly lipid-lowering activity in mice due to the presence of an isomeric mixture of cis and trans forms in the algal lycopene. Hence, the algae are an ideal candidate for lycopene production, and its several potentials are yet to address in the future.

Table 1. Pigment class of algal compounds and its uses.

Name of the Algae	Pigment compounds	Application	References
Undaria pinnatifida	Fucoxanthin	Strong radical scavenging activity	[51]
Cylindrotheca closterium	Fucoxanthin	Potent anti-oxidant activities	[55]
Porphyra sp.	Phycoerythrobilin	anti-oxidant activity	[56]
Spirulina platensis	Phycocyanin	Anti-oxidant, anti inflammatory and anti-cancer activity	[57]
Gracilaria chilensis		Anticancer	[58]
Porphyridium sp.	Phycoerythrin	immunomodulatory, anti-inflammatory, hypocholesterolemic, anti-microbial and anti-viral properties	[59]
Dunaliella salina	Carotenoids - β-Carotene	Anticancer Anti-oxidant immunomodulatory activity, hepatoprotective agent, and multivitamin preparations	[28]
Chlorella protothecoides	Lutein	Anti-inflammatory Anti-oxidant	
Chlorella zofingiensis	Canthaxanthin	Anti-oxidant and anti-cancer activities	
Nannochloropsis salina, Nannochloropsis oculata, Nannochloropsis gaditana	Canthaxanthin	Tanning cosmetics and cosmeceutics	[60]
Haematococcus pluvialis	Astaxanthin	Anti-oxidant Active against liver neoplasms Cardiovascular health	[3]
Dunaliella salina	β-carotene	Antioxidant	[61]
Chlorella marina, Anabaena vaginicola	Lycopene	Anti-oxidant, Anti-ageing Sunscreen	[40, 62]

Astaxanthin

Astaxanthin is a xanthophyll compound that is similar to the other carotenoids like zeaxanthin, β-carotene, and lutein, with many physiological and metabolic functions. The source of astaxanthin (3, 3'-dihydroxy- β, β'-carotene-4, 4'-dione) from unicellular freshwater microalga *Hematococcus pluvialis* and it is producing nearly about 4% of its dry wt. It is recognized as king of anti-oxidants with 65 times more potent than vitamin C and 10 times more than other carotenoids [41]. The biological activity of astaxanthin for human and animal benefits includes protection against UV light, singlet oxygen quenching, antilipid peroxidation activities, anti-cancer properties, cardioprotective, anti-Diabetic, anti-inflammatory, immune boosting activity and reproductive behavior in the aquatic animals along with its food and cosmetic applications [42, 43]. The growth inhibition of *Helicobacter pylori* in mice models using astaxanthin from *H. pluvialis* was reported by Kang and Kim [44]. It is widely used as anti-oxidant compound and has the potential to cure and prevent several health problems. Gammone *et al.* [45] reported cardioprotective activity of astaxanthin with considerably reduced blood pressure. This is achieved with ingestion of 50 mg/kg of astaxanthin for 14 days. The remarkable anti-diabetic activity of astaxanthin in mice fed with a high-fat, high-fructose diet can decrease triglycerides and blood glucose levels [46]. The US Food and Drug Administration (FDA) in 1987 approved the market sale of astaxanthin as a feed additive in aquaculture and subsequently approved it as a nutraceutical product. The global carotenoid sale in US increased from 1.5 billion in 2017 and is expected to reach US $2 billion by 2019. This is due to the health benefits offered by carotenoids. Astaxanthin seems to be a mandated bioactive compound whose cost ranges from 2500 to 7150 $ per kilogram based on its purity [47].

Fucoxanthin

Fucoxanthin ($C_{42}H_{58}O_6$) is a light-harvesting pigment molecule, xanthophyll of brown algae (Phaeophyceae) and diatoms (Bacillariophyceae) provides brown color to the algal group. A wide application of natural fucoxanthin in biomedical and pharmaceutical industries is rising, wherein the major resource is mainly achieved from naturally grown brown seaweeds and seaweed farming. The biological activities of fucoxanthin comprise of anti-microbial, anti-oxidant, anti-inflammatory, hepatoprotective, anti-cancer, anti-obesity, anti-diabetes, and cardiovascular protective effects [48 - 50]. This activity of fucoxanthin is affected by altered genetic expression at a cellular metabolic level. A report confirmed that 5-10% fucoxanthin content in *Undaria pinnatifida* acquired with high radical scavenging activity from its fucoxanthin metabolites, like fucoxanthinol and halocyn-thiaxanthin. The radical scavenging activity of fucoxanthin is because of the dual allenic bonds at the C-7' position of its structure [51]. The anti-angiogenic activity of fucoxanthin and the anti-cancer activity of fucoxanthin and fucoxanthinol is mainly mediated by the mechanism of cell cycle arrest followed by apoptosis, and metastasis will occur [52]. Of many *in vitro* studies, fucoxanthin was shown notable activity against Caco-2, DLD-1, and HT-29, human colorectal adenocarcinoma cell lines. Fucoxanthin inhibits the proliferation of cancer cells by limiting intercellular communications and intracellular signaling, thereby arresting cell growth [52]. Therefore, fucoxanthin is a potent anti-cancer agent. In addition, Fucoxanthin is a potent anti-diabetic agent, that can restore normal blood glucose levels by mRNA upregulation of glucose transporter 4 (GLUT4) proteins. Oral administration of fucoxanthin helps in reducing the level of abdominal white adipose tissue in obese mice through higher expression levels of uncoupling protein 1 (UCP1) and β3-adrenergic receptor (Adrb3) in fat tissues of the abdomen [45]. The few diatoms species also produce fucoxanthin, however, the cultivation and extraction are not up to the mark as a market requirement. Hence, the super-producing strain or strain improved diatom species should be developed for achieving renewable production.

Canthaxanthin

Canthaxanthin is a diketo-carotenoid with nine conjugated C=C bonds at the core of the molecule, and it is responsible for the orange/red colour of eggs, Salmon and rainbow trout flesh [53]. The high value of canthaxanthin is due to its wide applications in are human nutraceutical, cosmetic, and animal feed industries. It has high radical scavenging activity with unique anti-oxidant properties than other carotenes and xanthophylls. Canthaxanthin-rich algae are *Botryococcus braunii, Chlorella vulgaris, Chlorococcum, Haematococcus pluvialis, Nannochloropsis, Scenedesmus obliquus,* and *Chlorococcum* [54]. Canthaxanthin has the potency to enhance gap junction communication between cells so that it counteracts the cancer cells easily. The anti-cancer activity of canthaxanthin is reported in JB/MS, B16F10 and PYB6 cell lines, significantly reducing the growth rate by apoptosis. In most of the algae, canthaxanthin is produced inadequate when compared to other pigments. Thus, canthaxanthin production requires metabolic rearrangements in the algae to enhance the production, facilitating effective utilization.

Carbohydrates Based Compounds

Carbohydrates are biodegradable, water-soluble and functionally active compounds with biomedical applications. Seaweeds or marine macroalgae are majorly grouped into Chlorophyceae (green seaweeds), Phaeophyceae (brown algae, Chromophyta) and Rhodophyceae (red macroalgae). Seaweeds from marine resources are an abundant resource of Polysaccharides, which are classified into three groups, namely structural polysaccharides, intercellular mucilage, and storage polysaccharides. Microalgae have a negligible amount of polysaccharides production when compared to macroalgae, however, they are mostly exploited for their other biochemicals like protein, lipids, pigments and other secondary metabolites. Most of the microalgal groups (*Arthrospira/Spirulina*, *Chlorella vulgaris, Phaeodactylum tricornutum, Cylindrotheca closterium, Odontella aurita, Porphyridium purpureum, Anabaena, Aphanocapsa, Cyanothece* and *Cochlodinium polykrikoides*) produce polysaccharides. This exo- or extracellular polysaccharide, sPS, sulphate containing exopolysaccharide, proved to have several important biological activities. They are anti-viral agents, health foods, anti-oxidants, anti-inflammatory and immunomodulatory properties, and are also being used as important biomaterials like lubricants for bone joints, or even as drag-reducing substances [63].

Seaweeds or macroalgae are an excellent source of phycocolloids, and most of them being sulphated polysaccharides (sPS) reported for their antibacterial, anti-oxidant, anti-inflammatory, anti-viral, immune stimulant, immunomodulating, anticoagulant, antithrombotic and anti-tumor and cancer preventive activities [64, 65] (Table **2**). These biological activities show promising applications in pharmaceuticals, therapeutics, and regenerative medicines. According to Markets and Markets dot com, the seaweed cultivation market size is estimated to be valued at USD 16.7 billion in 2020 and is projected to reach USD 30.2 billion by 2025. The following macroalgal polysaccharides had superior bioactivities. Instead, polysaccharides were revolving in markets for various other commercial applications like cosmetics, toothpaste, soaps, shampoo and food products like milk, cheese, ice cream, dessert gels, candies, fruit juices, jams, jellies, sauces and processed meats, as well as in medical consumables like capsules and tablets, lotions and creams.

Fucans

Sulfated fucans are polysaccharides that possess neutral sugars and a trace amount of L-fucose with sulfated ester groups. It is commonly seen in the cell wall constituent of brown algae exists, along with alginates. It exhibits variegated activity against the treatment of anticoagulant, inflammation, radical scavenging, anti-viral, anti-oxidant, anticoagulant, antithrombotic, and anti-cancer effects [66].

Fucoidan

Fucoidan is a sulfated polysaccharide isolated from brown seaweeds. It is the constituent of the cell wall which is consumed as food supplements [67]. This oligosaccharide consists of sulfated fucose with galactose, glucoronic acid and xylose [68]. The structure of fucoidan consists of α-1, 3-linked sulphated L-fucose; a repeating sequence of alternating α (1→3) and α (1→4) glycosidic bonds is also possible [69]. The bioactivity of fucoidan purely depends upon its structure and sulphonation content. The rich source of fucoidan is *Laminaria digitata, Laminaria japonica, Fucus vesiculosus, Himanthalia elongata, Fucus serratus, Ascophyllum nodosum, Sargassum micracanthum, Ecklonia kurome, Lessonia nigrescence, Cladosiphon okamuranus* and *Turbinaria conoides* [70].

Commercially, it has been isolated on a large scale for various foods and cosmetics applications from *C. okamuranus* [71]. Fucoidan has been widely studied for its anti-oxidant, immunomodulatory, anti-inflammatory, anti-viral, anti-cancer and anti-oxidative effects, which find huge applications in food, cosmetic and medical industries [72, 73].

Research revealed that the sulfated polysaccharide of *Laminaria japonica* can act as a good anti-oxidant by inhibiting and removing the reactive oxygen radicals. Fucoidan from *Fucus vesiculosus* shows an immunomodulatory effect by oral intake due to its cell membrane binding ability. Thereby, activating the immune system and inhibiting viral growth [74]. Further, the anti-viral activities of fucoidan are studied against human immunodeficiency virus (HIV), Herpes simplex virus (HSV) and bovine viral diarrhea virus. The mechanism of anti-inflammatory activity of fucoidan underlies based on the radical scavenging property. Fucoidancan, as an alternative source to heparin, revealed an increase in clotting time [67]. The fucose from *Ecklonia cava* shows anti-tumor activity against murine colon carcinoma (CT-26), human leukemic monocyte lymphoma (U-937), and mouse melanoma (B-16) cell lines [75]. In another study, fucoidan from *Undaria pinnatifida* revealed anti-cancer properties against PC-3, HeLa, A549 and HepG2 cancer cell lines [76]. The mechanism of anti-tumor activity of fucoidan is proposed by inhibition of tumor growth along with tumor size reduction through mitochondria-mediated apoptosis induction by activating the caspases [69, 77]. Fucoidan has several anti-cancer activities against various cell lines however, detailed bioavailability and *in vivo* studies are yet to reveal the facts.

Laminarin Sulfate

Polysaccharide laminarins are a food source of brown algae composed of β-glucans, with Light molecular weight (LMW) linear polysaccharide chains consisting of D-glucopyranose linked by 1, 3 β-glycosidic linkages with an average of 20–25 disaccharides. The rich source of laminarin sulfate is found mostly in genera such as *Ascophyllum, Undaria,* and *Fucus*. Like the plant polysaccharides, laminarin is able to activate the macrophages, thus holding the strong immunostimulatory, anti-tumor and wound healing activities [78]. Other different biological activities such as anti-oxidant, antibacterial, lipid-lowering and anti-coagulant properties are also seen in Laminarin. Notably, laminarin has shown activities like stimulating, regenerating, energizing, and conditioning human keratinocytes and the dermis fibroblasts. Accordingly, Laminarin isolated from the *Laminaria coriacea* improves the binding of human skin fibroblast cells for wound healing. Holdt and Kraan [79] have reported that the presence of laminarin gives defense against strong irradiation and reduces the cholesterol levels in blood serum.

Alginate

The polysaccharide alginate is a rich source of brown seaweeds, namely *Ascophyllum nodosum*, *Laminaria hyperborea*, *Laminaria japonica*, *Macrocystis pyrifera,* and *Laminaria digitate* [80]. It is mainly composed of guluronate and mannuronate residues. The anti-oxidant property of Alginate oligosaccharides (AOS) has received significant attention, and possesses additional beneficial characteristics such as anti-inflammatory activity, and bacteriostatic function; the unique feature is its antineoplastic activity [81]. Interestingly the authors have revealed that AOS suppressed the tumorigenicity of prostate cancer cells via the Hippo/ YAP pathway in both *in vitro* and *in vivo* conditions. Apart from their bioactivity, the alginates have been widely used in making biomaterials like tablet capsules and scaffold preparation due to their low toxicity, biocompatibility, rapid gelation and the presence of calcium ions [80]. Alternatively, Selvaraj *et al.* [82] reported the use of Alginate as nanoparticle carriers to enhance anti-cancer activity against Human Leukemic Cancer U937 Cells. Moreover, a commercial algin-based wound dressing material available in the market is Sorbsan VR. The physical properties of Alginate pave the way for the multifaceted role in biomedical applications in both bioactive and biomaterial preparations.

Rhamnan Sulfate

The green seaweed *Monostroma nitidum* is a rich source of polysaccharide rhamnan. It consists of L-rhamnose sugars linked with the α-1,3 carbons. It exhibits anticoagulant, anti-viral and anti-tumor properties [83]. *In-vitro* studies reveal that the rhamnan sulfate was highly effective against a virus like HIV, Human cytomegalovirus and Herpex simplex virus. It averts obesity and reduces the elevation of LDL cholesterol and Blood Glucose levels [84]. Likewise, Suzuki and Terasawa [85] have elaborately discussed the biological activities of Rhamnan Sulfate and reported the Anti-hyaluronidase, Antitumor, Anti-obesity, Anti-thrombosis, Anti-hypercholesterolemia and Anti-hyperglycemia. Terasawa *et al.* [86] have recently reported that Rhamnan sulphate inhibits influenza virus infection by reducing the adsorption and promotes antibody production, suggesting that it is a potential candidate for the treatment of influenza virus infections.

Ulvan

The sulfated hetero polysacharide Ulvan consists of disaccharide units of D-glucuronic acid or L-iduronic acid attached with a sulfated 1,4 L-rhamnose residue with traces of D-xylose and D-glucose. About 18% of the carbohydrate content in green algae accounts for ulvan. The unique characteristic of the ulvan is the existence of rare sugars like sulphated rhamnose and iduronic acid in their backbone [87]. The rich source of ulvan is green algae *Ulva pertusa* and *Ulva lactuca* [88]. The bioactivity of ulvan includes antihyperlipidemic and anti-oxidant activities. It is widely known for its anti-viral activity against the Influenza virus and Herpes simplex virus 1. Ulvan can act as an antagonist against plant pathogens by activating the cellular signaling pathways of intracellular plant defense. In hypercholesterolemic rats, ulvan has reduced the total cholesterol (TC), triglycerides (TG), and total lipid levels by 46, 69, and 30%, respectively [89].

Agar

Gelidium and *Gracilaria* are marine red algae that possess a large amount of agar in their cell wall, as a linear polysaccharide. It has been used as a food source since 300 A.D. by the Chinese people. It is composed of alternating 3, 6-anhydr--L-galactose, and D-galactose units linked by alpha-(1,3) and beta-(1,4) glycosidic bonds [90]. This is widely used in food applications. The sulfated-galactan present in the agar exhibits bioactive properties. The anti-coagulant activity of agaropectin from *Gelidium amansii*, was reported at the concentrations of 3.125-25 mg/kg of rabbit [91]. It takes 690 sec for coagulation on heparin-treated blood, however, only 670 sec on agaropectin treatment. In general, agar as an individual does not possess any bioactivity thus, it is mostly utilized as gelling or texturizing agents in various microbiological mediums.

Carrageenan

Carrageenan is a linear polysaccharide extracted from red seaweed. It is composed of alternating units of D-galactose and 3, 6-anhydro-galactose form the core structure of carrageenan along with α-1,3- and β-1,4-glycosidic linkages [6]. It is classified into two groups, namely kappa and the lambda family. Based on the structure, they were further classified into various types known as iota, kappa, lambda, mu, nu, and theta carrageenans [88]. The rich sources are *Kappaphycus* sp., *Chondrus crispus*, *Gigartinam stellata, Euchema gigartina*, *Iridaea* sp., *Stellate iridaea*, *Hypnea*, and *Sarconema* [92]. In the food industry, Carrageenans are generally used as gelling, thickening, emulsifying and stabilizing agents [93]. Besides its commercial interest, the bioactivities of the compound include anti-hyperlipidemic, anti-coagulant, immuno-modulatory anti-viral, anti-oxidant and anti-cancer activities [94]. Carrageenan-administered rabbits showed reduced lipid and cholesterol in 18 weeks study period. The high sulfate content of carrageenan was reported as a potent anti-coagulant. Generally, phycocolloid carrageenan of red algae, have been extensively investigated for their vast array of bioactivities, however, the latest report shows their disaccharides (carrabioses and carrabiitols), obtained from extracts from *Hypnea musciformis*, *Iridaea undulosa* and *Euchema spinosum* as potential cytotoxic and anti-tumor compounds. Among the two disaccharides taken, κ-carrabiose was the most effective, showing high cytotoxic properties, killing the LM2 tumor cells through an apoptotic pathway [95].

Porphyran

Porphyran is a sulfated polygalactan available in red algae, and it consists of a linear backbone of 3-linked β-D-galactose units with alternating 4-linked α--galactose-6-sulfate or 3,6-anhydro-α-Lgalactose units. Porphyran can be extracted in large amounts from the cell walls of *Porphyra* sp., and it exhibits hypolipidemic, anti-cancer and anti-inflammatory activities in human beings. Jiang *et al.* [96] investigated the anti-inflammatory property of porphyran using mouse models and revealed that Porphyran impedes the production of nitric acid in macrophages by blocking the NF-B activation in the mouse macrophages of RAW264.7 cells. Furthermore, the cytotoxic study shows that porphyrin does not produce any toxicity against normal cells, whereas it is highly effective against cancer cells by inducing cell death. In this accordance, Dan *et al.* [97] reported that Porphyran isolated from *Pyropia yezoensis* is triple-helical sulfated hetero-galactopyranose, named PYP. On further degradation, PYP generates two low molecular weight derivatives, namely PYP-20 and PYP-50, respectively and one higher molecular weight PYP. All these porphyrans had no toxicity in normal human liver cells (HL-7702) and showed anti-tumor effects on Hep3B, HeLa and MDA-MB-231as reported. Adding to this, porphyran and oligo-porphyran have a series of pharmacological and biological functions, such as antioxidation, anti-cancer, antiaging, antiallergic, immunomodulatory, hypoglycaemic, and hypolipidemic effects [98].

Mycosporine Like Amino Acids (MAAs)

Mycosporine-like amino acids are small, water-soluble molecules with imino-carbonyl derivatives of cyclohexanone. It consists of amino acids bound to a chromophore with absorption maxima between 310 and 360 nm. The characteristic feature of MAA is the UV-absorbing property, which absorbs the UV radiation and dissipates the energy, thereby protecting from UV-induced damage. The MAA family includes mycosporine-glycine, palythinol, asterina-330, porphyra-334, palythine and shinorine types. Bioactive compounds like Porphyra-334 and shinorine were first identified in the marine red algae *Porphyra tenera* and *Chondrus yendoi* [108]. Shinorine is the most commonly seen and abundant MAA in several microalgae species. Approximately 40 MAAs are known to date and are widely seen in Rhodophyta. The microalgae that produce MAA are *Ankistrodesmus spiralis*, *Chlorella minutissima*, *Scenedesmus* sp. and *Scotiella nivalis*.The skin protective effect of MAAs in sunscreen against harmful effects of UV-A radiation. More recently, a chemical structure of a new MAA, called prasiolin, was reported in the terrestrial macroalga *Prasiola calophylla*, which is a member of the Chlorophyta (Trebouxiophyceae) [61]. Likewise, Hartmann *et al.* [109] isolated two novel MAAs from *Klebsormidium crenulatum,* namely klebsormidin A and klebsormidin B. This MAA is found to hit the cosmetic market in the near future because of its unique high molar UV absorption properties.

Table 2. Polysaccharide Class of Algae Compounds and their Applications

Types of Polysaccharide	Name of Algae	Biological Properties	References
Fucoidan	*Fucus vesiculosus*	Anti-metastatic, Wound contraction and healing properties, Anti-oxidant	[99]
	Laminaria japonica	Competitive inhibition of tyrosinase	[99]
	Canistrocarpus cervicornis	Anticoagulant, antioxidant, anti-proliferative	[100]
	Dictyopteris delicatula	Anticoagulant, antioxidant, antitumor, anti-proliferative	[101]
	D. polypodioides	Anti-tumor	[102]
	Pelvetia siliquosa	Anti-diabetic	[103]
	Eckloniasto lonifera	Anti-hypertensive activity	-
	Pelvetia fastigiata	Anti-viral	[104]
	Turbinaria conoides	Anti-oxidant	[105]
Rhamnnan	*Monostroma* latissimum	anti-HSV-1, anti-HCMV and anti–HIV-1 activity	[106]
Arabinans	*Codiumlatum* and *C. fragile, C. adhaerens*	anti-HSV-1 activity	-
Galactan	*C. fragile*	Inhibition of replication of HSV-2	-
Alginate	*Turbinaria conoides*	Anti-oxidant activity	[99]
	Sargassum horneri	Reduce macrophage activation	-
Sulfated galactan	*Codium cylindricum*	Anti-coagulant activity	[107]
Carrageenan	*Grateloupia filicina*	Antiangiogenic activity	-

Other Secondary Metabolites

Phenolics

Phenolics were sometime called as polyphenols, identified in a wide variety of marine algae [110]. The structures of polyphenols exhibit themselves as simple molecules, such as phenolic acids and other simple polyphenolic compounds, while the more complex is Phlorotannins, which consist of polymeric structures made up of phloroglucinol units (1,3,5-trihydroxybenzene), typically produced by marine brown algae or Phaeophyceae members. Phlorotanins are hydrophilic compounds that contain both phenolic and phenoxy groups in their structure and can be divided into four large subgroups: phlorethols and fuhalols (they have both bonds), fucols (they contain phenyl bonds), fucophloroethols (they have ether and phenyl bonds) and eckols (with dibenzodioxin bonds) [111]. Polyphenols are well known for their anti-oxidant property as it protects the human cells and organs from Reactive oxygen species (ROS) by neutralizing the free radicals. Algae comprise different types of phenolic metabolites like Tannins, Terpenoids and each compound shows different bioactivity in human health [112]. Phlorotannin isolated from the brown seaweed like *Ecklonia cava*, *Ecklonia stolonifera*, *Eisenia bicyclis*, *Sargassum thunbergii*, *Undaria pinnatifida*, and *Laminaria japonica* was known for its significant roles in algae which include UV protection, metal sequestration, deterrence from herbivory and defense against microbes and epiphytes. But from a bioactive perspective, it shows promising results in the studies like anti-HIV, anti-oxidant, anti-cancer, anti-bacteria, radio-protective, anti-diabetic, anti-inflammatory, enzyme inhibition activity and anti-allergic properties. Likewise, microalgae were revealed to have 4 simple phenolics (gallic, syringic, protocatechuic, and chlorogenic acids). Different strains of microalgae were reported to contain between 0.54 and 5.8 mg gallic acid g^{-1} dry weight. On the other hand, higher values of phenolics were reported from *Chlorella pyrenoidosa* (13.2-25.8 mg g^{-1} DW) and from *Arthrospira platensis* (17-43.2 mg g^{-1} DW) under similar circumstances. While the other study was done in crude extracts of several microalgae (*Ankistrodesmus* sp., *Spirogyra* sp., *Euglena cantabrica* and *Caespitella pascheri*) and cyanobacteria (*Nostoc* sp., *Nostoc commune*, *Nodularia spumigena*, *Leptolyngbya protospira*, *Phormidiochaete* sp., and *Arthrospira platensis*), *Euglena cantabrica* showed the highest concentration of phenolic compounds, particularly gallic and protocatechuic acids (5.87 and 2.97 mg per gram of dried biomass, resp. Unlike microalgal carotenoids, anti-oxidants of microalgal phenolics were not well established. Most of the microalgae contribute a huge presence of gallic acid, representing the strong anti-oxidant activity. In general, several extracts of different marine algae were rich in phenolic compounds like shinorine, porphyra-334, palythene, eckstolonol, eckol, sargachromenol, tetraprenyltoluquinol chromane meroterpenoid, scytonemin, and sargaquinoic acid, showing excellent photo-protectivecum anti-photoaging capacity [113]. Of the compounds, all showed multifunctional pharmacological roles like anti-diabetic, anti-microbial, antihypolipidemic, anti-inflammatory and anti-cancer activities.

Lectins

Lectins belong to the group of glycoproteins; it binds non-covalently with the carbohydrates to form agglutination. Lectins extracted from marine algae are monomeric in nature. The low molecular weight proteins comprise a large amount of acidic amino acids, and most of them show definite attraction towards glycoproteins. Lectins from several red marine algae like *Eucheuma serra, Acrocystis nana*, and *Solieria filiformis* show anti-cancer properties, and they can also act as drug delivery agents based on their carbohydrate specificity [114, 115]. Lectin shows some other medicinal properties like anti-viral, anti-inflammatory and anti-microbial activities [116, 117]. Besides, lectins isolated from the *Solieria robusta* act as mitogenic agents on mouse spleen lymphocytes and are useful for the detection of disease-related alterations of glycan synthesis and cell markers for diagnostic purposes, including infectious agents like viruses, bacteria, fungi and parasites [118]. Even though another activity of lectin was reported in literature, it may be ideally utilized for the theranostic purpose because of its unique binding properties.

CONCLUSION

Algae are emerging for their novel compounds because of their huge abundance, less toxic, renewable nature, edibility and eco perspectives. Despite, some of the faunas, like seaweeds, were long being consumed as a food material by human beings and their residues were also incorporated in most fast-moving consumer products. Algae are evolutionarily old organisms compared to plants but possess unique biochemical properties facilitated through their eco tolerance behavior. Though the present chapter discussed the potential bioactive compounds of the algae, nevertheless very few compounds were successfully taken up for commercialization. The fact that the high-value compounds exist in the algal groups having few disadvantages like poor water-solubility, chemical instability, and low bioavailability limits the competent usage in many food and beverage products which results in the futile of the product. Therefore, micro or nanoformulation is the future's promising tool to intensify the robustness of the original bioactive molecule and further which also improve the product formulation, thereby it becomes sustainable and functional food. Adding to this, algal-based products are well established in the international market because of their multiple applications, which pave the way to further novel insights on marine micro and macro algae since they are considered to be a repository for bioactive compounds. Hence, algae are significant contenders for holding a high-quality bioactive compound that is yet to discover in the near future.

CONSENT FOR PUBLICATION

Not applicable.

CONFLICT OF INTEREST

The author(s) declare no conflict of interest, financial or otherwise.

ACKNOWLEDGEMENT

Declared none.

REFERENCES

[1] Wang SK, Stiles AR, Guo C, Liu CZ. Microalgae cultivation in photobioreactors: An overview of light characteristics. Eng Life Sci 2014; 14(6): 550-9. [http://dx.doi.org/10.1002/elsc.201300170]

[2] Arunkumar K, Raja R, Kumar VBS, Joseph A, Shilpa T, Carvalho IS. Antioxidant and cytotoxic activities of sulfated polysaccharides from five different edible seaweeds. J Food Meas Charact 2021; 15(1): 567-76. [http://dx.doi.org/10.1007/s11694-020-00661-4]

[3] Raja R, Hemaiswarya S, Kumar NA, Sridhar S, Rengasamy R. A perspective on the biotechnological potential of microalgae. Crit Rev Microbiol 2008; 34(2): 77-88. [http://dx.doi.org/10.1080/10408410802086783] [PMID: 18568862]

[4] Raja R, Hemaiswarya S, Ganesan V, Carvalho IS. Recent developments in therapeutic applications of Cyanobacteria. Crit Rev Microbiol 2016; 42(3): 394-405. [PMID: 25629310]

[5] Yu G, Hu Y, Yang B, *et al.* Extraction, isolation and structural characterization of polysaccharides from a red alga Gloiopeltis furcata. J Ocean Univ China 2010; 9(2): 193-7. [http://dx.doi.org/10.1007/s11802-010-0193-7]

[6] Gómez-Ordóñez E, Jiménez-Escrig A, Rupérez P. Bioactivity of sulfated polysaccharides from the edible red seaweed Mastocarpus stellatus. Bioactive Carbohydrates and Dietary Fibre 2014; 3(1): 29-40. [http://dx.doi.org/10.1016/j.bcdf.2014.01.002]

[7] Shanab SMM, Hafez RM, Fouad AS. A review on algae and plants as potential source of arachidonic acid. J Adv Res 2018; 11: 3-13. [http://dx.doi.org/10.1016/j.jare.2018.03.004] [PMID: 30034871]

[8] Guedes AC, Meireles LA, Amaro HM, Malcata FX. Changes in lipid class and fatty acid composition of cultures of Pavlova lutheri, in response to light intensity. J Am Oil Chem Soc 2010; 87(7): 791-801.

[9] Cerón-García MC, Fernández-Sevilla JM, Sánchez-Mirón A, García-Camacho F, Contreras-Gómez A, Molina-Grima E. Mixotrophic growth of Phaeodactylum tricornutum on fructose and glycerol in fed-batch and semi-continuous modes. Bioresour Technol 2013; 147: 569-76. [http://dx.doi.org/10.1016/j.biortech.2013.08.092] [PMID: 24013294]

[10] Shanab SMM, Mostafa SSM, Shalaby EA, Mahmoud GI. Aqueous extracts of microalgae exhibit

antioxidant and anticancer activities. Asian Pac J Trop Biomed 2012; 2(8): 608-15. [http://dx.doi.org/10.1016/S2221-1691(12)60106-3] [PMID: 23569980]

[11] Kris-Etherton PM, Harris WS, Appel LJ. Fish consumption, fish oil, omega-3 fatty acids, and cardiovascular disease. circulation. 2002 2002; 106(21): 2747-57.

[12] Borowitzka MA. High-value products from microalgae—their development and commercialisation. J Appl Phycol 2013; 25(3): 743-56. [http://dx.doi.org/10.1007/s10811-013-9983-9]

[13] Ahmmed MK, Ahmmed F, Tian HS, Carne A, Bekhit AED. Marine omega□3 (n□3) phospholipids: A comprehensive review of their properties, sources, bioavailability, and relation to brain health. Compr Rev Food Sci Food Saf 2020; 19(1): 64-123. [http://dx.doi.org/10.1111/1541-4337.12510] [PMID: 33319514]

[14] Peltomaa ET, Aalto SL, Vuorio KM, Taipale SJ. The importance of phytoplankton biomolecule availability for secondary production. Front Ecol Evol 2017; 5: 128. [http://dx.doi.org/10.3389/fevo.2017.00128]

[15] Piironen V, Lindsay DG, Miettinen TA, Toivo J, Lampi AM. Plant sterols: biosynthesis, biological function and their importance to human nutrition. J Sci Food Agric 2000; 80(7): 939-66.

[16] Dufourc EJ. Sterols and membrane dynamics. J Chem Biol 2008; 1(1-4): 63-77. [http://dx.doi.org/10.1007/s12154-008-0010-6] [PMID: 19568799]

[17] Tanna B, Mishra A. Metabolites unravel nutraceutical potential of edible seaweeds: An emerging source of functional food. Compr Rev Food Sci Food Saf 2018; 17(6): 1613-24. [http://dx.doi.org/10.1111/1541-4337.12396] [PMID: 33350143]

[18] Seghiri R, Kharbach M, Essamri A. Functional composition, nutritional properties, and biological activities of Moroccan Spirulina microalga. J Food Qual 2019; 2019.

[19] Carbone DA, Pellone P, Lubritto C, Ciniglia C. Evaluation of Microalgae Antiviral Activity and Their Bioactive Compounds. Antibiotics (Basel) 2021; 10(6): 746. [http://dx.doi.org/10.3390/antibiotics10060746] [PMID: 34202941]

[20] Koyande AK, Chew KW, Rambabu K, Tao Y, Chu DT, Show PL. Microalgae: A potential alternative to health supplementation for humans. Food Sci Hum Wellness 2019; 8(1): 16-24. [http://dx.doi.org/10.1016/j.fshw.2019.03.001]

[21] Machado AR, da Silva Graça C, Machado MI, de Souza Soares LA. Anti-oxidant capacity of phenolic acids extracted from the biomass from Spirulina sp. LEB-18. Brazilian Journal of Development 2021; 7(3): 25956-70. [http://dx.doi.org/10.34117/bjdv7n3-345]

[22] Ou Y, Lin L, Pan Q, Yang X, Cheng X. Preventive effect of phycocyanin from Spirulina platensis on alloxan-injured mice. Environ Toxicol Pharmacol 2012; 34(3): 721-6. [http://dx.doi.org/10.1016/j.etap.2012.09.016] [PMID: 23121873]

[23] Liu Q, Huang Y, Zhang R, Cai T, Cai Y. Medical application of Spirulina platensis derived C-phycocyanin. Evid Based Complement Alternat Med 2016; 2016: 7803846. [PMID: 27293463]

[24] Thangam R, Suresh V, Asenath Princy W, *et al.* C-Phycocyanin from Oscillatoria tenuis exhibited an antioxidant and *in vitro* antiproliferative activity through induction of apoptosis and G0/G1 cell cycle arrest. Food Chem 2013; 140(1-2): 262-72. [http://dx.doi.org/10.1016/j.foodchem.2013.02.060] [PMID: 23578642]

[25] Mishra SK, Shrivastav A, Maurya RR, Patidar SK, Haldar S, Mishra S. Effect of light quality on the C-phycoerythrin production in marine cyanobacteria Pseudanabaena sp. isolated from Gujarat coast, India. Protein Expr Purif 2012; 81(1): 5-10. [http://dx.doi.org/10.1016/j.pep.2011.08.011] [PMID: 21906679]

[26] Senthilkumar N, Kurinjimalar C, Thangam R, *et al.* Further studies and biological activities of macromolecular protein R-Phycoerythrin from Portieria hornemannii. Int J Biol Macromol 2013; 62: 107-16.
[http://dx.doi.org/10.1016/j.ijbiomac.2013.08.004] [PMID: 23962717]

[27] Balić A, Mokos M. Do we utilize our knowledge of the skin protective effects of carotenoids enough? Antioxidants 2019; 8(8): 259.
[http://dx.doi.org/10.3390/antiox8080259] [PMID: 31370257]

[28] Raja R, Hemaiswarya S, Rengasamy R. Exploitation of Dunaliella for β-carotene production. Appl Microbiol Biotechnol 2007; 74(3): 517-23.
[http://dx.doi.org/10.1007/s00253-006-0777-8] [PMID: 17225103]

[29] Kawata A, Murakami Y, Suzuki S, Fujisawa S. Anti-inflammatory activity of β-carotene, lycopene and tri-n-butylborane, a scavenger of reactive oxygen species. *in vivo*. 2018; 32(2): 255-64.

[30] Christaki E, Bonos E, Giannenas I, Florou-Paneri P. Functional properties of carotenoids originating from algae. J Sci Food Agric 2013; 93(1): 5-11.
[http://dx.doi.org/10.1002/jsfa.5902] [PMID: 23044813]

[31] Novoveská L, Ross ME, Stanley MS, Pradelles R, Wasiolek V, Sassi JF. Microalgal carotenoids: A review of production, current markets, regulations, and future direction. Mar Drugs 2019; 17(11): 640.
[http://dx.doi.org/10.3390/md17110640] [PMID: 31766228]

[32] Jayappriyan KR, Rajkumar R, Rengasamy R. Unusual occurrence of non carotenogenic strains of Dunaliella bardawil and Dunaliella parva in India. J Basic Microbiol 2011; 51(5): 473-83.
[http://dx.doi.org/10.1002/jobm.201000384] [PMID: 21656806]

[33] Sluijs I, Cadier E, Beulens JWJ, van der A DL, Spijkerman AMW, van der Schouw YT. Dietary intake of carotenoids and risk of type 2 diabetes. Nutr Metab Cardiovasc Dis 2015; 25(4): 376-81.
[http://dx.doi.org/10.1016/j.numecd.2014.12.008] [PMID: 25716098]

[34] Borowitzka LJ, Borowitzka MA. β-Carotene (Provitamin A) Production with Algae. In: Vandamme EJ. Biotechnology of Vitamins, Pigments and Growth Factors. Elsevier Applied Biotechnology Series. Springer, Dordrecht 1989; 15-26.
[http://dx.doi.org/https://doi.org/10.1007/978-94-009-1111-6_2]

[35] Jayappriyan KR, Rajkumar R, Venkatakrishnan V, Nagaraj S, Rengasamy R. *in vitro* anticancer activity of natural β-carotene from Dunaliella salina EU5891199 in PC-3 cells. Biomedicine & Preventive Nutrition 2013; 3(2): 99-105.
[http://dx.doi.org/10.1016/j.bionut.2012.08.003]

[36] Guedes AC, Amaro HM, Malcata FX. Microalgae as sources of carotenoids. Mar Drugs 2011; 9(4): 625-44.
[http://dx.doi.org/10.3390/md9040625] [PMID: 21731554]

[37] Takaichi S. Carotenoids in algae: distributions, biosyntheses and functions. Mar Drugs 2011; 9(6): 1101-18.
[http://dx.doi.org/10.3390/md9061101] [PMID: 21747749]

[38] Tan JSL, Wang JJ, Flood V, Rochtchina E, Smith W, Mitchell P. Dietary antioxidants and the long-term incidence of age-related macular degeneration: the Blue Mountains Eye Study. Ophthalmology 2008; 115(2): 334-41.
[http://dx.doi.org/10.1016/j.ophtha.2007.03.083] [PMID: 17664009]

[39] Steiner BM, McClements DJ, Davidov-Pardo G. Encapsulation systems for lutein: A review. Trends Food Sci Technol 2018; 82: 71-81.
[http://dx.doi.org/10.1016/j.tifs.2018.10.003]

[40] Renju GL, Kurup GM, Saritha Kumari CH. Effect of lycopene from Chlorella marina on high cholesterol-induced oxidative damage and inflammation in rats. Inflammopharmacology 2014; 22(1): 45-54.

[http://dx.doi.org/10.1007/s10787-013-0178-4] [PMID: 23887896]

[41] Capelli B, Jenkins U, Cysewski GR. Role of astaxanthin in sports nutrition. In: Nutrition and enhanced sports performance. 2013; 465-71.
[http://dx.doi.org/10.1016/B978-0-12-396454-0.00048-5]

[42] Lorenz RT, Cysewski GR. Commercial potential for Haematococcus microalgae as a natural source of astaxanthin. Trends Biotechnol 2000; 18(4): 160-7.
[http://dx.doi.org/10.1016/S0167-7799(00)01433-5] [PMID: 10740262]

[43] Tanaka T, Shnimizu M, Moriwaki H. Cancer chemoprevention by carotenoids. Molecules 2012; 17(3): 3202-42.
[http://dx.doi.org/10.3390/molecules17033202] [PMID: 22418926]

[44] Kang H, Kim H. Astaxanthin and β-carotene in *Helicobacter pylori*-induced Gastric Inflammation: A Mini-review on Action Mechanisms. J Cancer Prev 2017; 22(2): 57-61.
[http://dx.doi.org/10.15430/JCP.2017.22.2.57] [PMID: 28698858]

[45] Maeda H, Hosokawa M, Sashima T, Murakami-Funayama K, Miyashita K. Anti-obesity and anti-diabetic effects of fucoxanthin on diet-induced obesity conditions in a murine model. Mol Med Rep 2009; 2(6): 897-902.
[http://dx.doi.org/10.3892/mmr_00000189] [PMID: 21475918]

[46] Bhuvaneswari S, Yogalakshmi B, Sreeja S, Anuradha CV. Astaxanthin reduces hepatic endoplasmic reticulum stress and nuclear factor-κB-mediated inflammation in high fructose and high fat diet-fed mice. Cell Stress Chaperones 2014; 19(2): 183-91.
[http://dx.doi.org/10.1007/s12192-013-0443-x] [PMID: 23852435]

[47] Kim DY, Vijayan D, Praveenkumar R, *et al.* Cell-wall disruption and lipid/astaxanthin extraction from microalgae: Chlorella and Haematococcus. Bioresour Technol 2016; 199: 300-10.
[http://dx.doi.org/10.1016/j.biortech.2015.08.107] [PMID: 26342788]

[48] Karpiński TM, Adamczak A. Fucoxanthin—An antibacterial carotenoid. Antioxidants 2019; 8(8): 239.
[http://dx.doi.org/10.3390/antiox8080239] [PMID: 31344844]

[49] Rajauria G, Abu-Ghannam N. Isolation and partial characterization of bioactive fucoxanthin from Himanthalia elongata brown seaweed: A TLC-based approach. Int J Anal Chem 2013; 2013: 1-6.
[http://dx.doi.org/10.1155/2013/802573] [PMID: 23762062]

[50] Stonik V, Stonik I. Low-molecular-weight metabolites from diatoms: structures, biological roles and biosynthesis. Mar Drugs 2015; 13(6): 3672-709.
[http://dx.doi.org/10.3390/md13063672] [PMID: 26065408]

[51] Sachindra NM, Sato E, Maeda H, *et al.* Radical scavenging and singlet oxygen quenching activity of marine carotenoid fucoxanthin and its metabolites. J Agric Food Chem 2007; 55(21): 8516-22.
[http://dx.doi.org/10.1021/jf071848a] [PMID: 17894451]

[52] Kumar S, Hosokawa M, Miyashita K. Fucoxanthin: a marine carotenoid exerting anti-cancer effects by affecting multiple mechanisms. Mar Drugs 2013; 11(12): 5130-47.
[http://dx.doi.org/10.3390/md11125130] [PMID: 24351910]

[53] Venugopalan V, Tripathi SK, Nahar P, Saradhi PP, Das RH, Gautam HK. Characterization of canthaxanthin isomers isolated from a new soil Dietzia sp. and their antioxidant activities. J Microbiol Biotechnol 2013; 23(2): 237-45.
[http://dx.doi.org/10.4014/jmb.1203.03032] [PMID: 23412067]

[54] Ferdous UT, Yusof ZNB. Medicinal Prospects of Antioxidants From Algal Sources in Cancer Therapy. Front Pharmacol 2021; 12: 593116.
[http://dx.doi.org/10.3389/fphar.2021.593116] [PMID: 33746748]

[55] Mattsson L, Sörenson E, Capo E, *et al.* Functional diversity facilitates stability under environmental changes in an outdoor microalgal cultivation system. Front Bioeng Biotechnol 2021; 9: 651895.
[http://dx.doi.org/10.3389/fbioe.2021.651895] [PMID: 33968914]

[56] Yabuta Y, Fujimura H, Kwak CS, Enomoto T, Watanabe F. Antioxidant activity of the phycoerythrobilin compound formed from a dried Korean purple laver (Porphyra sp.) during *in vitro* digestion. Food Sci Technol Res 2010; 16(4): 347-52.
[http://dx.doi.org/10.3136/fstr.16.347]

[57] Marzorati S, Schievano A, Idà A, Verotta L. Carotenoids, chlorophylls and phycocyanin from Spirulina: supercritical CO_2 and water extraction methods for added value products cascade. Green Chem 2020; 22(1): 187-96.
[http://dx.doi.org/10.1039/C9GC03292D]

[58] Contreras-Martel C, Matamala A, Bruna C, *et al.* The structure at 2 Å resolution of Phycocyanin from Gracilaria chilensis and the energy transfer network in a PC–PC complex. Biophys Chem 2007; 125(2-3): 388-96.
[http://dx.doi.org/10.1016/j.bpc.2006.09.014] [PMID: 17118524]

[59] Ardiles P, Cerezal-Mezquita P, Salinas-Fuentes F, Órdenes D, Renato G, Ruiz-Domínguez MC. Biochemical Composition and Phycoerythrin Extraction from Red Microalgae: A Comparative Study Using Green Extraction Technologies. Processes (Basel) 2020; 8(12): 1628.
[http://dx.doi.org/10.3390/pr8121628]

[60] Koller M, Muhr A, Braunegg G. Microalgae as versatile cellular factories for valued products. Algal Res 2014; 6: 52-63.
[http://dx.doi.org/10.1016/j.algal.2014.09.002]

[61] Hartmann A, Holzinger A, Ganzera M, Karsten U. Prasiolin, a new UV-sunscreen compound in the terrestrial green macroalga Prasiola calophylla (Carmichael ex Greville) Kützing (Trebouxiophyceae, Chlorophyta). Planta 2016; 243(1): 161-9.
[http://dx.doi.org/10.1007/s00425-015-2396-z] [PMID: 26358040]

[62] Mourelle M, Gómez C, Legido J. The potential use of marine microalgae and cyanobacteria in cosmetics and thalassotherapy. Cosmetics 2017; 4(4): 46.
[http://dx.doi.org/10.3390/cosmetics4040046]

[63] Raposo M, De Morais R, Bernardo de Morais A. Bioactivity and applications of sulphated polysaccharides from marine microalgae. Mar Drugs 2013; 11(1): 233-52.
[http://dx.doi.org/10.3390/md11010233] [PMID: 23344113]

[64] Qi H, Zhao T, Zhang Q, Li Z, Zhao Z, Xing R. Antioxidant activity of different molecular weight sulfated polysaccharides from Ulva pertusa Kjellm (Chlorophyta). J Appl Phycol 2005; 17(6): 527-34.
[http://dx.doi.org/10.1007/s10811-005-9003-9]

[65] Rodrigues JAG, Vanderlei ESO, Silva LMCM, *et al.* Antinociceptive and anti-inflammatory activities of a sulfated polysaccharide isolated from the green seaweed *Caulerpa cupressoides* Pharmacol Rep 2012; 64(2): 282-92.
[http://dx.doi.org/10.1016/S1734-1140(12)70766-1] [PMID: 22661177]

[66] Medeiros VP, Queiroz KCS, Cardoso ML, *et al.* Sulfated galactofucan from *Lobophora variegata*: Anticoagulant and anti-inflammatory properties. Biochemistry (Mosc) 2008; 73(9): 1018-24.
[http://dx.doi.org/10.1134/S0006297908090095] [PMID: 18976219]

[67] Hwang PA, Phan NN, Lu WJ, Hieu BTN, Lin YC. Low-molecular-weight fucoidan and high-stability fucoxanthin from brown seaweed exert prebiotics and anti-inflammatory activities in Caco-2 cells. Food Nutr Res 2016; 60(1): 32033.
[http://dx.doi.org/10.3402/fnr.v60.32033] [PMID: 27487850]

[68] Ponnan A, Ramu K, Marudhamuthu M, Marimuthu R, Siva K, Kadarkarai M. Antibacterial, antioxidant and anticancer properties of *Turbinaria conoides* (J. Agardh) Kuetz. Clinical Phytoscience 2017; 3(1): 5.
[http://dx.doi.org/10.1186/s40816-017-0042-y]

[69] Li B, Lu F, Wei X, Zhao R. Fucoidan: Structure and Bioactivity. Molecules 2008; 13(8): 1671-95.

[http://dx.doi.org/10.3390/molecules13081671] [PMID: 18794778]

[70] Ponnan A, Kulanthaiyesu A, Marudhamuthu M, Palanisamy K, Kadarkarai M. Protective effects of fucoidan against 4-nitroquinolin-1-oxide provoked genetic damage in mouse bone marrow cells. Environ Sci Pollut Res Int 2020; 27(25): 31760-6.
[http://dx.doi.org/10.1007/s11356-020-09472-y] [PMID: 32504434]

[71] Teruya T, Konishi T, Uechi S, Tamaki H, Tako M. Anti-proliferative activity of oversulfated fucoidan from commercially cultured Cladosiphon okamuranus TOKIDA in U937 cells. Int J Biol Macromol 2007; 41(3): 221-6.
[http://dx.doi.org/10.1016/j.ijbiomac.2007.02.010] [PMID: 17433432]

[72] Cumashi A, Ushakova NA, Preobrazhenskaya ME, *et al.* A comparative study of the anti-inflammatory, anticoagulant, antiangiogenic, and antiadhesive activities of nine different fucoidans from brown seaweeds. Glycobiology 2007; 17(5): 541-52.
[http://dx.doi.org/10.1093/glycob/cwm014] [PMID: 17296677]

[73] Gladyshev MI, Sushchik NN, Makhutova ON. Production of EPA and DHA in aquatic ecosystems and their transfer to the land. Prostaglandins Other Lipid Mediat 2013; 107: 117-26.
[http://dx.doi.org/10.1016/j.prostaglandins.2013.03.002] [PMID: 23500063]

[74] Hayashi K, Nakano T, Hashimoto M, Kanekiyo K, Hayashi T. Defensive effects of a fucoidan from brown alga Undaria pinnatifida against herpes simplex virus infection. Int Immunopharmacol 2008; 8(1): 109-16.
[http://dx.doi.org/10.1016/j.intimp.2007.10.017] [PMID: 18068106]

[75] Athukorala Y, Ahn GN, Jee YH, *et al.* Antiproliferative activity of sulfated polysaccharide isolated from an enzymatic digest of Ecklonia cava on the U-937 cell line. J Appl Phycol 2009; 21(3): 307-14.
[http://dx.doi.org/10.1007/s10811-008-9368-7]

[76] Synytsya A, Kim WJ, Kim SM, *et al.* Structure and antitumour activity of fucoidan isolated from sporophyll of Korean brown seaweed *Undaria pinnatifida.* Carbohydr Polym 2010; 81(1): 41-8.
[http://dx.doi.org/10.1016/j.carbpol.2010.01.052]

[77] Aisa Y, Miyakawa Y, Nakazato T, *et al.* Fucoidan induces apoptosis of human HS-Sultan cells accompanied by activation of caspase-3 and down-regulation of ERK Pathways. Am J Hematol 2005; 78(1): 7-14.
[http://dx.doi.org/10.1002/ajh.20182] [PMID: 15609279]

[78] Lee JY, Kim YJ, Kim HJ, Kim YS, Park W. Immunostimulatory effect of laminarin on RAW 264.7 mouse macrophages. Molecules 2012; 17(5): 5404-11.
[http://dx.doi.org/10.3390/molecules17055404] [PMID: 22569419]

[79] Holdt SL, Kraan S. Bioactive compounds in seaweed: functional food applications and legislation. J Appl Phycol 2011; 23(3): 543-97.
[http://dx.doi.org/10.1007/s10811-010-9632-5]

[80] Aderibigbe B, Buyana B. Alginate in wound dressings. Pharmaceutics 2018; 10(2): 42.
[http://dx.doi.org/10.3390/pharmaceutics10020042] [PMID: 29614804]

[81] Han Y, Zhang L, Yu X, *et al.* RETRACTED ARTICLE: Alginate oligosaccharide attenuates α2,6-sialylation modification to inhibit prostate cancer cell growth via the Hippo/YAP pathway. Cell Death Dis 2019; 10(5): 374.
[http://dx.doi.org/10.1038/s41419-019-1560-y] [PMID: 31076566]

[82] Selvaraj S, Shanmugasundaram S, Maruthamuthu M, Venkidasamy B, Shanmugasundaram S. facile synthesis and characterization of quercetin-loaded alginate nanoparticles for enhanced *in vitro* anticancer effect against human leukemic cancer U937 Cells. J Cluster Sci 2020; 1-2.

[83] Ropellato J, Carvalho MM, Ferreira LG, *et al.* Sulfated heterorhamnans from the green seaweed Gayralia oxysperma: partial depolymerization, chemical structure and antitumor activity. Carbohydr Polym 2015; 117: 476-85.

[http://dx.doi.org/10.1016/j.carbpol.2014.09.089] [PMID: 25498661]

[84] Kamimura Y, Hashiguchi K, Nagata Y, *et al.* Inhibitory effects of edible green algae Monostroma nitidum on glycemic responses. Nippon Shokuhin Kagaku Kogaku Kaishi=. Nippon Shokuhin Kagaku Kogaku Kaishi 2010; 57(10): 441-5.
[http://dx.doi.org/10.3136/nskkk.57.441]

[85] Suzuki K, Terasawa M. Biological activities of rhamnan sulfate extract from the green algae monostroma nitidum (Hitoegusa). Mar Drugs 2020; 18(4): 228.
[http://dx.doi.org/10.3390/md18040228] [PMID: 32344720]

[86] Terasawa M, Hayashi K, Lee JB, *et al.* Anti-influenza A virus activity of rhamnan sulfate from green algae Monostroma nitidum in mice with normal and compromised immunity. Mar Drugs 2020; 18(5): 254.
[http://dx.doi.org/10.3390/md18050254] [PMID: 32414158]

[87] Yapo BM. Rhamnogalacturonan-I: a structurally puzzling and functionally versatile polysaccharide from plant cell walls and mucilages. Polym Rev (Phila Pa) 2011; 51(4): 391-413.
[http://dx.doi.org/10.1080/15583724.2011.615962]

[88] Patil NP, Le V, Sligar AD, *et al.* Algal polysaccharides as therapeutic agents for atherosclerosis. Front Cardiovasc Med 2018; 5: 153.
[http://dx.doi.org/10.3389/fcvm.2018.00153] [PMID: 30417001]

[89] Borai IH, Ezz MK, Rizk MZ, *et al.* Hypolipidemic and anti-atherogenic effect of sulphated polysaccharides from the green alga Ulva fasciata. Int J Pharm Sci Rev Res 2015; 31(1): 1-2.

[90] Li J, Han D, Wang D, *et al.* Choreography of transcriptomes and lipidomes of Nannochloropsis reveals the mechanisms of oil synthesis in microalgae. Plant Cell 2014; 26(4): 1645-65.
[http://dx.doi.org/10.1105/tpc.113.121418] [PMID: 24692423]

[91] Qi H, Li D, Zhang J, Liu L, Zhang Q. Study on extraction of agaropectin from Gelidium amansii and its anticoagulant activity. Chin J Oceanology Limnol 2008; 26(2): 186-9.
[http://dx.doi.org/10.1007/s00343-008-0186-1]

[92] Pereira L. Population studies and carrageenan properties in eight Gigartinales (Rhodophyta) from Western Coast of Portugal. ScientificWorldJournal 2013; 2013: 1-11.
[http://dx.doi.org/10.1155/2013/939830] [PMID: 24288514]

[93] Campo VL, Kawano DF, Silva DB Jr, Carvalho I. Carrageenans: Biological properties, chemical modifications and structural analysis – A review. Carbohydr Polym 2009; 77(2): 167-80.
[http://dx.doi.org/10.1016/j.carbpol.2009.01.020]

[94] Wijesekara I, Kim SK. Angiotensin-I-converting enzyme (ACE) inhibitors from marine resources: prospects in the pharmaceutical industry. Mar Drugs 2010; 8(4): 1080-93.
[http://dx.doi.org/10.3390/md8041080] [PMID: 20479968]

[95] Calvo GH, Cosenza VA, Sáenz DA, *et al.* Disaccharides obtained from carrageenans as potential antitumor agents. Sci Rep 2019; 9(1): 6654.
[http://dx.doi.org/10.1038/s41598-019-43238-y] [PMID: 31040376]

[96] Jiang Z, Hama Y, Yamaguchi K, Oda T. Inhibitory effect of sulphated polysaccharide porphyran on nitric oxide production in lipopolysaccharide-stimulated RAW264.7 macrophages. J Biochem 2012; 151(1): 65-74.
[http://dx.doi.org/10.1093/jb/mvr115] [PMID: 21976706]

[97] He D, Wu S, Yan L, *et al.* Antitumor bioactivity of porphyran extracted from *Pyropia yezoensis* Chonsoo2 on human cancer cell lines. J Sci Food Agric 2019; 99(15): 6722-30.
[http://dx.doi.org/10.1002/jsfa.9954] [PMID: 31350864]

[98] Qiu Y, Jiang H, Fu L, Ci F, Mao X. Porphyran and oligo-porphyran originating from red algae Porphyra: Preparation, biological activities, and potential applications. Food Chem 2021; 349: 129209.
[http://dx.doi.org/10.1016/j.foodchem.2021.129209] [PMID: 33588184]

[99] Priyan Shanura Fernando I, Kim KN, Kim D, Jeon YJ. Algal polysaccharides: potential bioactive substances for cosmeceutical applications. Crit Rev Biotechnol 2019; 39(1): 99-113. [http://dx.doi.org/10.1080/07388551.2018.1503995] [PMID: 31690134]

[100] Robson SS, Maria Clara RV, Caroline SB, Claudio Cesar CS. Robson SS, Maria Clara RV, Caroline SB, Claudio Cesar CS. Anti-viral activity of Diterpenes from *Canistrocarpus cervicornis* against human Herpesvirus 1 and Bovine Herpes Virus 5. Arch Biomed Eng & Biotechnol 2019; 3(2): 000557.

[101] Costa LS, Fidelis GP, Cordeiro SL, *et al.* Biological activities of sulfated polysaccharides from tropical seaweeds. Biomed Pharmacother 2010; 64(1): 21-8. [http://dx.doi.org/10.1016/j.biopha.2009.03.005] [PMID: 19766438]

[102] de Jesus Raposo M, de Morais A, de Morais R. Marine polysaccharides from algae with potential biomedical applications. Mar Drugs 2015; 13(5): 2967-3028. [http://dx.doi.org/10.3390/md13052967] [PMID: 25988519]

[103] Waghmode AV, Khilare CJ. RP-HPLC profile of major phenolics from brown marine macro algae. J Appl Pharm 2018; 10(2): 1-5.

[104] Maria AG, Graziano R, Nicolantonio DO. Carotenoids: potential allies of cardiovascular health? Food Nutr Res 2015; 59(1): 26762. [http://dx.doi.org/10.3402/fnr.v59.26762] [PMID: 25660385]

[105] Chattopadhyay N, Ghosh T, Sinha S, Chattopadhyay K, Karmakar P, Ray B. Polysaccharides from Turbinaria conoides: Structural features and antioxidant capacity. Food Chem 2010; 118(3): 823-9. [http://dx.doi.org/10.1016/j.foodchem.2009.05.069]

[106] Hentati F, Tounsi L, Djomdi D, *et al.* Bioactive polysaccharides from seaweeds. Molecules 2020; 25(14): 3152. [http://dx.doi.org/10.3390/molecules25143152] [PMID: 32660153]

[107] Khalid S, Abbas M, Saeed F, Bader-Ul-Ain H, Suleria HA. Therapeutic potential of seaweed bioactive compounds. In: Maiti S, Ed. Seaweed Biomaterials. London: IntechOpen; 2018. [http://dx.doi.org/10.5772/intechopen.74060]

[108] Torres A, Enk CD, Hochberg M, Srebnik M. Porphyra-334, a potential natural source for UVA protective sunscreens. Photochem Photobiol Sci 2006; 5(4): 432-5. [http://dx.doi.org/10.1039/b517330m] [PMID: 16583025]

[109] Hartmann A, Glaser K, Holzinger A, Ganzera M, Karsten U. Klebsormidin A and B, two new UV-sunscreen compounds in green microalgal interfilum and Klebsormidium Species (Streptophyta) from terrestrial habitats. Front Microbiol 2020; 11: 499. [http://dx.doi.org/10.3389/fmicb.2020.00499] [PMID: 32292396]

[110] Li YX, Wijesekara I, Li Y, Kim SK. Phlorotannins as bioactive agents from brown algae. Process Biochem 2011; 46(12): 2219-24. [http://dx.doi.org/10.1016/j.procbio.2011.09.015]

[111] Jimenez-Lopez C, Pereira AG, Lourenço-Lopes C, *et al.* Main bioactive phenolic compounds in marine algae and their mechanisms of action supporting potential health benefits. Food Chem 2021; 341(Pt 2): 128262. [http://dx.doi.org/10.1016/j.foodchem.2020.128262] [PMID: 33038800]

[112] Okuda T, Ito H. Tannins of constant structure in medicinal and food plants—hydrolyzable tannins and polyphenols related to tannins. Molecules 2011; 16(3): 2191-217. [http://dx.doi.org/10.3390/molecules16032191]

[113] Wang HMD, Li XC, Lee DJ, Chang JS. Potential biomedical applications of marine algae. Bioresour Technol 2017; 244(Pt 2): 1407-15. [http://dx.doi.org/10.1016/j.biortech.2017.05.198] [PMID: 28697977]

[114] Anam C, Praseptiangga D, Nugraheni MA, *et al.* Preliminary characterization of crude lectin fraction of the red alga, Acrocystis nana from wediombo beach of the southern coast of Java island, Gunung kidul, Yogyakarta, Indonesia. InIOP Conference Series: Materials Science and Engineering 2017; 193(1): 012016.

[115] Chaves RP, Silva SR, Nascimento Neto LG, *et al.* Structural characterization of two isolectins from the marine red alga Solieria filiformis (Kützing) P.W. Gabrielson and their anticancer effect on MCF-7 breast cancer cells. Int J Biol Macromol 2018; 107(Pt A): 1320-9. [http://dx.doi.org/10.1016/j.ijbiomac.2017.09.116] [PMID: 28970169]

[116] Abreu T, Ribeiro N, Chaves H, *et al.* Antinociceptive and anti-inflammatory activities of the lectin from marine red alga Solieria filiformis. Planta Med 2016; 82(7): 596-605. [http://dx.doi.org/10.1055/s-0042-101762] [PMID: 27093245]

[117] Micewicz ED, Cole AL, Jung CL, *et al.* Grifonin-1: a small HIV-1 entry inhibitor derived from the algal lectin, Griffithsin. PLoS One 2010; 5(12): e14360. [http://dx.doi.org/10.1371/journal.pone.0014360] [PMID: 21179548]

[118] Conde E, Balboa EM, Parada M, Falqué E. Algal proteins, peptides and amino acids. In: Domínguez, Ed. Functional Ingredients from Algae for Foods and Nutraceuticals. Woodhead Publishing. 2013; 307-37. [http://dx.doi.org/10.1533/9780857098689.1.135]

CHAPTER 9

Bioactive Compounds from Edible Seaweeds and their Applications

Anuruddhika Udayangani Rathnayake[1,2], **Hee-Guk Byun**[1] and **Indira Wickramasinghe**[3,*]

[1] *Department of Marine Biotechnology, Gangneung-Wonju National University, Gangneung 25457, Republic of Korea*

[2] *Department of Science and Technology, Faculty of Applied Sciences, Uva Wellassa University, Badulla, Sri Lanka*

[3] *Department of Food Science & Technology, Faculty of Applied Sciences, University of Sri Jayewardenepura, Nugegoda, Sri Lanka*

Abstract: Edible seaweeds (macroalgae and microalgae) are considered superfoods of our waterways. Based on pigmentation seaweeds, macroalgae are classified into three groups, green seaweeds (Chlorophyta), brown seaweeds (Phaeophyta) and red seaweeds (Rhodophyta). Seaweeds are rich in macro and micro nutrients. They contain protein, dietary fiber, minerals, vitamins, polyphenols, peptides, sterols and polyunsaturated fatty acids. Apart from being a major nutritional source in the human diet, seaweed is considered a functional and nutraceutical due to its extra nutritional and physiological properties. Plethora of bioactive compounds with potential applications in food, medicinal, health and pharmaceutical industries are included in extensive research and industries. For example, phycocolloids from seaweeds are widely used in the food industry. Medicinal and pharmaceutical properties of edible seaweeds include anti-tumor property, anti-viral property, anti-coagulant property, anti-Alzheimer's property, and anti-oxidant property. Thus, a vast scope of investigating and understanding bioactive compounds from edible seaweeds is well-timed.

Keywords: Agar, Alginate, Anti-Oxidant, Anti-Tumor, Anti-Viral, Bioactive Compounds, Carrageenan, Edible, Food Industry, Fucoidan, Human Diet, Hydrocolloids, Laminarin, Medicinal Uses, Peptides, Phycocolloids, Polysaccharide, Seaweeds, Ulvan.

* **Corresponding author Indira Wickramasinghe:** Department of Food Science & Technology, Faculty of Applied Sciences, University of Sri Jayewardenepura, Sri Lanka; Tel: +94716861908; E-mail: indiraw@sjp.ac.lk

Avinash Mishra (Ed.)

INTRODUCTION

The oceans have been one of Earth's most valuable sources of food. Seaweed has inhabited the oceans for over 200 million years. Since the seventh century, Asian people have been using seaweeds as a food source. Human migration from countries like China, Korea, Japan, Indonesia, and Malaysia, where edible seaweed has been used as food, brought this custom. Therefore, consumption of seaweed is not unusual in many more countries all over the world, including Europe [1]. The appropriate human diet is important in maintaining good health and preventing certain diseases resulting from inappropriate dietary intake and lifestyle. Many of the healthy properties of the diet are due to a series of bioactive compounds that are not necessarily nutrients. Bioactive compounds have a positive effect on certain physiological functions or diseases. They are the functional ingredients obtained from different sources. Seaweed can produce a wide range of secondary metabolites that are not found in other species. Besides being a low-calorie healthy food, high fiber and mineral content in edible seaweeds help them to act as a potential functional ingredient [2].

On the coastlines of the US and Canada, some companies are growing macroalgae for human consumption and exportation. Some other countries are also showing renewed interest in macro algae with the increasing global utilization of macroalgae. Other than the direct consumption of seaweed as sea vegetables, alginate extracted from brown macroalgae, carrageenan and agar extracted from red macroalgae, and microalgae have been served as texture or viscosity modifiers in prepared food industries. Phycocolloids from seaweed have a wide range of uses in the food, cosmetics, medicinal and pharmaceutical industries. Medicinal and pharmaceutical properties of edible seaweeds include anti-tumour, anti-viral, anti-coagulant, anti-Alzheimer's, and anti-oxidant properties. Therefore, this chapter focuses on bioactive compounds derived from their applications [3 - 5].

BIOACTIVE COMPOUNDS PRESENT IN EDIBLE SEAWEEDS

Bioactive Compounds in Macroalgae

Carrageenans and Agarans

Carrageenans and agarans are natural sulphated polysaccharides extracted from edible red marine algae. They consist of D-galactose and its derivatives alternatively linked by α (1→3) and β (1→4) glycosidic linkages. There are six basic forms of carrageenans depending on solubility, source of extraction, and sulphate content (Kappa (κ) carrageenan, Iota (ι) carrageenan, Lambda (λ) carrageenan, Mu (μ) carrageenan, Nu (ν) carrageenan, and Theta (θ) carrageenan).

Among six types, κ, ι and λ carrageenans (Fig. **1**) are commercially important due to their thickening and gelling properties [6, 7].

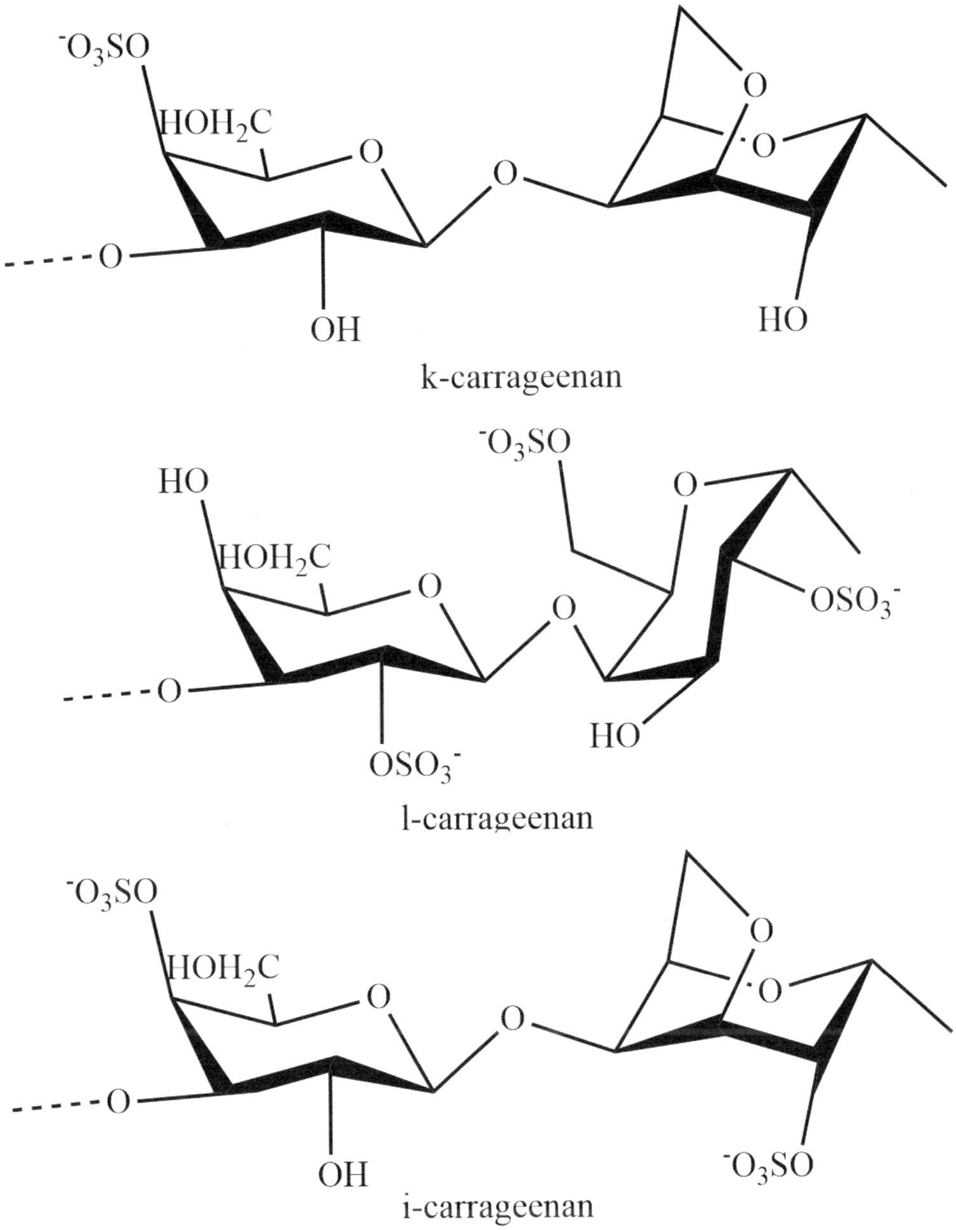

Fig (1). Repeating disaccharide units of commercially important carrageenans.

Carrageenan is an excellent gelling agent with a long history of seaweed-derived hydrocolloids in the food industry. Carrageenans are used mainly as a food additive due to their biological and physicochemical properties such as thickening, gelling, emulsifying, and stabilizing abilities. Carrageenans are incorporated into dairy desserts, cottage cheese and puddings to improve texture and stability. In the meat processing industry of hamburger, sausage and patties production, carrageenans are used as texture improvers and binder [8]. Industrial use of κ carrageenan also includes immobilization of biocatalysts in bioprocessing. Biocatalyst employed in biotransformation and fermentation process, immobilized in κ carrageenan gel beds. For instance, *Acetobacter suboxydans* cells and yeast cells (*Saccharomyces* sp.) immobilized in κ carrageenan gel beds have been used in acetic acid fermentation and ethanol production [8].

Carrageenan-based hydrogels have a broad range of uses in biomedical industries. Biomedical applications include tissue engineering, wound healing and drug delivery. Incorporation of κ carrageenan into collagen-hydroxyapatite composite gel acted as an efficient bone repair material by increasing its compressive strength [9]. Combination of biodegradable polyhydroxybutyrate or polyhydroxybutyrate valerate with the anionic sulfated polysaccharide κ-carrageenan pronounced bioactive and osteogenic properties in tissue engineering [10]. κ -Carrageenan beads have been identified as a useful drug carrier due to their high encapsulation efficiency and good stability [7]. Additionally, carrageenan is used in modern non-food items such as beer, culture media, toothpaste, air fresheners, textiles, tissues, hair shampoos, bactericides, and fungicides [7]. The sulphated polysaccharide of the agaran group contributes more to the bioactivities coming from *Gracilaria* species. Examples include anti-oxidant activities, anti-viral activities, antinociceptive activities, and anti-inflammatory activities [9 - 11].

Agar

Phycocolloids are gelling agents extracted from marine algae, and agar was the first phycocolloid used as a food additive in our civilization. Agarophyte seaweed used for agar production is *Gelidium* and *Gracilaria* species ((Rhodophyceae phylum). The chemical structure of agar is characterized by repeating units of D-galactose and 3-6,anhydro-L-galactose with their derivatives. Agar has a wide range of uses in food, pharmaceutical, and biomedical industries due to its excellent rheological properties. Agar is a substitute for gelatin in a vegetarian diet and is also a universal food additive. Agar is used as a thickener in soups, and different types of desserts, including jelly and ice cream. Further, agar acts as a clarifying agent in the brewing industry. Other uses include culture media for antibacterial activity, packaging material, paper industry, photography, dairy industry, cosmetic industry, leather industry and pharmaceutical industry [12].

Alginate

Alginate is a cell wall component of brown seaweed with structural and functional similarities to carrageenan and agar (Fig. **2**). Alginate is also used in a wide range of industries, including food additives, pharmaceutical, wound dressing, paper coating, and textile printing. Industrial applications of alginate are based on three main properties (thickening ability of aqueous solution, gel formation ability, and film formation ability). Industrial use of alginate as a food additive includes frozen desserts, frozen foods, pastry fillings, syrups, bakery products and icings [7].

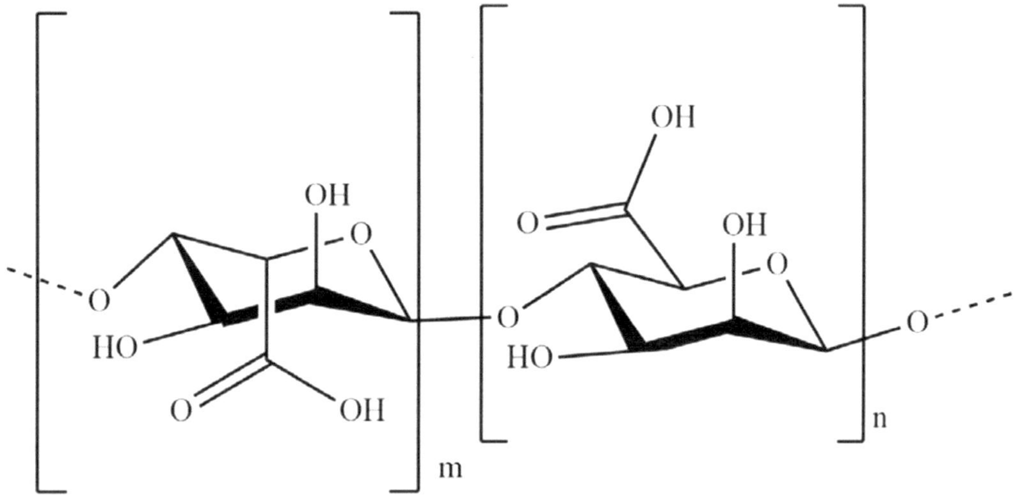

Alginic acid

Fig. (2). Structure of alginic acid (salt of alginic acids are known as alginates).

Laminarin

Laminarin is a low molecular weight polysaccharide mainly found in brown seaweed. Laminarin content in brown algae is greater compared to red and green algae. Examples of laminarin-containing algae are *Saccharina latissimi, Laminaria hyperborean, Laminaria digitate, Fucus vesiculosus, Undaria pinnatifida*, and *Ascophyllum nodosum*. Laminarins are composed of (1,3)-β-D-glucan and (1,3)-β-D -glucopyranose residues with some 6-*O*- branching and some β-(1,6)-intra-chain links in the main chain. The molecular weight of laminarin is approximately 5 kDa. The level of branching present in the laminarin structure affects solubility. There are two types of laminarin chains depending on the reducing end (M chains end with 1-*O*-substituted D -mannitol, and G chains end with glucose; (Fig. **3**) [13, 14]. Laminarin has been identified as a good source of dietary fibers. Biofunctional activities of laminarin include anticoagulant activity, anti-oxidant activity, antitumor activity, anti-inflammatory activity and anti-apoptotic activity [13].

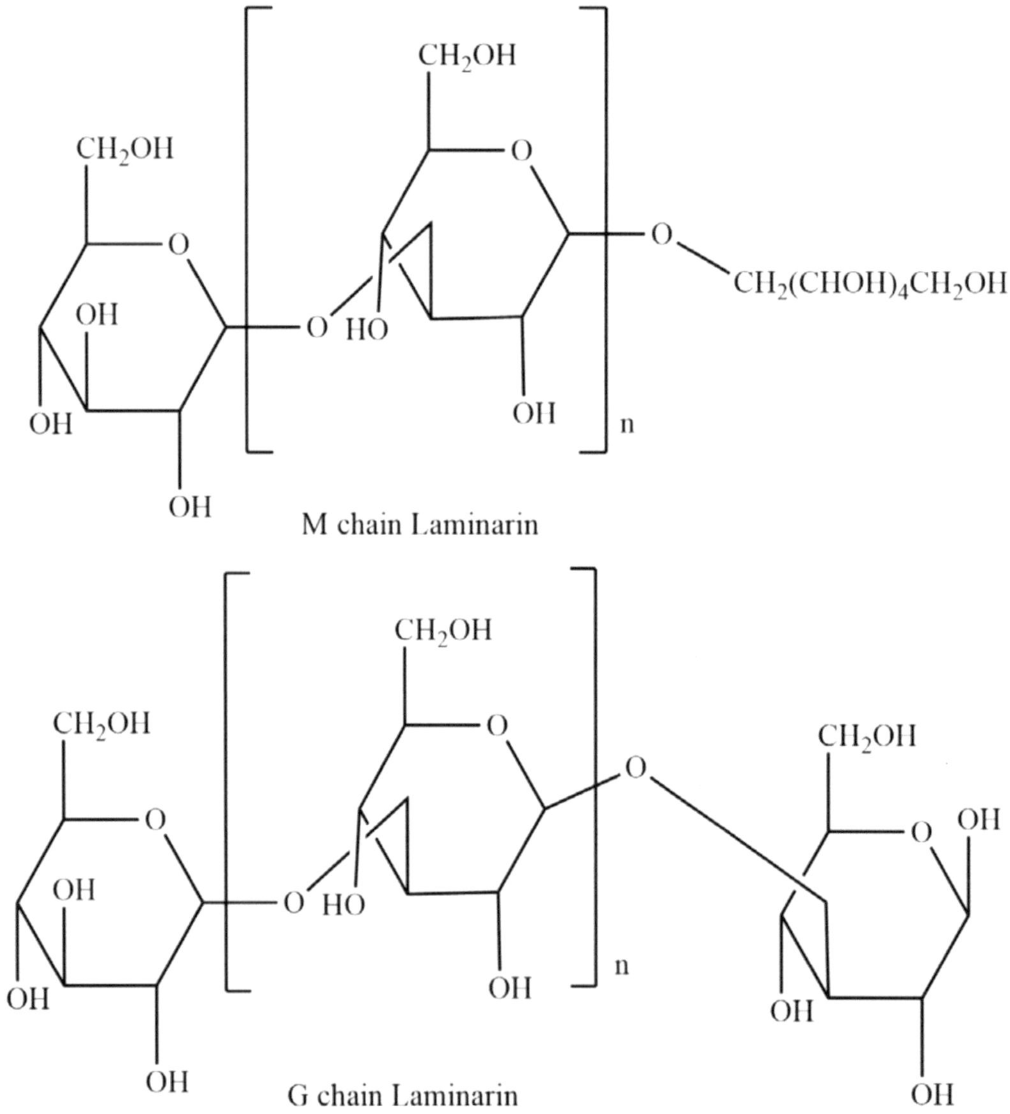

Fig. (3). Chemical structure of M and G chain Laminarin one unit.

Ulvan

Ulvan is the major water-soluble sulfated polysaccharide (soluble dietary fibers) found in green seaweed of order Ulvales. Ulvan content in green seaweeds can vary 18%-29% on the dry weight. Green seaweeds *Ulva* spp. and *Enteromorpha* spp are the major sources of ulvan [15]. Studies have identified chemical composition and structure of ulvan from green seaweeds that are complex and composed of glucose, rhamnose, glucuronic acid, and xylose. Ulvan can act as a chemo-preventive and chemo-therapeutic agent against human breast cancer (MCF7), hepatocellular carcinoma (HepG2), and cervical cancer (HeLa) by showing significant cancer cell cytotoxicity. Further, identified biological activities of ulvan include anti-oxidant activity against oxidative stress, antihyperlipidemic activity to regulate dyslipidemia, and anticoagulant activity. Other biological activities of ulvan include anti-viral activity *in vitro* against human and avian influenza viruses. Green seaweeds are normally consumed raw in salads, cooked, in soups, dried, and garnish. As an example, *Ulva* sp. known as aosa nori in Japan used in dried form to flavor foods [15, 16].

Fucoidan

Fucoidans are high molecular weight fucose-rich sulfated polysaccharides found in brown macro algae and echinoderms (Fig. **4**). Some fucoidan-containing brown algae are *Undaria pinnatifida, Fucus vesiculosus, Himanthalia lorea, Bifurcaria bifurcate, Padina pavonia, Ecklonia kurome, Sargassum* sp., and *Hizikia fusiforme* [17]. Chemical composition of fucoidans is complex. Other than fucose and sulfate, fucoidans contain other monosaccharides such as glucose, galactose, mannose, and xylose. Fucoidans have been studied for a wide range of biological activities (Fig. **5**). At present, fucoidans are available for use in different areas such as functional foods, dietary supplements, cosmetics, and for inclusion in aquaculture, livestock, and pet feed supplements.

Bioactive Compounds in Microalgae

Microalgae is comprised of a highly diverse group of 200,000–800,000 species. They are unicellular algae with a diameter of 1 and 50 µm. Microalgae contain a high amount of protein, whereas some microalgae species even contain a similar amount of protein found in meat, eggs and milk. Macronutrients and bioactive compounds present in edible microalgae are proteins, lipids (EPA, 20:5 n-3 & DHA, 22:6 n-3), dietary fiber, minerals (P, Mg, K, Ca, Fe, Zn), Vitamins (E, B1, B2, B3, B6, B9, C, E, B12), and pigments (phycocyanin, carotenoids, chlorophylls). Microalgae are a rich source of high-quality protein and bioactive compounds. They are even low in fat content. Therefore, supplementation of microalgae in humans and other mammals is associated with the anti-oxidant, antihypertensive, immunomodulatory, antidiabetic and antihyperlipidemic effects [18 - 20].

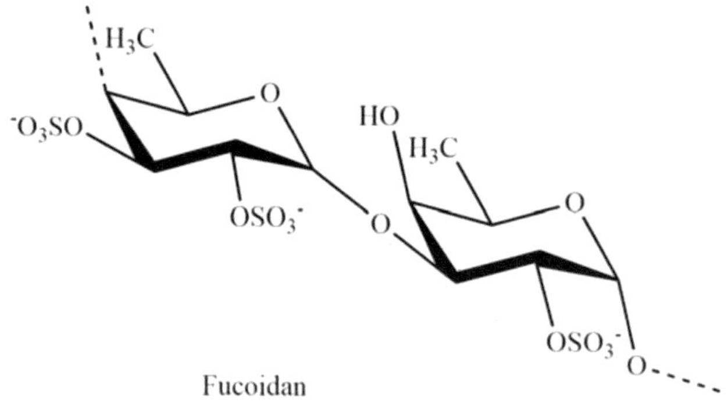

Fig. (4). Chemical structure of fucoidan.

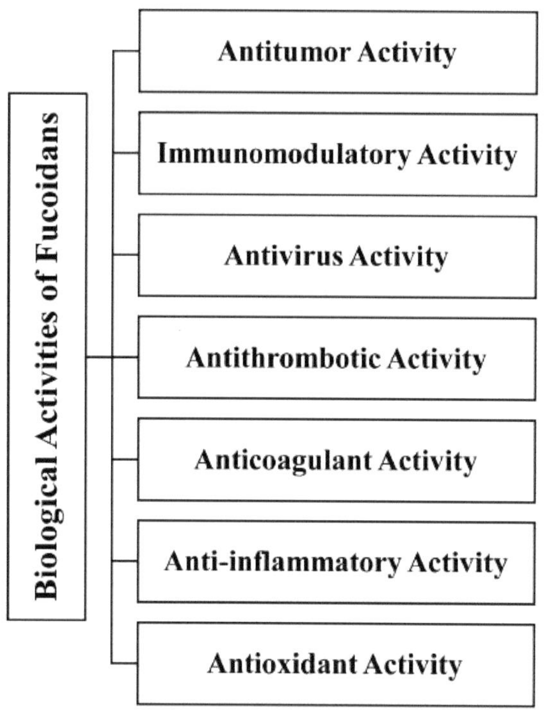

Fig. (5). Biological activities of Fucoidans.

CONCLUDING REMARKS

Edible macro and micro algae are a reliable source of functional foods and nutraceuticals. There are several reports showing the nutritional and medicinal benefits and importance of seaweeds. Phycocolloids from edible seaweeds are widely used in the food industry and have become an essential part of the vegetarian diet. Many bioactive compounds in seaweeds, such as laminarin, ulvan and fucoidan, have been studied for their therapeutic applications in inflammation, oxidative stress, cancer, hypertension and neurodegeneration, yet clinical trials are recommended to check the applicability in pharmaceutical industries.

CONSENT FOR PUBLICATION

Not applicable.

CONFLICT OF INTEREST

The author(s) declare no conflict of interest, financial or otherwise.

ACKNOWLEDGEMENTS

The authors would like to acknowledge the University of Sri Jayewardenepura, Sri Lanka, Uva Wellassa University, Sri Lanka, 2019 Academic Research Support Program in Gangneung-Wonju National University and Ministry of Oceans and Fisheries, Republic of Korea (No. 20140441) for their support and continuous encouragement.

REFERENCES

[1] Farrer J, Wang C. Who owns a cuisine? The grassroots politics of Japanese food in Europe. Asian Anthropol 2021; 20(1): 12-29. [http://dx.doi.org/10.1080/1683478X.2020.1774960]

[2] Ferraces-Casais P, Lage-Yusty MA, Rodríguez-Bernaldo de Quirós A, López-Hernández J. Evaluation of bioactive compounds in fresh edible seaweeds. Food Anal Methods 2012; 5(4): 828-34. [http://dx.doi.org/10.1007/s12161-011-9321-2]

[3] Gupta S, Abu-Ghannam N. Bioactive potential and possible health effects of edible brown seaweeds. Trends Food Sci Technol 2011; 22(6): 315-26. [http://dx.doi.org/10.1016/j.tifs.2011.03.011]

[4] Tanna B, Mishra A. Metabolites unravel nutraceutical potential of edible seaweeds: An emerging source of functional food. Compr Rev Food Sci Food Saf 2018; 17(6): 1613-24. [http://dx.doi.org/10.1111/1541-4337.12396] [PMID: 33350143]

[5] Abdul Khalil HPS, Saurabh CK, Tye YY, *et al.* Seaweed based sustainable films and composites for food and pharmaceutical applications: A review. Renew Sustain Energy Rev 2017; 77: 353-62. [http://dx.doi.org/10.1016/j.rser.2017.04.025]

[6] Udayangani RM, Somasiri GD, Wickramasinghe I, Kim SK. Potential Health Benefits of Sulfated Polysaccharides from Marine Algae. Encyclopedia of Marine Biotechnology 2020; 1: 629-35. [http://dx.doi.org/10.1002/9781119143802.ch22]

[7] Janarthanan M, Senthil Kumar M. The properties of bioactive substances obtained from seaweeds and their applications in textile industries. J Ind Text 2018; 48(1): 361-401. [http://dx.doi.org/10.1177/1528083717692596]

[8] Ghanbarzadeh M, Golmoradizadeh A, Homaei A. Carrageenans and carrageenases: versatile polysaccharides and promising marine enzymes. Phytochem Rev 2018; 17(3): 535-71. [http://dx.doi.org/10.1007/s11101-018-9548-2]

[9] Torres P, Santos JP, Chow F, dos Santos DYAC. A comprehensive review of traditional uses, bioactivity potential, and chemical diversity of the genus Gracilaria (Gracilariales, Rhodophyta). Algal Res 2019; 37: 288-306. [http://dx.doi.org/10.1016/j.algal.2018.12.009]

[10] Yegappan R, Selvaprithiviraj V, Amirthalingam S, Jayakumar R. Carrageenan based hydrogels for drug delivery, tissue engineering and wound healing. Carbohydr Polym 2018; 198: 385-400. [http://dx.doi.org/10.1016/j.carbpol.2018.06.086] [PMID: 30093014]

[11] Goonoo N, Khanbabaee B, Steuber M, *et al.* κ-Carrageenan enhances the biomineralization and osteogenic differentiation of electrospun polyhydroxybutyrate and polyhydroxybutyrate valerate fibers. Biomacromolecules 2017; 18(5): 1563-73. [http://dx.doi.org/10.1021/acs.biomac.7b00150] [PMID: 28346782]

[12] Olatunji O. Agar. In: Aquatic Biopolymers. Springer, Cham: 2020; 145-68. [http://dx.doi.org/10.1007/978-3-030-34709-3_7]

[13] Kadam SU, Tiwari BK, O'Donnell CP. Extraction, structure and biofunctional activities of laminarin from brown algae. Int J Food Sci Technol 2015; 50(1): 24-31. [http://dx.doi.org/10.1111/ijfs.12692]

[14] Kadam S, O'Donnell C, Rai D, *et al.* Laminarin from Irish brown seaweeds Ascophyllum nodosum and Laminaria hyperborea: Ultrasound assisted extraction, characterization and bioactivity. Mar Drugs 2015; 13(7): 4270-80. [http://dx.doi.org/10.3390/md13074270] [PMID: 26184235]

[15] Udayangani C, Wijesekara I, Wickramasinghe I. Characterization of sea lettuce (*Ulva lactuca*) from Matara, Sri Lanka and development of nutribars as a functional food. Ruhuna J Sci 2019; 10(2): 96-107.

[16] Wijesekara I, Pangestuti R, Kim SK. Biological activities and potential health benefits of sulfated polysaccharides derived from marine algae. Carbohydr Polym 2011; 84(1): 14-21. [http://dx.doi.org/10.1016/j.carbpol.2010.10.062]

[17] Rathnayake AU, Abuine R, Kim YJ, Byun HG. Anti-Alzheimer's materials isolated from marine bio-resources: A review. Curr Alzheimer Res 2019; 16(10): 895-906. [http://dx.doi.org/10.2174/1567205016666191024144044] [PMID: 31647396]

[18] Koyande AK, Chew KW, Rambabu K, Tao Y, Chu DT, Show PL. Microalgae: A potential alternative to health supplementation for humans. Food Sci Hum Wellness 2019; 8(1): 16-24. [http://dx.doi.org/10.1016/j.fshw.2019.03.001]

[19] Batista AP, Gouveia L, Bandarra NM, Franco JM, Raymundo A. Comparison of microalgal biomass profiles as novel functional ingredient for food products. Algal Res 2013; 2(2): 164-73. [http://dx.doi.org/10.1016/j.algal.2013.01.004]

[20] Ramos-Romero S, Torrella JR, Pagès T, Viscor G, Torres JL. Edible microalgae and their bioactive compounds in the prevention and treatment of metabolic alterations. Nutrients 2021; 13(2): 563. [http://dx.doi.org/10.3390/nu13020563] [PMID: 33572056]

CHAPTER 10

Microalgae as a Sustainable Source of Vegan Protein and Omega-3 Fatty Acids

Luke J. Webster[1], **Xuan Ji**[1] and **Peer M. Schenk**[1,*]

[1] *Algae Biotechnology Laboratory, School of Agriculture and Food Sciences, The University of Queensland, Brisbane, Queensland 4105, Australia*

Abstract: As global populations increase, sustainable food production has become concerning, with a massive proportion of the population suffering from malnutrition, and the gradual worsening of negative environmental impacts. Microalgae offer an alternative to traditional food sources, being able to meet nutritional requirements while not competing for arable land, using large amounts of water, or polluting the atmosphere. Algae can be used to produce key nutrients of the human diet – amino-acid-rich protein and healthy fats and oils, the benefits of which have been clearly documented previously. Mass cultivation of algae as a food product would be a step in the correct direction for sustainable food production and can provide significant improvements over traditional agriculture and crop farming industries in terms of health and sustainability.

Keywords: Agriculture, Aquaculture, Food Security, Microalgae, Nutraceuticals, Protein, Sustainability.

INTRODUCTION

Microalgae form a large proportion of the world's biomass [1], and for a good reason. Many microalgae offer very high nutritional value and are rich in essential amino acids, polyunsaturated fatty acids, vitamins, minerals and antioxidants, as well as numerous other medically bioactive compounds. The consumption of many algal species is documented to improve the immune system, reduce the risk of heart disease, and act in anti-inflammation. Among the many useful compounds produced by microalgae, vegan protein and omega-3 oils are held primarily in focus for this chapter.

Current sources of protein and omega-3 fatty acid-rich oils face similar issues, with large proportions of protein and omega-3 sourced from animal agriculture

* **Corresponding author Peer M. Schenk:** Algae Biotechnology Laboratory, School of Agriculture and Food Sciences, The University of Queensland, Brisbane, Queensland, Australia; Tel: +61 7 33651142; Fax: +61 7 33651699; E-mail: p.schenk@uq.edu.au

that in turn also require protein-rich feeds. With approx. 45% of global arable land used for feed production, the rising surge in the popularity of a vegan diet offers a new opportunity to address food security for the future. Animal agriculture also contributes significantly to pollution, greenhouse gas emissions and usage of clean water [2]. Production of vegan proteins and omega-3-rich oils from microalgae does not require massive consumption of fresh water, fixes atmospheric carbon in its production, and many systems have been developed to minimise land usage.

VEGAN PROTEIN

Across the world, various food products are used as the main source of protein. Most of the world sources protein for consumption from grains and legumes, however, this varies depending on the country. Notably, wealthier countries tend to consume much higher levels of protein overall, and the fraction of protein from meat products becomes more significant.

Meat products are significantly high in protein and contain a diverse range of amino acids readily bioavailable when cooked. However, meat products are also associated with severe negative health effects as a result of over-consumption. In fact, processed meat has been labelled as "carcinogenic to humans" by The International Agency for Research on Cancer [3]. In addition to posing a relative risk to humans, the process of meat production also takes a very serious toll on the environment. Approximately 18% of global carbon emissions stem from livestock agriculture, the second largest contributor of emissions behind the energy industry [4]. Animal husbandry is also estimated to already cover more than two-thirds of the world's agricultural land [5], while also being the largest cause of water pollution and wastage.

Based on the above reasons, combined with the moral questionability of meat-sourced protein, many people are turning towards vegan sources of protein as an alternative. Currently, however, many vegan protein choices come with significant drawbacks that limit their large-scale use. Most plant-based proteins, such as those derived from grains, often lack essential amino acids for human nutritional requirements, and, therefore must be consumed in conjunction with other protein sources. Some of the more common non-meat protein sources include soy protein, nuts, seeds, and high-protein vegetables such as peas. Soy protein production is 6-20 times more energy efficient than animal protein production. Production of soy and legume-based meat replacement uses significantly less water than beef, with a production of a 150-gram burger using a mere 158 litres of water compared to 2350 for beef [6]. However, these sources of protein also compete for arable land where soybean is mass-cultivated (albeit mostly for animal feed), and have led to deforestation of large areas in Brazil, Argentina and the US [7]. There have also been unproven health concerns surrounding the consumption of soy proteins, which limit customer approval.

Algae-based proteins could overcome many of these issues. The cultivation of microalgae does not need to compete with traditional crops for arable land. Algae can accumulate protein to a similar level as meat and soy and boast a protein yield of 4-15 tons/ha/year, compared to 0.6-1.2 tons/ha/year for soybean [8]. Marine microalgae do not require the use of freshwater in cultivation, and some species of microalgae can be cultivated in wastewater [9]. Algal protein is also regarded as a viable source of protein, meeting the amino acid requirements of the Food and Agriculture Organisation of the United Nations (FAO) under certain conditions.

Table 1. Essential amino acid fractions from microalgal protein samples compared with traditional protein sources and the FAO guidelines.

-	-	**Essential Amino Acids g/100 g Protein**								
-	**Source**	**Ile**	**Leu**	**Val**	**Lys**	**Phe + Tyr**	**Met + Cys**	**Trp**	**His**	**Thr**
Guidelines	**FAO/WHO [10]**	**4.0**	**7.0**	**5.0**	**5.5**	**6.0**	**3.5**	**1.0**	—	**2.0**
Red Meat	Beef [11]	5.1	8.4	5.7	8.4	7.2	3.7	1.1	2.9	4.0
Protein Alternative	Egg [10]	6.6	8.8	7.2	*5.3*	10.0	5.5	1.7	2.4	5.0
	Soybean [10]	5.3	7.7	5.3	6.4	8.7	*3.2*	1.4	2.6	4.0
	Corn [10]	*3.4*	11.0	*4.9*	*3.2*	8.9	*3.3*	4.4	—	3.7
Microalgae	*Chlorella Vulgaris* [8]	3.8	8.8	5.5	8.4	8.4	3.6	2.1	2.0	4.8
	Dunaliella bardawil [8]	4.2	11.0	5.8	7.0	9.5	3.5	*0.7*	1.8	5.4
	Nannochloropsis oceanica	*3.9*	9.5	5.4	6.9	9.5	*3.4*	1.4	2.0	5.4
	Scenedesmus obliquus [8]	*3.6*	7.3	6.0	5.6	8.0	*2.1*	*0.3*	2.1	5.1
	Arthrospira maxima [8]	6.0	8.0	6.5	*4.6*	8.8	*1.8*	1.4	1.8	4.6
	Arthrospira platensis [12]	6.7	9.8	7.1	*4.8*	10.6	3.4	*0.3*	2.2	6.2
	Aphanizomenon sp. [8]	*2.9*	*5.2*	*3.2*	*3.5*	*2.5*	*0.9*	*0.7*	0.9	3.3

Incidences, where amino acid fractions are lower than FAO guidelines are indicated by italics.

Apart from red meat and fish, none of the traditional protein sources meet the recommended guidelines, with eggs covering the essential amino acids but lacking slightly in lysine (Table **1**). There is a large discrepancy in amino acid fractions from different species of microalgae. This is due in part to large differences between algal species. For example, *Arthrospira platensis*, also known as spirulina, is a cyanobacterium and thus distinctly different from eukaryotic algae. The amino acid profiles of algae also differ significantly depending on culture/processing methods. For example, Tibbetts *et al.* [13] found that supercritical fluid CO_2 extraction increased free amino acid content, and that lysine content was reduced when the process was performed at higher temperatures (90°C). This has indicated that the amino acid profile can be manipulated by varying the processing conditions, producing algal protein samples that meet the FAO guidelines. The most promising microalgae are *Chlorella vulgaris* with only Ile slightly lower than recommended by FAO (3.8% instead of 4.0%), *Dunaliella bardawil* with only Trp lower (0.7% instead of 1%), and *Nannochloropsis oceanica* with only two amino acid groups failing the FAO recommendation by 0.1%. Considering that *N. oceanica* can be readily cultivated in seawater, it is likely the most promising species to address current and future protein demand.

When comparing protein nutrition, it can be useful to consider the EAAI (essential amino acids index), which is calculated by the following:

$$EAAI = 10^{\log EAA} \tag{1}$$

for

$$logEAA = 0.1\left[log\left(\frac{a_1}{a_{1s}}\right) + log\left(\frac{a_2}{a_{2s}}\right) + log\left(\frac{a_n}{a_{ns}}\right)\right] \tag{2}$$

Where $a_{1....}a_n$ are the amino acid content (g/100 g protein), and $a_{1s....}a_{ns}$ are the recommended levels of each amino acid. The EAAI is therefore the geometric mean of the ratio of all essential amino acids to their reference comparison.

Similarly, the chemical score index can be calculated by the following equation:

$$CSI = 100 * \frac{a_{limiting}}{a_s} \tag{3}$$

The EAAI is useful for showing the average nutritional value for every amino acid in a sample, while the CSI is useful for determining the severity of the deficiency in the limiting amino acid.

Table 2. EAAI and CSI Values of Various Protein Sources in Reference to the FAO Guidelines.

-	Source	EAAI	CSI
Guidelines	**FAO/WHO**	**1.00**	**100**
Red Meat	Beef	1.22	110
Protein Alternatives	Egg	1.41	96
	Soybean	1.21	91
	Corn	1.24	58
Microalgae	*C. vulgaris*	1.31	95
	N. oceanica	1.26	97
	D. bardawil	1.22	70
	A. maxima	1.16	51
	A. platensis	1.17	30
	S. obliquus	0.97	30
	Aphanizomenon sp.	0.70	26

As shown in Table **2**, after beef, protein from eggs matches the recommended guidelines most appropriately, with an average ratio of 1.41 essential amino acid content and a CSI of 96.4, indicating that 96.4% of the limiting amino acid (lysine) is present. Most forms of algal protein investigated offer decent EAAI values but have lower CSI values. This deficiency of essential amino acid would mean that protein needs to be acquired from other sources to balance the amino acid intake. Most of these low CSI values are caused by a deficiency in tryptophan.

However, two species of microalgae stand out as the most nutritious: *C. vulgaris* has comparable EAAI and CSI values to egg protein and presents superior nutritional value to soy protein. *C. vulgaris* is also renowned for high protein accumulation in the cell (51-58% DCW), thus offering a highly viable vegan protein alternative. *N. oceanica* is a marine microalga well-renowned for producing high lipid content, including omega-3 fatty acids, but it also contains highly nutritious protein. Both organisms can produce a protein that can compete with beef, egg and soy, while remaining much more sustainable to produce. Notably, these values were taken from purified protein samples of algae cultured under standard conditions – there is a significant improvement in protein quality possible if optimised for this purpose.

Sustainability of Vegan Protein from Microalgae

The major limiting factor for the use of microalgae for large-scale nutritional use is the relatively high cost of production when compared to more traditional sources such as animal agriculture. However, the traditional methods, while engrained in society currently, are not sustainable and actual costs from indirect effects, such as pollution, land degradation and greenhouse gas emissions, are typically not taken into consideration. Taking the US as an example: the exploitation of the land for agricultural purposes is causing a soil erosion rate 13 times higher than the recommended sustainable level of 1 ton/ha/yr [14]. With the world population expected to approach 10 billion by 2050, protein and nutrient production would need to be ramped up to support another 2 billion, yet land exploitation resulting in nutrient depletion would mean that traditional farming approaches would no longer be viable to support this.

Alarmingly, meat produced by animals in feedlots requires protein-rich feed sources (*e.g.*, soybean), which covers 45% of arable land and is the main reason for deforestation in the Amazon. Additionally, the energy costs and associated fossil fuel usage are simply not acceptable and cannot meet the demands of halving carbon emissions by 2030, as established by the Paris agreement. In fact, beef production requires a ratio of 40 kcal fossil fuel energy per 1 kcal of nutritional energy, as compared to an input of 2.2-5 kcal for 1 kcal of plant-based energy.

Production of proteins from microalgae, however, offers a far more sustainable solution. The growth of algae fixes carbon dioxide from the atmosphere or can use waste carbon gases from other production processes. Furthermore, the algal growth can be paired with an anaerobic digester to produce biogas from algal residues for energy use and to recycle nutrients [15]. Microalgae utilise significantly less land than traditional agriculture, and our own measurements from large-scale cultivation have shown that protein yields from *Scenedesmus* microalgae are 72 times higher on a per area basis than soybean. Microalgal protein could completely substitute soy protein by using less than 7% of the total land used for soybean growth [16]. Algal culture does not significantly impact soil erosion, and does not typically require the use of pesticides, thereby posing a reduced risk to human health or the environment.

It can be concluded that certain species of microalgae present a promising alternative for vegan protein production, both in terms of nutrition and environmental impact. However, wide-spread use of microalgae as a food product has not yet been achieved. This is mainly associated with the relatively high costs of algae cultivation and processing using current methods, but these expenses are gradually being reduced as the technology matures.

OMEGA-3 FATTY ACIDS

While microalgae do pose an excellent alternative for vegan protein production, many marine microalgae also produce high levels of lipids. Of interest specifically for nutritional purposes are omega-3 polyunsaturated fatty acids (PUFAs). These fats are an essential component of the cellular membrane and cannot be produced by human metabolism. Omega-3 PUFAs have been well documented to substantially improve brain development [17], lower the risk of heart disease and pancreatic cancers [18], and possess anti-inflammatory properties by reducing cytokine liberation [19].

On the other hand, high levels of omega-6 PUFAs have been demonstrated to increase the risk of dangerous diseases, such as heart disease and depression. It is postulated that the factor determining risk is most likely to be the ratio of these two fatty acids, as humans likely evolved on a diet with an omega-6/omega-3 ratio of approximately 1, compared to the modern diet with a ratio of 15:1 [20]. Reducing the ratio to 4 demonstrated a reduction of 70% in overall mortality [21] over a 5-year period with 600 participants. Dropping the ratio to 2:1 yielded a reduction in risk for breast cancer, lowered inflammation in arthritis patients, and lowered the severity of asthma symptoms. Thus, the incorporation of more omega-3 oils in the modern diet would provide beneficial changes to public health.

Currently, the primary sources of omega-3 oil in the diet are fish products – yet the fish do not produce these oils themselves. The high levels of omega-3 fatty acids in fish are due to the consumption of significant amounts of marine algae in their diet. These fatty acids are eicosapentaenoic acid (EPA) and docosahexaenoic acid (DHA). However, global fish stocks have declined by approx. 90% and high consumption of fish has been linked to a risk of mercury poisoning, so management of the amounts and types of fish would be necessary to optimise omega-3 intake from fish. Cutting out the "middlefish" and directly consuming omega-3 oil from microalgae would help to reduce processing costs, lower the risk of mercury poisoning, and halt the ravaging of the oceans due to current fishing methods.

Omega-3 oils can also be consumed via nuts and seeds in the form of alpha-linolenic acid (ALA). While ALA does present benefits for human consumption, DHA and EPA have been highly linked to improvements in brain health and function, thus a diet consisting of all three would be preferable [22]. DHA and EPA are both at high levels in algae and fish oil, however, the growth and processing conditions of algal oil production can be manipulated to produce optimised ratios of these fatty acids, providing an advantage over traditionally farmed fish oil.

Table **3** highlights the fatty acid profiles present in some fish and microalgae-based oils. To meet the nutritional guidelines, oil supplements should include high levels of omega-3-rich oil, with a small amount of omega-6. To obtain the full health benefits, omega-3-rich oil would optimally be in the form of EPA and DHA. As can be seen, algal oil offers a significantly high quantity of DHA and/or EPA – however, this can vary greatly between products depending on the algal species and strain. Generic fish oil contains a decent but much lower level of both EPA and DHA, and tuna oil can compete nutritionally with some algal oils (but this process requires enzymatic extraction and is generally more expensive). Overall, microalgal oil seems to be nutritionally superior, while also being more readily controlled via strain selection and growth conditions.

Table 3. Fatty acid compositions from various fish and microalgae-based omega-3 oil supplements.

Fatty Acid (g/100g)	Microalgal Oil [23]	Microalgal Oil [24]	Fish Oil [23]	Tuna Oil [25]
16:0 (PALM)	24.34	2.67	32.65	20.09
16:1 (PALMO)	0.37	17.12	13.99	5.53
18:0 (SA)	1.32	0.45	2.81	6.99
18:1n9 (OA)	0.08	0.59	10.37	19.44
18:2n6 (LA)	0.17	0.43	3.23	1.50
20:3n6 (DGLA)	0.30	0.00	0.10	0.00
20:4n6 (AA)	0.25	4.40	0.70	0.00
18:3n3 (ALA)	0.23	0.37	1.56	0.00
20:5n3 (EPA)	0.49	26.67	6.68	6.71
22:6n3 (DHA)	50.22	4.20	11.65	23.73
Total n-6	0.72	4.83	4.03	1.50
Total n-3	50.94	31.24	19.95	30.44
ω6/ω3 ratio	0.01	0.15	0.20	0.05

As with vegan proteins, the levels of omega-3 PUFAs produced differ dramatically based on the microalgal species. *N. oceanica,* for example, can typically accumulate 30% lipids DCW, yet can reach 65-70% under nitrogen deprivation. Of this, approx. 30% is comprised of EPA [26]. By comparison, *C. reinhartii* has only 6% of its lipids constituting EPA or DHA, with 42% containing ALA [27], and some heterotrophic algae, such as *Shizochytrium* sp., produce large amounts of DHA. Therefore, it is clear that microalgal-based omega-3-rich oil can be tailored to produce very different ratios of fatty acids.

Sustainability of Omega-3-Rich Oil from Microalgae

Approximately 20-25% of wild-caught fish are caught to produce fish meal and oil. Fish oil is the traditional oil product high in omega-3 and was thus touted for its health benefits. However, there are rising concerns regarding the sustainability of industrial fishing, declining fish populations and potential ocean pollution [28]. Over-fishing has led to massive changes in predator diversity, which has had cascading effects such as altering marine biomass levels, and pest population levels and has consequently led to the extinction of many marine species [29]. Plant oils, while high in omega-3 ALA, do not contain high levels of EPA or DHA and thus are not as beneficial. The production of these oils also requires high cropland use, the use of pesticides and can be energy intensive.

Omega-3-rich oil produced from microalgae does not require the exploitation of natural resources, and the land use of microalgae culture is also very low – rendering microalgal oil to be the most viable renewable source of omega-3.

CO-PRODUCTION OF VEGAN PROTEIN AND OMEGA-3 PUFAS

Typically, the optimal conditions for the production of algal oil or vegan protein can be achieved for either product, whereas co-production requires a sacrifice in the yield of either product. For example, strains that typically produce high levels of omega-3-rich oils, such as *N. oceanica* may not yield a similar protein quantity as *C. vulgaris*.

A common way to combat this is to prioritise the production of one product to maximise productivity, and to draw the other as a by-product of the main process. Many algal refinery processes are designed around oil extraction. These processes generally use the following workflow: cultivation, dewatering and oil extraction. The defatted biomass is often highly rich in protein, and can be used directly for human consumption or animal feed, purified to produce protein isolates, or modified into various food products. We have shown that with this co-production method, it is possible to produce 15,000 L/ha/year of oil, and 30 tons/ha/year of protein-rich biomass. In fact, with the current economic challenges faced by algal farming, it often becomes significantly more profitable to use a co-production method.

Another method is to skip the extraction steps entirely – utilising whole algal cells in products such as food additives and supplements. This is regularly done with spirulina, a cyanobacterium that is widely regarded as a superfood even in its unprocessed form. Incorporating algae into items such as "snack bars" or other healthy foods is becoming more frequent as the nutritional value, and sustainable production becomes more widely known. Avoiding the extraction steps and utilising unprocessed algae also ensures that other beneficial compounds in microalgae find their way into the diet. Compounds such as essential vitamins, anti-oxidants (including carotenoids), minerals, fibre and enzymes are highly touted for their nutritional benefits and are readily available in unprocessed algae food products.

CONCLUDING REMARKS

The cultivation of microalgae for sustainable production of vegan protein and omega-3-rich oils clearly has incredible potential yet is rarely taken advantage. The impact of traditional food sources, both on global health and the environment becomes clearer each passing year, and it becomes more apparent that the use of microalgae could help to remediate these issues. As more techniques become available for optimisation of algae culture, extraction and processing, this technology will hopefully become more widespread – and the consumption of algae-based products will become as common as other dietary staples.

CONSENT FOR PUBLICATION

Not applicable.

CONFLICT OF INTEREST

The author(s) declare no conflict of interest, financial or otherwise.

ACKNOWLEDGEMENT

We wish to acknowledge Dr. Skye Thomas-Hall for providing amino-acid and yield data for *N. oceanica.*

REFERENCES

[1] Hemantkumar J, Rahimbhai M. Hemantkumar J, Rahimbhai M. Microalgae and its use in nutraceuticals and food supplements. In: Vítová M, Ed. Microalgae - From Physiology to Application. London: IntechOpen; 2019. [http://dx.doi.org/10.5772/intechopen.90143]

[2] Henchion M, Hayes M, Mullen A, Fenelon M, Tiwari B. Future protein supply and demand: strategies and factors influencing a sustainable equilibrium. Foods 2017; 6(7): 53. [http://dx.doi.org/10.3390/foods6070053] [PMID: 28726744]

[3] WHO (2015). IARC monographs evaluate consumption of red meat and processed meat: 240.

[4] Ilea RC. Intensive livestock farming: global trends, increased environmental concerns, and ethical solutions. J Agric Environ Ethics 2009; 22(2): 153-67. [http://dx.doi.org/10.1007/s10806-008-9136-3]

[5] Dopelt K, Radon P, Davidovitch N. Environmental effects of the livestock industry: the relationship between knowledge, attitudes, and behavior among students in Israel. Int J Environ Res Public Health 2019; 16(8): 1359.
[http://dx.doi.org/10.3390/ijerph16081359] [PMID: 31014019]

[6] Ercin E, Martinez-Aldaya M, Hoekstra A. Ercin E, Martinez-Aldaya M, Hoekstra A. The water footprint of soy milk and soy burger and equivalent animal products. 49: Unesco-IHE Institute for Water Education; 2011; 36.

[7] Barona E, Ramankutty N, Hyman G, Coomes OT. The role of pasture and soybean in deforestation of the Brazilian Amazon. Environ Res Lett 2010; 5(2): 024002.
[http://dx.doi.org/10.1088/1748-9326/5/2/024002]

[8] Bleakley S, Hayes M. Algal Proteins: extraction, application, and challenges concerning production. Foods 2017; 6(5): 33.
[http://dx.doi.org/10.3390/foods6050033] [PMID: 28445408]

[9] Abou-Shanab RAI, Ji MK, Kim HC, Paeng KJ, Jeon BH. Microalgal species growing on piggery wastewater as a valuable candidate for nutrient removal and biodiesel production. J Environ Manage 2013; 115: 257-64.
[http://dx.doi.org/10.1016/j.jenvman.2012.11.022] [PMID: 23270891]

[10] Becker EW. Micro-algae as a source of protein. Biotechnol Adv 2007; 25(2): 207-10.
[PMID: 17196357]

[11] Anaeto M, Adeyeye J, Chioma G, Olarinmoye A, Tayo G. Goat products: Meeting the challenges of human health and nutrition. Agric Biol J N Am 2010; 1(6): 1231-6.
[http://dx.doi.org/10.5251/abjna.2010.1.6.1231.1236]

[12] Amorim M, Soares J. 'Coimbra J, Liete M, Albino L, Martins M. Microalgae proteins: production, separation, isolation, quantification, and application in food and feed. Crit Rev Food Sci Nutr 2020; 1-27.

[13] Tibbetts SM, Bjornsson WJ, McGinn PJ. Biochemical composition and amino acid profiles of *Nannochloropsis granulata* algal biomass before and after supercritical fluid CO_2 extraction at two processing temperatures. Anim Feed Sci Technol 2015; 204: 62-71.
[http://dx.doi.org/10.1016/j.anifeedsci.2015.04.006]

[14] Pimentel D, Pimentel M. Sustainability of meat-based and plant-based diets and the environment. Am J Clin Nutr 2003; 78(3) (Suppl.): 660S-3S.
[PMID: 12936963]

[15] González-González LM, Zhou L, Astals S, *et al.* Biogas production coupled to repeat microalgae cultivation using a closed nutrient loop. Bioresour Technol 2018; 263: 625-30.
[PMID: 29800924]

[16] Brune D, Lundquist T, Benemann J. Algal production and harvest for food, feed and biofuels. n.d.

[17] McNamara RK, Carlson SE. Role of omega-3 fatty acids in brain development and function: potential implications for the pathogenesis and prevention of psychopathology. Prostaglandins Leukot Essent Fatty Acids 2006; 75(4-5): 329-49.
[PMID: 16949263]

[18] Fiala M. Curcumin and omega-3 fatty acids enhance NK cell-induced apoptosis of pancreatic cancer cells but curcumin inhibits interferon-γ production: benefits of omega-3 with curcumin against cancer. Molecules 2015; 20(2): 3020-6.
[http://dx.doi.org/10.3390/molecules20023020] [PMID: 25685909]

[19] Alagawany M, Elnesr SS, Farag MR, *et al.* Nutritional significance and health benefits of omega-3, -6 and -9 fatty acids in animals. Anim Biotechnol 2021; 1-13.
[http://dx.doi.org/10.1080/10495398.2020.1869562] [PMID: 33470155]

[20] Simopoulos AP. The importance of the ratio of omega-6/omega-3 essential fatty acids. Biomed Pharmacother 2002; 56(8): 365-79.
[http://dx.doi.org/10.1016/S0753-3322(02)00253-6] [PMID: 12442909]

[21] de Lorgeril M, Renaud S, Salen P, *et al.* Mediterranean alpha-linolenic acid-rich diet in secondary prevention of coronary heart disease. Lancet 1994; 343(8911): 1454-9.
[http://dx.doi.org/10.1016/S0140-6736(94)92580-1] [PMID: 7911176]

[22] Lane K, Derbyshire E, Li W, Brennan C. Bioavailability and potential uses of vegetarian sources of omega-3 fatty acids: a review of the literature. Crit Rev Food Sci Nutr 2014; 54(5): 572-9.
[http://dx.doi.org/10.1080/10408398.2011.596292] [PMID: 24261532]

[23] Feng J, Long S, Zhang H, Wu S, Qi G, Wang J. Comparative effects of dietary microalgae oil and fish oil on fatty acid composition and sensory quality of table eggs. Poult Sci 2020; 99(3): 1734-43.
[http://dx.doi.org/10.1016/j.psj.2019.11.005] [PMID: 32115040]

[24] Zhao C, Brück T, Lercher JA. Catalytic deoxygenation of microalgae oil to green hydrocarbons. Green Chem 2013; 15(7): 1720.
[http://dx.doi.org/10.1039/c3gc40558ç]

[25] Oliveira D, Licodiedoff S, Furigo A, Ninow J, Bork J, Podesta R, *et al.* Enzymatic extraction of oil from yellowfin tuna (*Thunnus albacares*) by-products: a comparison with other extraction methods. Food Sci Technol (Campinas) 2017; 52(3): 699-705.

[26] Lai YJ. Omega-3 fatty acid obtained from *Nannochloropsis oceanica* cultures grown under low urea protect against Abeta-induced neural damage. J Food Sci Technol 2015; 52(5): 2982-9.
[http://dx.doi.org/10.1007/s13197-014-1329-3] [PMID: 25892799]

[27] Darwish R, Gedi MA, Akepach P, Assaye H, Zaky AS, Gray DA. *Chlamydomonas reinhardtii* is a potential food supplement with the capacity to outperform *Chlorella* and *Spirulina.* Appl Sci (Basel) 2020; 10(19): 6736.
[http://dx.doi.org/10.3390/app10196736]

[28] Vázquez-Rowe I. A fine kettle of fish: the fishing industry and environmental impacts. Curr Opin Environ Sci Health 2020; 13: 1-5.
[http://dx.doi.org/10.1016/j.coesh.2019.08.004]

[29] Bruno JF, O'Connor MI. Cascading effects of predator diversity and omnivory in a marine food web. Ecol Lett 2005; 8(10): 1048-56.
[http://dx.doi.org/10.1111/j.1461-0248.2005.00808.x]

CHAPTER 11

Micro-Algae as a Source of Food and Bioactive Compounds for Human Health

Sonal Mishra[1], **Neha Kumari**[1] and **Rajeshwar P. Sinha**[1,*]

[1] *Laboratory of Photobiology and Molecular Microbiology, Centre of Advanced Study in Botany, Institute of Science, Banaras Hindu University, Varanasi-221005, India*

Abstract: Micro-algae are a rich natural source of highly valuable bioactive compounds like mycosporine-like amino acids (MAAs), scytonemin, carotenoids, vitamins, essential amino acids (EAAs), polyunsaturated fatty acids (PUFAs), minerals, polysaccharides, lectins, halogenated compounds, polyketides, agar-agar, alginic acid, carrageenan, enzymes and fibers. These compounds are involved in several defense strategies helping organisms to survive in various extreme environmental conditions, such as high salinity, high and low temperatures and deleterious solar radiation. Furthermore, they are promising biocatalysts and can be used in the field of "white biotechnology" to improve the sustainable production of metabolites. Micro-algae have the potential to become one of the most promising and innovative sources for novel food and functional products. They play a major role as functional ingredients for the enhancement of the nutritional value of foods and thus improving human health, well-being and quality of life. Micro-algal-derived bioactive compounds have UV-screening, anti-inflammatory, anti-microbial, antioxidant, antibiotics, immunosuppressants, anticancer and antiviral properties. These bioactive compounds have many biotechnological, industrial, pharmaceutical and cosmetic uses that are discussed in this chapter with their emerging techniques of harvesting to enhance their production on a commercial scale. The diverse metabolites from indigenous sources facilitate the promising effect, enabling researchers and pharmaceutical companies for the production of novel non-toxic and cost-effective medicines related to various applicable fields.

Keywords: Anti-Microbial, Bioactive Compounds, Food, Micro-Algae, UV-Screening.

INTRODUCTION

Micro-algae are one of the earliest forms of life on Earth. When the Earth's environment formed over 3 billion years ago, micro-algae existed in Earth's

* **Corresponding author Rajeshwar P. Sinha:** Laboratory of Photobiology and Molecular Microbiology, Centre of Advanced Study in Botany, Institute of Science, Banaras Hindu University, Varanasi, India; Tel: +91 542 2307147; Fax: +91 542 2366402; E-mail: rpsinhabhu@gmail.com

oceans. The diversity of micro-algae is vast, but this diversity has not yet been fully exposed [1, 2]. There are more than 50,000 different types of micro-algal species present in oceans and fresh water (lakes, ponds and rivers). Among these species, only 30,000 have been studied. Micro-algae have been used as food by humans for thousands of years [3]. Micro-algae can convert solar energy to chemical energy by fixing CO_2, and their efficiency is ten times greater than terrestrial plants. The commercial production of micro-algae is approximately 5,000 tons/year of dry matter [4]. Much of the microalgal biomass has been an attractive source for producing a wide range of highly valuable products, including MAAs, scytonemin, carotenoids, EAA, steroids, polysaccharides, lectins, halogenated compounds, polyketides, toxins, agar-agar, alginic acid, carrageenan, enzymes and fiber. Besides these compounds, microalgae are also a source of several useful compounds with health benefits such as proteins, carbohydrates, PUFA, essential minerals, and vitamins which can enhance the nutritional value of food products. In nutraceutical industries, *Arthrospira* (*Spirulina*) and *Chlorella* are commercialized as sources of healthy foods and nutritional supplements with several health benefits, including boosting immune system activity, anti-tumor effects, and animal growth promotion, because of the presence of abundant proteins, vitamins, active polysaccharides, and other important compounds [5, 6]. Micro-algal carotenoids like β-carotene obtained from *Dunaliella* and astaxanthin derived from *Haematococcus* are commercially produced in large-scale processes. Micro-algal-derived products are currently successfully produced for use in cosmetics and pharmaceutical products [7]. Examples include cyanobacterial polysaccharides used in personal skincare products and oligopeptides derived from extracts of *Chlorella* sp., which can promote firmness of the skin [8]. Various drug candidates with anti-inflammatory, anticancer and anti-infective activities have been identified in the pharmaceutical industries [9]. Adenosine obtained from *Phaeodactylum tricornutum* can act as an anti-arrhythmic agent for the treatment of tachycardia, and the green algal metabolite caulerpin has anti-tuberculosis activities [10, 11]. Production of oleaginous micro-alga is becoming attractive as an alternative source of biofuels with the potential to meet global demand for renewable bioenergy [12]. This chapter focuses on the production of potentially valuable bioactive compounds derived from micro-algae and their respective applications in the food and pharmaceutical industries for the benefit of human health.

BIOACTIVE COMPOUNDS DERIVED FROM MICRO-ALGAE

Micro-algae are the main source of bioactive compounds. Most of these compounds derived from micro-algae have antimicrobial, antiviral, anti-inflammatory, anti-cancer and antioxidant properties (Table **1**), which may be categorized as follows:

Polysaccharides

Polysaccharides act as carbon and energy sources and are excreted during normal as well as stressful physiological processes. They function as thickening or gelling agents [13, 14]. They are immunomodulatory, antibacterial, anticoagulant, antimutagenic, radioprotective, anti-oxidative, antiulcer, anticancer, and anti-inflammatory agents [14 - 16] (Fig. **1**). Various types of polysaccharides are derived from micro-algae like phycocolloids, carrageenan, agar, alginate, lectins *etc*.

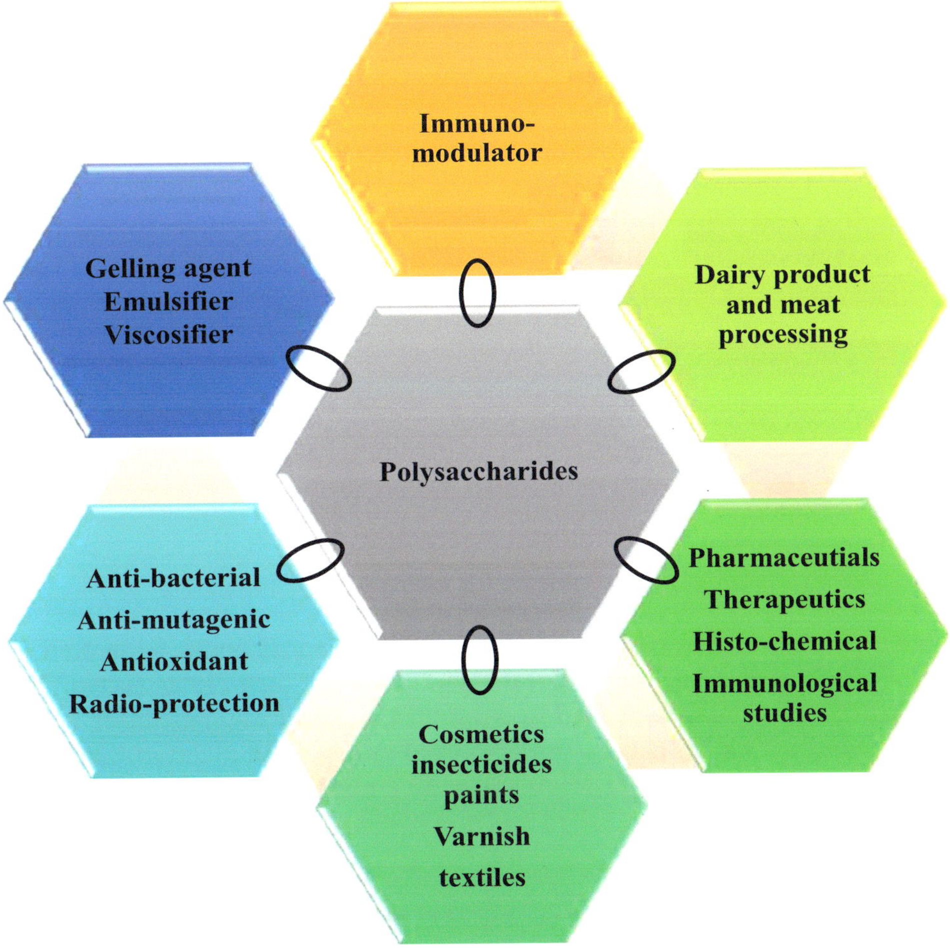

Fig. (1). Significance of polysaccharides derived from micro-algae (modified from Singh *et al.* [18]).

Phycocolloids

Phycocolloids are unique type of polysaccharides synthesized by various species of seaweeds. Among different types of phycocolloids, carrageenan, agar and alginates have great importance due to their multifunctional uses. Sulfated polysaccharides, agar and carrageenan, are extracted from red algae, while alginates are binary polyurodine containing guluronic and mannuronic acids obtained from brown algae [17]. Phycocolloids are used as an emulsifier, viscosifier and gelling agents.

Carrageenan

Carrageenans are high molecular weight compounds that are formed through replicating disaccharide units with modifications in 4-linked α-galactopyranose and/or 3,6- anhydro-D-galactopyranose and 3-linked β-D-galactopyranose. These are synthesized by several species of seaweeds. α-carrageenan is produced by red algae *Chondrus* and *Gigartina* [19]. It plays an important role in the dairy industry, meat processing and other miscellaneous products like toothpaste, air freshener gels, and pet food due to its viscosity properties.

Agar

Agar is a seaweed galactan consisting of α (1→4)-3,6-anhydro-L-galactose and β(1→3)-D-galactose [20]. Agar is used as emulsifying, stabilizing, and thickening agent. Due to its solidifying nature, it is also used in growth mediums for bacteria and fungi.

Alginate

It is a linear polysaccharide consisting of α-L-guluronic acid and 1,4-linked β-D-annuronic [21]. It is also known as alginic acid or algin. They are extracted from brown algae like *Laminaria* and used in cosmetics, pharmaceuticals, insecticides, paints, and printers' ink [22]. They are also used as a gelling agent in the textile industry for sizing cotton yarns. Brown algae are a good source of alginates (Fig. **2**) and have importance in several food and pharmaceutical industries [22].

Lectins

Lectins or agglutinins are synthesized in cyanobacteria and algae and are the complex form of proteins, having the ability to bind directly with carbohydrates without altering the property of carbohydrates to which they are bound [22]. They have various applications in histo-chemical and immunological studies due to their specificity of carbohydrates binding and are also used in identifying sugar type on the cell surface. Lectins have therapeutic properties due to their ability of binding the epithelium of intestine and enhance the diffusion of drugs. Due to their therapeutic importance, numerous lectins, i.e., scytovirin, microvirin, agglutinin, and cyanovirin-N, have been isolated from several cyanobacteria such as *Scytonema varium*. Lectins are used in prophesying transmission of HIV due to the interaction of glycans with HIV gp120 and play an important role in antiviral activities [23, 24].

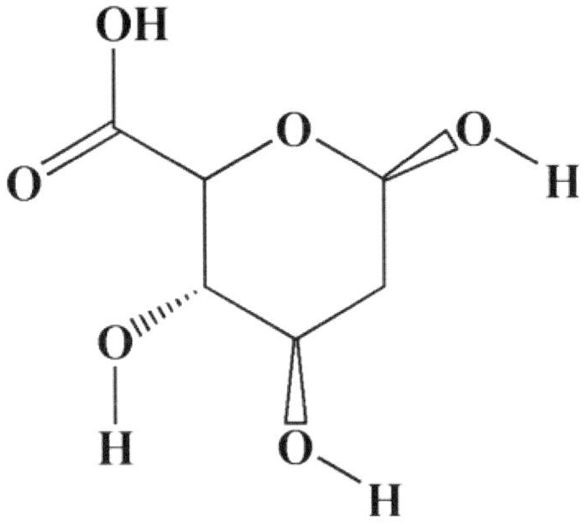

Fig. (2). Structure of alginate.

Table 1. Bioactive compounds from micro-algae.

Micro-Algae	Bioactive Compounds	References
Spirulina sp.	Polysaccharides, C-phycocyanin, Allophycocyanin	[25]
Spirulina platensis	Phycocyanin, C-phycocyanin, Phenolic acids, tocopherols (vitamin E), neophytadiene, phytol, PUFAs (*n*-3) fatty acids, oleic acid, linolenic acid, palmitoleic acid	[26 - 28]
Spirulina maxima	Phenolic, Flavonoids	[29]
Spirulina fusiformis	Diacylglycerols, C-phycocyanin	[28]
Haematococcus pluvialis	Astaxanthin, lutein, zeaxanthin, canthaxanthin, lutein, β-carotene, oleic acid	[27, 30]
Chlorella sp.	Carotenoids, sulfated polysaccharides, sterols, PUFAs (*n*-3) fatty acids	[27]
Chlorella vulgaris	Canthaxanthin, astaxanthin, peptide, oleic acid	[25, 30, 31]
Chlorella minutissima	Eicosapentaenoic acid (EPA)	[28]

(Table 1) cont.....

Micro-Algae	Bioactive Compounds	References
Chlorella ellipsoidea	Zeaxanthin, violaxanthin	[25]
Dunaliella salina	*trans*-Betacarotene, cis-carotene, β-carotene, oleic acid, linolenic acid, palmitic acid	[28, 30]
Dunaliella	Diacylglycerols	[28]
Dunaliella tertiolecta	Sterols	[32]
Botryococcus braunii	Linear alkadienes (C25, C27, C29, and C31), triene (C29)	[33]
Chlorella zofingiensis	Astaxanthin	[27]
Chlorella protothecoides	Lutein, zeaxanthin, canthaxanthin	[33]
Chlorella pyrenoidosa	Lutein, sulfated polysaccharide	[30]
Nostoc linckia and *Nostoc spongiaeforme*	Borophycin	[28]
Nostoc sp.	Cryptophycin, Phycocyanin	[28]
Nostoc muscorum	Phenolic, Terpenoids, Alkaloids, Phycobilins	[34]
Nostoc humifusum	Phenolic, Terpenoids, Alkaloids, Phycobilins	[34]
Chlorella protothecoides	Lutein	[27]
Chlorella zofingiensis	Astaxanthin	[27]
Chlorella vulgaris	Phenolic, Terpenoids, Alkaloids	[34]
Chlorella minutissima	Phytol, Phenol	[35]
Nannochloropsis gaditana	Eicosapentaneoic acid	[36]
Haematococcus pluvalis	Carotenoid, astaxanthin,	[37]

PIGMENTS AND VITAMINS

Micro-algal pigments such as chlorophyll, carotenoids and phycobiliproteins exhibit different colors like green, yellow and brown to red. Some natural colorants obtained from different micro-algae such as blue pigment-phycocyanin from *Spirulina,* yellow pigment-β-carotene from *Dunaliella* and yellow to red pigment-astaxanthin from *Haematococcus* are gaining importance over synthetic as they are non-toxic and non-carcinogenic [38]. Becker [39] reported that micro-algae are an abundant source of vitamins, such as vitamins A, B1, B2, B12, C, E, nicotinate, biotin, folic acid, pantothenic acid and niacin. Vitamin B12 is synthesized by *Chlorella* sp. and *Pleurochrysis carterae, Dunaliella tertiolecta, Tetraselmis suecica, Nannochloropsis oculata, Chaetoceros calcitrans* and *Porphyridium cruentum* synthesize vitamin E. Smerilli *et al.* [40] reported that the vitamin C content in micro-alga *Skeletonema marinoi* depends on the spectral composition and ultraviolet radiation intensity. Several micro-algal pigments are briefly discussed below:

Phycobiliproteins

Phycobiliproteins are hydrophilic proteins bonded to phycobilins (Fig. **3a**), a photosynthetic pigment and are mainly found in cyanobacteria and some red algae [27]. Based on their maximum UV-visible absorption spectra, phycobiliproteins are classified as phycocyanin (blue pigment), phycoerythrin (red pigment), and allophycocyanin (pale-blue pigment). The major sources of phycobiliproteins are *Spirulina* sp., *Arthrospira platensis* and *Amphanizomenon flos-aquae. Spirulina* sp. is a rich source of c-phycocyanins and allophycocyanins and contains relatively low nucleic acid contents. Phycocyanin produced by *Arthrospira platensis* is extensively used as a food colorant, while *Porphyridium* is used for the production of erythrin, a fluorescent agent used in flow cytometers and immunoassay assays, in microscopy and DNA assays as nonradioactive markers. Phycobiliproteins are commercially used as natural dyes and fluorescent agents and in the pharmaceutical sectors as antioxidants, anti-inflammatory, neuroprotective and hepatoprotective agents [41]. Due to its stability, phycocyanin is also used in cosmetic formulations [33].

Astaxanthin

It is keto- carotenoid extracted from *Haematococcus pluvialis* (*H. pluvialis*). It has 100 times more antioxidant activity than α-tocopherol. It is also produced by *Chlorella zofingiensis,* and its production rate is less than that of *H. pluvialis*. Several other genera, such as *Chlorella, Chlorococcum* and *Scenedesmus,* are also potential producers of astaxanthin [42]. Astaxanthin (Fig. **3b**) plays a major role in protecting the alga against UV light damage and photooxidation of the poly-unsaturated astaxanthin. It shows strong free-radical scavenging activity, and protects against lipid peroxidation and oxidative damage of low-density lipoprotein (LDL)-cholesterol, cell membranes, cells and tissues. It also shows protective effects against diseases like cancers, inflammatory diseases, metabolic syndrome, diabetes, diabetic nephropathy, neurodegenerative diseases and eye diseases [43]. Tso and Lam [44] reported that astaxanthin is effective at ameliorating retinal injury and it is also effective in protecting photoreceptors from degeneration in rats. The intake of astaxanthin has been shown to have a protective effect against ethanol-induced gastric ulcers caused by *Helicobacter pylori* and affects cholesterol levels in the blood, which could help to cure heart disease, and prevent obesity [45] and age-related cognitive function in humans [46] (Table **2**).

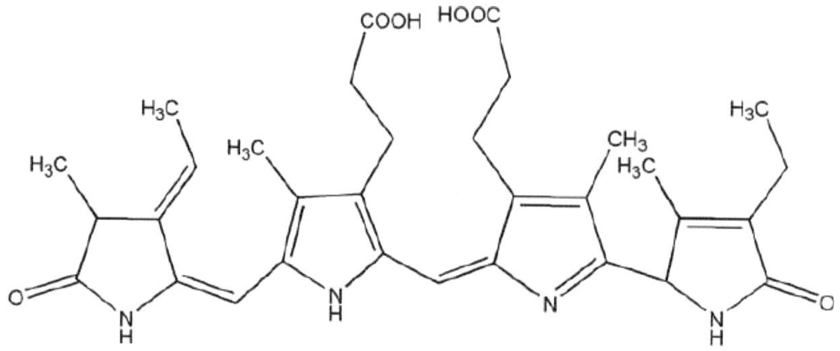

(a) Phycobilins

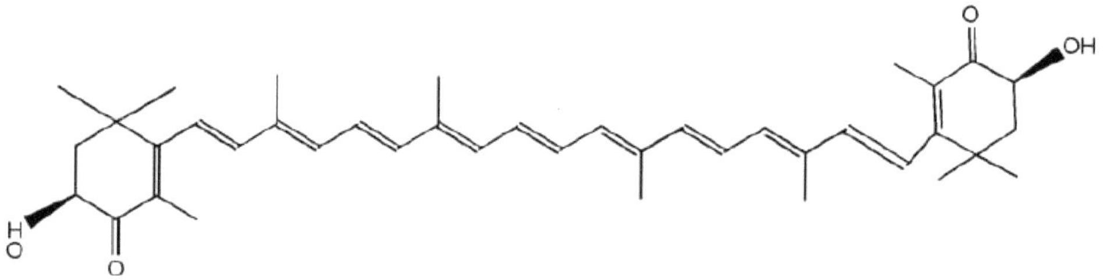

(b) Astaxanthin

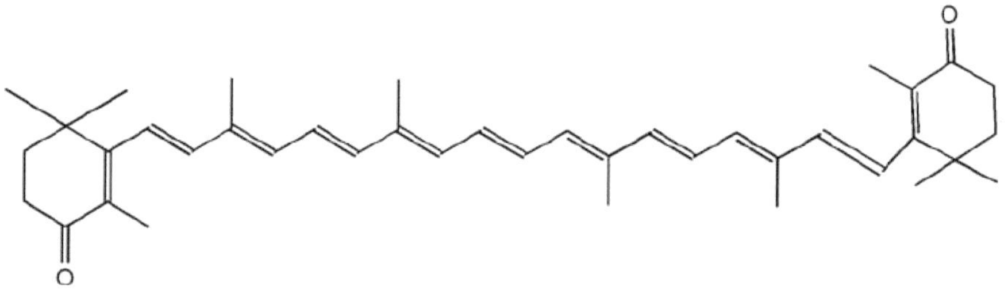

(c) Canthaxanthin

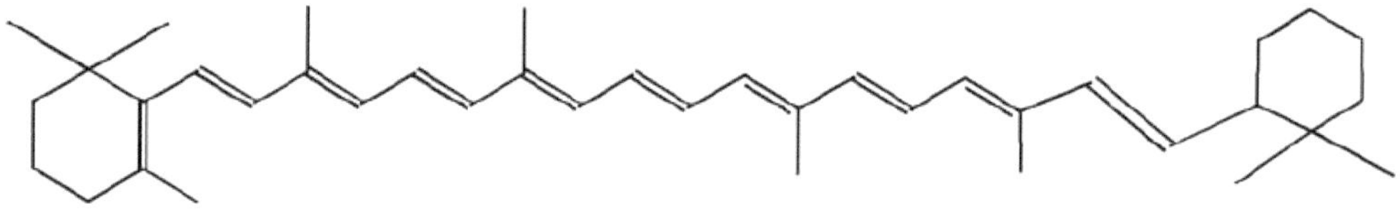

(d) β-carotene

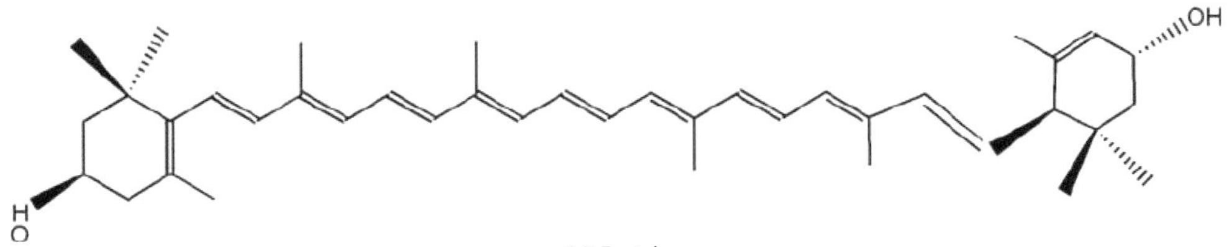

(e) Lutein

Fig. (3). Chemical structure of **(a)** Phycobilins, **(b)** Astaxanthin, **(c)** Canthaxanthin, **(d)** β-carotene and **(e)** Lutein, obtained from different micro-algae.

Canthaxanthin

It is a carotenoid used as a food dye, giving color to egg yolks and chicken skin (Table **2**). It has antioxidative, anti-inflammatory and neuroprotective effects [47], and it is produced by *Coelastrella striolata* and *C. zofingiensis*. It reduces the excess amount of light that reaches to PSII complex of the photosystem and thus providing passive photoprotection to the photosynthetic apparatus. Canthaxanthin (Fig. **3c**) supplemented diet is used as fodder for aquatic animals such as shrimp, lobster, and ornamental fishes to enhance their pigmentation.

β -Carotene

β-carotene is a high demanding carotene because of its diverse variety of commercial applications, including food colorant, provitamin A (retinol) in food and animal feed, additives to cosmetics and multivitamin preparations and a powerful antioxidant in health food products. β-carotenes are naturally produced by micro-algae *Dunaliella*. β-carotene, a component of the photosynthetic reaction center in *Dunaliella*, is accumulated in lipid globules in the interthylakoid spaces of the chloroplast [42]. Plasma high-density lipoprotein-cholesterol is enhanced in fibrate-treated patients by β-carotene (Fig. **3d**) that is rich in *Dunaliella* powder while reducing plasma triglyceride levels. Lin *et al.* [48] reported that β-carotenes reduced lipid peroxidation and also inhibited the activation of hepatic stellate cells.

Lutein

It is an essential component of pigment present in the macula lutea (or yellow spot) in the retina and lens of the eyes. Lutein (Fig. **3e**) is used as a colorant in cosmetics and food products [49] and is responsible for pigmentation in fish and poultry. It is extracted from *Muriellopsis* sp. and *Scenedesmus almeriensis*. Micro-alga synthesizes other carotenoids, such as Zeaxanthin from *D. salina* mutants [49], echinenone from *Botryococcus braunii* (*B. braunii*) [50], fucoxanthin from *Phaeodactylum tricornutum* (*P. tricornutum*) [51], and the colorless carotenoids phytoene and phytofluene from *D. salina* [52]. They also play a major role in protection against oxidative stress caused by a wide spectrum of disease and aging.

Table 2. Biotechnological applications of micro-algal metabolites.

Algae	Compound	Uses	References
Spirulina platensis	Phycocyanins	Nutraceuticals, Cosmetics	[30]
Chlorella vulgaris	Ascorbic acid	Nutraceuticals, Cosmetics	[30]
Hematococcus pluvialis	Carotenoids, astaxanthin	Nutraceuticals, pharmaceuticals, additives	[30]
Odontella aurita	Fatty acids	Pharmaceuticals, Cosmetics	[30]
Porphyridium cruentum	Polysaccharides	Pharmaceuticals, Cosmetics	[30]
Dunaliella salina	Carotenoids	Nutraceuticals, food supplement	[30]
Isochrisis galbana	Fatty acids	Nutrition	[161]
Phaedactylum tricornutum	Lipids, fatty acids	Human nutrition	[158, 159]
Nannochloropsis sp., *Phaeodactylum tricornutum, Porphyridium cruentum*	Eicosapentaenoic acid	Cardiovascular benefits, mental development and support, anti-inflammatory, protection against atherosclerosis	[29, 160, 161]
Porphyridium purpureum, P. cruentum, Parietochloris incisa	Arachidonic acid	Improves normal growth, visual and functional development in infants	[162 - 164]
Crypthecodinium cohnii, Schizochytrium sp., *Ulkenia* sp.	Docosahexaenoic acid	Cardiovascular benefits, improves nervous system development and function of the brain	[165, 166]
Dunaliella salina	Bixin	Food additives, Cosmetics	[167, 168]
Chlorella sp., *Nannochlropsis oculata, Euglena gracilis*	Tocopherol (α-Tocopherol)	Vitamin E, food additive E306, E307, E308 antioxidant in cosmetics and foods	[169, 170]
Haematococcus pluvialis, Chlorella Zofingiensis	Canthaxanthin	Food additive E161 g, farming of salmonids and chicken, tanning pills	[171]
Botryococcus braunii, Dunaliella tetriolecta	Violaxanthin	Food additive E161e (approved in Australia & New Zealand)	[167]
Phaeodactylum tricornutum	Zeaxanthin	Food additive E161 h, animal feed, pharmaceutical (anti-colon cancer, eye health)	[172, 173]
Phaeodactylum tricornutum	Fucoxanthin	Anti-adipositas	[174]

MICROALGAL FATTY ACIDS

Crypthecodinium, Schizochytrium, Thraustochytrids and Ulkenia are used for the production of essential fatty acids (EFAs) [53, 54]. Docosahexaenoic acid (DHA), rich in oil from *Crypthecodinium cohnii* is available on the market and contains 40-50% DHA and no eicosapentaenoic acid (EPA) or other long-chain polyunsaturated fatty acids (PUFAs) [55]. Some PUFAs producing micro-algae are *A. platensis*, *Isochrysis galbana*, *Odontella*, *Porphyridium cruentum*, *Cryptothecodinium cohnietc*. Micro-alga of the genus *Phaeodactylum* and diatoms are rich in EPA (Fig. **4a**) while *P. lutheri* has been shown to contain large amounts of DHA (Fig. **4b**) [56]. DHA is mainly found in *Cryptothecodinium cohnii,* which is a non-photosynthetic, marine dinoflagellate [57]. *Schizochytrium* strains that produce DHA and EPA are currently used as an adult dietary supplement in food and beverages, health foods, animal feeds and maricultural products, including cheeses, yogurts, spreads and dressings, and breakfast cereals. These important EFAs in this micro-alga are used as supplements in foods for pregnant and nursing women and cardiovascular patients [55]. EPA is produced from *Chlorella minutissima,* and its production is enhanced by lowering the temperature and increasing salinity [58]. A high amount of gamma-Linolenic acid (GLA) is produced from *Spirulina* and *Nostoc* strains.

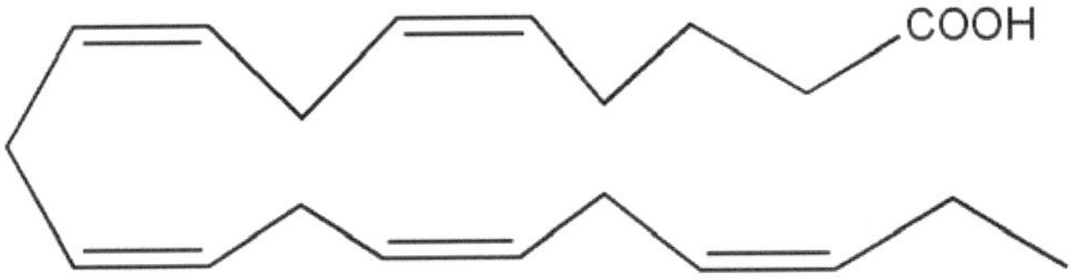

(a) EPA

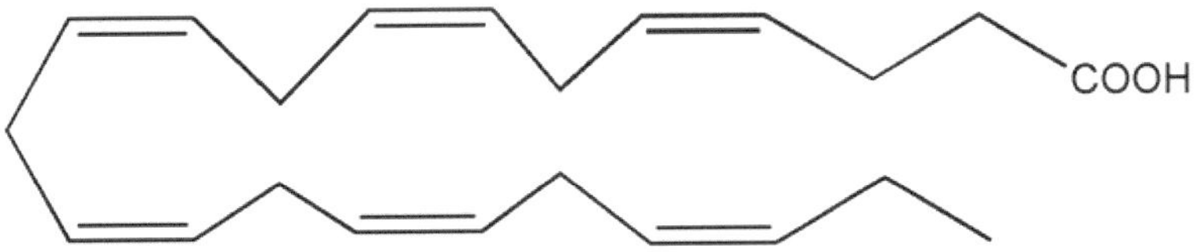

(b) DHA

Fig. (4). Chemical structure of fatty acids **(a)** EPA and **(b)** DHA.

OTHER BIOACTIVE COMPOUNDS

The micro-alga *Chlorella* sp. is known to produce β-1,3-glucan, a bioactive compound important in human health. It is routinely used in tumor immunotherapy since it can stimulate recovery of the bone marrow following chemotherapy. It also has the potential to inhibit antiviral activity in HIV infected patients and accelerate hematopoietic recovery in both sub-lethally and lethally radiated mice [59, 60]. Some micro-algal toxins are also produced like ciguatoxin which is a heat tolerant toxin produced by the dinoflagellate *Gambierdiscus toxicus* [33], a cyanotoxin, okadaic acid is produced by *Prorocentrum lima*, *Dinophysis acuminata* and *Dinophysis acuta* [61 - 63] and diatom *Pseudonitzchia* produces a neurotoxin domoic acid by which amnesic shellfish poisoning is caused [64]. Recently, two novel compounds Karatungiol A and B were isolated by Washida *et al.* [65] from cultivated symbiotic marine dinoflagellate *Amphidinium* sp. Karatungiol showed antifungal property against *Aspergillus niger* at 12 µg/disc and antiprotozoan activity against *Trichomonas foetus* at 1 µg/ml. Chlorophyll is the main photosynthetic pigment found in all photoautotrophic organisms and liver function recovery was stimulated and bile secretion was increased [66]. It also repairs cells, enhances blood hemoglobin and promotes rapid cell growth. Being a natural pigment, chlorophyll is used in the food industry as a colorant due to the increasing consumer demands for natural foods. Micro-alga such as the cyanobacterium *Aphanizomenon* contains a significant amount of chlorophyll (approx.1-2% dry weight) which stimulates liver function and increases bile secretion [66].

UV-SCREENING COMPOUNDS

In response to the harmful UV radiation, micro-algae have various defense mechanisms to cope with the radiation effects (Fig. **5**). In aquatic environments, micro-alga frames prominently as a source of UV-absorbing compounds (Table **3**), namely sporopollenin, scytonemin and MAAs have been entrenched.

Sporopollenin

Sporopollenin resistant to acetolysis is an inert biopolymer retaining complicated structure among aliphatic, especially aromatic and isoprenoid components present infrequently in some algal cell walls, spores and pollens [67 - 69]. Members of chlorophytes, such as *Enallax coelastroides, Coelastrum microporum, Characium terrestre, Scenedesmus* sp., *Scotiella chlorelloidea, Scotiellopsis rubescens,* and *Spongiochloris spongiosa* which contain large amounts of sporopollenin are highly resistant to UV-B radiation [70]. Sporopollenin was also reported in the zygotes of *Dunaliella salina* [71]. According to Rozema *et al.* [72], extracts containing sporopollenin have UV-B absorbing properties. In chlorophytes, increased tolerance to the UV-B radiation is due to the presence of sporopollenin, which increases in response to UV-B exposure, which is evident by the optical density of UV-B. UV-B-absorbing ability of sporopollenin makes it natural, innocuous and sheltered material and because of this ability, it may be used in sunscreens/cosmetics.

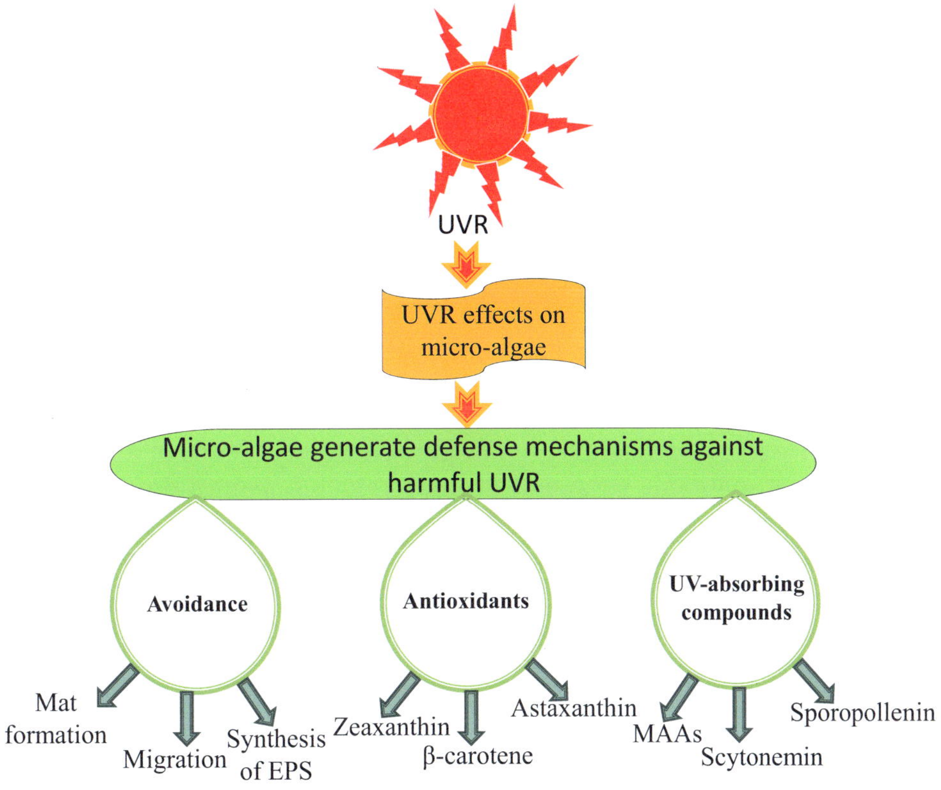

Fig. (5). Micro-algae have various defense mechanisms in response to the UV radiation effects.

Scytonemin

Scytonemin is a yellowish-brown, lipid-soluble, dimeric pigment with a molecular mass of 544 kDa. Its structure consists of phenolic and indolic subunits [73]. It has an *in vivo* and *in vitro* absorption maximum of 370 nm and 386 nm, respectively. Purified scytonemin also has smaller peaks at 252 nm, 278 nm and 300 nm [73]. Scytonemin (Fig. **6a**) can be easily extracted in non-polar solvents like acetone or ethyl acetate and validated through the techniques such as spectrophotometry and chromatography. Micro-algal scytonemin seems confined to cyanobacteria, especially in the extracellular polysaccharide sheath of *Scytonema* sp., *Rivularia* sp., *Chlorogloeopsis* sp., *Calothrix* sp., and *Nostoc commune* [74]. Scytonemin is the principal UV-absorbing compound in *Lyngbya* cf. *aestuarii,* and its area content appears to observe the seasonal alteration of solar intensity [75]. More than 30 species of cyanobacterial cultures retain scytonemin in their extracellular sheath and natural populations exposed to intense UV-radiation [76]. UV-A-induced photoinhibition and growth delay are seen in the *Chlorogloeopsis* sp. under UV-A exposure, and the process reverses after the synthesis of scytonemin in the extracellular envelopes [77]. Correlation between scytonemin content and UV-exposed photosynthetic resistance has been exhibited in *Calothrix* sp. [78, 79]. The adaptive strategy of photoprotection against UV irradiation is strongly suggested by Garcia-Pichel and Castenholz [76]. In *Nostoc punctiforme,* scytonemin synthesis was also enhanced by the incident periodic desiccation [80]. Scytonemin has strong stability against temperature and UV-A radiation and has a capacity for UV-screening even after prolonged physiological inactivity like desiccation. Due to all these characteristics, scytonemin can be used in various welfare of human beings [81].

Mycosporine-Like Amino Acids

Micro-algae (including cyanobacteria) are treated as extensive biomass producers having an important role in retaining the trophic energy dynamics of an ecosystem [82, 83]. They are a notable source of various valuable natural products having economic and ecological importance. In the past few decades, increased UV-radiations (280-400) on the Earth's surface [84] have inflamed enormous concern about its harmful effects on aquatic as well as terrestrial lifeforms, including cyanobacteria and other micro-algae [85]. These organisms synthesize bioactive compound MAAs to cope up with the extreme UV-radiations [86 - 88]. MAAs (Fig. **6b**) are low molecular weight (<400 Da), intracellular, colorless, hydrophilic compounds consisting of either aminocyclohexenone or an aminocyclohexinimine chromophore conjugated with the nitrogen substituent of an amino acid or its imino alcohol. There are approximately 40 MAAs known so far, with an absorption maximum ranging from 309-365 nm [89 - 92]. Including members of cyanophyceae, MAAs have also been reported in various micro-algae such as Chlorophyceae [93], Bacillariophyceae [94] and Chrysophyceae [95]. MAAs have high molar extinction coefficients (ε=28,100-50,000 M^{-1} cm^{-1}) and the ability to prevent UV-induced formation of thymine dimers [83, 96]. MAAs also contain antioxidant properties because they can disperse absorbed solar energy as heat into their surroundings without reactive oxygen species (ROS) formation [97]. The loop betwixt UV-radiation and nascent life is inseparable evidence, strengthened by the corroboration for biogenicity of ~3.5 Giga-annum (Ga) old euphotic stromatolitic formations [98] of apparently photosynthetically active, still UV-C bathed, microbial colonies of cyanobacteria like organisms [99]. The oldest stromatolite fossil (~3.7 Ga) from an Eoarchean shallow marine environment provides the evidence for life flourishing on Earth under extreme UV-radiation [100]. There are also effects of temperature and pH on micro-algae [101]. At high temperatures and low pH organisms are prone to demean before the synthesis of photoprotective /bioactive compounds. In this way, these compounds are affected indirectly at various temperatures and pH [102]. Cyanobacteria, along with micro-algae were collected from historical Kunds of Varanasi, India, was examined for their ability to synthesize MAAs [101, 102].

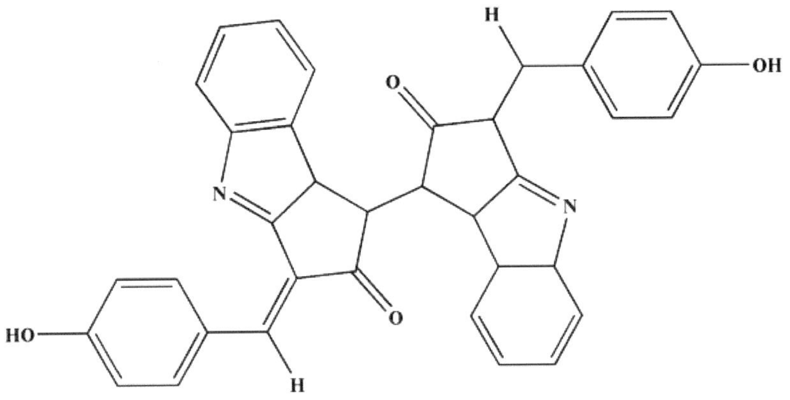

(a) Scytonemin

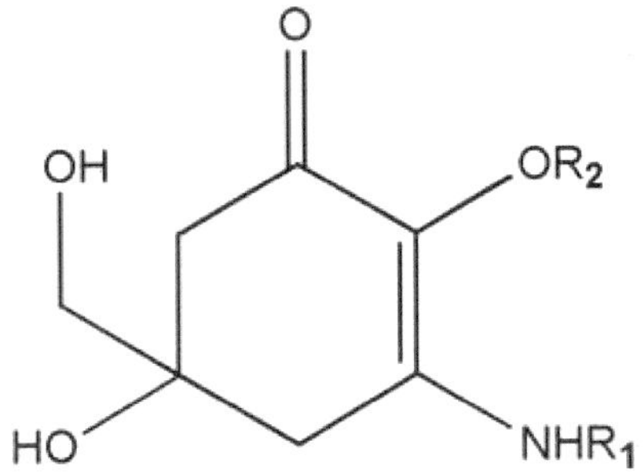

(b) Micosporine-like amino acid

Fig. (6). Chemical structures of photoprotective compounds **(a)** scytonemin and **(b)** MAA [R_1 and R_2 represent COOH and CH_3, respectively].

Table 3. UV-Screening compounds in micro-algae.

UV-Screening Compound	Algae	References
Sporopollenin	*Characium terrestre, Coelastrum microporum, Scenedesmus* sp., *Dunaliella slina, Chlorella fusca*	[37]
Scytonemin	*Chlorogloeopsis* sp., *Calothrix, Nostoc commune, Nostoc punctiforme*	[37]

(Table 3) cont.....

UV-Screening Compound	Algae	References
Mycosporine-like amino acids	*Chlorella minutissima, Chlorella sorokiniana, Dunaliella tertiolecta, Isochrysis* sp., *Corethron criophilum, Stellarima microtrias, Alexandrium catenella*	[37]

MICRO-ALGAE IN THE FOOD INDUSTRY

The functional food concept defeats the world and inspires science and industry to discover avant-garde ingredients capable of certain physiological effects for incorporation in food products. Seawater creates an environment for very different species to exist, many of them exceptional and rich in desired ingredients with anticancer, antioxidant and anti-inflammatory activity [103, 104]. The exploitation of microorganisms is considered relevant to increase the bioactivity of food products and their chemical components increase the nutritional value of food products. Micro-algae have launched into food merchandise in the form of food supplements, mainly tablets, capsules and sometimes liquid suspensions. They are instantly included in the super-food or nutritious-food category, which find their place in the growing consciousness of consumers aware of the benefits of a healthy diet. The nutrient composition of micro-algae contains particularly high levels of proteins with EAAs, hydrocolloids of the polysaccharide group, antioxidant compounds such as carotenoids and chlorophyll, phycobiliproteins, active enzymes, monounsaturated fatty acids (MUFAs) and PUFAs and minerals. The role of phycobiliproteins in the food industry has already been discussed above. Micro-algae have lower carbon food material than any other food raw material resource, which makes micro-algae a more sustainable future food [103, 105, 106]. Other than supplements traditional forms of food products such as pasta, snack foods, candy bars, gums, drink mixes and beverages have also been recently introduced into the market. There are many other ways of micro-algae utilization due to the richness in proteins, polysaccharides or fatty acids in specific parts of micro-algal biomass. The inclusion of micro-algae in various food products attracts the attention of researchers due to its techno-functional and nutritional benefits (Table **4**). Among the macro-algae, many species have only received recognition and appreciation, and some of them are cultivated in the European seas. *Porphyra* (a red alga) is a cold-water species that grows in shallow seawater. Its content of 40% - 45% carbohydrate, 30% - 35% protein and lots of vitamins make it a very popular food ingredient. In Japan, a marine vegetable product called "amanori" is manufactured. It is one of the most commonly used seaweeds for food, as almost all human populations have access to *porphyra* and have included it in their diet in some way.

Table 4. Micro-algae suitable for food products incorporation and their applications.

Micro-algae sp.	Product	Purpose	References
Aporosa fusiformis	bread	Nutritional properties (proteins and mineral content)	[107]
Arthrospira maxima	vegetarian food gels, pasta	Techno-functional and nutritional properties (antioxidative activity, ω-3 PUFAs)	[108 - 111]
Arthrospira platensis	dairy products (fermented milk), natural and probiotic yogurt, biscuits, bread, gluten-free bread, pasta	Techno-functional and nutritional properties (protein, fiber, mineral content and antioxidative activity)	[112 - 116]
Arthrospira sp.	frozen yogurt, bread, extruded snacks	Techno-functional properties and nutritional properties (proteins content)	[117 - 119]
Chlorella vulgaris	cookies, biscuits, vegetarian food gels	Colouring agent, techno-functional and nutritional properties (antioxidative activity, ω -3 PUFAs)	[116, 120]
Chlorella vulgaris green and *Chlorella vulgaris* orange (before and after carotenogenesis)	oil/water emulsionspasta, oil/water emulsions	Techno-functional properties, coloring and nutritional properties (antioxidative activity)	[121, 122, 126]
Chlorella sp.	yogurt, processed cheese	Techno-functional properties and nutritional properties	[123, 124]
Dunaliella salina	pasta	Techno-functional properties and nutritional properties	[125]
Diacronema vikianum	vegetarian food gels, pasta	Techno-functional and nutritional properties (antioxidative activity, ω -3 PUFAs)	[109, 126]
Dunaliella sp.	bread	Nutritional properties (protein content)	[127]
Haematococcus pluvialis	cookies, vegetarian food gels	Techno-functional and nutritional properties (antioxidative activity, ω -3 PUFAs)	[108, 109]
Haematococcus pluvialis (after carotenogenesis)	oil/water emulsions	Coloring and nutritional properties (antioxidative activity)	[122]
Isochrysis galbana	pasta, bread, biscuits	Techno-functional properties and nutritional properties (ω -3 PUFAs)	[109, 126, 128]
Nannochloropsis gaditana	bread	Techno-functional properties	[128]
Oscilltoria amphebia	bread	Techno-functional properties and nutritional properties	[115]

(Table 4) cont.....

Micro-algae sp.	Product	Purpose	References
Phaeodactylum tricornutum	biscuits	Techno-functional properties and nutritional properties (antioxidative activity)	[116]
Scenedesmus almeriensis	bread	Techno-functional properties	[128]
Tetraselmis suecica	biscuits, bread	Techno-functional properties and nutritional properties (antioxidative activity)	[116, 128]

Palmaria palmata is another red alga commonly called dulce, red dulce, or sea lettuce flakes and it occurs naturally on the northern coasts of both the Atlantic and Pacific Oceans. It is known to have high vitamin content and a high protein content, although dulse proteins lack EAAs. Dried *Palmaria palmata* is offered as a salty cocktail snack in the bars of Nova Scotia and Maine, and in Ireland, it is eaten raw or cooked [129]. The market demand for *Laminaria* is huge because it forms the main ingredient of food products known as "kombu" or "konbu", which are used extensively in Japanese cuisines, such as dashi, one of the basic ingredients required to make soup stock. The Japanese food product "sarumen" is based on the *Alaria* genus seaweeds, and *Alaria esculenta* specifically provides both excellent quality protein and iodine [129]. Ulva is particularly rich in iron and protein and is also known in Europe, especially in Scotland, where *Ulva lactuca* was utilized for salads and soups. *Caulerpa racemosa* seaweed is very rich in proteins, fiber, minerals such as calcium and magnesium, and vitamins such as folic acid, ascorbic acid, vitamin A and vitamin B1 [129]. The Chinese first used over 2000 years ago *Nostoc* sp. as food, and later, the commercial forms of other micro-algae *Chlorella* sp. and *Spirulina* sp., were consumed as healthy foods in Japan, Taiwan and Mexico [130, 131]. Currently, most commercial products of micro-algae are available in the market as a health food, tablets, capsules and liquids [132], and their products are mixed with pastes, snacks, candies, gums, noodles, wine, beverages, and breakfast cereals [133]. *Aphanizomenon flos-aquae*, *Chlorella* sp., *Dunaliella salina*, *Dunaliella tertiolecta* and *Spirulina plantensis* are some of the micro-algae species widely used as a human food source because they are rich in protein content and have high nutritive value [134, 135]. Among these micro-algae, *Spirulina* (*Arthrospira*) and *Chlorella* are currently dominating the market of micro-algae. In addition, *Scenedesmus* sp. is also recognized as a food source of nutritional micro-alga, but its commercial production is limited. Their extracts are assimilated into commonly used foods such as desserts, fruit puddings, noodles, soups and ravioli. In 1961, various studies showed that the extracts of *Scenedesmus* sp. and *Chlorella* sp. can be used for human consumption. More than 50% of the world's *Spirulina* production is used as feed supplement. In 1999, the micro-algae production rate for aquaculture was 10^3 tons, of which 62% for molluscs, 21% for shrimps and 16% for fish and global world production of 43 x 10^6 tons of plants and animals.

APPLICATIONS OF MICRO-ALGAE IN HUMAN HEALTH

Antimicrobial Compounds from Micro-Algae

Micro-algae have antimicrobial properties due to the production of α- and β-ionone, β-cyclocitral, neophytadiene, and phytol [136]. Micro-algae have shown antimicrobial activity against human pathogens, such as *Escherichia coli, Pseudomonas aeruginosa, Staphylococcus aureus,* and *Staphylococcus epidermidis* due to the production of γ-linolenic acid, eicosapentaenoic acid, hexadecatrienoic acid, docosahexaenoic acid, palmitoleic acid, lauric acid, oleic acid, lactic acid, and arachidonic acid [137]. Micro-algal-derived sulfated polysaccharides inhibit viral infection, such as encephalomyocarditis virus, Herpes simplex virus types 1 and 2 (HSV1, HSV2), human immunodeficiency virus (HIV), hemorrhagic septicemia in salmonid virus, swine fever virus, and varicella virus [138, 139].

Micro-Algal Antioxidants

Dimethylsulfoniopropionate and mycosporine amino acids are isolated from micro-algae and act as potent UV radiation protectants [140]. Carotenoids and phycocyanins are also prominently found. It is used as natural food colorants and in cosmetics. It is obtained from *Spirulina* [141] and *Porphyridium cruentum* [142] and *Synechococcus* [136]. β-carotene and astaxanthin (Fig. **5**) are the most important among these carotenoids. They have major importance in pharmaceutical industries and food because of their antioxidant properties and pigmentation ability. Major natural source of β-carotene is *Dunaliella salina,* and more than 14% is produced in dry biomass.

Antitumor Compounds from Micro-Algae

Skeletonema marinoi was effective against human melanoma cells A2058 when cultured in nitrogen deficient conditions [143]. Sulfated polysaccharides were extracted from filamentous micro-algae *Tribonema* sp. and their activity was tested on liver cancer cells, HepG2. The MTT test showed that the sulfated polysaccharide from *Tribonema* sp. possessed antitumor activity, causing induced cell apoptosis. The antitumor activity of the alcoholic fatty acid ester synthesized by the micro-algae *Phaeodactylum tricornutum* by Samarakoon *et al.* [144]. Amphidinolide-A is an antitumor agent that is produced by *Amphidinium* sp., Acutiphycins produced by *Oscillatoria acutissima* showed activity against murine and Lewis lung carcinoma [145]. Glycoprotein produced from *Chlorella vulgaris* showed protective activity against tumor metastasis and chemotherapy-induced immunosuppression in mice. Aeruginosins isolated from *Microcystis aeruginosa* has been shown to have inhibitory properties against thrombin, plasmin and trypsin. Other depsipeptides, such as micropeptin, microcystilide, cyanopeptolin, oscillapeptin and nostocyclin are inhibitors of enzymes such as trypsin, plasmin, thrombin and chymotrypsin.

Anti-inflammatory Activity and Antiplatelet Activity

Spirulina (*Arthrospira platensis*) is a highly nutritious blue-green micro-algae used worldwide as a nutraceutical food supplement and also contains anti-inflammatory activity. *Spirulina* containing phycocyanin can inhibit the formation of pro-inflammatory cytokines like TNFα, reduce the prostaglandin E(2) production and inhibit the expression of cyclooxygeanase-2 (COX-2) [145]. *Spirulina* contains β-carotene, which is responsible for the suppression of the transcription of IL-1β, IL-6 and IL-12, inflammatory cytokines, in the macrophage cell line stimulated by lipopolysaccharide or IFN [146]. *Spirulina* biomass also enhances the growth of probiotic bacteria such as *Lactobacillus casei*, *Streptococcus thermophilus*, *Bifidobacteria* and *Lactobacillus acidophilus*. Anti-inflammatory activity is also exhibited by *Phaeodactylum tricornutum* and *Chlorella stigmatophora* [147]. Villar *et al.* [148] reported that crude aqueous extract of the micro-alga *Dunaliella tertiolecta* has a strong inhibitory effect on human platelet aggregation caused by thrombin, arachidonic acid and ionomycin isolated from micro-algae.

Micro-algal Compounds Against Degenerative Diseases

Dunaliella salina produces β-carotene, which has the potential to decrease the risk of cancer and degenerative diseases in humans. Lutein is effective against several diseases like cataracts and macular degeneration and in the early stages of atherosclerosis. β-carotene and lutein extracted from *Chlorella* sp. Extracts of *Chlorella* sp. prevented the cognitive disability that accompanies Alzheimer's disease in rats. It was also shown that lutein obtained from *Chlorella* reduced the incidence of cancer. Growth of colon cancer was inhibited by carotenoids obtained from *Chlorella ellipsoidea* and *Chlorella vulgaris* [149]. Micro-algal lycopene obtained from *Chlorella marina* reduced the proliferation of prostate cancer in mice [150] and also inhibited total cholesterol and low-density lipoprotein (LDL) levels [150] and ameliorated rheumatoid arthritis.

OTHER APPLICATIONS OF MICRO-ALGAE

Micro-Algae in Biofuels

Micro-algae are usually harvested during the stationary phase of growth for maximal recovery of oil. Harvesting of micro-algal biomass is technically difficult due to its small size and association with epiphytes in dilute aqueous suspension. Various techniques have been developed for the harvesting of micro-algae like flocculation, centrifugation, filtration, agglomeration and pressure dewatering [151]. Of these processes, flocculation is considered the most promising for large-scale production of micro-algae. *Botryococcus, Chlorella*, *Scenedesmus*, *Chlamydomonas*, *Dunaliella* and *Nannochloropsis* may provide raw materials for the production of biodiesel.

Micro-Algae as a Feedstock for Bioethanol Production

Micro-algal biomass can be used as a potential alternative feedstock for bioethanol production because they are highly efficient photosynthetic organisms, can accumulate a high amount of starch and cellulose that can be used for bioethanol production, does not compete with food crops for precious arable land and water, and contain no lignin, making saccharification an easier process [152]. Micro-algae like *Chlorella, Dunaliella, Chlamydomonas, Scenedesmus and Spirulina* contain a large amount (450% of the dry weight) of starch and glycogen and are used as raw materials for ethanol production [153]. Guo *et al*. [154] used micro-algal strains, *M. afer* and *S. abundans,* for saccharification and subsequent fermentative bioethanol production that were isolated from coastal waters of Pearl River Delta, Ethanol yield from the algal system is very high, which is 5,000–15,000 gal/acre or 46,760–140,290 L/ha. A high amount of carbohydrate accumulated in the biomass of *Scenedesmus*, *Chlorella* [152] and *Chlamydomonas* biomass contains a high amount of carbohydrate [155]. *Chlorella vulgaris* is a good source of ethanol with 65% conversion efficiency [156] due to the large amount of starch.

Micro-Algal Bioactive Compounds in Cosmetic Industry

Moisturizers

Micro-algae can be effectively used in the cosmetics due to the presence of bioactive compounds that help in moisturization. Micro-algae *Thalassiosira* is rich in amino acids mainly serine which facilitates moisturization. *Monodus subterraneus* containing PUFAs can be used for the production of moisturizing products [157] and *Nannochloropsis* is rich in linolenic acid, an omega-6 fatty acid [150]. It has been reported that protein and its hydrolysates derived from

Chlorella, *Spirulina*, and *Porphyra* sp. contain essential moisturizing (Face lotion) properties (Fig. **7**). Micro-algal species *Thraustochytrium*, *Aurantiochytrium*, and *Schizochytrium* contain significant amounts of squalene, which is used in cosmetics as emollient. Other than cosmetics micro-algal products are used in various industries, such as nutraceuticals, food additives *etc*. (Table **2**).

Skin Whitening and Hair Care

Melanin plays a major role in the pigmentation of the skin. Although melanin plays an important role in protection against UV damage but its overproduction results in hyperpigmentation. Tyrosinase is a major enzyme responsible for melanin synthesis, and it catalyzes the hydroxylation of levodopa (L-DOPA). L-DOPA is oxidized to dopaquinone and finally converted to melanin. Tyrosinase inhibitory activity is reported in several micro-algae species. It has been reported that ethanol extracts of *Arthrospira platensis* have shown inhibiting activity with an IC50 value of 1.4×10^{-3} g/mL [175]. *N. oculata* derived Zeaxanthin acts as an efficient tyrosinase inhibitor. *Nannochloropsis oculata* extracts are also patented to be used in formulations of cosmetics. Micro-algal extracts are also being used in hair care treatments because it helps in hair growth and improves the structure by promoting the differentiation of keratinocytes. Some micro-algal species have the ability to soften the hair and make it flexible because they produce a high amount of EFAs [176].

Anti-Aging

Aging of the skin is a complex biological phenomenon that is inevitable and occurs through cellular senescence. Deposition of ROS in the cells leads to cell death (either by apoptosis or necrosis), leading to the wrinkling of the skin and can also result in skin cancer. UV radiations (UVA and UVB) cause extrinsic aging resulting in the production of ROS. UV radiation destroys elastin and collagen, which are the essential fibers for the skin [177]. Carotenoids act as quenching agents and inhibit ROS formation. So, carotenoids can be used as active ingredients in anti-aging components (Fig. **7**).

Micro-alga is a significant source of antioxidant compounds such as β-carotene, astaxanthin, lutein, zeaxanthin, and canthaxanthin. Astaxanthin is an effective antioxidant molecule and possesses high antioxidant properties as compared to other antioxidants. Japanese cosmetic product "Astablank," containing astaxanthin in its composition, can combat wrinkles and spots. Astaxanthin is a major carotenoid and helps in preventing the oxidation of essential lipids by increasing SOD and catalase activity [178]. Melanogenesis of skin is reduced by fucoxanthin and hence is used in cosmeceuticals [179]. β-Carotene also helps in the prevention of free radicals formation and helps in avoiding premature aging. Lutein is another important molecule present in the epidermal layer protecting skin damage induced by UV radiation. Micro-algal species *Scenedesmus* and *Chlorella* contain a significant amount of lutein which can help in the rejuvenation of the skin. Polysaccharides derived from *Porphyridium* sp. stimulate elastin synthesis and help in cell renewal. Extracts from *Chlorella* species are the main source of anti-aging components in cosmetics because it increases the synthesis of collagen via promoting tissue regeneration and wrinkle reduction [180]. Ocean Pharma (Germany) and Pentapharm (Switzerland) produced cosmetics based on algal extracts of *Arthrospira maxima* and *N. oculata* respectively. These products contain significant anti-aging properties by improving skin regeneration capability [181].

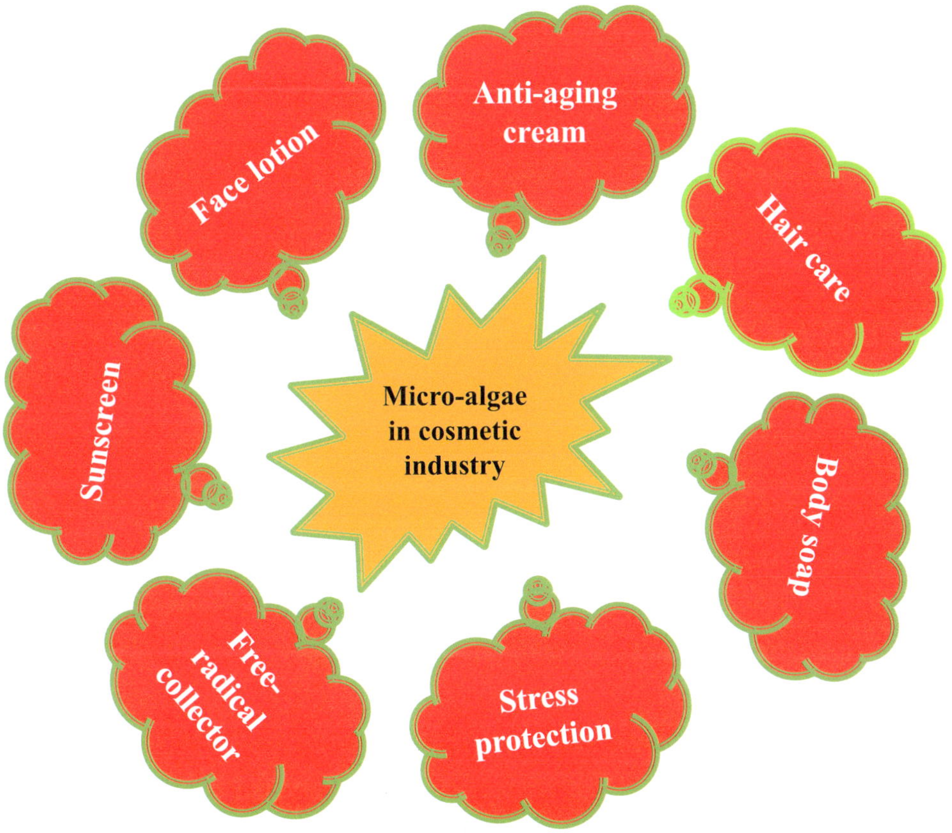

Fig. (7). Uses of micro-algal bioactive compounds in cosmetic industry.

CONCLUSION

Micro-algae are a valuable source of health-promoting and disease-conquering metabolites. The sundry groups of bioactive compounds like MAAs, scytonemin, carotenoids, fatty acids, and amino acids in micro-algae increase the nutritional value of human diets. Micro-algae have wide applications in various fields in the success of defeating barriers in the way of human health. Mass culture and commercial production of micro-algae have increased in recent decades. They are useful for aquaculture, poultry and piggery, thereby increasing their production. The medicinal use of micro-algae has helped to cure assorted diseases in humans. This review accentuates the importance of micro-algal bioactive compounds for future developments in the food and pharmaceutical industries.

CONSENT FOR PUBLICATION

Not applicable.

CONFLICT OF INTEREST

The author(s) declare no conflict of interest, financial or otherwise.

ACKNOWLEDGEMENTS

S. Mishra and N. Kumari are thankful to the University Grants Commission (Joint CSIR-UGC JRF-2019/NTA Ref. No.: 191620046790) and the Council of Scientific and Industrial Research (09/013(0819)/2018-EMR-I), New Delhi, India, respectively, for the financial support in the form of JRF.

REFERENCES

[1] Fehling J, Stoecker DK, Baldauf SL. Photosynthesis and the eukaryote tree of life. In: Falkowski PG, Knoll AH, Eds. Evolution of Primary Producers in the Sea 2007; 75-107. [http://dx.doi.org/10.1016/B978-012370518-1/50007-2]

[2] Stern RF, Horak A, Andrew RL, *et al.* Environmental barcoding reveals massive dinoflagellate diversity in marine environments. PLoS One 2010; 5(11): e13991. [http://dx.doi.org/10.1371/journal.pone.0013991] [PMID: 21085582]

[3] Milledge JJ. Commercial application of microalgae other than as biofuels: a brief review. Rev Environ Sci Biotechnol 2011; 10(1): 31-41. [http://dx.doi.org/10.1007/s11157-010-9214-7]

[4] Raja R, Hemaiswarya S, Kumar NA, Sridhar S, Rengasamy R. A perspective on the biotechnological potential of microalgae. Crit Rev Microbiol 2008; 34(2): 77-88. [http://dx.doi.org/10.1080/10408410802086783] [PMID: 18568862]

[5] Kang HK, Salim HM, Akter N, *et al.* Effect of various forms of dietary *Chlorella* supplementation on growth performance, immune characteristics, and intestinal microflora population of broiler chickens. J Appl Poult Res 2013; 22(1): 100-8. [http://dx.doi.org/10.3382/japr.2012-00622]

[6] Choi W, Kang D, Lee H. Enhancement of immune activation activities of *Spirulina maxima* grown in deep-sea water. Int J Mol Sci 2013; 14(6): 12205-21. [http://dx.doi.org/10.3390/ijms140612205] [PMID: 23743830]

[7] Guerin M, Huntley ME, Olaizola M. *Haematococcus* astaxanthin: applications for human health and nutrition. Trends Biotechnol 2003; 21(5): 210-6. [http://dx.doi.org/10.1016/S0167-7799(03)00078-7] [PMID: 12727382]

[8] Jaspars M, De Pascale D, Andersen JH, Reyes F, Crawford AD, Ianora A. The marine biodiscovery pipeline and ocean medicines of tomorrow. J Mar Biol Assoc U K 2016; 96(1): 151-8. [http://dx.doi.org/10.1017/S0025315415002106]

[9] Romano G, Costantini M, Sansone C, Lauritano C, Ruocco N, Ianora A. Marine microorganisms as a promising and sustainable source of bioactive molecules. Mar Environ Res 2017; 128: 58-69. [http://dx.doi.org/10.1016/j.marenvres.2016.05.002] [PMID: 27160988]

[10] Blunt JW, Copp BR, Keyzers RA, Munro MHG, Prinsep MR. Marine natural products. Nat Prod Rep 2016; 33(3): 382-431. [http://dx.doi.org/10.1039/C5NP00156K] [PMID: 26837534]

[11] Lauritano C, Andersen JH, Hansen E, *et al.* Bioactivity screening of microalgae for antioxidant, anti-inflammatory, anticancer, anti-diabetes, and antibacterial activities. Front Mar Sci 2016; 3: 68. [http://dx.doi.org/10.3389/fmars.2016.00068]

[12] Chen X, He G, Deng Z, Wang N, Jiang W, Chen S. Screening of microalgae for biodiesel feedstock. Adv Microbiol 2014; 4(7): 365-76. [http://dx.doi.org/10.4236/aim.2014.47044]

[13] Delattre C, Fenoradosoa TA, Michaud P. Galactans: an overview of their most important sourcing and applications as natural polysaccharides. Braz Arch Biol Technol 2011; 54(6): 1075-92. [http://dx.doi.org/10.1590/S1516-89132011000600002]

[14] Kraan S. Algal polysaccharides, novel applications and outlook. In: Chang CF, Ed. Carbohydrates - Comprehensive Studies on Glycobiology and Glycotechnology 2012; 489-532. [http://dx.doi.org/10.5772/51572]

[15] Misurcova L, Orsavova J, Ambrozova JV. Algal polysaccharides and health. In: Ramawat KG, Merillon JM, Eds. Polysaccharides Bioactivity and Biotechnology 2015; 110-44. [http://dx.doi.org/10.1007/978-3-319-16298-0_24]

[16] de Jesus Raposo M, de Morais A, de Morais R. Marine polysaccharides from algae with potential biomedical applications. Mar Drugs 2015; 13(5): 2967-3028. [http://dx.doi.org/10.3390/md13052967] [PMID: 25988519]

[17] Kılınç B, Cirik S, Turan G, Tekogul H, Koru E. Seaweeds for food and industrial applications.In: Muzzalupo I, Ed. Food Industry. London: IntechOpen; 2013. [http://dx.doi.org/10.5772/53172]

[18] Singh R, Parihar P, Singh M, *et al.* Uncovering potential applications of cyanobacteria and algal metabolites in biology, agriculture and medicine: current status and future prospects. Front Microbiol 2017; 8: 515. [http://dx.doi.org/10.3389/fmicb.2017.00515] [PMID: 28487674]

[19] Zhou G, Sheng W, Yao W, Wang C. Effect of low molecular λ-carrageenan from *Chondrus ocellatus* on antitumor H-22 activity of 5-Fu. Pharmacol Res 2006; 53(2): 129-34. [http://dx.doi.org/10.1016/j.phrs.2005.09.009] [PMID: 16310373]

[20] Cardozo KHM, Guaratini T, Barros MP, *et al.* Metabolites from algae with economical impact. Comp Biochem Physiol C Toxicol Pharmacol 2007; 146(1-2): 60-78. [http://dx.doi.org/10.1016/j.cbpc.2006.05.007] [PMID: 16901759]

[21] Raja A, Vipin C, Aiyappan A. Biological importance of marine algae-An overview. Int J Curr

Microbiol Appl Sci 2013; 2: 222-7.

[22] Lam SK, Ng TB. Lectins: production and practical applications. Appl Microbiol Biotechnol 2011; 89(1): 45-55. [http://dx.doi.org/10.1007/s00253-010-2892-9] [PMID: 20890754]

[23] Bewley CA, Gustafson KR, Boyd MR, *et al.* Solution structure of cyanovirin-N, a potent HIV-inactivating protein. Nat Struct Biol 1998; 5(7): 571-8. [http://dx.doi.org/10.1038/828] [PMID: 9665171]

[24] Huskens D, Férir G, Vermeire K, *et al.* Microvirin, a novel α(1,2)-mannose-specific lectin isolated from Microcystis aeruginosa, has anti-HIV-1 activity comparable with that of cyanovirin-N but a much higher safety profile. J Biol Chem 2010; 285(32): 24845-54. [http://dx.doi.org/10.1074/jbc.M110.128546] [PMID: 20507987]

[25] Amaro HM, Barros R, Guedes AC, Sousa-Pinto I, Malcata FX. Microalgal compounds modulate carcinogenesis in the gastrointestinal tract. Trends Biotechnol 2013; 31(2): 92-8. [http://dx.doi.org/10.1016/j.tibtech.2012.11.004] [PMID: 23260440]

[26] Ibañez E, Cifuentes A. Benefits of using algae as natural sources of functional ingredients. J Sci Food Agric 2013; 93(4): 703-9. [http://dx.doi.org/10.1002/jsfa.6023] [PMID: 23339029]

[27] Markou G, Nerantzis E. Microalgae for high-value compounds and biofuels production: A review with focus on cultivation under stress conditions. Biotechnol Adv 2013; 31(8): 1532-42. [http://dx.doi.org/10.1016/j.biotechadv.2013.07.011] [PMID: 23928208]

[28] Singh NK, Dhar DW. Microalgae as second generation biofuel. A review. Agron Sustain Dev 2011; 31(4): 605-29. [http://dx.doi.org/10.1007/s13593-011-0018-0]

[29] El-Baky HH, El Baz FK, El-Baroty GS. Production of phenolic compounds from *Spirulina maxima* microalgae and its protective effects. Afr J Biotechnol 2009; 8(24): 7059-67.

[30] Plaza M, Herrero M, Cifuentes A, Ibáñez E. Innovative natural functional ingredients from microalgae. J Agric Food Chem 2009; 57(16): 7159-70. [http://dx.doi.org/10.1021/jf901070g] [PMID: 19650628]

[31] Priyadarshani I, Rath B. Commercial and industrial applications of micro algae–A review. J Algal Biomass Util 2012; 3(4): 89-100.

[32] Francavilla M, Trotta P, Luque R. Phytosterols from *Dunaliella tertiolecta* and *Dunaliella salina*: A potentially novel industrial application. Bioresour Technol 2010; 101(11): 4144-50. [http://dx.doi.org/10.1016/j.biortech.2009.12.139] [PMID: 20110165]

[33] de Jesus Raposo MF, de Morais RMSC, de Morais AMMB. Health applications of bioactive compounds from marine microalgae. Life Sci 2013; 93(15): 479-86. [http://dx.doi.org/10.1016/j.lfs.2013.08.002] [PMID: 23994664]

[34] Mostafa SS. Microalgal biotechnology: prospects and applications. Plant Sci 2012; 12: 276-314.

[35] Custódio L, Justo T, Silvestre L, *et al.* Microalgae of different phyla display antioxidant, metal chelating and acetylcholinesterase inhibitory activities. Food Chem 2012; 131(1): 134-40. [http://dx.doi.org/10.1016/j.foodchem.2011.08.047] [PMID: 26434272]

[36] Asgharpour M, Rodgers B, Hestekin J. Eicosapentaenoic acid from *Porphyridium cruentum*: Increasing growth and productivity of microalgae for pharmaceutical products. Energies 2015; 8(9): 10487-503. [http://dx.doi.org/10.3390/en80910487]

[37] Priyadarshani I, Biswajit R. Commercial and industrial applications of micro algae – A review. J Algal Biomass Util 2012; 3(4): 89-100.

[38] Dufossé L, Galaup P, Yaron A, *et al.* Microorganisms and microalgae as sources of pigments for food

use: a scientific oddity or an industrial reality? Trends Food Sci Technol 2005; 16(9): 389-406. [http://dx.doi.org/10.1016/j.tifs.2005.02.006]

[39] Becker W. Microalgae in human and animal nutrition. In: Richmond A, Ed. Handbook of microalgal culture: biotechnology and applied phycology Microalgal Culture Handbook 2004; 312-51.

[40] Smerilli A, Orefice I, Corato F, Gavalás Olea A, Ruban AV, Brunet C. Photoprotective and antioxidant responses to light spectrum and intensity variations in the coastal diatom *Skeletonema marinoi*. Environ Microbiol 2017; 19(2): 611-27. [http://dx.doi.org/10.1111/1462-2920.13545] [PMID: 27712003]

[41] Eriksen NT. Production of phycocyanin—a pigment with applications in biology, biotechnology, foods and medicine. Appl Microbiol Biotechnol 2008; 80(1): 1-14. [http://dx.doi.org/10.1007/s00253-008-1542-y] [PMID: 18563408]

[42] Del Campo JA, García-González M, Guerrero MG. Outdoor cultivation of microalgae for carotenoid production: current state and perspectives. Appl Microbiol Biotechnol 2007; 74(6): 1163-74. [http://dx.doi.org/10.1007/s00253-007-0844-9] [PMID: 17277962]

[43] Yuan JP, Peng J, Yin K, Wang JH. Potential health-promoting effects of astaxanthin: A high-value carotenoid mostly from microalgae. Mol Nutr Food Res 2011; 55(1): 150-65. [http://dx.doi.org/10.1002/mnfr.201000414] [PMID: 21207519]

[44] Tso MO, Lam TT. Method of retarding and ameliorating central nervous system and eye damage. Board of Trustees of the University of Illinois United States Patent 5527533 1996.

[45] Ikeuchi M, Koyama T, Takahashi J, Yazawa K. Effects of astaxanthin in obese mice fed a high-fat diet. Biosci Biotechnol Biochem 2007; 71(4): 893-9. [http://dx.doi.org/10.1271/bbb.60521] [PMID: 17420580]

[46] Satoh A, Tsuji S, Okada Y, *et al.* Preliminary clinical evaluation of toxicity and efficacy of a new astaxanthin-rich *Haematococcus pluvialis* extract. J Clin Biochem Nutr 2009; 44(3): 280-4. [http://dx.doi.org/10.3164/jcbn.08-238] [PMID: 19430618]

[47] Chan K, Mong M, Yin M. Antioxidative and anti-inflammatory neuroprotective effects of astaxanthin and canthaxanthin in nerve growth factor differentiated PC12 cells. J Food Sci 2009; 74(7): H225-31. [http://dx.doi.org/10.1111/j.1750-3841.2009.01274.x] [PMID: 19895474]

[48] Lin J-T, Lee Y-C, Hu C-C, *et al.* Evaluation of carotenoid extract from *Dunaliella salina* against cadmium-induced cytotoxicity and transforming growth factor beta 1 induced expression of smooth muscle alpha-actin with rat liver cell lines. Yao Wu Shi Pin Fen Xi 2010; 18: 301-6.

[49] Jin ES, Polle JE, Lee HK, Hyun SM, Chang M. Xanthophylls in microalgae: from biosynthesis to biotechnological mass production and application. J Microbiol Biotechnol 2003; 13(2): 165-74.

[50] Matsuura H, Watanabe MM, Kaya K. Echinenone production of a dark red-coloured strain of *Botryococcus braunii.* J Appl Phycol 2012; 24(4): 973-7. [http://dx.doi.org/10.1007/s10811-011-9719-7]

[51] Kim SM, Jung YJ, Kwon ON, *et al.* A potential commercial source of fucoxanthin extracted from the microalga *Phaeodactylum tricornutum.* Appl Biochem Biotechnol 2012; 166(7): 1843-55. [http://dx.doi.org/10.1007/s12010-012-9602-2] [PMID: 22371063]

[52] Von Oppen-Bezalel L, Shaish A. Application of the colourless carotenoids, phytoene and phytofluene in cosmetics, wellness, nutrition, and therapeutics. In: Ben-Amotz A, Polle JEW, Subba Rao DV, Eds. The alga Dunaliella: biodiversity, physiology, genomics and biotechnology 2009; 423-44. [http://dx.doi.org/10.1201/b10300-19]

[53] Barclay W, Weaver C, Metz J, Hansen J. Development of a docosahexaenoic acid production technology using *Schizochytrium*: a historical perpective. In: Cohen Z, Ratledge C, Eds. Single Cell Oils: Microbial and Algal Oils 2010; 36-52. [http://dx.doi.org/10.1016/B978-1-893997-73-8.50008-6]

[54] Wynn J, Behrens P, Sundararajan A, Hansen J, Apt K. Production of single cell oils from dinoflagellates. In: Cohen Z, Ratledge C, Eds. Single cell oils Microbial and algal oils 2010; 115-29. [http://dx.doi.org/10.1016/B978-1-893997-73-8.50010-4]

[55] Ward OP, Singh A. Omega-3/6 fatty acids: Alternative sources of production. Process Biochem 2005; 40(12): 3627-52. [http://dx.doi.org/10.1016/j.procbio.2005.02.020]

[56] Guedes AC, Amaro HM, Malcata FX. Microalgae as sources of carotenoids. Mar Drugs 2011; 9(4): 625-44. [http://dx.doi.org/10.3390/md9040625] [PMID: 21731554]

[57] Silva TL, Mendes A, Mendes RL, *et al.* Effect of n-dodecane on *Crypthecodinium cohnii* fermentations and DHA production. J Ind Microbiol Biotechnol 2006; 33(6): 408-16. [http://dx.doi.org/10.1007/s10295-006-0081-8] [PMID: 16501933]

[58] Seto A, Wang HL, Hesseltine CW. Culture conditions affect eicosapentaenoic acid content of *Chlorella minutissima.* J Am Oil Chem Soc 1984; 61(5): 892-4. [http://dx.doi.org/10.1007/BF02542159]

[59] Amparyup P, Sutthangkul J, Charoensapsri W, Tassanakajon A. Pattern recognition protein binds to lipopolysaccharide and β-1,3-glucan and activates shrimp prophenoloxidase system. J Biol Chem 2012; 287(13): 10060-9. [http://dx.doi.org/10.1074/jbc.M111.294744] [PMID: 22235126]

[60] Vetvicka V, Terayama K, Mandeville R, *et al.* Pilot study: orally administered yeast β-1, 3-glucan prophylactically protects against anthrax infection and cancer in mice. J Am Neutraceut Assn 2002; 5: 1-5.

[61] Hu W, Xu J, Sinkkonen J, Wu J. Polyketides from marine dinoflagellates of the genus *Prorocentrum*, biosynthetic origin and bioactivity of their okadaic acid analogues. Mini Rev Med Chem 2010; 10(1): 51-61. [http://dx.doi.org/10.2174/138955710791112541] [PMID: 20105116]

[62] MacKenzie L, Beuzenberg V, Holland P, McNabb P, Suzuki T, Selwood A. Pectenotoxin and okadaic acid-based toxin profiles in *Dinophysis acuta* and *Dinophysis acuminata* from New Zealand. Harmful Algae 2005; 4(1): 75-85. [http://dx.doi.org/10.1016/j.hal.2003.12.001]

[63] Rundberget T, Sandvik M, Larsen K, *et al.* Extraction of microalgal toxins by large-scale pumping of seawater in Spain and Norway, and isolation of okadaic acid and dinophysistoxin-2. Toxicon 2007; 50(7): 960-70. [http://dx.doi.org/10.1016/j.toxicon.2007.07.003] [PMID: 17825866]

[64] Trainer VL, Bates SS, Lundholm N, *et al.* Pseudo-nitzschia physiological ecology, phylogeny, toxicity, monitoring and impacts on ecosystem health. Harmful Algae 2012; 14: 271-300. [http://dx.doi.org/10.1016/j.hal.2011.10.025]

[65] Washida K, Koyama T, Yamada K, Kita M, Uemura D. Karatungiols A and B, two novel antimicrobial polyol compounds, from the symbiotic marine dinoflagellate *Amphidinium* sp. Tetrahedron Lett 2006; 47(15): 2521-5. [http://dx.doi.org/10.1016/j.tetlet.2006.02.045]

[66] Bishop W, Zubeck H. Evaluation of microalgae for use as nutraceuticals and nutritional supplements. J Nutr Food Sci 2012; 2: 3-6. [http://dx.doi.org/10.4172/2155-9600.1000147]

[67] Pickett-Heaps JD, Staehelin LA. The ultrastructure of *Scenedesmus* (Chlorophyceae). II. Cell division and colony formation. J Phycol 1975; 11(2): 186-202. [http://dx.doi.org/10.1111/j.1529-8817.1975.tb02766.x]

[68] Schulze Osthoff K, Wiermann R. Phenols as integrated compounds of sporopollenin from *Pinus*

pollen. J Plant Physiol 1987; 131(1-2): 5-15.
[http://dx.doi.org/10.1016/S0176-1617(87)80262-6]

[69] Guilford WJ, Schneider DM, Labovitz J, Opella SJ. High resolution solid state C NMR spectroscopy of sporopollenins from different plant taxa. Plant Physiol 1988; 86(1): 134-6.
[http://dx.doi.org/10.1104/pp.86.1.134] [PMID: 16665854]

[70] Xiong F, Komenda J, Kopecky J, Nedbal L. Strategies of ultraviolet-B protection in microscopic algae. Physiol Plant 1997; 100(2): 378-88.
[http://dx.doi.org/10.1111/j.1399-3054.1997.tb04796.x]

[71] Komaristaya VP, Gorbulin OS. Sporopollenin in the composition of cell walls of *Dunaliella salina* Teod. (Chlorophyta) zygotes. Int J Algae 2006; 8(1): 43-52.
[http://dx.doi.org/10.1615/InterJAlgae.v8.i1.40]

[72] Rozema J, Broekman RA, Blokker P, *et al.* UV-B absorbance and UV-B absorbing compounds (para-coumaric acid) in pollen and sporopollenin: the perspective to track historic UV-B levels. J Photochem Photobiol B 2001; 62(1-2): 108-17.
[http://dx.doi.org/10.1016/S1011-1344(01)00155-5] [PMID: 11693361]

[73] Proteau PJ, Gerwick WH, Garcia-Pichel F, Castenholz R. The structure of scytonemin, an ultraviolet sunscreen pigment from the sheaths of cyanobacteria. Experientia 1993; 49(9): 825-9.
[http://dx.doi.org/10.1007/BF01923559] [PMID: 8405307]

[74] Sinha RP, Klisch M, Gröniger A, Häder DP. Ultraviolet-absorbing/screening substances in cyanobacteria, phytoplankton and macroalgae. J Photochem Photobiol B 1998; 47(2-3): 83-94.
[http://dx.doi.org/10.1016/S1011-1344(98)00198-5]

[75] Karsten U, Maier J, Garcia-Pichel F. Seasonality in UV-absorbing compounds of cyanobacterial mat communities from an intertidal mangrove flat. Aquat Microb Ecol 1998; 16: 37-44.
[http://dx.doi.org/10.3354/ame016037]

[76] Garcia-Pichel F, Castenholz RW. Characterization and biological implications of scytonemin, a cyanobacterial sheath pigment. J Phycol 1991; 27(3): 395-409.
[http://dx.doi.org/10.1111/j.0022-3646.1991.00395.x]

[77] Garcia-Pichel F, Sherry ND, Castenholz RW. Evidence for an ultraviolet sunscreen role of the extracellular pigment scytonemin in the terrestrial cyanobacterium *Chlorogloeopsis* sp. Photochem Photobiol 1992; 56(1): 17-23.
[http://dx.doi.org/10.1111/j.1751-1097.1992.tb09596.x] [PMID: 1508978]

[78] Brenowitz S, Castenholz RW. Long-term effects of UV and visible irradiance on natural populations of a scytonemin-containing cyanobacterium (*Calothrix* sp.). FEMS Microbiol Ecol 1997; 24(4): 343-52.
[http://dx.doi.org/10.1111/j.1574-6941.1997.tb00451.x]

[79] Dillon JG, Castenholz RW. The synthesis of the UV-screening pigment, scytonemin, and photosynthetic performance in isolates from closely related natural populations of cyanobacteria (*Calothrix* sp.). Environ Microbiol 2003; 5(6): 484-91.
[http://dx.doi.org/10.1046/j.1462-2920.2003.00436.x] [PMID: 12755715]

[80] Fleming ED, Castenholz RW. Effects of periodic desiccation on the synthesis of the UV-screening compound, scytonemin, in cyanobacteria. Environ Microbiol 2007; 9(6): 1448-55.
[http://dx.doi.org/10.1111/j.1462-2920.2007.01261.x] [PMID: 17504482]

[81] Sinha RP, Häder DP. UV-protectants in cyanobacteria. Plant Sci 2008; 174(3): 278-89.
[http://dx.doi.org/10.1016/j.plantsci.2007.12.004]

[82] Abed RMM, Dobretsov S, Sudesh K. Applications of cyanobacteria in biotechnology. J Appl Microbiol 2009; 106(1): 1-12.
[http://dx.doi.org/10.1111/j.1365-2672.2008.03918.x] [PMID: 19191979]

[83] Rastogi RP, Sinha RP. Biotechnological and industrial significance of cyanobacterial secondary

metabolites. Biotechnol Adv 2009; 27(4): 521-39.
[http://dx.doi.org/10.1016/j.biotechadv.2009.04.009] [PMID: 19393308]

[84] Williamson CE, Zepp RG, Lucas RM, *et al.* Solar ultraviolet radiation in a changing climate. Nat Clim Chang 2014; 4(6): 434-41.
[http://dx.doi.org/10.1038/nclimate2225]

[85] Häder DP, Williamson CE, Wängberg SÅ, *et al.* Effects of UV radiation on aquatic ecosystems and interactions with other environmental factors. Photochem Photobiol Sci 2015; 14(1): 108-26.
[http://dx.doi.org/10.1039/C4PP90035A] [PMID: 25388554]

[86] Garcia-Pichel F, Castenholz RW. Occurrence of UV-absorbing, mycosporine-like compounds among cyanobacterial isolates and an estimate of their screening capacity. Appl Environ Microbiol 1993; 59(1): 163-9.
[http://dx.doi.org/10.1128/aem.59.1.163-169.1993] [PMID: 16348839]

[87] Rastogi RP, Richa , Sinha RP, Singh SP, Häder DP. Photoprotective compounds from marine organisms. J Ind Microbiol Biotechnol 2010; 37(6): 537-58.
[http://dx.doi.org/10.1007/s10295-010-0718-5] [PMID: 20401734]

[88] Rastogi RP, Incharoensakdi A. Characterization of UV-screening compounds, mycosporine-like amino acids, and scytonemin in the cyanobacterium *Lyngbya* sp. CU2555. FEMS Microbiol Ecol 2014; 87(1): 244-56.
[http://dx.doi.org/10.1111/1574-6941.12220] [PMID: 24111939]

[89] Dunlap WC, Chalker BE. Identification and quantitation of near-UV absorbing compounds (S-320) in a hermatypic scleractinian. Coral Reefs 1986; 5(3): 155-9.
[http://dx.doi.org/10.1007/BF00298182]

[90] Rosic NN, Dove S. Mycosporine-like amino acids from coral dinoflagellates. Appl Environ Microbiol 2011; 77(24): 8478-86.
[http://dx.doi.org/10.1128/AEM.05870-11] [PMID: 22003012]

[91] Wada N, Sakamoto T, Matsugo S. Mycosporine-like amino acids and their derivatives as natural antioxidants. Antioxidants 2015; 4(3): 603-46.
[http://dx.doi.org/10.3390/antiox4030603] [PMID: 26783847]

[92] Řezanka T, Temina M, Tolstikov AG, Dembitsky VM. Natural microbial UV radiation filters — Mycosporine-like amino acids. Folia Microbiol (Praha) 2004; 49(4): 339-52.
[http://dx.doi.org/10.1007/BF03354663] [PMID: 15530001]

[93] Shih MF, Cherng JY. Potential protective effect of fresh grown unicellular green algae component (resilient factor) against PMA- and UVB-induced MMP1 expression in skin fibroblasts. Eur J Dermatol 2008; 18(3): 303-7.
[PMID: 18474460]

[94] Riegger L, Robinson D. Photoinduction of UV-absorbing compounds in Antarctic diatoms and *Phaeocystis antarctica.* Mar Ecol Prog Ser 1997; 160: 13-25.
[http://dx.doi.org/10.3354/meps160013]

[95] Llewellyn CA, Airs RL. Distribution and abundance of MAAs in 33 species of microalgae across 13 classes. Mar Drugs 2010; 8(4): 1273-91.
[http://dx.doi.org/10.3390/md8041273] [PMID: 20479978]

[96] Misonou T, Saitoh J, Oshiba S, *et al.* UV-absorbing substance in the red alga *Porphyra yezoensis* (Bangiales, Rhodophyta) block thymine photodimer production. Mar Biotechnol (NY) 2003; 5(2): 194-200.
[http://dx.doi.org/10.1007/s10126-002-0065-2] [PMID: 12876656]

[97] Conde FR, Churio MS, Previtali CM. The photoprotector mechanism of mycosporine-like amino acids. Excited-state properties and photostability of porphyra-334 in aqueous solution. J Photochem Photobiol B 2000; 56(2-3): 139-44.

[http://dx.doi.org/10.1016/S1011-1344(00)00066-X] [PMID: 11079474]

[98] Vankranendonk M, Philippot P, Lepot K, Bodorkos S, Pirajno F. Geological setting of Earth's oldest fossils in the ca. 3.5Ga Dresser Formation, Pilbara Craton, Western Australia. Precambrian Res 2008; 167(1-2): 93-124.
[http://dx.doi.org/10.1016/j.precamres.2008.07.003]

[99] Westall F, de Ronde CEJ, Southam G, *et al.* Implications of a 3.472–3.333 Gyr-old subaerial microbial mat from the Barberton greenstone belt, South Africa for the UV environmental conditions on the early Earth. Philos Trans R Soc Lond B Biol Sci 2006; 361(1474): 1857-76.
[http://dx.doi.org/10.1098/rstb.2006.1896] [PMID: 17008224]

[100] Nutman AP, Bennett VC, Friend CRL, Van Kranendonk MJ, Chivas AR. Rapid emergence of life shown by discovery of 3,700-million-year-old microbial structures. Nature 2016; 537(7621): 535-8.
[http://dx.doi.org/10.1038/nature19355] [PMID: 27580034]

[101] Mishra S, Pandey A, Sinha RP. Characterization of photoprotective and pharmaceutically important compounds from cyanobacteria and algal assemblages from historical kunds of Varanasi, India. J Chem Pharm Res 2019; 11(9): 24-36.

[102] Mishra S, Pandey A, Ahmed H, Kumar D, Singh V, Sinha RP. Mycosporine-like amino acids (MAAs) profile of cyanobacteria from different historical kunds of Varanasi, India. Int J Appl Sci Biotechnol 2019; 7(3): 317-26.
[http://dx.doi.org/10.3126/ijasbt.v7i3.25712]

[103] Batista AP, Gouveia L, Bandarra NM, Franco JM, Raymundo A. Comparison of microalgal biomass profiles as novel functional ingredient for food products. Algal Res 2013; 2(2): 164-73.
[http://dx.doi.org/10.1016/j.algal.2013.01.004]

[104] Nuño K, Villarruel-López A, Puebla-Pérez AM, Romero-Velarde E, Puebla-Mora AG, Ascencio F. Effects of the marine microalgae *Isochrysis galbana* and *Nannochloropsis oculata* in diabetic rats. J Funct Foods 2013; 5(1): 106-15.
[http://dx.doi.org/10.1016/j.jff.2012.08.011]

[105] Goh LP Jr, Loh SP, Fatimah MY, Perumal K. Bioaccessibility of Carotenoids and Tocopherols in Marine Microalgae, *Nannochloropsis* sp. and *Chaetoceros* sp. Malays J Nutr 2009; 15(1): 77-86.
[PMID: 22691807]

[106] Tokuşoglu O, üunal MK. Biomass nutrient profiles of three microalgae: *Spirulina platensis, Chlorella vulgaris*, and *Isochrisis galbana.* J Food Sci 2003; 68(4): 1144-8.
[http://dx.doi.org/10.1111/j.1365-2621.2003.tb09615.x]

[107] Achour HY, Doumandji A, Sadi S, Saadi S. Evaluation of nutritional and sensory properties of bread enriched with *Spirulina.* Ann Food Sci Technol (Valahia Univ Târgoviste) 2014; 15: 270-5.

[108] Batista AP, Nunes MC, Raymundo A, *et al.* Microalgae biomass interaction in biopolymer gelled systems. Food Hydrocoll 2011; 25(4): 817-25.
[http://dx.doi.org/10.1016/j.foodhyd.2010.09.018]

[109] Gouveia L, Coutinho C, Mendonça E, *et al.* Functional biscuits with PUFA-ω3 from *Isochrysis galbana.* J Sci Food Agric 2008; 88(5): 891-6.
[http://dx.doi.org/10.1002/jsfa.3166]

[110] Fradique M, Batista AP, Nunes MC, Gouveia L, Bandarra NM, Raymundo A. Incorporation of *Chlorella vulgaris* and *Spirulina maxima* biomass in pasta products. Part 1: Preparation and evaluation. J Sci Food Agric 2010; 90(10): 1656-64.
[http://dx.doi.org/10.1002/jsfa.3999] [PMID: 20564448]

[111] Varga L, Szigeti J, Kovács R, Földes T, Buti S. Influence of a *Spirulina platensis* biomass on the microflora of fermented ABT milks during storage (R1). J Dairy Sci 2002; 85(5): 1031-8.
[http://dx.doi.org/10.3168/jds.S0022-0302(02)74163-5] [PMID: 12086036]

[112] Molnár N, Gyenis B, Varga L. Influence of a powdered *Spirulina platensis* biomass on acid production

of lactococci in milk. Milchwissenschaft 2005; 60(4): 380-2.

[113] Singh P, Singh R, Jha A, Rasane P, Gautam AK. Optimization of a process for high fibre and high protein biscuit. J Food Sci Technol 2015; 52(3): 1394-403. [http://dx.doi.org/10.1007/s13197-013-1139-z] [PMID: 25745207]

[114] Ak B, Avsaroglu E, Isik O, Özyurt G, Kafkas E, Etyemez M. Nutritional and physicochemical characteristics of bread enriched with microalgae *Spirulina platensis.* Int J Eng Res Appl 2016; 6: 30-8.

[115] Saleh AM, El-Fouly MM, Shaheen AB, El-Malky W, Khorshid AM. Blaugrüne Algen bei der Brotherstellung. Getreide. Mehl und Brot 1987; 41(1): 19-22.

[116] Batista AP, Niccolai A, Fradinho P, *et al.* Microalgae biomass as an alternative ingredient in cookies: Sensory, physical and chemical properties, antioxidant activity and in vitro digestibility. Algal Res 2017; 26: 161-71. [http://dx.doi.org/10.1016/j.algal.2017.07.017]

[117] Beheshtipour H, Mortazavian AM, Mohammadi R, Sohrabvandi S, Khosravi-Darani K. Supplementation of *Spirulina platensis* and *Chlorella vulgaris* algae into probiotic fermented milks. Compr Rev Food Sci Food Saf 2013; 12(2): 144-54. [http://dx.doi.org/10.1111/1541-4337.12004]

[118] Dinu M, Vlasceanu G, Dune A, Rotaru G. Researches concerning the growth of nutritive value of the bread products through the *Spirulina* adding. J Environ Prot Ecol 2012; 13(2): 660-5.

[119] Lucas BF, Morais MG, Santos TD, Costa JAV. *Spirulina* for snack enrichment: Nutritional, physical and sensory evaluations. Lebensm Wiss Technol 2018; 90: 270-6. [http://dx.doi.org/10.1016/j.lwt.2017.12.032]

[120] Gouveia L, Batista AP, Miranda A, Empis J, Raymundo A. *Chlorella vulgaris* biomass used as colouring source in traditional butter cookies. Innov Food Sci Emerg Technol 2007; 8(3): 433-6. [http://dx.doi.org/10.1016/j.ifset.2007.03.026]

[121] Raymundo A, Gouveia L, Batista AP, Empis J, Sousa I. Fat mimetic capacity of *Chlorella vulgaris* biomass in oil-in-water food emulsions stabilized by pea protein. Food Res Int 2005; 38(8-9): 961-5. [http://dx.doi.org/10.1016/j.foodres.2005.02.016]

[122] Gouveia L, Raymundo A, Batista AP, Sousa I, Empis J. *Chlorella vulgaris* and *Haematococcus pluvialis* biomass as colouring and antioxidant in food emulsions. Eur Food Res Technol 2006; 222(3-4): 362-7. [http://dx.doi.org/10.1007/s00217-005-0105-z]

[123] Cho EJ, Nam ES, Park SI. Keeping quality and sensory properties of drinkable yoghurt with added *Chlorella* extract. Korean J Food Nutr 2004; 17(2): 128-32.

[124] Jeon JK. Effect of *Chlorella* addition on the quality of processed cheese. Journal of the Korean Society of Food Science and Nutrition 2006; 35(3): 373-7. [http://dx.doi.org/10.3746/jkfn.2006.35.3.373]

[125] El-Baz FK, Abdo SM, Hussein AM. Microalgae *Dunaliella salina* for use as food supplement to improve pasta quality. Int J Pharm Sci Rev Res 2017; 46(2): 45-51.

[126] Fradique M, Batista AP, Nunes MC, Gouveia L, Bandarra NM, Raymundo A. *Isochrysis galbana* and *Diacronema vlkianum* biomass incorporation in pasta products as PUFA's source. Lebensm Wiss Technol 2013; 50(1): 312-9. [http://dx.doi.org/10.1016/j.lwt.2012.05.006]

[127] Finney KF, Pomeranz Y, Bruinsma BL. Use of algae *Dunaliella* as a protein supplement in bread. Cereal Chem 1984; 61(5): 402-6.

[128] García-Segovia P, Pagán-Moreno MJ, Lara IF, Martínez-Monzó J. Effect of microalgae incorporation on physicochemical and textural properties in wheat bread formulation. Food Sci Technol Int 2017;

23(5): 437-47.
[http://dx.doi.org/10.1177/1082013217700259] [PMID: 28675970]

[129] Mouritsen OG, Rhatigan P, Pérez-Lloréns JL. The rise of seaweed gastronomy: phycogastronomy. Bot Mar 2019; 62(3): 195-209.
[http://dx.doi.org/10.1515/bot-2018-0041]

[130] Durand-Chastel H. Production and use of *Spirulina* in Mexico. In: Shelef G, Soeder CJ, Eds. Algae Biomass 1980; 51-64.

[131] Soong P. Production and development of *Chlorella* and *Spirulina* in Taiwan. In: Shelef G, Soeder CJ, Eds. Algae Biomass 1980; 97-113.

[132] Pulz O, Gross W. Valuable products from biotechnology of microalgae. Appl Microbiol Biotechnol 2004; 65(6): 635-48.
[http://dx.doi.org/10.1007/s00253-004-1647-x] [PMID: 15300417]

[133] Liang S, Liu X, Chen F, Chen Z. Current microalgal health food R & D activities in China. In: Ang PO, Ed. Asian Pacific Phycology in the 21st Century: Prospects and Challenges 2004 2004; 45-8.
[http://dx.doi.org/10.1007/978-94-007-0944-7_7]

[134] Soletto D, Binaghi L, Lodi A, Carvalho JCM, Converti A. Batch and fed-batch cultivations of *Spirulina platensis* using ammonium sulphate and urea as nitrogen sources. Aquaculture 2005; 243(1-4): 217-24.
[http://dx.doi.org/10.1016/j.aquaculture.2004.10.005]

[135] Rangel-Yagui CO, Danesi EDG, de Carvalho JCM, Sato S. Chlorophyll production from *Spirulina platensis* : cultivation with urea addition by fed-batch process. Bioresour Technol 2004; 92(2): 133-41.
[http://dx.doi.org/10.1016/j.biortech.2003.09.002] [PMID: 14693445]

[136] Cazzonelli CI. Carotenoids in nature: insights from plants and beyond. Funct Plant Biol 2011; 38(11): 833-47.
[http://dx.doi.org/10.1071/FP11192] [PMID: 32480941]

[137] Amaro HM, Guedes AC, Malcata FX. Antimicrobial activities of microalgae: an invited review. In: M'endez-Vilas A, Ed. Science against Microbial Pathogens: Communicating Current Research and Technological Advances 2011; 1272-80.

[138] Smelcerovic A, Knezevic-Jugovic Z, Petronijevic Z. Microbial polysaccharides and their derivatives as current and prospective pharmaceuticals. Curr Pharm Des 2008; 14(29): 3168-95.
[http://dx.doi.org/10.2174/138161208786404254] [PMID: 19075698]

[139] Chen X, Song L, Wang H, *et al.* Partial characterization, the immune modulation and anticancer activities of sulfated polysaccharides from filamentous microalgae *Tribonema* sp. Molecules 2019; 24(2): 322.
[http://dx.doi.org/10.3390/molecules24020322] [PMID: 30658396]

[140] Mata TM, Martins AA, Caetano NS. Microalgae for biodiesel production and other applications: A review. Renew Sustain Energy Rev 2010; 14(1): 217-32.
[http://dx.doi.org/10.1016/j.rser.2009.07.020]

[141] Viskari PJ, Colyer CL. Rapid extraction of phycobiliproteins from cultured cyanobacteria samples. Anal Biochem 2003; 319(2): 263-71.
[http://dx.doi.org/10.1016/S0003-2697(03)00294-X] [PMID: 12871721]

[142] Bermejo Román R, Alvárez-Pez JM, Acién Fernández FG, Molina Grima E. Recovery of pure B-phycoerythrin from the microalga *Porphyridium cruentum.* J Biotechnol 2002; 93(1): 73-85.
[http://dx.doi.org/10.1016/S0168-1656(01)00385-6] [PMID: 11690696]

[143] Samarakoon KW, Ko JY, Lee JH, Kwon ON, Kim SW, Jeon YJ. Apoptotic anticancer activity of a novel fatty alcohol ester isolated from cultured marine diatom, *Phaeodactylum tricornutum.* J Funct Foods 2014; 6: 231-40.
[http://dx.doi.org/10.1016/j.jff.2013.10.011]

[144] Kobayashi J, Ishibashi M, Nakamura H, *et al.* Amphidinolide-A, a novel antineoplastic macrolide from the marine dinoflagellate *Amphidinium* sp. Tetrahedron Lett 1986; 27(47): 5755-8.
[http://dx.doi.org/10.1016/S0040-4039(00)85318-6]

[145] Khanra S, Mondal M, Halder G, Tiwari ON, Gayen K, Bhowmick TK. Downstream processing of microalgae for pigments, protein and carbohydrate in industrial application: A review. Food Bioprod Process 2018; 110: 60-84.
[http://dx.doi.org/10.1016/j.fbp.2018.02.002]

[146] Andrade LM, Andrade CJ, Dias M, Nascimento CA, Mendes MA. *Chlorella* and *Spirulina* microalgae as sources of functional foods, nutraceuticals, and food supplements. MOJ Food Processing & Technology 2018; 6(1): 45-58.
[http://dx.doi.org/10.15406/mojfpt.2018.06.00144]

[147] Raposo M, De Morais R, Bernardo de Morais A. Bioactivity and applications of sulphated polysaccharides from marine microalgae. Mar Drugs 2013; 11(1): 233-52.
[http://dx.doi.org/10.3390/md11010233] [PMID: 23344113]

[148] Villar R, Laguna R, Martínez D, Núñez L, Jiménez C. Anti-aggregant effects on human platelets of the crude aqueous extract and polar fractions of the microalga *Dunaliella tertiolecta.* Phytother Res 1997; 11(1): 70-2.
[http://dx.doi.org/10.1002/(SICI)1099-1573(199702)11:1<70::AID-PTR36>3.0.CO;2-U]

[149] Guedes AC, Amaro HM, Barbosa CR, Pereira RD, Malcata FX. Fatty acid composition of several wild microalgae and cyanobacteria, with a focus on eicosapentaenoic, docosahexaenoic and α-linolenic acids for eventual dietary uses. Food Res Int 2011; 44(9): 2721-9.
[http://dx.doi.org/10.1016/j.foodres.2011.05.020]

[150] Renju GL, Kurup GM, Saritha Kumari CH. Effect of lycopene from *Chlorella marina* on high cholesterol-induced oxidative damage and inflammation in rats. Inflammopharmacology 2014; 22(1): 45-54.
[http://dx.doi.org/10.1007/s10787-013-0178-4] [PMID: 23887896]

[151] Halim R, Danquah MK, Webley PA. Extraction of oil from microalgae for biodiesel production: A review. Biotechnol Adv 2012; 30(3): 709-32.
[http://dx.doi.org/10.1016/j.biotechadv.2012.01.001] [PMID: 22266377]

[152] Yen HW, Hu IC, Chen CY, Ho SH, Lee DJ, Chang JS. Microalgae-based biorefinery – From biofuels to natural products. Bioresour Technol 2013; 135: 166-74.
[http://dx.doi.org/10.1016/j.biortech.2012.10.099] [PMID: 23206809]

[153] Ho SH, Huang SW, Chen CY, Hasunuma T, Kondo A, Chang JS. Bioethanol production using carbohydrate-rich microalgae biomass as feedstock. Bioresour Technol 2013; 135: 191-8.
[http://dx.doi.org/10.1016/j.biortech.2012.10.015] [PMID: 23116819]

[154] Guo H, Daroch M, Liu L, Qiu G, Geng S, Wang G. Biochemical features and bioethanol production of microalgae from coastal waters of Pearl River Delta. Bioresour Technol 2013; 127: 422-8.
[http://dx.doi.org/10.1016/j.biortech.2012.10.006] [PMID: 23138065]

[155] Kim M, Baek J, Yun Y, Junsim S, Park S, Kim S. Hydrogen production from *Chlamydomonas reinhardtii* biomass using a two-step conversion process: Anaerobic conversion and photosynthetic fermentation. Int J Hydrogen Energy 2006; 31(6): 812-6.
[http://dx.doi.org/10.1016/j.ijhydene.2005.06.009]

[156] Hirano A, Ueda R, Hirayama S, Ogushi Y. CO_2 fixation and ethanol production with microalgal photosynthesis and intracellular anaerobic fermentation. Energy 1997; 22(2-3): 137-42.
[http://dx.doi.org/10.1016/S0360-5442(96)00123-5]

[157] Couteau C, Coiffard L. Microalgal application in cosmetics. In: Levine I, Fleurence J, Eds. Microalgae in health and disease prevention 2018; 317-23.
[http://dx.doi.org/10.1016/B978-0-12-811405-6.00015-3]

[158] Yongmanitchai W, Ward OP. Growth of and omega-3 fatty acid production by *Phaeodactylum tricornutum* under different culture conditions. Appl Environ Microbiol 1991; 57(2): 419-25. [http://dx.doi.org/10.1128/aem.57.2.419-425.1991] [PMID: 2014989]

[159] Acién Fernández FG, Hall DO, Cañizares Guerrero E, Krishna Rao K, Molina Grima E. Outdoor production of *Phaeodactylum tricornutum* biomass in a helical reactor. J Biotechnol 2003; 103(2): 137-52. [http://dx.doi.org/10.1016/S0168-1656(03)00101-9] [PMID: 12814873]

[160] Molina Grima E, Garcia Camacho F, Ácien Fernandez FG. Production of EPA from *Phaeodactylum tricornutum*. In: Cohen Z, Ed. Chemicals from Microalgae 1999; 57-92.

[161] Sukenik A. Production of eicosapentaenoic acid by the marine eustigmatophyte *Nannochloropsis*. In: Cohen Z, Ed. Chemicals from Microalgae 1999; 41-56.

[162] Su G, Jiao K, Chang J, *et al.* Enhancing total fatty acids and arachidonic acid production by the red microalgae *Porphyridium purpureum.* Bioresour Bioprocess 2016; 3(1): 33. [http://dx.doi.org/10.1186/s40643-016-0110-z]

[163] Solovchenko AE, Khozin-Goldberg I, Didi-Cohen S, Cohen Z, Merzlyak MN. Effects of light intensity and nitrogen starvation on growth, total fatty acids and arachidonic acid in the green microalga *Parietochloris incisa.* J Appl Phycol 2008; 20(3): 245-51. [http://dx.doi.org/10.1007/s10811-007-9233-0]

[164] Giménez Giménez A, Ibáñez González MJ, Robles Medina A, Molina Grima E, García Salas S, Esteban Cerdán L. Downstream processing and purification of eicosapentaenoic (20: 5n-3) and arachidonic acids (20: 4n-6) from the microalga *Porphyridium cruentum.* Bioseparation 1998; 7(2): 89-99. [http://dx.doi.org/10.1023/A:1008021330785]

[165] Chu WL. Biotechnological applications of microalgae. International e-Journal of Science, Medicine & Education 2012; 6(1) (Suppl. 1): S24-37. [http://dx.doi.org/10.56026/imu.6.Suppl1.S24]

[166] Horrocks LA, Yeo YK. Health benefits of docosahexaenoic acid (DHA). Pharmacol Res 1999; 40(3): 211-25. [http://dx.doi.org/10.1006/phrs.1999.0495] [PMID: 10479465]

[167] Koller M, Muhr A, Braunegg G. Microalgae as versatile cellular factories for valued products. Algal Res 2014; 6: 52-63. [http://dx.doi.org/10.1016/j.algal.2014.09.002]

[168] Pour Hosseini SR, Tavakoli O, Sarrafzadeh MH. Experimental optimization of SC-CO_2 extraction of carotenoids from *Dunaliella salina.* J Supercrit Fluids 2017; 121: 89-95. [http://dx.doi.org/10.1016/j.supflu.2016.11.006]

[169] Pangestuti R, Kim SK. Biological activities and health benefit effects of natural pigments derived from marine algae. J Funct Foods 2011; 3(4): 255-66. [http://dx.doi.org/10.1016/j.jff.2011.07.001]

[170] Millao S, Uquiche E. Antioxidant activity of supercritical extracts from *Nannochloropsis gaditana*: Correlation with its content of carotenoids and tocopherols. J Supercrit Fluids 2016; 111: 143-50. [http://dx.doi.org/10.1016/j.supflu.2016.02.002]

[171] Esatbeyoglu T, Rimbach G. Canthaxanthin: From molecule to function. Mol Nutr Food Res 2017; 61(6): 1600469. [http://dx.doi.org/10.1002/mnfr.201600469] [PMID: 27687695]

[172] Nwachukwu ID, Udenigwe CC, Aluko RE. Lutein and zeaxanthin: Production technology, bioavailability, mechanisms of action, visual function, and health claim status. Trends Food Sci Technol 2016; 49: 74-84. [http://dx.doi.org/10.1016/j.tifs.2015.12.005]

[173] Eilers U, Dietzel L, Breitenbach J, Büchel C, Sandmann G. Identification of genes coding for functional zeaxanthin epoxidases in the diatom *Phaeodactylum tricornutum.* J Plant Physiol 2016; 192: 64-70.
[http://dx.doi.org/10.1016/j.jplph.2016.01.006] [PMID: 26851888]

[174] Gómez-Loredo A, Benavides J, Rito-Palomares M. Growth kinetics and fucoxanthin production of *Phaeodactylum tricornutum* and *Isochrysis galbana* cultures at different light and agitation conditions. J Appl Phycol 2016; 28(2): 849-60.
[http://dx.doi.org/10.1007/s10811-015-0635-0]

[175] Sahin SC. The potential of *Arthrospira platensis* extract as a tyrosinase inhibitor for pharmaceutical or cosmetic applications. S Afr J Bot 2018; 119: 236-43.
[http://dx.doi.org/10.1016/j.sajb.2018.09.004]

[176] Ariede MB, Candido TM, Jacome ALM, Velasco MVR, de Carvalho JCM, Baby AR. Cosmetic attributes of algae - A review. Algal Res 2017; 25: 483-7.
[http://dx.doi.org/10.1016/j.algal.2017.05.019]

[177] Lan CCE, Hung YT, Fang AH, Ching-Shuang W. Effects of irradiance on UVA-induced skin aging. J Dermatol Sci 2019; 94(1): 220-8.
[http://dx.doi.org/10.1016/j.jdermsci.2019.03.005] [PMID: 30956032]

[178] Pan JL, Wang HM, Chen CY, Chang JS. Extraction of astaxanthin from *Haematococcus pluvialis* by supercritical carbon dioxide fluid with ethanol modifier. Eng Life Sci 2012; 12(6): 638-47.
[http://dx.doi.org/10.1002/elsc.201100157]

[179] Thomas N, Kim SK. Beneficial effects of marine algal compounds in cosmeceuticals. Mar Drugs 2013; 11(12): 146-64.
[http://dx.doi.org/10.3390/md11010146] [PMID: 23344156]

[180] Wang HMD, Chen CC, Huynh P, Chang JS. Exploring the potential of using algae in cosmetics. Bioresour Technol 2015; 184: 355-62.
[http://dx.doi.org/10.1016/j.biortech.2014.12.001] [PMID: 25537136]

[181] Ryu B, Himaya SW, Kim SK. Applications of microalgae-derived active ingredients as cosmeceuticals. In: Kim SK, Ed. Handbook of marine microalgae 2015; 309-16.
[http://dx.doi.org/10.1016/B978-0-12-800776-1.00020-0]

CHAPTER 12

Dietary Component and Composition of Algal Functional Food

Soraya Paz[1,*], **Carmen Rubio**[1], **Ángel J. Gutiérrez**[1], **Dailos González-Weller**[1,2] and **Arturo Hardisson**[1]

[1] *Area of Toxicology, School of Medicine, Universidad de La Laguna, 38071 Santa Cruz de Tenerife, Tenerife, Canary Islands, Spain*

[2] *Health Inspection and Laboratory Service, Canary Health Service, 38006 Santa Cruz de Tenerife, Tenerife, Canary Islands, Spain*

Abstract: Algae, formerly used in food in countries on the Asian continent, are increasingly common in Western countries. These are foods with great nutritional value due to their composition in minerals and vitamins, as well as bioactive substances. Seaweed is a valuable source of antioxidants that, with its consumption, would help reduce oxidative stress. Likewise, it should be noted that the production of algae is sustainable and that, due to its own growth factors and low requirements, it reduces economic costs. This chapter will study the nutritional properties of the consumption of edible algae and the possible differences between edible species of algae and its contribution to the recommended daily requirements.

Keywords: Algal Foods, Antioxidant, Calcium, Chlorophyta, Cobalt, Copper, Dietary Intake, Dietary Reference Intake (Dri), Edible Seaweeds, Iron, Magnesium, Manganese, Minerals, Molybdenum, Phaeophyta, Potassium, Rhodophyta, Sodium, Trace Elements, Vitamins, Zinc.

INTRODUCTION

Seaweed is marketed as a source of certain nutrients. Since ancient times, algae have been a staple in the diet of Asian countries, such as China, Japan and Thailand. However, these are also of relevant importance in European countries, such as Spain, Galicia being one of the largest producers of algae, or in the Republic of Ireland, where algae play an important role in the Irish diet and customs. Currently, due to migratory flows, globalization and new eating habits, algae are increasingly common in European markets and shops.

* **Corresponding author Soraya Paz:** Area of Toxicology, School of Medicine, Universidad de La Laguna, Santa Cruz de Tenerife, Tenerife, Spain; Tel: +34 922318903; Fax: +34 922319307; E-mail: spazmont@ull.edu.es

The consumption of algae has increased notably in Western countries due, in particular, to vegetarian and vegan diets [1, 2], which are followed by more and more people who seek a supposedly healthier diet, avoiding the consumption of products of animal origin. However, the general population is unaware of the nutritional value and chemical composition of these products. Algae contain proteins, lipids, sugars, fatty acids, minerals, vitamins and bioactive compounds responsible for their biological activity [3, 4]. The nutrients present in algae can be divided into two large groups, the group of minerals formed by macronutrients and trace elements, on the other hand, other compounds, such as antioxidants, fiber, *etc*.

PRODUCTION AND CONSUMPTION OF ALGAE IN EUROPE

In Europe, algae are more and more frequent due to new trends in food. They constitute a staple food in vegan and vegetarian diets, in which you can find a large number of by-products made with algae (snacks, cookies, tofu, *etc*.). Although there are no official data on the average consumption of algae, there are data on the sales and production of these products in Europe. A report on the European market for sea vegetables carried out in 2014 shows that the countries that consume and produce the most algae in Europe are France, the United Kingdom, Germany and Spain [5].

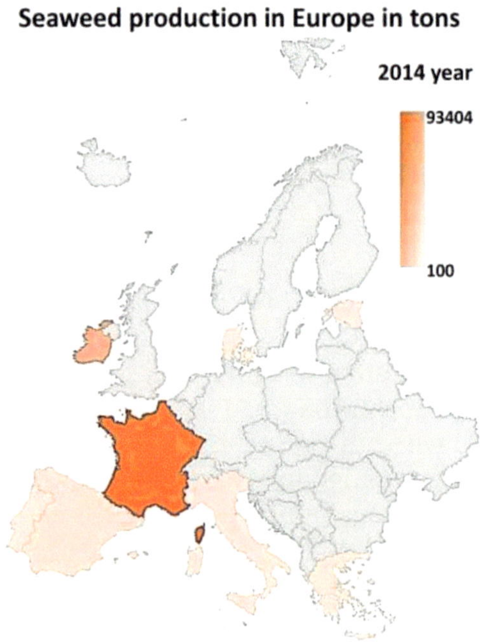

Fig. (1). Seaweed production in Europe in the 2014 [6].

The algae sector has grown exponentially in recent years; growth of 10% per year is expected. Most of the seaweed consumed in Europe comes mainly from Asia. However, Spain stands out as one of the main producers of European wakame and kombu algae (Fig. **1**) [5]. It is estimated that around 60% of the known algae species are used for food, with Asian countries being the largest consumers of these [7]. The consumption of seaweed in the Japanese population is 8.5 g/day, according to data from the Korean National Health and Nutrition Survey [7]. However, due to the multiple beneficial effects of algae, western countries have become interested in its consumption.

To date, the consumption of algae is not stipulated, and the average consumption data in the European population is unknown. However, in view of the greater availability of edible algae in various commercial presentations on all commercial surfaces in Western countries, as well as its presence in restaurants, it is reasonable to estimate that the consumption of algae in Western countries has increased exponentially.

ALGAE AND ITS NUTRIENTS

The Nutrient Database of the American organization USDA (United States, Department of Agricultural), collects nutritional reports of different foods and a great variety of algae.

Classified in the group of vegetables, the nutritional value of algae varies markedly between species. Algae are low-calorie foods, with remarkable protein content, as well as a high content of dietary fiber, minerals, and vitamins. However, in order to compare the content of macronutrients, micronutrients and minerals, it is necessary to take into account other foods of plant origin (Table **1**).

Table 1. Comparison of the nutritional value of different algae with some vegetable foods given for 100 grams of each product.

-	Wakame[a]	Irish Moss[b]	Kelp[c]	Brussels Sprouts[d]	Green Peas[e]	Spinach[f]
Energetic value(Kcal)	45	49	43	36	81	23
Protein (g)	3.03	1.51	1.68	2.55	5.42	2,86
Carbohydrates (g)	9.14	12.29	9.57	7.10	14.45	3,63
Sugar (g)	0.65	0.61	0.60	1.74	5.67	0.42
Fat (g)	0.64	0.16	9.57	0.50	0.40	0.39
- Saturated fats (g)	0.130	0.033	0.247	0..102	0.071	0.063
- Monounsaturated (g)	0.058	0.015	0.098	0.038	0.035	0.010
- Polyunsaturated (g)	0.218	0.055	0.047	0.255	0.187	0.165

(Table 1) cont.....

-	Wakame[a]	Irish Moss[b]	Kelp[c]	Brussels Sprouts[d]	Green Peas[e]	Spinach[f]
Dietary fiber (g)	0.5	1.3	1.3	2.6	5.7	2.2
Cholesterol (mg)	0	0	0	0	0	10
Calcium (mg)	150	72	168	36	25	99
Iron (mg)	2.18	8.90	2.85	1.20	1.47	2.71
Magnesium (mg)	107	144	121	20	33	79
Potassium (mg)	50	63	89	317	108	558
Sodium (mg)	872	67	233	21	5	79
Zinc (mg)	0.38	1.95	1.23	0.33	1.24	0.53
Copper (mg)	0.284	0.149	0.130	0.083	0.176	0.130
Manganese (mg)	1.40	0.370	0.200	0.227	0.410	0.897
Vitamin C (mg)	3.0	3.0	3.0	62.0	40.0	28.1
Niacin (mg)	1.60	0.593	0.470	0.607	2.090	0.724
Vitamin E (mg)	1.00	0.87	0.87	0.43	0.13	2.03
Vitamin K (µg)	5.3	5.0	66.0	140.3	24.8	482.9

[a]USDA [8], [b]USDA [9], [c]USDA [10], [d]USDA [11], [e]USDA [12], [f]USDA [13].

In Table **1**, the values of different nutrients in algae and other vegetable foods are collected considering a portion of 100 grams of each food. Seaweed is a food rich in vitamins and fiber that can be considered a source of essential elements such as phosphorus, potassium, iron, and sodium [14].

Red algae such as Irish Moss (*Chondrus crispus*) contain higher levels of minerals, such as Fe, Mg, K and Zn, while brown algae such as Wakame (*Undaria spp.*) have higher Ca, Na, Cu and Mn contents. Comparing the content of these nutrients between algae and other vegetables, such as peas, spinach or Brussels sprouts, shows that these terrestrial vegetables, in general, have a higher content of dietary fiber, and a higher content of vitamin C, which can favor the absorption of minerals such as Fe and a lower sodium (Na) content. Taking into account that Na is present in all foods, the total daily intake of this element can be exceeded in cases of high consumption of algae, since they have high Na content.

Although the Fe content of the three types of algae that are presented in Table **1** is remarkable, it is necessary to take into account that the bioavailability and, therefore, the absorption of this iron from plant foods, is a lot lower compared to the absorption of iron present in foods of animal origin. This is due to the chemical form of Fe, which is found as non-heme iron (Fe^{3+}) in plant products, while, in products of animal origin, it is found in the heme form (Fe^{2+}) [15].

Dietary Fiber

As mentioned above, algae are a valuable source of dietary fiber [16, 17]. Rasmussen *et al.* [18] estimated that the content of dietary fiber in algae can range between 36 and 60% of its dry matter, of that percentage, about 55-70% is soluble dietary fiber.

Considering the various published studies on the fiber content in algae, it is found that the soluble to insoluble dietary fiber content varies considerably between the different species of algae. However, studies carried out by Dawczynski *et al.* [19] did not detect significant differences in the content of dietary fiber between the red and brown algae analyzed. Table **2** collects the dietary fiber content recorded by various authors. As can be seen in the table, the *Grateloupia turuturu* alga contains the highest average level of soluble dietary fiber (48.1 g/100 g), while it is the *Durvillaea antárctica* alga that registers the highest concentration of insoluble dietary fiber (43.7 g/100 g).

Table 2. Soluble and insoluble dietary fiber content in different algae species.

Specie	Soluble Dietary Fiber (g/100g Dry Weight)	Insoluble Dietary Fiber (g/100g dw)	References
Gracilaria	23.8 ± 1.9	36.0 ± 3.7	[20]
Macrocystis	6.4 ± 0.3	43.6 ± 1.2	
Ulva	18.2 ± 0.8	36.0 ± 1.2	
Ulva clathrata	21.9 ± 0.9	18.7 ± 2.1	[21]
Ulva lactuca	20.53 ± 0.28	34.37 ± 0.67	[22]
Grateloupia turuturu	48.1 ± 1.0	12.3 ± 1.2	[23]
Himanthalia elongate	23.63± 0.48	13.51 ± 0.45	[24]
Laminaria saccharina	17.12 ± 0.84	13.11 ± 0.56	
Ulva lactuca	27.2 ± 1.2	33.3 ± 0.3	[25]
Durvillaea antarctica	27.7 ± 1.2	43.7 ± 0.3	

Therefore, algae are a valuable source of both soluble and insoluble dietary fiber; this makes algae a food capable of reducing the glycemic response and blood cholesterol [7].

Protein Content

The proteins that can be found in different species of algae can vary between species. The glycine, arginine, alanine and glutamic acid content stand out. Likewise, considering the recommended values indicated by the World Health

Organization (WHO) together with the Food and Agricultural Organization (FAO), it is necessary that the algae contain sufficient essential amino acids to cover the indicated requirements. However, it should be noted that its limiting amino acids are lysine and cysteine [26, 27].

Among the different algae, the amino acid taurine is found in red algae. This amino acid is of great interest since it participates in multiple physiological processes such as osmoregulation, immunomodulation, *etc* (Fig. **2**). Likewise, taurine plays an important role in eye development and the nervous system. The fact that taurine is found in high contents in red algae, favors its use as a reinforcement of foods intended for children, who need a higher content of taurine in the first months of life.

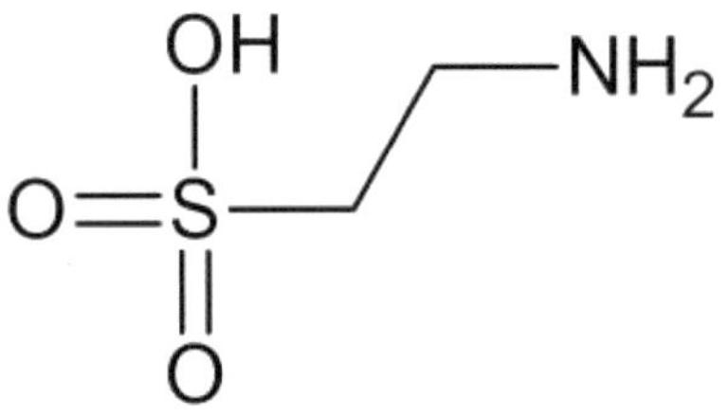

Fig. (2). Taurine chemical structure.

Dawczynski *et al.* [19] showed that brown algae contain high concentrations of the amino acid phosphoserine. Phosphoserine is a chelator that has a certain antioxidant capacity [28].

Essential Elements

Algae are organisms whose most of their matter is made up of minerals. Minerals or essential elements are elements required daily by the body since they participate in physiological reactions. The essential elements can be divided into two groups, the macroelements formed by alkali and alkaline earth metals (Fig. **3**): calcium (Ca), potassium (K), magnesium (Mg) and sodium (Na). These are required in large quantities.

On the other hand, there are the essential trace elements (Fig. **4**), these being iron (Fe), copper (Cu), cobalt (Co), zinc (Zn), manganese (Mn), molybdenum (Mo) and iodine. (I). Although these elements are vital for the proper functioning of the body, these elements are required in smaller quantities [29 - 31].

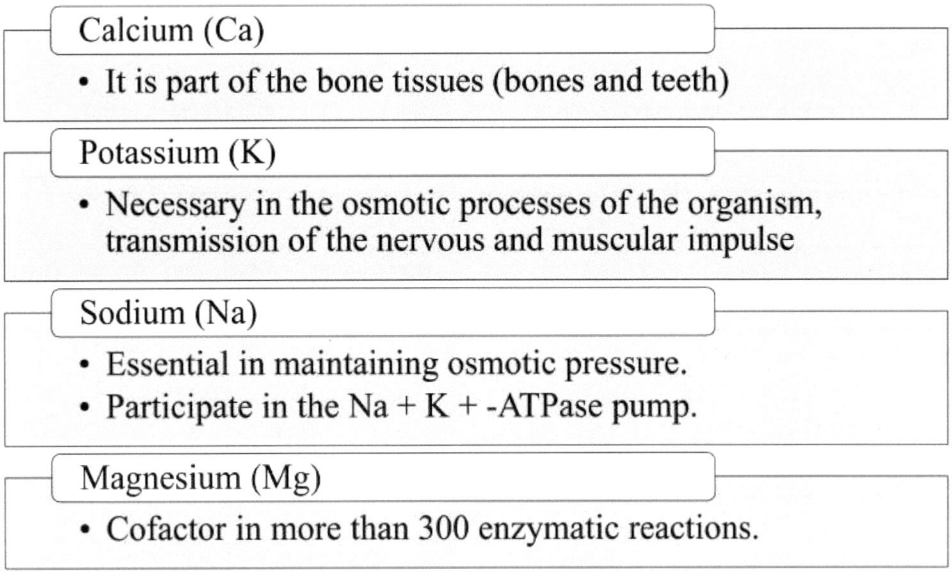

Fig. (3). Main functions of the macroelements in the organism.

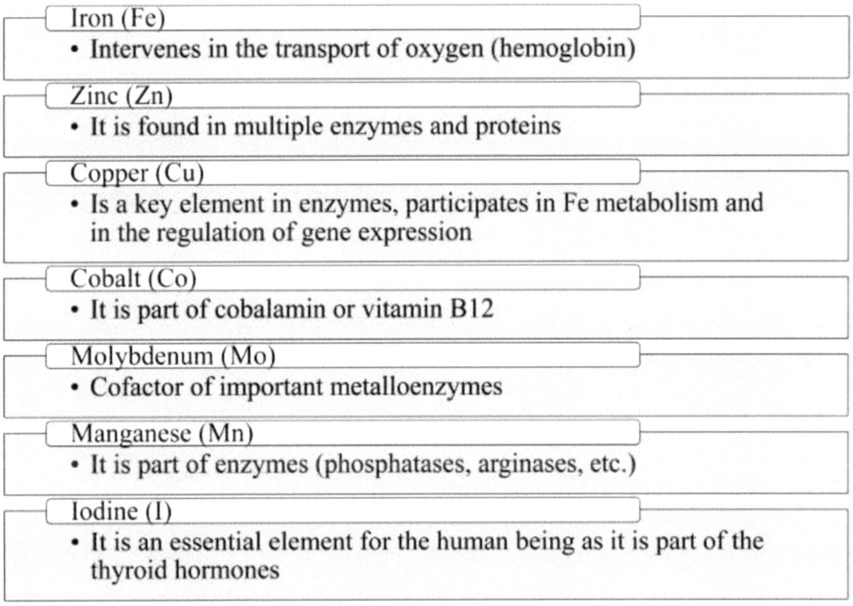

Fig. (4). Main functions of the trace elements in the organism.

Studies carried out by Paz *et al.* [2] in which the content of macroelements and essential trace elements in algae commercialized in the Canary Islands was analyzed, recorded concentrations of these that varied significantly between species as well as in the geographical origin, which directly influences the content of macroelements and trace elements of algae (Table **3**).

Table 3. Macroelements and trace elements levels (mg/kg) in commercialized edible seaweeds determined by icp-oes (inductively coupled plasma optical emission spectrometry) [2].

-	***Undaria Pinnatifida***		***Himanthalia Elongata***	***Laminaria Ochroleuca***	**Mixed**	
	European	Asian	European			Asian
Macroelements						
Ca	3314±1303	3927±1222	3469±1526	2910±1267	2227±987	1327±623
Na	19783±9625	28246±2712	25085±7924	15031±10684	18277±8149	17199±3157
K	33714±13121	8640±2691	57480±19976	38410±15553	12580±5970	5259±2352
Mg	2768±1527	4662±1223	3537±1497	1653±1386	1111±582	525±534
Trace elements						
Co	0.12±0.05	0.08±0.01	0.65±0.14	0.16±0.31	0.08±0.03	0.02±0.04
Cu	1.8±1.2	2.0±0.4	2.2±0.9	2.5±1.1	1.6±0.5	21.2±10.1
Fe	22.6±14	58.8±17	17.8±3.3	27.8±8.9	27±4.4	21.8±12
Mn	2.71±1.1	5.64±0.70	14.1±12	8.61±13	10.2±4.6	3.45±3.2
Mo	0.11±0.03	0.08±0.02	0.08±0.03	0.08±0.04	0.06±0.02	0.06±0.01
Zn	40.3±32	18.1±4.3	21.3±13	10.3±7.1	20.1±6.5	9.27±7.1

As can be seen in the table above, the algae analyzed contain high concentrations of the macroelements Ca, Na, K and Mg. They can be considered an important source of these elements that are required in large quantities on a daily basis.

In the case of trace elements, the content of Fe and Zn stands out considerably, which makes algae a food that can be considered a source of both elements. However, it is necessary to consider the limitations of Fe and Zn present in plant foods, since their bioavailability is lower compared to foods of animal origin.

ALGAE AS A FUNCTIONAL FOOD

The consumption of algae has increased in western countries because this new food stands out as a functional food. This is due to its composition, algae contain bioactive substances of great value. This has caused thousands of laboratories around the world to study the content of bioactive substances in different species of algae and their possible beneficial effects. Algae stand out for containing high concentrations of bioactive substances such as polyphenols, peptides and carotenoids [32 - 34].

Carotenoids

Carotenoids, found in photosynthetic organisms like algae, play an important role in the human organism due to their physiological effects [35]. In addition, diet is the main route of entry for carotenoids into the human body since human beings are not capable of synthesizing these compounds [36].

Carotenoids are lipophilic pigments that can be found in algae in different concentrations. These natural pigments have beneficial properties due to their antioxidant, anticancer, anti-inflammatory capacity due to modulation on macrophage function, among other beneficial factors [37 - 43].

Antioxidants are recognized for their ability to reduce the risk of diseases, especially due to their ability to reduce oxidative stress [43, 44]. The extraction of carotenoids from algae has acquired great importance due to the beneficial effects attributed to them. Teramukai *et al.* [36] have recently published the results obtained from the extraction of carotenoids from brown algae with edible oils.

Fucoxanthin

One of the most studied carotenoids for their antioxidant properties is fucoxanthin ($C_{42}H_{58}O_6$), which is found in brown or brown algae (Fig. **5**). The brown color of brown algae is due to the presence of fucoxanthin. This compound belonging to the xanthophyll family has shown anti-cancer, anti-inflammatory and anti-obesity effects [33, 45]. The wakame algae or Undaria pinnatifida are one of the macroalgae with the highest levels of fucoxanthin [45].

Sargasso algae, abundant on the coasts of different geographical locations such as the Caribbean, the Canary Islands, *etc*., also stands out for its fucoxanthin content [35].

Studies on the properties of fucoxanthin are increasing, discovering new beneficial properties of this carotenoid, such as, for example, its activity as a neuroprotective, photoprotective and preventive of osteoporosis [46 - 50]. Although fucoxanthin is a lipophilic carotenoid, its absorption in the body depends on multiple factors such as the amount and type of lipids ingested, and the content of dietary fiber, among other factors that are still unknown.

Bhat *et al.* [51] determined the fucoxanthin content in different species of algae, registering the following concentrations: *Ulva prolifera* (0.69±0.09 µg/kg dry weight), *Sargassum whightii* (3.13±0.09 µg/kg dw), *Bryopsis sp.* (3.44±0.11 µg/kg dw), *Grateloupia filicina* (3.45±0.06 µg/kg dw), *Ceramium sp.* (4.85±0.04 µg/kg dw) and *Gracilaria corticata* (3.06±0.05 µg/kg dw).

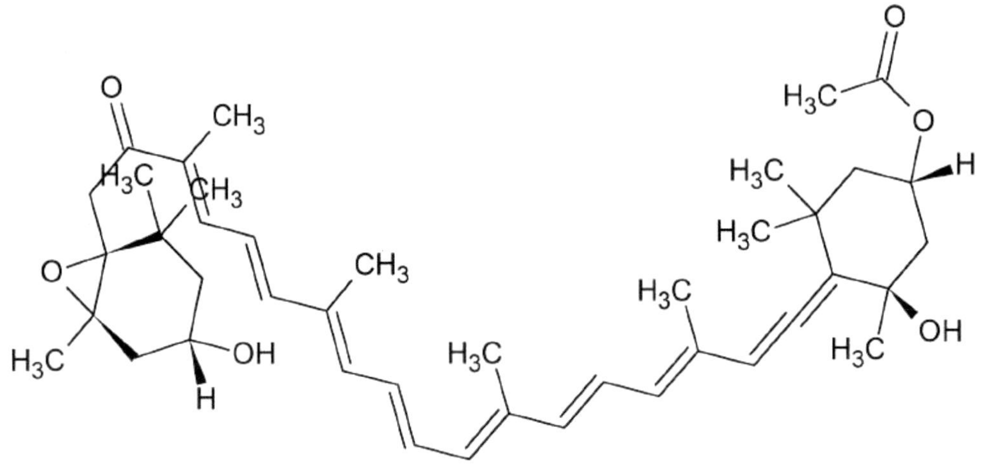

Fig. (5). Chemical structure of fucoxanthin.

β-carotene

B-carotene is a fat-soluble carotenoid (Fig. **6**) that provides a red, orange and/or yellow coloration in photosynthetic organisms. It is also known as a provitamin A carotenoid, being the most abundant and most efficient of those found in food.

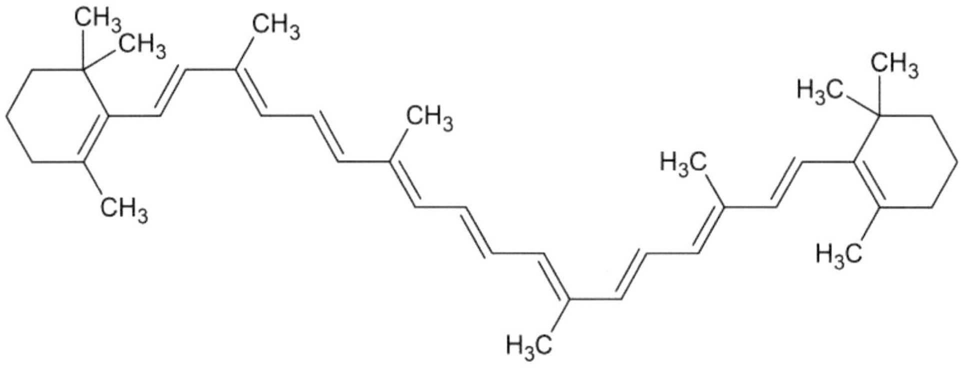

Fig. (6). Chemical structure of β-carotene.

The functions of this carotenoid in the human body are mainly due to its activity as a provitamin A or a precursor of Vitamin A [52, 53]. Vitamin A is necessary for proper growth and development, for eye and vision health, and the immune system. Likewise, β-carotene is an effective antioxidant against reactive oxygen species [54 - 56]. A study carried out by Kasperczyk *et al.* [57] showed that β-carotene reduced oxidative stress, improved glutathione metabolism and, in addition, modified the antioxidant defense system in workers exposed to high levels of lead.

Bhat *et al.* [51] determined the beta-carotene content in algae from India, registering concentrations: *Gracilaria corticata* (4.13±0.07 µg/kg dw), *Grateloupia filicini* (9.40±0.05 µg/kg dw), *Bryopsis sp.* (5.46. ±0.08 µg/kg dw), *Ulva prolifera* (10.85±0.11 µg/kg dw). Meanwhile, in another study by Silva-Marinho *et al.* [43], the beta-carotene content in *Saccharina latissima* algae was determined in different stations, registering a content of between 3.06±3.12 and 23.02±7.14 µg/kg dw, confirming that the beta-carotene content varied between stations. Of the year, the highest concentration is registered between the months of November to January.

Polyphenols

Algae contain water-soluble compounds, such as polyphenols. Polyphenols are water-soluble antioxidant compounds. There are numerous studies carried out on the content of polyphenols in algae and their antioxidant activity (Table **4**). It should be noted that brown algae contain higher concentrations of polyphenols than red and green algae [7].

Table 4. Total polyphenols concentration by different algae species.

Specie	Algae Type	Total Polyphenols (mg GAE/100 g dw)	References
Stypocaulon scoparium	Brown	328±2.87	[58]
Ulva lactuca	Green	2.86±0.04	[59]
Alaria esculenta	Brown	2.80±0.05	
Sargassum marginatum		0.29	[37]

GAE, gallic acid equivalent.

Swiriwardhana *et al.* [60] studied the content of polyphenols in the aqueous, organic and alcoholic phases extracted from Hijiki algae (brown algae), finding a higher antioxidant activity linked to the aqueous phase, that is, the phase containing the highest concentration of polyphenols is the one that is related to the highest antioxidant activity. Another study carried out by Vidal *et al.* [61], in which they determined the content of polyphenols in the aqueous fraction of two species of green algae (*Halimeda opuntia* and *Halimeda monile*), concluded that of the different polyphenols studied, the salicylic acid content in both species stood out considerably. Salicylic acid and its activity as an antioxidant have been studied in multiple investigations. This chemical compound is found in plants and plays an important role in the defense system of plant organisms against oxidative stress [61 - 65]. However, the antioxidant effect may be due not only to salicylic acid, but also to all the different polyphenols found in algae. The polyphenols that stand out in algae and can be determined in algae today are salicylic, cinnamic, p-coumaric, gallic, pyrogallic, ferulic, and caffeic (Fig. **7**) [61].

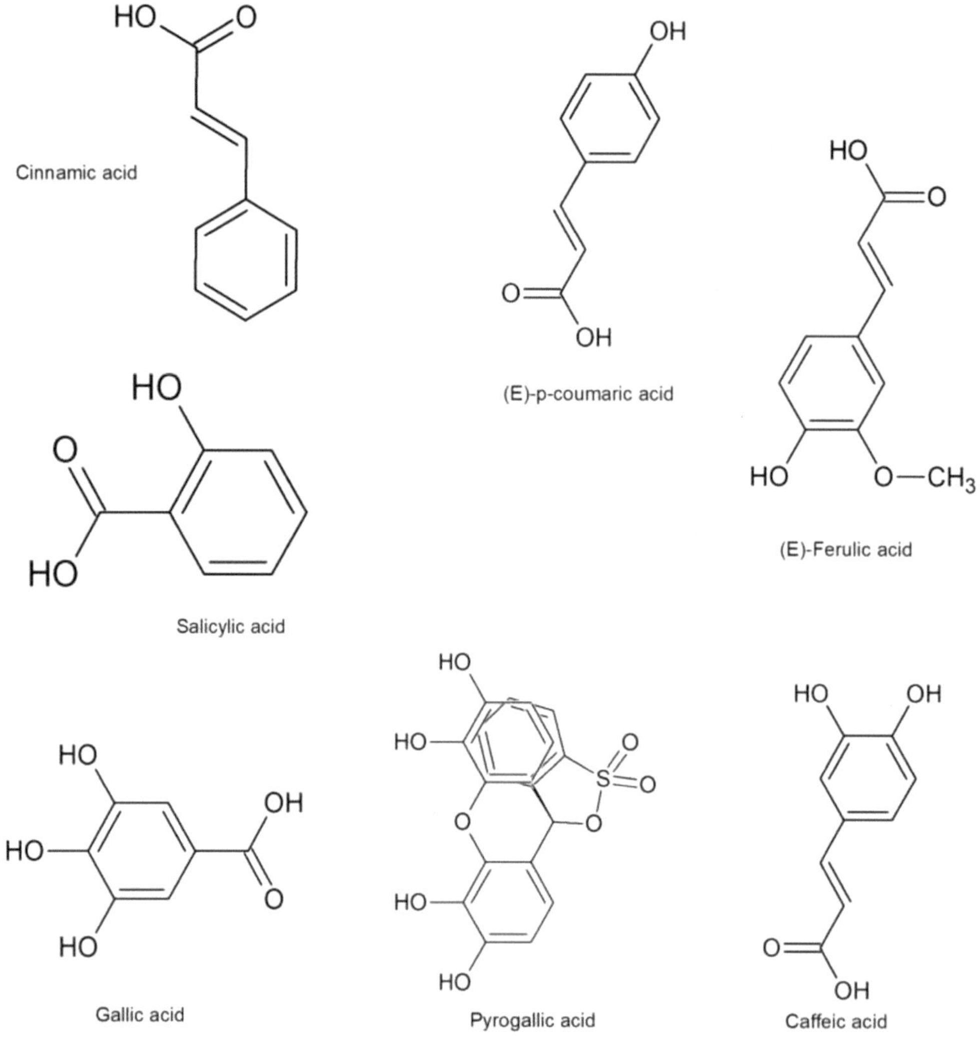

Fig. (7). Chemical structure of polyphenols found in algae.

CONCLUSION

Seaweed, consumed for centuries in Asian countries, has become a functional and attractive food in Western countries. Studies on the nutritional content of the different species of edible algae show that these foods are an important source of soluble and insoluble dietary fiber, macroelements and essential trace elements, such as Ca, Na, K, Mg, Fe, Zn or I, and sufficient vitamins considering the daily requirements. Likewise, it is a food with a low calorie and fat content. Although these are plant foods, studies have shown that algae, depending on the species in question, can be a valuable source of amino acids, such as glycine, arginine, alanine and glutamic acid. However, lysine and cystine are limiting amino acids not found in algae.

Algae, in addition to their valuable nutritional profile, stand out for their composition in bioactive substances that exert benefits on health. The bioactive compounds that stand out in algae are polyphenols, peptides and carotenoids. Among the carotenoids that stand out in algae, is fucoxanthin, a pigment that gives the characteristic brown color to brown algae or Phaeophyta. Fucoxanthin is a lipophilic compound belonging to the group of xanthophylls; it has anti-cancer, anti-inflammatory and anti-obesity effects. Another carotenoid that stands out, especially in red algae and gives the characteristic color to red algae or Rhodophyta, is β-carotene, with antioxidants effects. Another bioactive components of algae are the water-soluble polyphenols present in the aqueous fraction of algae. Salicylic acid stands out, whose antioxidant activity has been widely studied. Other polyphenols that stand out in algae are cinnamic, p-coumaric, gallic, pyrogallic, ferulic and caffeic acid, with more powerful antioxidant activity. In general, algae are functional and nutritious foods that, together with a balanced diet, can provide an important supply of nutrients. Likewise, the presence of bioactive compounds in algae gives it a beneficial role in the health of consumers.

CONSENT FOR PUBLICATION

Not applicable.

CONFLICT OF INTEREST

The author(s) declare no conflict of interest, financial or otherwise.

ACKNOWLEDGEMENT

Declared none.

REFERENCES

[1] Paz S, Rubio C, Frías I, *et al.* Metal Concentrations in Wild-Harvested Phaeophyta Seaweed from the Atlantic Ocean (Canary Islands, Spain). J Food Prot 2018; 81(7): 1165-70. [http://dx.doi.org/10.4315/0362-028X.JFP-18-038] [PMID: 29939794]

[2] Paz S, Rubio C, Frías I, *et al.* Toxic metals (Al, Cd, Pb and Hg) in the most consumed edible seaweeds in Europe. Chemosphere 2019; 218: 879-84. [http://dx.doi.org/10.1016/j.chemosphere.2018.11.165] [PMID: 30609492]

[3] MacArtain P, Gill CIR, Brooks M, Campbell R, Rowland IR. Nutritional value of edible seaweeds. Nutr Rev 2007; 65(12): 535-43.
[http://dx.doi.org/10.1111/j.1753-4887.2007.tb00278.x] [PMID: 18236692]

[4] Gupta VK, Tuohy MG. Biotechnology of Bioactive Compounds: Sources and Applications. WILEY United Kingdom. Blackwell 2015.
[http://dx.doi.org/10.1002/9781118733103]

[5] Monitor Organic. The European Market for Sea Vegetables. 2014; pp. 1-36.

[6] FAO (Food and Agriculture Organization of the United Nations) The State of World Fisheries and Aquaculture. Opportunities and challenges. Roma FAO Publications 2014.

[7] Quitral R V, Morales G C, Sepúlveda L M, Schwartz M M. Propiedades nutritivas y saludables de algas marinas y su potencialidad como ingrediente funcional. Rev Chil Nutr 2012; 39(4): 196-202.
[http://dx.doi.org/10.4067/S0717-75182012000400014]

[8] USDA (United States Department of Agriculture). National Nutrient Database for Standard Reference Release 28. Basic Report: 11669, Seaweed, wakame, raw 2017. a

[9] USDA (United States Department of Agriculture). National Nutrient Database for Standard Reference Release 28. Basic Report: 11444, Seaweed, Irish moss, raw; 2017. b

[10] USDA (United States Department of Agriculture). National Nutrient Database for Standard Reference Release 28. Full Report (All Nutrients): 11445, Seaweed, kelp, raw 2017. c

[11] USDA (United States Department of Agriculture). National Nutrient Database for Standard Reference Release 28. Full Report (All Nutrients): 11099, Brussels sprouts, cooked, boiled, drained, without salt; 2017. d

[12] USDA (United States Department of Agriculture). National Nutrient Database for Standard Reference Legacy Release. Full Report (All Nutrients): 11304, Peas, green, raw; 2017. e

[13] USDA (United States Department of Agriculture). National Nutrient Database for Standard Reference Release 28. Full Report (All Nutrients): 11457, Spinach, raw; 2017. f

[14] Astorga-España MS, Rodríguez Galdón B, Rodríguez Rodríguez EM, Díaz Romero C. Mineral and trace element concentrations in seaweeds from the sub-Antarctic ecoregion of Magallanes (Chile). J Food Compos Anal 2015; 39: 69-76.
[http://dx.doi.org/10.1016/j.jfca.2014.11.010]

[15] Abbaspour N, Hurrell R, Kelishadi R. Review on iron and its importance for human health. J Res Med Sci 2014; 19(2): 164-74.
[PMID: 24778671]

[16] Ajanth Praveen M, Karthika Parvathy KR, Jayabalan R, Balasubramanian P. Dietary fiber from Indian edible seaweeds and its in-vitro prebiotic effect on the gut microbiota. Food Hydrocoll 2019; 96: 343-53.
[http://dx.doi.org/10.1016/j.foodhyd.2019.05.031]

[17] Peñalver R, Lorenzo JM, Ros G, Amarowicz R, Pateiro M, Nieto G. Seaweeds as a functional ingredient for a healthy diet. Mar Drugs 2020; 18(6): 301.
[http://dx.doi.org/10.3390/md18060301] [PMID: 32517092]

[18] Rasmussen RS, Morrissey MT. Marine biotechnology for production of food ingredients. Adv Food Nutr Res 2007; 52: 237-92.
[http://dx.doi.org/10.1016/S1043-4526(06)52005-4] [PMID: 17425947]

[19] Dawczynski C, Schubert R, Jahreis G. Amino acids, fatty acids, and dietary fibre in edible seaweed products. Food Chem 2007; 103(3): 891-9.
[http://dx.doi.org/10.1016/j.foodchem.2006.09.041] [PMID: 26065750]

[20] Sanz-Pintos N, Pérez-Jiménez J, Buschmann AH, Vergara-Salinas JR, Pérez-Correa JR, Saura-Calixto

F. Macromolecular antioxidants and dietary fiber in edible seaweeds. J Food Sci 2017; 82(2): 289-95. [http://dx.doi.org/10.1111/1750-3841.13592] [PMID: 28152188]

[21] Peña-Rodríguez A, Mawhinney TP, Ricque-Marie D, Cruz-Suárez LE. Chemical composition of cultivated seaweed Ulva clathrata (Roth) C. Agardh. Food Chem 2011; 129(2): 491-8. [http://dx.doi.org/10.1016/j.foodchem.2011.04.104] [PMID: 30634256]

[22] Yaich H, Garna H, Besbes S, Paquot M, Blecker C, Attia H. Chemical composition and functional properties of *Ulva lactuca* seaweed collected in Tunisia. Food Chem 2011; 128(4): 895-901. [http://dx.doi.org/10.1016/j.foodchem.2011.03.114]

[23] Denis C, Morançais M, Li M, *et al.* Study of the chemical composition of edible red macroalgae Grateloupia turuturu from Brittany (France). Food Chem 2010; 119(3): 913-7. [http://dx.doi.org/10.1016/j.foodchem.2009.07.047]

[24] Gómez-Ordóñez E, Jiménez-Escrig A, Rupérez P. Dietary fibre and physicochemical properties of several edible seaweeds from the northwestern Spanish coast. Food Res Int 2010; 43(9): 2289-94. [http://dx.doi.org/10.1016/j.foodres.2010.08.005]

[25] Ortiz J, Romero N, Robert P, *et al.* Dietary fiber, amino acid, fatty acid and tocopherol contents of the edible seaweeds *Ulva lactuca* and Durvillaea antarctica. Food Chem 2006; 99(1): 98-104. [http://dx.doi.org/10.1016/j.foodchem.2005.07.027]

[26] Dhargalkar VK, Verlecar XN. Southern Ocean seaweeds: A resource for exploration in food and drugs. Aquaculture 2009; 287(3-4): 229-42. [http://dx.doi.org/10.1016/j.aquaculture.2008.11.013]

[27] Rajapakse N, Kim SK. Nutritional and digestive health benefits of seaweed. Adv Food Nutr Res 2011; 64: 17-28. [http://dx.doi.org/10.1016/B978-0-12-387669-0.00002-8] [PMID: 22054935]

[28] Oliva Y, Vega S. Péptidos bioactivos derivados de las proteínas lácteas: propiedades y aplicaciones principales. Rev Salud Anim 2004; 26(3): 151-62.

[29] Mertz W. The essential trace elements. Science 1981; 213(4514): 1332-8. [http://dx.doi.org/10.1126/science.7022654] [PMID: 7022654]

[30] Blanco A. Química Biológica. Octava edición. Madrid: Editorial El Ateneo; 2006.

[31] (Federación Española de Sociedades de Nutrición, Alimentación y Dietética). Ingestas Dietéticas de Referencia (IDR) para la población española. Acta Diet 2010; 14(4): 196-7.

[32] Tokudome S, Kuriki K, Moore MA. Seaweed and cancer prevention. Jpn J Cancer Res 2001; 92(9): 1008-10. [http://dx.doi.org/10.1111/j.1349-7006.2001.tb01192.x] [PMID: 11831193]

[33] Fung A, Hamid N, Lu J. Fucoxanthin content and antioxidant properties of Undaria pinnatifida. Food Chem 2013; 136(2): 1055-62. [http://dx.doi.org/10.1016/j.foodchem.2012.09.024] [PMID: 23122162]

[34] López-Hidalgo AM. Las macroalgas, ¿pueden ayudarnos a resolver problemas de índole mundial? Universitarios Potosinos 2017; 14(215): 4-10.

[35] Araiza-Macías MJ, Balandrano Fernández AL, Hernández Contreras JP. Alga Sargazo como posible fuente de materias primas para la extracción de carotenoides. Memorias del XXI Concurso Las Allistas de Investigación, Desarrollo e Innovación CLIDi SAL-25.

[36] Teramukai K, Kakui S, Beppu F, Hosokawa M, Miyashita K. Effective extraction of carotenoids from brown seaweeds and vegetable leaves with edible oils. Innov Food Sci Emerg Technol 2020; 60: 102302. [http://dx.doi.org/10.1016/j.ifset.2020.102302]

[37] Chandini SK, Suresh PV, Bhaskar N. Seaweeds as source of nutritionally beneficial compounds – A review. J Food Sci Technol 2008; 45: 1-13.

[38] Wang T, Jónsdóttir R, Kristinsson HG, *et al.* Enzyme-enhanced extraction of antioxidant ingredients from red algae Pal-maria palmate. LWT - Food. Sci Tech (Paris) 2010; 43: 1387-93.

[39] Souza BWS, Cerqueira MA, Martins JT, *et al.* Antioxidant potential of two red seaweeds from the Brazilian coasts. J Agric Food Chem 2011; 59(10): 5589-94.
[http://dx.doi.org/10.1021/jf200999n] [PMID: 21491929]

[40] Cho M, Lee HS, Kang IJ, Won MH, You S. Antioxidant properties of extract and fractions from Enteromorpha prolifera, a type of green seaweed. Food Chem 2011; 127(3): 999-1006.
[http://dx.doi.org/10.1016/j.foodchem.2011.01.072] [PMID: 25214089]

[41] Pangestuti R, Kim SK. Biological activities and health benefit effects of natural pigments derived from marine algae. J Funct Foods 2011; 3(4): 255-66.
[http://dx.doi.org/10.1016/j.jff.2011.07.001]

[42] Wijesinghe WAJP, Jeon YJ. Enzyme-assistant extraction (EAE) of bioactive components: A useful approach for recovery of industrially important metabolites from seaweeds: A review. Fitoterapia 2012; 83(1): 6-12.
[http://dx.doi.org/10.1016/j.fitote.2011.10.016] [PMID: 22061659]

[43] Marinho GS, Sørensen A-DM, Safafar H, Pedersen AH, Holdt SL. Antioxidant content and activity of the seaweed Saccharina latissima: a seasonal perspective. J Appl Phycol 2019; 31(2): 1343-54.
[http://dx.doi.org/10.1007/s10811-018-1650-8]

[44] Shahidi F, Zhong Y, Chandrasekara A. Antioxidants and human health. In: Yu LL, Tsao R, Shahidi F, Eds. Cereal and Pulses: Nutraceutical Properties and Health Benefits 1st ed. 2012; 273-308.
[http://dx.doi.org/10.1002/9781118229415.ch19]

[45] Prabhasankar P, Ganesan P, Bhaskar N, *et al.* Edible Japanese seaweed, wakame (Undaria pinnatifida) as an ingredient in pasta: Chemical, functional and structural evaluation. Food Chem 2009; 115(2): 501-8.
[http://dx.doi.org/10.1016/j.foodchem.2008.12.047]

[46] Sasaki K, Ishihara K, Oyamada C, *et al.* Effects of fucoxanthin addition to ground chicken breast meat on lipid and colour stability during chilled storage, before and after cooking. Asian-Australas J Anim Sci 2008; 21(7): 1067-72.
[http://dx.doi.org/10.5713/ajas.2008.70670]

[47] Maeda H, Hosokawa M, Sashima T, Miyashita K. Chapter 32. Antiobesity Effect of Fucoxanthin from Edible Seaweeds and Its Multibiological Functions. In ''Functional Food Health'', T. Shibamoto, K. Kanazawa, F. Shahidi, and C. T. Ho (eds). Washington: ACS Publications 2008; 376-88.
[http://dx.doi.org/10.1021/bk-2008-0993.ch032]

[48] Heo SJ, Yoon WJ, Kim KN, *et al.* Evaluation of anti-inflammatory effect of fucoxanthin isolated from brown algae in lipopolysaccharide-stimulated RAW 264.7 macrophages. Food Chem Toxicol 2010; 48(8-9): 2045-51.
[http://dx.doi.org/10.1016/j.fct.2010.05.003] [PMID: 20457205]

[49] Das SK, Ren R, Hashimoto T, Kanazawa K. Fucoxanthin induces apoptosis in osteoclast-like cells differentiated from RAW264.7 cells. J Agric Food Chem 2010; 58(10): 6090-5.
[http://dx.doi.org/10.1021/jf100303k] [PMID: 20420432]

[50] Shimoda H, Tanaka J, Shan SJ, Maoka T. Anti-pigmentary activity of fucoxanthin and its influence on skin mRNA expression of melanogenic molecules. J Pharm Pharmacol 2010; 62(9): 1137-45.
[http://dx.doi.org/10.1111/j.2042-7158.2010.01139.x] [PMID: 20796192]

[51] Bhat I, Haripriya G, Jogi N, Mamatha BS. Carotenoid composition of locally found seaweeds of Dakshina Kannada district in India. Algal Res 2021; 53: 102154.
[http://dx.doi.org/10.1016/j.algal.2020.102154]

[52] Clark S. Beta Carotene. xPharm: The Comprehensive Pharmacology Reference 2007; 1-3.

[53] Pattani KM. Chemoprevention of Head and Neck Cancer. Early Diagnosis and Treatment of Cancer Series: Head and Neck cancers Netherlands: Elsevier; 2010. [http://dx.doi.org/10.1016/B978-1-4160-5202-9.50011-9]

[54] Lustgarten M, Muller FL, Van Remmen H. Chapter 8 - An Objective Appraisal of the Free Radical Theory of Aging. In: Masoro EJ, Austad SN. Handbook of the Biology of Aging (Seventh Edition). Academic Press, Elsevier 2011; 177-202. [http://dx.doi.org/10.1016/B978-0-12-378638-8.00008-7]

[55] Tanumihardjo SA. Carotenoids: health effects. Encyclopedia of Human Nutrition 3rd ed. 2013; 292-7. [http://dx.doi.org/10.1016/B978-0-12-375083-9.00045-3]

[56] Padmanabhan P, Cheema A, Paliyath G, *et al.* Solanaceous fruits including tomato, eggplant, and peppers. Ency Food Health 2016; 2016: 24-32. [http://dx.doi.org/10.1016/B978-0-12-384947-2.00696-6]

[57] Kasperczyk S, Dobrakowski M, Kasperczyk J, Ostałowska A, Zalejska-Fiolka J, Birkner E. Beta-carotene reduces oxidative stress, improves glutathione metabolism and modifies antioxidant defense systems in lead-exposed workers. Toxicol Appl Pharmacol 2014; 280(1): 36-41. [http://dx.doi.org/10.1016/j.taap.2014.07.006] [PMID: 25038314]

[58] López A, Rico M, Rivero A, Suárez de Tangil M. The effects of solvents on the phenolic contents and antioxidant activity of solvents on the phenolic contents and antioxidant activity of Stypocaulon scoparium algae extracts. Food Chem 2011; 125: 1104-9. [http://dx.doi.org/10.1016/j.foodchem.2010.09.101]

[59] Nwosu F, Morris J, Lund VA, Stewart D, Ross HA, McDougall GJ. Anti-proliferative and potential anti-diabetic effects of phenolic-rich extracts from edible marine algae. Food Chem 2011; 126(3): 1006-12. [http://dx.doi.org/10.1016/j.foodchem.2010.11.111]

[60] Siriwardhana N, Lee KW, Jeon Y-J, Kim S-H, Haw J-W. Antioxidant activity of Hizikia fusiforms on reactive oxygen species scavenging and lipid peroxidation inhibition. Food Sci Technol Int 2003; 9(5): 339-46. [http://dx.doi.org/10.1177/1082013203039014]

[61] Vidal A, de Andrade-Wartha ES, De Oliveira AM, Pavan R, Lima A, Fallarero A, *et al.* Ars Pharm 2009; 50(1): 24-31.

[62] Wang Z, Ma L, Zhang X, Xu L, Cao J, Jiang W. The effect of exogenous salicylic acid on antioxidant activity, bioactive compounds and antioxidant system in apricot fruit. Sci Hortic (Amsterdam) 2015; 181: 113-20. [http://dx.doi.org/10.1016/j.scienta.2014.10.055]

[63] Agarwal S, Sairam RK, Srivastava GC, Tyagi A, Meena RC. Role of ABA, salicylic acid, calcium and hydrogen peroxide on antioxidant enzymes induction in wheat seedlings. Plant Sci 2005; 169(3): 559-70. [http://dx.doi.org/10.1016/j.plantsci.2005.05.004]

[64] Díaz Gutierrez D, Méndez Ortega W, Oliveira e Silva AM, Zaldívar Muñoz C, Mancini-Filho J, Vidal Novoa A. Comparación de las propiedades antioxidantes y contenido de polifenoles de extractos acuosos de las algas marinas Bryothamnion triquetrum y Halimeda opuntia. Ars Pharmaceutica (Internet) 2015; 56(2): 89-99. [http://dx.doi.org/10.4321/S2340-98942015000200003]

[65] Vidal A, Fallarero A, Silva de Andrade-Wartha ER. de Oliveira e Silva AM, de Lima A, Pavan Torres R, Vuorela P, Mancini-Filho J. Composición química y actividad antioxidante del alga marina roja Bryothamnion triquetrum (S.G.Gmelin) Howe. Rev Bras Ciê Farm 2006; 42(4): 590-600.

CHAPTER 13

Metabolomics of Different Seaweeds in Functional Food Prospects

Ninian P.P. Pabbathi[1], **Aditya Velidandi**[1], **Prashakha J. Shukla**[2], **Neelam M. Nathani**[3,*] and **Chandrashekar Mootapally**[3,*]

[1] *Department of Biotechnology, National Institute of Technology, Warangal, Telangana, India*

[2] *Parul Institute of Applied Sciences, Parul University, Vadodara, Gujarat, India*

[3] *School of Applied Sciences & Technology (SAST-GTU), Gujarat Technological University, Chandkheda Campus, Ahmedabad, Gujarat, India*

Abstract: The well-being and improvement of an individual's health, is directly linked to food. Acknowledging the well-studied interrelation between health and nutritional food determines the need to cultivate, maintain and advance a novel branch of food science, known as functional foods. Research interest has extensively increased in the last three decades for the exploitation of seaweeds as functional foods or as nutraceuticals, which provide health and dietary benefits beyond their nutritional content. Additionally, seaweeds have been studied for their metabolites which exhibit biological activity, to manufacture therapeutically superior products. This chapter focuses on providing information related to omics-based seaweed metabolite identification and their applications in human health as a food ingredient.

Keywords: Bioactive Compounds, Macroalgae, Metabolomics, Microalgae, Seaweeds.

INTRODUCTION

As stated by Maimonides, a Jewish philosopher, "No disease that can be treated by diet should be treated with any other means". Given this, at present, the notion of optimum nutritional diet, *i.e.,* nutrition science, is being promoted because of the role of diet in the prevention of deficiencies and diseases as well as in supporting body growth [1 - 4]. The significance of diet in promoting health is in the spotlight once again. It is worth mentioning that even though consumer expectations towards convenience and canned foods are increasing rapidly, the

* **Corresponding authors Neelam M. Nathani and Chandrashekar Mootapally:** School of Applied Sciences & Technology (SAST-GTU) Gujarat Technological University, Gujarat, India; Tel: +91 8473836647; Fax: +91-79-23267580; E-mails: neelam.nathani@yahoo.com (NMN), chandu.avi@gmail.com (CM)

Avinash Mishra (Ed.)

need for maintaining and improving the health of an individual remains a basic factor in the consumer food market [3, 5]. In the well-being and improvement of the health of an individual, food plays a vital role, and acknowledging the well-studied interrelation between health and nutritional food determines the need to cultivate, maintain and advance a novel branch of food science, known as functional foods. This helps in combating health issues arising from increased material comfort and urbanization, which are related to everyday life routine that involves low physical activity. Also, the occurrence of chronic non-communicable diseases, which are considered to be the characteristic of the modern age lifestyle, such as dental problems, allergies, diabetes, cancer, osteoporosis, and wildly present obesity, can be addressed by a change in the diet [6 - 8].

The purpose of functional foods is to prevent disease and provide clinical care by considering the differences in the environment, lifestyle, molecular phenotype, and genetics of an individual. Employing clinical genomics in the treatment of cancer provides an option to choose therapies based on information provided and ultimately foreseeing the outcomes of the chosen therapies. Molecular genomics is a powerful tool that provides fundamental information by which one can observe and study individual variability, however it will not offer information on other biomolecules (for example, metabolites) that can also define phenotypes at the molecular level. Preferably, food science should also help us to understand the biochemistry of the metabolites involved in human health improvement. Extensive research has been done to use the analyses of metabolites for developing disease diagnostic tools and in providing necessary clinical care, such as developing test strips for blood glucose measurement for diabetic patients or estimating phenylalanine concentrations in infants for phenylketonuria diagnosis [9, 10]. Undeniably, unlike genomic changes, variations in the levels of metabolites that occur during monogenic disorders and complex diseases can be quantifiable. These variations could reveal temporal dynamics and tissue specificity. Metabolomics is a progressively evolving branch of science that can be broadly defined as the complete measurement of all the metabolites as well as the low molecular weight biomolecules present in an individual organism. In comparison with the typical biochemical-clinical laboratory techniques, metabolomics can profile and identify a considerably large number of metabolites, thus can provide a complete understanding of metabolic pathways and biological processes. It also helps us to study and understand the biochemical interactions of the metabolites with the biological specimen.

Most of the required elements or molecules to sustain life can be found in seawater. Planet Earth's surface is covered by about 72% water; life itself has emerged from seawater. Around 5 lakh species inhabit the ocean, which means that nearly ~75% of known species live in seawater. Algae are one such species found majorly inhabiting saltwater but can also sustain in freshwater. Phycology is a branch of science that deals with scientific studies related to algae. Nearly 30,000 species of algae (both microalgae and macroalgae) have been identified (Fig. **1**). As per literature, macroalgae, also known as seaweed, are classified into 3 sub-groups: green algae (Chlorophyta), red algae (Rhodophyta), and brown algae (Ochrophytas) [11].

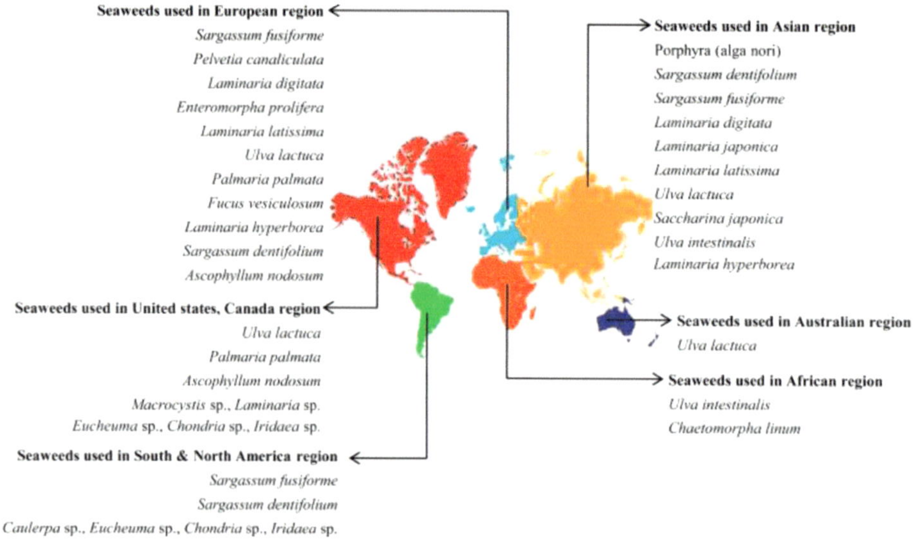

Fig. (1). Global distribution of seaweeds.

Research interest has extensively increased in the last three decades for the exploitation of seaweeds as functional foods or as nutraceuticals, which provide health and dietary benefits beyond their nutritional content [12]. Additionally, seaweeds have been studied for their metabolites which exhibit biological activity, to manufacture therapeutically superior products [13, 14]. In contradiction with malnutrition (vitamin and mineral deficiency), obesity has become a paradox and escalated to become a global epidemic by affecting many nations around the world, such as Australia, USA, and Europe [15]. In conjunction, the emergence of lifestyle and diet-associated diseases, particularly cancer, metabolic syndrome, and diabetes type-II, has increased. According to the global dietary survey, countries that consume seaweed on a regular basis are reported to face considerably less diet and obesity-associated diseases [16, 17]. However, a vast number of diet-related parameters might be associated with this complex relationship. Hence, the exploitation of seaweeds for their vast medical and dietary benefits has gained significant momentum [18]. This chapter focuses on providing information related to omics-based seaweed metabolite identification and their applications in human health as a food ingredient.

WORKFLOW OF METABOLOMICS

Sample Collection

The massive diversity of metabolites present in the seaweed specimen along with their dynamic variations caused due to the delicate changes in their micro-environment, presents a daunting task for analysts during the identification of metabolites. Metabolite profiling depends on various factors such as accessibility of nutritional elements, diurnal fluctuations, variations during photosynthesis, and respiration depending on the light [19 - 22]. Hence, it is vital to choose the appropriate condition and time for seaweed collection before extraction of metabolites. Extreme precautions should be taken such that the entire sampling is done exactly under the same environmental conditions. Due to the dynamic exchange rates of the metabolites from one form to another, fixing the state of seaweed specimen right after harvesting is essential, and the best possible way to do that is through flash freezing. Lyophilisation of the harvested sample is yet another vital step before extraction of metabolites since differences in the water content would result in biased quantification of metabolites. Later, optimization of sample extraction depending on the type of seaweed, such as entire or part of seaweed, and also based on the analytical tool to be employed for analysis is required [23].

Sample Preparation

Homogenization is a very crucial method employed to achieve effective extraction, which breaks down the rigid cell wall, in turn releasing the metabolites. Various methods are used for matrix disruption or reduction of particle size. Disruption of matrix results in achieving maximum extraction of metabolites in huge amounts, whereas smaller particle size reduces the solvent diffusion time [24]. Among the methods employed for homogenization and matrix disruption, mechanical shear is widely acknowledged as the most effective, which can be done by using mortar and pestle. However, this is a laborious approach. Various research groups employ automated grinders (such as cryomills or balls) as well as acidic solvents (such as trichloroacetic acid or perchloric acid) to achieve proper homogenization [25]. Nonetheless, certain precautions (regarding temperature, *etc.*) are necessary such that whatever the approach used, the conditions should not trigger enzymatic activities, since it can impact the metabolite profiling [26]. Various extraction techniques and solvent combination systems are already available, and the proper choice of an effective approach is crucial in obtaining the maximum amount of metabolites from the specimen. Extraction techniques such as solid-phase extraction, supercritical fluid extraction, solvent extraction, and sonication are available, each having its advantages and disadvantages. Likewise, various solvent systems are present, and certain features such as pH, chemical reactivity, dissolution rate, selectivity, solubilization power, polarity, and toxicity should be considered while choosing an appropriate solvent system [27]. Perchloric acid is a suitable solvent for the extraction of hydrophilic, polar, and ionizable metabolites. Additionally, it causes protein denaturation and inhibits all enzymatic reactions, which can impede metabolite analysis [28]. Organic solvents are chiefly employed to extract hydrophobic, nonpolar, or polar metabolites. However, for the extraction of polar and hydrophilic as well as nonpolar metabolites, two-phase solvent systems such as water-methano--chloroform are frequently employed. Hence they are much suitable for seaweed metabolomics. So far, no such method or approach has been reported that can provide maximum extraction of entire metabolites present in a biological specimen along with reproducibility and robustness [29].

Techniques Used for Metabolomics Analysis

As the metabolome is more intricate with various metabolites with varied properties, there is no single analytical approach to understand the entire metabolome of any sample completely. It has already been proven that metabolites that can be unravelled through multiple analysis techniques are much higher than employing a single technique [30]. As per existing literature, various techniques have been employed for the analysis and profiling of seaweed metabolites (Fig. **2**), such as liquid/gas chromatography (LC/GC), mass spectrometry (MS), capillary electrophoresis (CE), Fourier transform–ion cyclotron resonance–mass spectrometry (FT-ICR-MS), Ultra performance liquid chromatography-tandem mass spectrometer (UPLC-MS) and Nuclear magnetic resonance (NMR) [31]. Each approach is extremely sophisticated however they have an inevitable inherent biased nature towards the particular metabolite profiling [32]. In metabolomics, reproducibility should be the primary criteria for the selection of the technique. Among the various techniques employed, NMR is considered the appropriate approach; however, the sensitivity is relatively less (1-1000 μM in an NMR tube) in comparison with chromatography and MS-based approaches in metabolite profiling [29, 33]. Moreover, LC/GC-MS separates individual chemical molecules, thus determining the chemical composition of the sample, whereas NMR only offers a chemical fingerprint [34]. NMR can detect non-volatile metabolites which are present in major proportion, whereas GC-MS can analyse only thermostable and volatile metabolites [34]. Depending on the type of metabolites targeted, different techniques are employed. Each method has its benefits and drawbacks, however so far, no single technique has the capacity to achieve non-targeted metabolite profiling. As discussed earlier, to achieve the maximum as well as reproducible metabolite profiling, various techniques are employed in combination. One study reported the application of GC-MS, LC-MS and UPLC coupled with a time-of-flight mass spectrometer (TOF-MS) for exo-metabolomic profiling in the green seaweed species (*Ulva mutabilis*) [35].

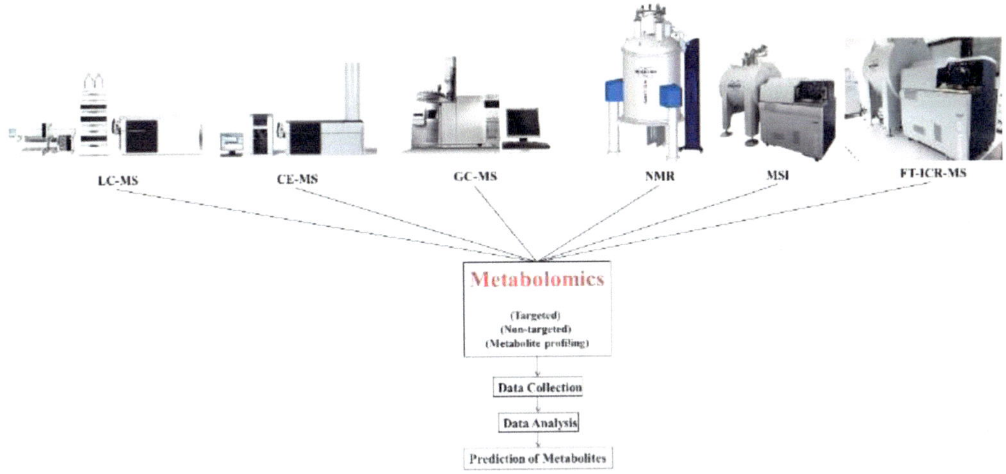

Fig. (2). Various techniques used for metabolomics.

Mass Spectroscopy

MS is the most widely employed detector system due to its speed, sensitivity and broader applicability. The choice of method for separation of the metabolites depends upon the extract preparation, and LC/GC in combination with MS is the extensively employed separation technique [36]. Recently, CE and FT-ICR coupled with MS are also being used for metabolite profiling [37].

LC-MS

In LC-MS, separation of metabolite molecules which are dissolved in the liquid mobile phase is achieved based on their interaction with the solid stationary phase, followed by passing into the mass analyser [36]. This technique is appropriate for chemically unstable structures and heat-labile functional groups as well as high-molecular-weight and high-vapor-point secondary structures (>500 kDa) (such as steroids, sterols, fatty acids, phospholipids and non-polar metabolites) [38]. Mainly due to the advancement in the column dimensions and matrix particle sizes, speedy analysis of a wider range of metabolites has become possible, along with obtaining a higher efficiency during separations and peak resolutions [29]. In general, reverse-phase (RP) stationary phases are employed, which consist of silica particles, which are covalently bound with hydrophobic alkyl chains. Octade-cylsilane is the widely used stationary phase, in which the C18 functional group is covalently attached to the surface of silica particles [39]. In such cases, mobile phases with different polarities are used in combinations such as water:methanol or water:acetonitrile. Therefore, low polar molecules such as primary metabolites (like glycolytic intermediates and carbohydrates) can be retained in the non-polar stationary phase [20].

Among the various ionization approaches employed, atmospheric pressure ionization (API) and electron spray ionization (ESI) are most widely used because these approaches allow the direct analysis of metabolites from the liquid mobile phase and, therefore, are most appropriate for LC coupling. API and ESI are known as soft ionization methods as they offer only partial structural information of metabolites because they inject low internal energy, which is essential for fragmentation thus, it generates limited fragments [40]. In-depth structural information can be attained by the occurrence of a sufficient number of fragments, which can be possible by the collision of metabolite ions with the inert gas molecules (such as N_2, Ar and He), known as the collision-induced dissociation (CID) method. In general, CID is employed in a tandem MS instrument, which allows for two or more consecutive mass spectrometry analyses (tandem MS^n experiments). In LC-MS-based metabolomics, tandem-in-space and tandem-i--time are regularly used tandem MS^n instrument types. Ion trapping instruments such as FT-ICR-MS, Orbitrap and quadrupole ion trap mass spectrometers (QIT-MS) are categorized under tandem-in-time MS. In such instruments, fragmentation is carried out in various stages, among which first-generation fragment offers structural elucidation, specifically about unknown metabolites, additional fragmentation stages provide much-needed information about metabolites structure (MS^n analysis). Whereas, two consecutive mass spectrometers are present for analysis, separated by a collision cell in a tandem-i--space MS [29, 41].

However, analysis through LC-MS has certain drawbacks, such as the occurrence of several metabolites could affect the exact quantification of the metabolites through ionization and also the transmission of analytes from liquid to gaseous phase [42]. Therefore, optimization is necessary for the proper separation of metabolites in LC, which further reduces the effect of matrix-induced impact on the mass analyser. Metabolites having higher similarity can be eluted at the same time by forming a complex metabolite matrix [43]. Additionally, metabolite analysis using LC-MS requires authenticated metabolite molecules as a reference for the quantification and identification of the targeted metabolite. This, in turn, restricts the analysis of metabolites that are not available commercially [44]. In a study by Gaubert *et al.* 2019, it was observed that out of the three techniques (GC-MS, NMR and LC-MS), LC-MS was observed to be the best for distinguishing metabolites. In the same study, with the combination of all the above-stated techniques, chemo markers were identified [45].

GC-MS

GC-MS is employed to separate and detect the naturally occurring derivatives of volatile metabolites such as esters, monoterpenes and alcohols. In general, this approach is used for primary metabolites, which are polar and non-volatile, such as amino acids and sugars. By applying chemical derivatization, these metabolites are transformed into thermostable and volatile molecules. GC-MS is the most widely employed analytical method for metabolite analysis due to its robust separation and electron impact spectrometry approach. In addition, GC-MS provides a single chromatogram for important primary metabolites. Even at low concentrations of known metabolites, GC-MS offers highly efficient metabolite quantification, detection, resolution and separation with outstanding reproducibility [29, 30]. The identification of the detected metabolites is generally confirmed by employing available databases and commercial standards as reference molecules [46]. However, this approach is not viable for thermolabile metabolites such as high-molecular-weight (>500 Da) metabolites, larger oligosaccharides, as well as di- and triphosphates [47].

Derivatization is the first and foremost essential step in GC-MS analysis. This is a technical process in which analytes (metabolites) are modified at one or more functional groups, thus producing chemically similar derivatives by the application of a derivatizing reagent [48]. In general, the derivatization process for metabolites analysis occurs in two stages: (1) methoxyamination, in this keto or aldehyde functional groups are transformed to oxides in the presence of methoxyamine; Z- and E- isomers of the resultant oximes are detected in GC-MS and followed by (2) trimethylsilylation step, this increases the volatile nature of the metabolite derivatives. Active hydrogens (such as -SH, -NH, -COOH and -OH) react with metabolite derivatives and convert them into more volatile and thermostable as well as less polar trimethylsilyl (TMS) sulfide, TMS-amine, TMS-ester and TMS-ether groups respectively. Other silylation reagents, such as N,O-bis-trimethyl-silyl-trifluor-oacet-amide and N-methyl-trime-thylsily-ltrif-uor-oacet-amide are also widely used in metabolomics investigations using GC-MS [49].

High reproducibility is partially attained by the application of the electron ionization (EI) process in GC-MS analysis. In the EI process, electrons which are kinetically activated with an acceptable average standard energy of 70 eV interact with sample metabolites. At present, orthogonal time of flight (TOF) and quadrupole (Q) are the most widely employed mass analysers, however, TOF-MS analysers offer speedy data acquisition times, superior duty cycles and mass accuracy compared to Q-MS analysers. This approach is more feasible for metabolite profiling, because it can precisely deconvolute coinciding chromatogram peaks which are attained from multiple metabolites extract, thus reducing the time required for the overall metabolite analysis [29, 50, 51].

CE-MS

CE offers superior resolving potential and detection limits which are in turn dependent on the metabolite charge to size ratio. The detection limits are dependent on the type of metabolites to be analysed, since not all the types of metabolites are similarly responsive to CE-MS. CE is performed in a capillary tube fused with silica; the tube ends are immersed in buffer reservoirs across which high voltages (10-40 kV) are present. The occurrence of variations in the electrophoretic mobility of the various metabolites is responsible for the metabolite separation [52].

Compared to traditional LC/GC-MS, CE-MS is CE-MS is superior in terms of speed and efficiency [53]. This is often employed in the analysis of primary metabolites. Due to the very minimal flow of applied voltage for electrophoretic separation and effluent from the capillary, a complex interface was observed in between the ESI process and CE-MS. Additionally, a continuous electrical supply has to be kept, and MS-amenable background electrolytes are required in the running buffer. Due to the above-stated reasons, CE-MS provides very poor reproducibility, and highly skilled personnel are mandatory to use this instrument efficiently. CE-MS is not easy to set up and maintain as well as due to the lack of services in this aspect makes this approach very hard to employ [29].

FT-ICR-MS

FT-ICR-MS uses an ion trapping approach which provides highly precise mass measurements. Compared to other MS analysers, FT-ICR-MS offers superior mass accuracy and resolution while determining the empirical formula [54]. Owing to the technological advancement, the recently developed Orbitrap FT-MS analytical instrument offers higher sensitivity [55] and is also less expensive in comparison with the cyclotron FT-MS instrument [56]. However, this instrument cannot separate structural isomers which have identical mono-isomeric masses, which is a major drawback [29].

FT-ICR-MS coupled with ESI method is the most commonly used approach for metabolite profiling [57]. FT-ICR-MS coupled with direct injection high-resolution MS is considered a powerful ion-trapping instrument that has a lower detection limit. FT-ICR-MS also offers superior compatibility with other MS analysers. Major limitations such as complicated ion-ion associations, high cost and occurrence of high magnetic field restrict the widespread application of the instrument [58].

UPLC-MS

Compared to traditional HPLC, a UPLC coupled with MS is a powerful instrument that offers high-throughput capabilities, superior specificity and higher peak capacity. In general, UPLC columns contain porous particles which have <2 mm internal diameter [29]. UPLC instrument offers 9- and 3-fold decrease in analysis time when compared with HPLC instrument using 5- and 3-mm particle size analytical columns without compromising the overall separation [59]. UPLC-MS provides faster quantifications without the loss of resolution due to its wider range of optimal linear velocity. Hence, it is more appropriate for metabolomics studies. UPLC, in combination with Q-TOF-MS, is a powerful instrument and can provide superior and precise mass measurements at concentration levels <5 ppm with efficient resolution [29]. Therefore, it can even be employed in the identification of the trace metabolites in the complex extracts. One study reported the application of an untargeted UPLC-Q-TOF metabolite fingerprinting approach to investigate the differences in the metabolites of four brown seaweed species (*Lobophora sonderii*, *L. rosacea*, *L. obscura* and *L. monticola*) [60].

NMR

In the metabolomics field, NMR has been extensively employed for various applications, such as metabolic flux analysis, metabolite profiling and fingerprinting [61]. NMR is capable for the identification and quantification of a diverse number of metabolites. NMR is a highly sophisticated automated instrument with a simple non-destructive sample preparation process. Unlike MS-based approaches, NMR does not require sample (metabolites) derivatization or separation using chromatographic techniques [61]. It is widely acknowledged that high-resolution NMR can be employed as an ideal instrument for metabolite profiling, but the sensitivity is low in comparison with MS-based approaches. Additionally, it is highly sensitive to pH variations [62]. However, NMR has immense potential for metabolite analysis due to its direct isotopomer distribution analysis, identification of metabolite structures, does not require standard/reference molecules, huge isotope editing potential and superior quantitative accuracy [63]. 1H NMR approach is simple yet provides quick analysis; hence it is most widely used for metabolite profiling. However, its main limitation is the overlapping of the signal during the analysis of complex metabolites. In such instances, other advanced NMR-based approaches such as high-resolution magic angle spinning (HRMAS)-NMR spectroscopy, heteronuclear multiple quantum coherence and nuclear overhauser effect spectroscopy, total correlation spectroscopy, heteronuclear single quantum coherence (HSQC) spectroscopy and two-dimensional (2D) 1H J-resolved NMR spectroscopy are employed [26]. Each approach has its own advantages and limitations.

SEAWEED CLASSIFICATION

As discussed above, seaweeds are broadly classified into green (Chlorophyta), red (Rhodophyta) and brown (Ochrophyta) algae. Both green and red algae are included in the Plantae kingdom, whereas brown algae are classified under the Phaeophyceae class. Brown algae belong to both the Protista kingdom and microalgae kingdom. This algal classification is based on the pigment present in the algal species (Table **1**), which is used for photosynthesis [11]. Around 10-40% of the total species that live in the marine habitat belong to macroalgae [64]. Inter-species and inter-genus seaweed metabolite variations were defined depending on solely the carbohydrate metabolites. In brown algae, it was observed that mannitol played an important role, whereas it was glucose, fructose and sucrose in green algae. Contrastingly, different species of red algae were observed to have distinctive metabolites. For example, in *Pyropia pseudolinearis,* the major carbohydrate metabolite was sorbitol, and similarly, *Dasya sessilis* has panose [65]. In the natural habitat, seaweeds experience both biotic and abiotic stress. They acclimatise to the stresses by altering their physiological profiles, particularly by altering their nitrogen and carbon metabolisms. One of the important metabolites for carbon-nitrogen metabolism is γ-aminobutyrate (GABA). It was observed that GABA was more in red algae, followed by green algae and, lastly, brown algae. It was observed that red algae have an abundance of nitrogen-containing amino acids (ornithine, glutamate, *etc.*) when compared to brown algae. It has been reported that the red algae are more stress tolerant for the above reasons when compared to the other two algae.

Table 1. Pigment content of three groups of seaweeds [66].

S. No	Seaweeds	Example	Pigments		
			Chlorophyll	Carotinoids	Phycobiliproteins
1	Green Algae	 *Ulva lactuca*	Chlorophyll-a, b & their derivatives	β-carotene, Xanthophyll	...

(Table 1) cont.....

S. No	Seaweeds	Example	Pigments		
			Chlorophyll	Carotinoids	Phycobiliproteins
2	Red Algae	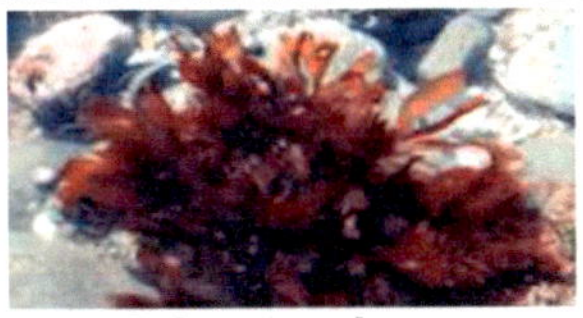 *Palmaria palmata*	Chlorophyll-a, d & their derivatives	Xanthophyll	Phycoerythrin & Phycocyanin
3	Brown Algae	 *Ascophyllum nodosum*	Chlorophyll-c, b & their derivatives	Fucoxanthin & xanthophyll, β-carotene	...

SEAWEED CHEMICAL COMPOSITION

The seaweed chemical composition is dependent on various factors such as type of species, cultivation place, harvesting, and environmental conditions. In the nutritional aspect (Fig. **3**), seaweeds are a major source of lipids and proteins. Generally, red and green seaweeds contain higher protein content (dry weight (DW) of 10-47%) in comparison to brown seaweeds (DW of 5-24%). Seaweeds contain around 0.79-7.87% (DW) of lipids; among this, polyunsaturated fatty acids (PUFAs) such as ω-3 and ω-6 constitute a major portion [67 - 73].

The occurrence of various bioactive compounds in the seaweeds offers them several biological properties related to the treatment and prevention of diseases such as neuroprotective, fat-lowering, antivirus, antitumor, antimicrobial, anti-inflammatory, antihypertensive and anti-diabetic agents [75 - 77]. Both primary and secondary metabolites could also be involved in these biological applications. Primary metabolites such as lipids, polysaccharides, and proteins are found to be involved in several physiological processes. Fibres and polysaccharides are the major bioactive molecules that positively impact several chronic diseases such as obesity, diabetes, cardiovascular diseases, and cancer. However, secondary metabolites such as small peptides, terpenes, sterols, halogenated and phenolic compounds are minor molecules that are produced in seaweeds during stressful conditions such as environmental pollutants, salinity, ultraviolet radiation, and temperature variations [78]. In a study conducted by Belghit *et al.*, 21 seaweeds (red, green and brown alga) were studied for variation in 391 metabolites. They have identified significant metabolite differences among them. The differences were observed in carbohydrate metabolites, peptide metabolites, fatty acid & sterol metabolites and even few secondary metabolites [79]. They acclimatise to the stresses by altering their physiological profiles, particularly by altering their nitrogen and carbon metabolisms. One of the important metabolite for carbon-nitrogen metabolism is γ-aminobutyrate (GABA). It was observed that GABA was more in red algae followed by green algae and lastly brown algae. It was observed that red algae have abundance of nitrogen containing amino acids (ornithine, glutamate *etc.*) when compared to brown algae. It has been reported that the red algae are more stress tolerant for the above reasons when compared to other two algae [79].

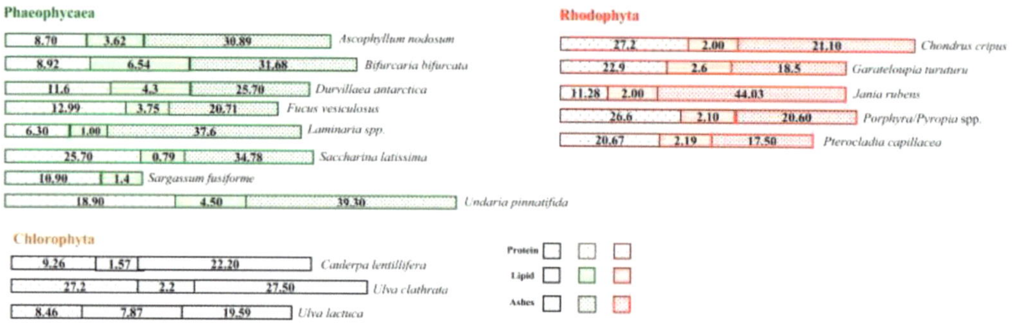

Fig. (3). Chemical composition of seaweeds [74].

Polysaccharides

Seaweeds possess more amount carbohydrates in the form of functional, structural as well as storage polysaccharides. The carbohydrate content varies from species to species, and the range is 20-70%. Despite their high carbohydrate content, they cannot act as a good source of carbohydrates due to the presence of high dietary fibre (about 70%) [74, 80]. Outcomes of various metabolomics and 16S rDNA sequencing studies are tabulated below (Table **2**).

Table 2. Metabolomics and 16s rdna sequencing studies in seaweeds [101].

Seaweed	Substrate	Method of Microbial Enumeration	Microbial Variations			Technique used	Metabolomic analysis			Refs.
			Increase	Decrease	No Change		Metabolites Increased	Metabolites Decreased	No Change	
Eckloniaradiate	Crude fraction (C)	qPCR	*F. prausnitzii, C. coccoides, Firmicutes, Bacteroidetes, E. coli, Enterococcus* (C)	...	...	GC-FID	Acetate, Propionate, Butyrate, Total SCFA (C)	...	...	[84]
	Phlorotannin-enriched fraction (P)		*Bacteroidetes, E. coli, Enterococcus*(P)	...	...		Propionate, Butyrate, Total SCFA (L)	...	...	
	-									
	-									

(Table 2) cont.....

Seaweed	Substrate	Method of Microbial Enumeration	Microbial Variations			Technique used	Metabolomic analysis			Refs.
			Increase	Decrease	No Change		Metabolites Increased	Metabolites Decreased	No Change	
-	Low-molecular weight polysaccharide fraction (L)	-	*Bifidobacterium Lactobacillus,F. prausnitzii, C. coccoides, Firmicutes, Bacteroidetes, E. coli* (L)	… … …	… … …	GC-FID	Butyrate, Total SCFA (H)	… … …	… … …	[84]
	High-molecular weight polysaccharide fraction (H)		*Bacteroidetes, E. coli* (H)	…	…			…	…	
…	Alginate	16S rRNA NGS	*Bacteroides*	…	…	GC-FID	Propionate, Butyrate, Total SCFA	…	…	[102]
Sargassumthunbergii	Polysaccharide extract	16S rRNA NGS	*Bacteroidetes, Roseburia, Parasutterella, Fusicatenibacter, Coprococcus, Fecalibacterium*	…	…	GC-MS	Acetate, Propionate, Butyrate, Valerate, Total SCFA	…	…	[82]
Ascophyllum nodosum	Sulphated polysaccharide extract	16S rRNA NGS	*Bacteroides, Phascolarctobacterium, Oscillospira, Fecalibacterium*	…	…	GC-FID	Acetate, Propionate, Butyrate, Total SCFA	…	…	[103]
Laminaria digitate	Crude polysaccharide extract (CPE)	16S rRNA NGS	*Parabacteroides, Fibrobacter, Dialister, γ B38UC* (CPE)	*Streptococcus, Ruminococcus* (CPE)	…	GC-FID	Acetate, Propionate, Butyrate, Total SCFA	…	…	[101]
	Depolymerised crude polysaccharide extract (DPE)		*Parabacteroides, Lachnospiraceae UC, Dialister*(DPE)	*Peptostreptococcaceae* IS, *Streptococcus, Ruminococcus* (DPE)	…		Acetate, Propionate, Butyrate, Total SCFA	…	…	
Osmundeapinnatifida	*Osmundeapinnatifida* Viscozyme extract	FISH	…	…	*Bifidobacterium, Lactobacillus, Clostridium histolticum*	HPLC	Total SCFA	…	…	[104]
Kappaphycus alvarezii	Whole Seaweed	FISH	*Bifidobacterium*	*Clostridium coccoides/ Eubacteriumrectale*		HPLC	Total SCFA	…	…	[105]
Gracilariarubra	Polysaccharide extract	16S rRNA NGS	*Bacteroides, Prevotella, Phascolarctobacterium*	*Firmicutes:Bacteroidetes*		GC-FID	Acetate, Propionate, Isobutyrate, Total SCFA	…	…	[106]
…	Porphyran	qPCR	*Bifidobacterium, Bacteroides*	…	…	HPLC	…	…	Acetate, Propionate, Butyrate, Total SCFA	[107]
Enteromorphaprolifera	Polysaccharide extract	Microbial Culturing	*Enterobacter*	…	*Enterococcus, Lactobacillus, Bifidobacterium*	GC-FID	…	…	Acetate, Butyrate, Lactate	[90]
…	Ulvan	qPCR	*Bifidobacterium, Lactobacillus*			HPLC	…	…	Acetate, Lactate	[107]

Table 3. Seaweed polysaccharides and their biological applications [99].

Seaweed	Polysaccharide	Composition	Dry Weight	Biological Applications	Ref.
Green	**Ulvan**	Sulfated 1,4-β-D-Glucuronic acid α-l-Rhamnose 1,4-β-D-xyloglucan	8–29%	Antioxidant, anti-proliferative, hypocholesterolaemic	[97]

(Table 3) cont.....

Seaweed	Polysaccharide	Composition	Dry Weight	Biological Applications	Ref.
Red	Agar	1,3-β-D-galactose 1,4-3,6-anhydro-α-L-galactose	...	...	...
	Carrageenan	1,4-β-D-galactose 1,3-α-D-galactose 3,6-anhydro-D-galactose	30-75%	Antiviral, anticoagulant, antithrombotic	[100]
	Porphyran	Sulfated 1,3-β-D-galactose 1,4-α-l-galactose-6-sulfate 3,6-anhydro-α-L-galactose	...	Antitumor, immunomodulatory, hypolipidaemic, hypotensive	...
	Xylan	1,3-1,4-β-D-xylose	...	...	
Brown	Alginate	1,4-β-D-mannuronic acid, α-L-guluronic acid	17-45%	...	[100]
	Laminarin	1,3-1,6-β-glucose	10–35%	Antitumor, decreases liver triglyceride, cholesterol and phospholipid levels; serum hypocholesterolaemic, hypotensive, antibacterial, immunomodulator, anticoagulant	[100]
	Fucoidan	Sulfated 1,2-1,3-1,4-α-L-fucose	5–20%	Antioxidant, anticoagulant, antithrombotic, anti-adhesive, anti-proliferative, anti-inflammatory, anti-angiogenic, anti-metastatic	[100]

Brown Seaweed Polysaccharides

Brown seaweeds are commonly used as food ingredients owing to their commercial abundance [81]. When alginate, along with fucoidan, was fermented, they affected the hydrolase activity of microbial bile salts suggesting an anti-obesogenic effect. It is reported that the brown seaweed as well as its extracted polysaccharides, when fermented by fecal microbes, displayed elevated levels of metabolites like butyrate, propionate, and acetate, along with total SCFAs. In correspondence to the increased metabolites, few microbial populations like *Roseburia, Bacteroides, Fusicatenibacter, Bifidobacterium, Fecalibacterium, Coprococcus, Lactobacillus,* and *Parasutterella* were also observed to be high in number after the fermentation [82 - 84]. Biological applications of seaweed polysaccharides are tabulated in Table **3**.

Alginate

Alginates are extensively used in the food industry, biotechnology, and medicinal industries [85]. In Japan, digested sodium alginate is used in the diet as a treatment to control cholesterol levels in the blood [86]. In a study, eight individuals belonging to Japan were supplemented with 10 grams of alginate daily and the differential populations in the fecal microbiome were analysed [87]. After one and two weeks, it was observed that alginate supplementation has increased the population of *Bifidobacterium* and metabolomics analysis after two weeks revealed that there was a considerable increase in propionic as well as acetic acid. The metabolomics analysis after two weeks also revealed that harmful metabolites like p-cresol, sulfide, indole, skatole, and ammonia have significantly reduced suggesting the capability of alginate to be used as a functional food.

Laminarin

Laminarin is a type of storage polysaccharide consisting of 1,3- or 1,6- β-glucose units, and it has high water solubility. An *in vivo* study on rats was performed where 143 mg of laminarin was supplemented for weeks. The results suggested that *Lactobacillus* and *Bifidobacterium* were not involved in the selective fermentation of laminarin, whereas it has modified mucus secretion and composition of the caecum, jejunum as well as colon to defend against bacterial infections [88, 89].

Fucoidan

These are structural polysaccharides of brown seaweeds and are constituted by 1,2- or 1,3- or 1,4-α-L-fucose. These are highly soluble in water. In two different studies performed on rats, fucoidans from different sources (*Ascophyllum nodosum* & *Laminaria japonica*) were supplemented, and it was observed that the caecal microbiota had higher populations of *Bifidobacterium, Ruminococcaceae*, and *Lactobacillus*. In one of the above studies, it was also reported that fucoidan had decreased a load of antigen, thereby decreasing the inflammatory response [90, 91].

Red Algae Polysaccharides

Galactans

Red seaweed species like *Gracilaria* spp., *Gelidium* spp., *etc*., are already being exploited commercially for food emulsification & encapsulation agents, agar, and other food additives [81]. Galactans can be grouped into three different types. Carrageenans, agar, and poryphyrans. Carrageenans are chiefly made of sulfated 3,6-anhydro-D-galactose, 1,3-α-D-galactose, and 1,4-β-D-galactose, whereas agar and porphyran are made up of sulfated 3,6-1,4-anhydro-α-L-galactose, 1,3-β-D-galactose and sulfated 3,6-anhydro-α-L-galactose,1,4-α-L-galactose-6-sulfate, 1,3-β-D-galactose respectively [92 - 94].

Xylan

Xylan is made up of 1,4-1,3-β-D-xylose. In a study conducted by Lahaye *et al*, a six-hour *in vitro* fecal fermentation of xylan from *Palmariapalmata* demonstrated an increase in metabolites butyrate, acetate as well as propionate [95]. It has already been reported that the human gut has a range of populations that produce CAZymes like xylosidases, xylanases, *etc* [96]; studies are being done to investigate the capability of human gut microbes to digest xylooligosaccharides from seaweeds.

Green Algae Polysaccharides

Ulvan

These are cell wall polysaccharides, mainly made up of sulfated 1,4-β-D-glucuronic acid, 1,3-α-L-rhamnose, and 1,4-β-D-xyloglucan. These are highly water-soluble compounds [97]. In a study, mice were induced constipation by loperamide, and the effect of *Enteromorpha* and polysaccharides derived from *Enteromorpha* was analysed. The analysis showed that supplementation of *Enteromorpha* and its associated polysaccharides had reduced inflammation and even improved the gut microbiota [98].

Proteins

Microalgae from all three classes can fulfill the human essential amino acid requirement. Cereals are the major source of amino acids in human diets, out of which wheat, corn and rice are mostly used. Regrettably, none of the three cereals is a good protein source compared to microalgae. The major amino acid that lacks in terrestrial cereals diet is lysine, which is nine times more in microalgae [108]. The protein composition varies greatly amongst the different groups of seaweeds (green, red, and brown). In general, the protein composition in brown seaweeds is low (around 5-24% in DW) in comparison with green and red seaweeds, which contain high protein content (around 10-47% in DW) [109]. Same as other nutrient elements of seaweeds, the composition of amino acids, peptides, and proteins are also impacted by several parameters, in particular seasonal variations [110]. For instance, during the period between February-May months, brown seaweeds such as (*Laminaria* and *Saccharina*) produced the highest concentration of protein [110]. Similar variation was observed in red seaweed species, where the maximum was seen in summer and a significant drop in winter [109]. Generally, seaweed proteins are abundant with essential amino acids such as glutamic acid, alanine, arginine, and glycine at concentrations equivalent to the requirements specified by WHO/FAO [80, 111]. However, certain amino acids are observed in low concentrations, such as cysteine and lysine. Several other amino acids observed in seaweeds are domoic acid, kainic, kainoids, laminin, taurine, and few mycosporin-related amino acids [112, 113]. In humans, taurine is involved in several physiological mechanisms such as nervous system [114], ocular development, membrane stabilization, and immunomodulation. Additionally, domoic and kainic acids are essential in regulating several neurophysiological mechanisms [115]. It has been reported that *Palmaria palmate* (red algae) has been considered one of the best alternatives for edible cereals [108].

Lipids

The lipid content in the seaweeds is present in low concentrations (around 1-5% in DW), among which glycolipids and neutral lipids being present abundantly. The concentration of essential fatty acids observed in seaweeds is greater than what is observed in terrestrial plants since they produce long-chain PUFAs, majorly the docosahexaenoic acid (DHA) and eicosapentaenoic acid (EPA), which belong to the ω-3 fatty acids family [116]. Generally, red seaweeds contain high arachidonic acid, oleic acid, palmitic acid, and EPA concentrations compared to brown seaweeds which have higher concentrations of α-linolenic acid, linoleic acid, and oleic acid but lower EPA concentrations. Higher concentrations of DHA, oleic, palmitic, α-linolenic, and linoleic acids are observed in green seaweeds [117, 118]. Both brown and red seaweeds are a major source of omega-6 and omega-3 fatty acids [119, 120]. According to a metabolomic study conducted on four species of brown algae (*L. rosacea, L. monticola, L. obscura and L. sonderii*) with the help of Ultra high performance liquid chromatography-high resolution mass spectrometry (UHPLC-HRMS) fingerprinting technique, it was observed that the metabolome varies in every species with the altering climatic conditions as well as substrates. Under such conditions, polyolefins, as well as derivatives of fatty acids, are recognised as potential chemomarkers. A significant variation in the metabolome was observed every month in four of them and the variation might be related to the salinity and the temperature of the sea surface [60].

Dietary Fibre

Nowadays, the fibre content in food consumption is considered essential to maintaining a healthy diet. Conversely, still, there is no well-defined or practiced analytical technique which can measure the food components that exert the physiological effects of dietary fibre [74]. However, there is an agreement to take into account the physiological role of dietary fibre while defining the nutritional composition. Dietary fibre comprises of diverse compounds which contain a comprehensive blend of polymers and carbohydrates, which include both oligo- and poly-saccharides such as inulin, resistant starch, gums, pectin, hemicellulose, and cellulose, which might be in combination with lignin or other non-carbohydrate elements such as resistant protein, cutins, saponins, waxes, and polyphenols [121].

The biological properties of dietary fibre are primarily defined by the ratio of two fractions (insoluble and soluble). The formation of the viscous gel upon contact with water in the gastrointestinal tract is a well-defined characteristic feature of soluble fibre. In contrast, insoluble fibre does not form gels when in contact with water; however, it can retain water in its structural matrix, which results in an increased fecal quantity that quickens intestinal transit. These dissimilarities in the behavior of the fibre in the intestinal transit will result in diverse physiological features. A major proportion of the soluble fibre is fermented. Its primary features are the reduction of glucose and cholesterol concentrations in blood and the development of microbiota in the intestine. Whereas insoluble fibre is meagrely fermented, it provides an intestinal regulating effect since it contains a marked laxative [74].

The majority of these are fibres, which cannot be digested by the human intestinal enzymes, however, a small fraction of them can be degradable by various enzymes produced from colonic bacteria [122]. This fraction of dietary fibre is present in considerate amounts ranging from 36-60% in DW [123]. Whereas, compared to vegetables cultivated on land, the soluble dietary fibre present in seaweeds is in a higher proportion (around 55-70% in DW; (Fig. **4**)) [80].

Vitamins and Minerals

In humans, vitamin deficiency causes a wide range of diseases like pernicious anaemia caused by the deficiency of cobalamin (vitamin B12), beriberi due to the deficiency of thiamine (vitamin B1), pellagra due to niacin deficiency, scurvy due to ascorbic acid, vitamin C deficiency and many more. All these vitamins are crucial for many metabolic functions as they act as precursors for vital enzyme cofactors. Many mammals have lost their capability to produce these vital cofactors, so they depend on the diet to obtain them [124]. Seaweeds are also an outstanding source of vitamins [111]; however, the vitamin content is dependent on variations in their taxonomy, seasons, and habitat. Seaweeds spend a lot of time being exposed to sunlight directly in a watery ecosystem. Hence, seaweeds contain several types of anti-oxidative molecules, which include various protective pigments and vitamins, both fat- and water-soluble [125]. Based on this, seaweeds are an abundant source of vitamins A (retinoids), B1 (thiamin), B2 (Riboflavin), B3 (niacin), B5 (pantothenic acid), B9 (folic acid), B12 (cobalamin), C (ascorbic acid), D (calciferol) and E (tocopherols) [111]. Vitamin C, a water-soluble vitamin, is present in huge concentrations in brown algae *Ecklonia arborea* (previously known as *Eisenia arborea*), *Crassiphycus changii* (previously known as *Gracilaria changii*), sea spaghetti (*Himanthalia elongata*), and laver (*Porphyra umbilicalis*) [126 - 128]. Seaweeds are an excellent alternative for providing vitamin B12 for individuals who follow a strict vegetarian diet. Since vitamin B12 is not essential or produced by higher taxa plant species, thus vegetables and fruits do not provide the required daily dose of vitamin B12. Hence, vitamin B12 deficiency is commonly observed among individuals who follow vegan or vegetarian diets [129 - 131].

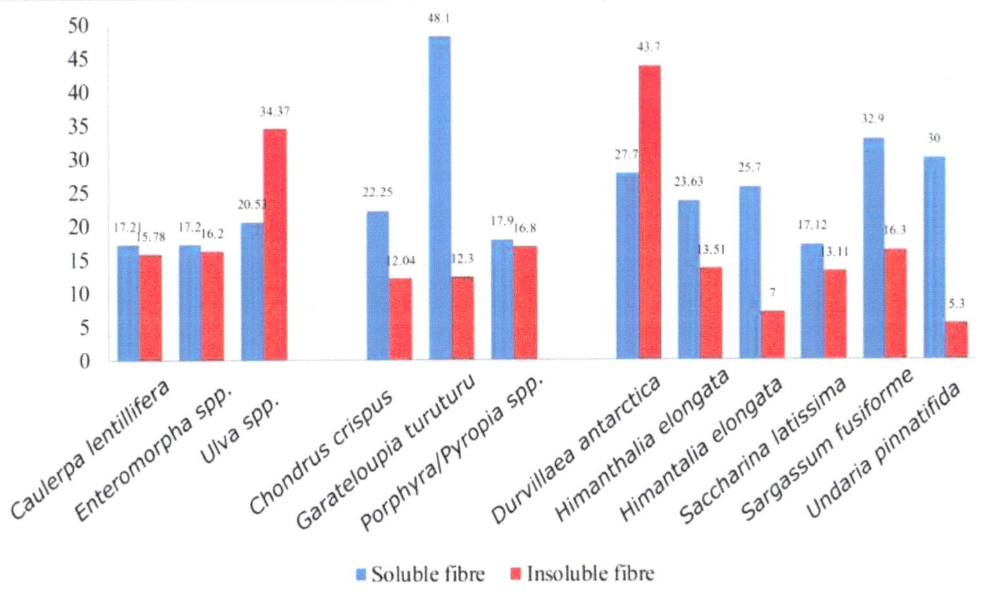

Fig. (4). Graph representing dietary fibre content in various seaweeds (*Caulerpa lentillifera, Enteromorpha spp., Ulva spp.*belong to Chlorophyta, *Chondrus crispus, Garateloupia turuturu, Porphyra/Pyropia spp.*belong to Rhodophyta, *Durvillaea antarctica, Himanthalia elongate, Himantali aelongate, Saccharina latissimi, Sargassum fusiforme, Undaria pinnatifida* belong to Phaeophyceae [74].

The seaweeds which inhabit the marine ecosystem also contain high mineral content ranging from 8-40% in DW of seaweed (Fig. **5**). Necessary mineral elements such as phosphorous, sulfate, chloride, potassium, magnesium, calcium, sodium, and micronutrient elements such as cobalt, nickel, boron, manganese, fluoride, molybdenum, selenium, copper, zinc, iron, and iodine are available abundantly [132]. The composition of these minerals will vary depending on the seaweed classification, physiological, seasonal, and geographical variations, as well as the type of mineralization and processing approach, employed [120]. Seaweeds are a major source of iodine, thus providing the much-needed daily iodine dose (150 µg/day). Due to the presence of high mineral content, seaweeds can be employed as nutraceuticals or as dietary supplements to reach the daily dose recommended for a few macro-and micro-nutrients.

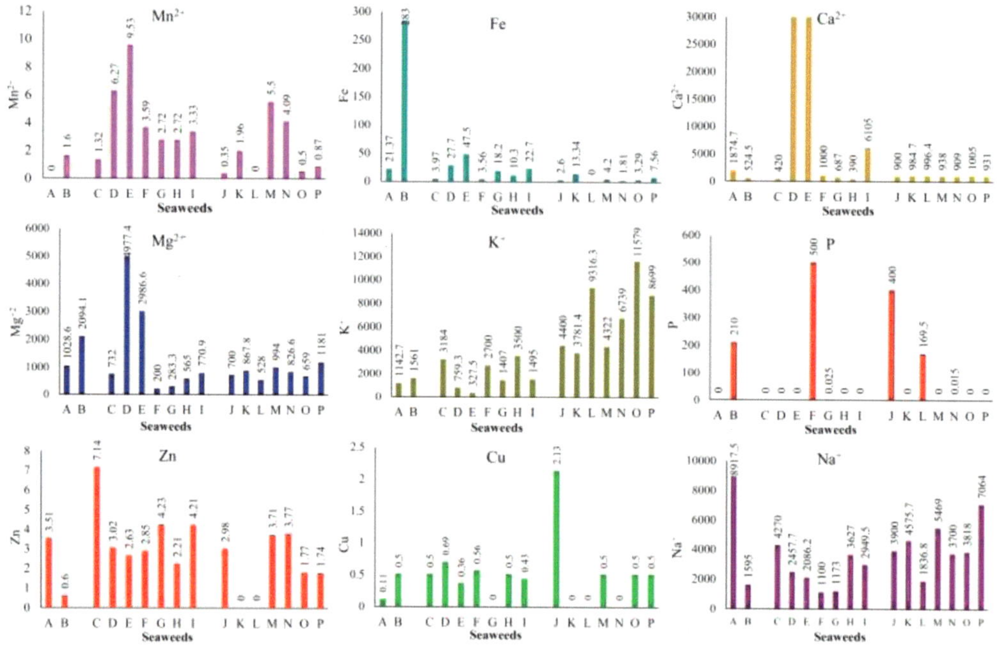

Fig. (5). Mineral content of various seaweeds (A is *Caulerpa lentillifera*, B is *Ulva rigida,* C is *Chondrus crispus*, D is *Ellisolandia elongate* (formerly *Corallina mediterranea*), E is *Janiarubens*, F is *Palmaria palmate,* G is *Porphyra umbilicalis*, H is *Pyropia tenera* (formerly *Porphyra tenera*), I is *Pterocladiella capillacea* (formerly *Pterocladia capillacea*), J is *Alaria esculenta*, K is *Ascophyllum nodosum*, L is *Bifurcaria bifurcate,* M is *Fucus vesiculosus*, N is *Himanthalia elongate*, O is *Laminaria digitate*, P is *Undaria pinnatifida*) [74].

Polyphenols

Polyphenols are found as a minor component in seaweeds. Low concentrations of polyphenols are found in red and green seaweeds (around 1% in DW) in comparison with brown seaweeds [120, 133]. However, in *Fucus* and *Ascophyllum* genera, around 14% (in DW) can be observed [110]. Among the several polyphenols found in the brown seaweeds, phlorotannins are most extensively reported, particularly among the species belonging to the *Ecklonia* genus [110, 134]. Phlorotannins are usually produced from phloroglucinol (1, 3, 5-trihydroxybenzene) oligomeric structures [135]. Phlorotannins are usually observed in concentrations ranging from 20-250 mg/g (in DW) among *Cystophora retroflexa*, *Sargassums pinuligerum*, *Fucus vesiculosus,* and *Ascophyllum nodosum* species [136]. According to Ford *et al* 2020, phlorotannins have antibacterial activity and the potency of the bacterial growth inhibition depends on the chemical structure of phlorotannins. The chemical structures of these compounds vary from species to species. Accordingly the antibacterial activity of *Ascophyllum nodosum* has been observed to be more than *Fucus serratus* [137]. Other polyphenols reported are flavonoids and their derivatives, such as epicatechin and catechin, fucol, and its derivatives [110].

BIOACTIVE COMPOUNDS IN ALGAE

Aside from macro- and micro-nutrient elements, seaweeds also contain several bioactive molecules that exhibit superior anti-oxidative potential, such as polyphenols and carotenoids [70, 134, 138 - 142]. Most of the natural pigments extracted from seaweeds are investigated for their anti-inflammatory (primarily related to modulating macrophage functions), anticancer, and antioxidant activities, among others [143]. Fucoxanthin, a carotenoid, is among the naturally available pigment primarily found in various species of brown seaweeds, which is reported by various research groups for their osteoporosis preventive, photoprotective, neuroprotective, anti-obesity, anti-inflammatory, anti-cancer, and anti-oxidative effects [75, 144 - 148].

Few research reports have indicated that phycobiliproteins extracted from red seaweeds (phycoerythrin) can be used in the treatment or prevention of diseases related to neurodegeneration, which are associated with oxidative stress (such as Parkinson's and Alzheimer) because of their anti-oxidative property [149]. The lectins extracted from *Bryothamnion* sp. (Rhodophyta) exhibited growth inhibition activity against the *Streptococcus* sp., hence they can also be employed as anti-bacterial agents [150]. Among the peptides observed in the seaweeds, which contain 2-20 amino acids, can be cyclic, linear, peptides or depsipeptides with one or more amide bonds substituted by ester-kahalalides- bonds, dipeptides (almazole D, carnosine), tripeptides (glutathione), pentapeptides (galaximide), hexapeptides, oligopeptides, and phycobiliproteins [151]. These extracted peptides are studied for their immunostimulatory, anticoagulatory, antihypertensive, antimicrobial, antiviral, antitumor, antioxidant properties [152]. Specifically, the P- and Q-kahalalides found in green seaweeds exhibit cytotoxicity against HL-60 cell lines, whereas F-kahalalide extracted from *Bryopsis* sp. (Chlorophyta) exhibited the potential to reduce the cell viability of non-metastatic prostate tumor cells [115]. Glycoproteins, agglutinins, phycobiliproteins, and lectins are found in great abundance in seaweeds [113].

BIOLOGICAL APPLICATIONS

The occurrence of various bioactive compounds in the seaweeds offers them several biological properties related to the treatment and prevention of diseases (Fig. **6**) such as neuroprotective, fat-lowering, antivirus, antitumor, antimicrobial, anti-inflammatory, antihypertensive and anti-diabetic agents [75 - 77].

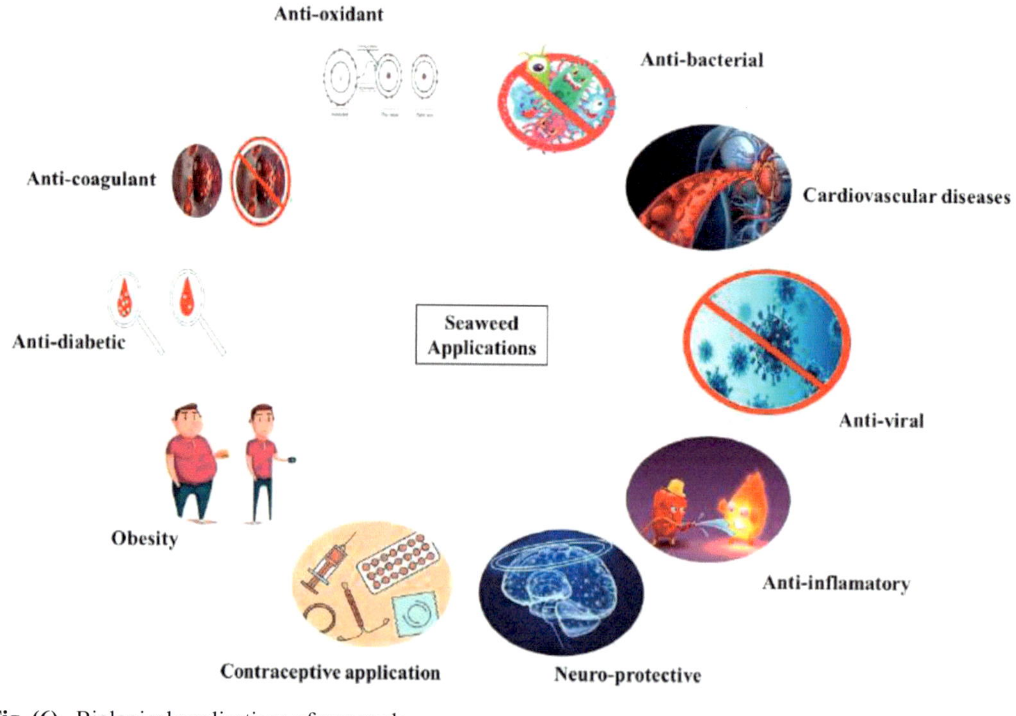

Fig. (6). Biological applications of seaweeds.

Antibacterial Activity

In present scenario, evolution of bacteria by developing resistance against several antibiotics is forcing researchers and scientists to search for new natural bactericidal compounds [153]. Natural compounds will offer higher efficiency and low or no toxicity [154]. Thus, seaweeds offer several such bio-active compounds, which could be useful in treating several diseases. As these compounds are extracted from sources available naturally, the adverse effects would be minimal and these could be primary treatment choice [29]. The chemical investigative studies carried out on few types of seaweed have demonstrated that they synthesize a wholly new variety of bio-active secondary metabolites, which have distinctive molecular structures not synthesized by other species. The three groups of seaweeds (such as green, red and brown) have demonstrated pharmacological activity and antimicrobial activity by inhibiting the growth of several virus, fungi and bacteria [155]. Based on the study carried out with few species of brown seaweed (such as *Sargssum* sp., *Laminaria* sp. and *Ascophyllumnodosum*),it was established that they can be employed to fight against the infectious diseases caused by harmful bacteria [156]. So far, several investigative studies based solely on the antimicrobial activity of the seaweed extracts collected from various parts across the world against broad spectrum of infectious bacteria (both gram negative and gram positive), fungi and viruses. Thus, extractions from seaweeds can be employed as potential antimicrobial agents.

The bactericidal potential of three different solvent extracts (such as ethyl acetate, methanol and water) of marine seaweed *Amphiroa anceps* was evaluated against *Enterobacter* sp. and ethyl acetate extract exhibited maximum inhibitory zone [157]. In addition, the chemical compositional analysis showed the occurrence of terpenoids, amino acids and flavonoids in methanol extract. Terpenoids, flavonoids, phenols and protein in ethyl acetate extract, whereas aqueous extract showed only tannins. In similar study done by using methanol extract of *Ulva lactuca* exhibited good bactericidal potential against harmful gram negative (such as *Salmonella typhi*, *Pseudomonas aeruginosa*, *Klebsiella sp.* and *Escherichia coli*) and gram positive (such as *Bacillus* sp., *Staphylococcus epidermidis*, *Staphylococcus aureus* and *Bacillus subtilis*) bacterial species [158]. In some studies, the microalgae were given physical treatments like cutting and grinding further they were induced by bacterial cultures to produced antibacterial compounds. For instance, In *Halimeda gracilis*, the induction by bacterial cultures (*Stenotrophomonas maltophilia* and *Vibrio harveyi*) has reduced its antibacterial activity but contrastingly the physical treatment has increased its activity. Similarly in *Halimeda macroloba* the induction enhanced the antibacterial activity; on the other hand the physical treatment decreased it [159].

Antioxidant Activity

As a result of seaweeds inhabitation in extremely oxidative environmental conditions they have developed robust anti-oxidative protective mechanisms. Similar to photosynthetic organisms, seaweeds grow in combination of high oxygen and light concentrations, which supports the development of efficient anti-oxidative compounds and free radicals [160]. Thus, the lack of oxidative damage to the thylakoid membranes of the chloroplasts demonstrates that the seaweeds have developed efficient protective processes [161]. Based on the chemical compositional analysis seaweeds contain a broad variety of biologically active compounds such as amino acids, peptides, unsaturated fats, sulphated polysaccharides and polyphenols, which display several anti-oxidative features. Among them, the most essential are carotenoids, fucoidnas and phlorotannins [74]. In a study by Walailuck *et al* 2011, four seaweed species (*Turbinaria conoides* (J. Agardh) Küzting, *Sargassum binderi* (Sonder), *Halimeda macroloba* (Decaisne) and *Amphiroa sp.*) were extracted (in both aqua and ethanol) for phenolic compounds and more phenolics were observed in aqueous extraction than ethanol. Among the four *T. conoides* displayed the best anti-oxidative activity. So the authors suggest that dehydrated *T. conoides* could be used as a hot beverage like tea for human consumption and can be a potent nutraceutical product [162].

Seaweeds such as *Chnoospora maxima*, *Sargassum wightii*, *Amphyroa anceps*, *Spyridia hypnoides*, *Chaetmorpha antennia* and *Ulva fasciata* collected from Kanyakumari coast (India) and their methonalic extracts were analysed for their biochemical compositions and *in vitro* anti-oxidative potentials [163]. Among the 6 seaweeds studied, *S. wightii* showed highest protein content, whereas highest lipid, phenolic and carbohydrate content were observed in *C. maxima*. Additionally, highest anti-oxidative activity was observed in *C. maxima* and *S. wightii*, both belonging to brown seaweeds, which supports the fact that brown seaweeds contain high phenolic concentrations. Immunity associated gene expression in *Litopenaeus vannamei* was observed to be improved by the presence of a polysaccharide extracted from *Grateloupia filicina* [164].

In general, the bio-active compounds found in brown seaweeds contain maximum number of isomers because of the notable variations in their branching positions occur between their phlorogucinol units (PGUs). The presence of PGUs oligomerization and free hydroxyl groups in their structure seems to be the main reason for the anti-oxidative potential of phlorotannins [165]. Additionally, occurrence of polyphenols with definitive biological activity in various seaweeds can impact gene expression [166, 167]. Hence, there is a huge research interest in exploring the polyphenols features associated with the prevention of cancer and cardiovascular diseases as well as aging [168, 169].

The polysaccharides such as alginic acid, laminarans and fucoidans found in brown algae, exhibit good *in vitro* anti-oxidative potential and can be considered as excellent potential anti-oxidative compounds [170]. The anti-oxidative potential of sulphated polysccharies is dependent on various factors such as glycosidic bonds, sugar type, molecular weight and degree of sulfation [171, 172]. Compared to high molecular weight polysaccharides, lower molecular weight polysaccharides exhibit high anti-oxidative potential [173]. This is due to the easy incorporation of low molecular weight polysaccharides into the cells and donation of protons more effectively than high molecular weight polysaccharides [171]. Additionally, a positive correlation was established in between the free radical scavenger potential and sulfate content of fucoidan fractions of *Saccharina japonica,* a brown seaweed (previously known as *Laminaria japonica*) [174]. In another report, Sargachromanol E extracted from *S. horneri* (brown seaweed) is studied for its scavenging potential against ROS, shielding the cell membrane against oxidative modification in human dermal fibroblasts exposed to ultra violet light [175]. Bio-active compounds known as carotenoids are also proficient anti-oxidative compounds found in seaweeds [74]. Tocopherols and xanthophyll are most abundantly observed carotenoids. Among xanthophylls, fucoxanthin is the most effective quenchers of singlet oxygen, whereas tocopherols are commonly employed in the food industries because of their effective radical scavenging potential.

Anti-viral Activity

Anti-viral activities against respiratory syncytial virus (RSV), Herpes simplex virus (HSV) and human immunodeficiency virus (HIV) were reported using the seaweeds. Sulphated polysaccharides such as sulfated rhamnogalactans, fucoidans and carrageenans were studied for their inhibitory potential against to cell damage caused by HIV-1, which were extracted from *U. lactuca, Nothogeniafastigiata* sp. and *Agardhiella subulata* (previously known as *Agardhi ellatenera*) [176 - 179]. In another study, the chemical characterization of extracts of the bio-active compounds from *Schizymenia dubyi* (red seaweed) showed differences in the activity against HIV-1 virus, which was due to the seasonal variations. *S. dubyi* contains high levels of glucuronic acid, a polysaccharide called glucurono-galacto-sulfate which is accountable for its anti-viral potential [74, 180].

Antiviral potential of sulphated polysaccharides, which are extracted from various groups of seaweeds such as green (*Caulerpa racemosa*), red (*Gracilaria caoticata*, *Scinaia hatei* and *Grateloupia indica*) and brown (*Cystoseira indica* and *Stoechospermum marginatum*) were investigated against dengue virus. All the seaweed extracts demonstrated anti-viral activity against DENV-2 with IC_{50} values in the range from 0.12 to 20 µg/mL. The maximum anti-viral activity was exhibited by sulfate-galactan extracted from *G. indica*, with an IC_{50} value of 0.1 µg/mL. Additionally, extracts of *G. indica* exhibited anti-viral activity against HSV [181].

Anti-inflammatory Activity

Extract of edible red seaweed (*Eucheuma cottonii*) exhibited several anti-inflammatory properties such as eosinophil infiltration, mucin synthesis and MMP-9 expression in rat model [182]. The reverse-phase high performing liquid chromatography (RP-HPLC) and liquid chromatography-mass spectrophotometry (LC-MS) analysis revealed the absence of quercetin, rutin, phlorogucinol and phlorotannins in water-methanol (50% *v/v*) solvent extract. However, Fourier Transform Infrared Spectroscopy (FT-IR) analysis indicated the presence of prominent biologically active compounds, peptides or simple proteins. The results also revealed that the monocyte, segmented neutrophil, band neutrophil, eosinophil and total white blood cells numbers were greatly declined. Additionally, extract of 300 mg /kg reduced the lung inflammation and airway smooth muscle constrictions in comparison with loratadine treatment of 10 mg/kg in rat model studies. Another study reported the tumor suppressor potential of polyphenolic rich extract from *E. cottonii* [183].

Anticoagulant Activity

The anticoagulant potential of the sulfated polysaccharides extracted from seaweeds is one of the extensively investigated properties, in order to discover an alternative for heparin, a naturally occurring anticoagulant. The anticoagulant potential of the sulfated polysaccharides extracted from seaweeds is dependent of several factors such as distribution as well as degree of sulfate groups, sugar composition and molecular weight of the compound [176]. Several studies reported the extraction, isolation and characterization of sulfated polysaccharides which have significant anticoagulation potential from seaweeds [171]. Among them, sulfatedgalactans or carrageenans found in red seaweeds and sulphate fucoidans found in brown seaweeds are two main types [171, 184, 185]. Several research groups have studied the relation between the anticoagulation potential and molecular structure of few sulfated polysaccharides extracted from seaweeds [184, 186, 187]. Results indicated that the occurrence of sulfated groups determined their potential for anticoagulation. Based on the literature evidence available so far indicates that sulfated polysaccharides extracted from seaweeds might be used as a substitute for heparin because of its natural origin and superior anticoagulation capacity in pharmaceutical industries [171, 176].

Anticancer Activity

Among the developed countries, cancer is one of the main contributors in mortality [188]. Numerous treatment and patient care options have been developed to in the hope to cure cancer. However, these treatments have several adverse health impacts. Hence, new treatment options should be discovered to help reduce these adverse effects. The polysaccharides extracted from seaweeds can be an alternative choice, as they have exhibited growth inhibitory potential against few types of cancer cell lines in *in vitro* conditions [189]. Their anticancer potential can be because of several mechanisms of action such as free radical production, DNA damage, depolarization of mitochondrial membrane and cell cycle arrest [188].

In a study, anti-allergic phlorotannins (such as eckol, 0,0-bieckil, 0,2-bieckol, 2,2-bieckol, phlorofucofuroeckol-A and phlorofucofuroeckol-B) were extracted, purified and studied for their activity against RBL-H cells from edible brown seaweed (*Eisenia arborea*). All purified phlorotannins inhibited the release of b-Hex from RHL-H cells [190]. Another study reported the biological activities of *S. marginatum* extract as well as their chemical compositional analysis using GC-MS [191]. Polysaccharide (k-carrageenan), which was extracted, purified and isolated from marine seaweed (*Kappaphycus alvarezii*) was studied for anticancer potential against human colon cancer cell lines (HCT116) [192]. Results indicated the exposure of cancer cells to the polysaccharide resulted in cell death by apoptosis, formation of apoptosome, fragmentation of nucleus, up-regulation of mRNA expression levels of Bcl-xL, Bcl-2 and caspase3 and down-regulation of PARP-1 and XIAP.

In another study, methanol extract of seaweed (*Solieria robusta*) exhibited cell death of oral cancer cells (Ca9-22) through oxidative stress and apoptosis [193]. NMR spectral analysis showed the occurrence of fatty acid and sugar moiety. Flow cytometry analysis of the extract exposed cancer cells treated with annexin V/PI stain showed that the mortality of the cancer cells increased in a concentration-dependent manner. Cell cycle analysis using propidium iodide (PI) based assay revealed the drastic change in the cell cycle distribution, primarily in subG1 accumulation. Oxidative stress analysis revealed the occurrence of reactive oxygen species (ROS) in a time- and concentration-dependent manner, resulting in depolarization of the mitochondrial membrane.

The ethanolic extracts of 13 seaweeds (such as *Melanothamnus afaqhusainii*, *Botryocladia leptopoda*, *Stokeyia indica*, *Stoechospermum marginatum*,

Sargassum variegatum, *S. swartzii*, *S. binderi*, *Spatoglossum variabile*, *Padina pavonia*, *Jolynala linarioides*, *Dictyota intrica*, *Rhizoclonium implexum* and *Halimeda tuna*) were studied for their hypo-lipidemic potential using normal and triton-induced rats [194]. Among them 8 (*S. robusta*, *S. variabile*, *H. tuna*, *S. swartzii*, *S. binderi*, *M. afaqhusainii*, *J. laminarioides*, *D. intrica*) were studied foe their efficacy and ultimately 3 (*M. afaqhusainii, S. binderi and J. laminarioides*) were further investigated for their potential in rats with high-fat diet-induced hyperlipidemia. Some seaweed extracts have showed a similar activity compared to standard drugs such as fenofibrate and benzafibrate (Lipocor). The ethyl acetate extract of seaweed (*Ulva intestinalis*) collected from the Zhejiang coast (China), has exhibited excellent algicidal potential against pathogenic red tide microalgae (*Prorocentrum micans* and *Heterosigma akashiwo*) [195]. The analysis using NMR spectroscopy showed the occurrence of three (such as hexadecanoic acid, (6E,9E,12E)-(2-acetoxy-β-d-glucose)-octadecatrienoic acid ester and 15-ethox--(6z,9z,12z)-hexadecatrienoic acid) algicidal molecules. Among them, the second compound is found to be highly effective against microalgae.

Seaweeds synthesize a broad spectrum of chemically active metabolites to defend themselves from other settlement species [196]. Preliminary investigation has revealed the presence of a few anti-oxidative compounds, in particular, β-carotene, which can be beneficial for treating oral leukoplakia, a precancerous condition, which is a possible precursor for oral cancer. Polysaccharides extracted from seaweeds play a significant part in the regulation of immune responses and few of them are well-studied anti-inflammatory compounds. Among them, porphyran, phlorotannins, peptides, laminaram, carrageenans, and alginate are associated with various potential applications. Phlorotannins are essential for cancer chemoprevention. Thus, phlorotannins extracted from the brown seaweed (*Ecklonia cava*) defend the cells from oxidative stress and radiation-induced injuries [197]. Among the various bioactive metabolites studied in seaweeds, fucoxanthin and fucoidan are essential metabolites well-recognized for their anticancer potential and are also employed as potential chemopreventive and chemotherapeutic agents.

Fucoidan, a sulfated polysaccharide, is extracted from brown seaweeds which is rich in L-fucose. Several studies reported the application of fucoidan as anticancer agent resulting in cell death through inhibition of tumor-induced angiogenesis and apoptosis induction [198, 199]. It is also being reported for its anticancer potential in the prostate lung, liver, colon, breast, bladder and melanoma cells [200, 201]. In one study, the crude extract of brown seaweed (*S. wightii*) was studied for its anticancer potential. The results indicated the inhibition of breast cancer cells, which is dependent on the concentration of the extract applied [202]. Anti-proliferative activity was also reported from *Halidrys siliquosa, Dictyota dichotoma, Desmarestia ligulata, Carpodesmia tamariscifolia* (previously known as *Cystoseira tamariscifolia*) and *Bifurcaria bifurcate* crude extracts [203].

Neuro-protective Activity

Seaweeds are extensively studied for their neuro-protective potential, particularly brown seaweeds [204, 205]. Parkinson's disease is one of the common neurodegenerative diseases related to difficulties in movement and characterized by the reduction of dopamine (a neurotransmitter), loss of dopaminergic neurons in the substantia nigra of pars compacta (SNpc) and by the aggregation of Lewy bodies. So far, no cure has been discovered and even the mechanisms associated with the disease are not fully understood. Hence, it is essential to develop novel control and preventive treatment approaches. The red seaweed, *Cistus crispus* is well-studied for its various biological properties because of the presence of several functional metabolites. In addition, it showed excellent results in a transgenic *Caenorhabditis elegans* (Nematoda) PD model by reducing the α-synuclein accumulation and defended nemotodes from the neurotoxic 6-hydroxydopamine (6-OHDA) and caused induction of dopaminergic neurodegeneration. Thus, such favourable outcomes can allow its application in pharmaceuticals for the designing of better anti-neurodegenerative agents [206].

Enzymatic hydrolysates from brown seaweed, *Saccharina japonica* (previously known as *Laminaria japonica*) exhibited the anti-oxidative, anti-inflammatory and anti-inflammatory properties [207]. The anti-oxidative potential was dependent on the carbohydrases and proteases used. However, hydrolysates acquired by using Celluclast (60%) and Flavourzyme (90%) showed maximum inhibition for acetylcholinesterase activity. Additionally, hydrolysates exhibited excellent cell viable percentage, thus allowing its application in pharmaceuticals. The bioactive metabolites (such as polysaccharides) observed in seaweeds play a significant part in neurodegenerative diseases [208]. Phlorotannins, specific to brown seaweeds, have a tremendous capacity to be used as therapeutics for human health. The factors such as solid/liquid ratio during upon extraction, extraction time, temperature, particle size, drying temperature, solvent type and pre-treatment choice affect the quality as well as efficiency of the compound. Phloroeckol and a tetrameric phloroglucinol observed in brown seaweed (*Macrocystis pyrifera*) are accountable for its anti-allergic, anti-alzheimer's and anti-diabetic effects.

The edible brown seaweed (*Ecklonia cava*), which is well known for its therapeutic activity against the prevention of Parkinson's disease, was first noticed by Cha *et al.* [209]. Its anti-neurodegenerative potential can be associated with the anti-oxidative potential of dieckol in dopaminergic neuronal cells. The increase in oxidative stress makes the cells vulnerable to intracellular toxic events, in turn leading to aggregation of protein (α-synuclein), the occurrence of cell death and finally resulting in Parkinson's disease. The dieckol, a polyphenol extracted from the seaweed avoids the aggregation of α-synuclein. The algal pigments present naturally in the seaweeds also exhibit neuroprotective effects. Fucoxanthin, extracted from brown seaweed (*Undaria pinnatifida*) exhibited a potential effect; this can reduce the neuronal cell damage in cortical neurons under re-oxygenation and hypoxia [210]. This might be because of the occurrence of intramolecular oxygen atoms and allenic bonds in their molecular structure [211]. Promising effects were also observed in the xanthophyll astaxanthin, such as 6-OHDA-induced apoptosis suppression, inflammatory mediators' production inhibition, effects against cerebral ischemia and H_2O_2-induced toxicity [205]. These effects can be related to the occurrence of several conjugated double bonds in the molecule and to the strong anti-oxidative properties which are unique to this pigment.

Type-II Diabetes

In the world, around108 million people had been diagnosed with diabetes by 1980 (~4.7% of total population). This reached to 422 million by 2014 (~8.5% of total population). Among these, 90-95% belong to type-II diabetes [212]. Currently, in India, 8.7% of adults have been suffering from type-II diabetes, whereas 7.7 and 9.1% in UK and USA, respectively [213 - 215]. Type-II diabetes mellitus, commonly known as adult-onset diabetes, is a dietary-associated metabolic disorder that responds to diet change. On the other hand, type-I diabetes, which is inherited genetically, is an autoimmune disorder, which damages pancreatic beta cells, and regular injections of insulin should be administered for treatment. Acarbose is the generally used drug for treating type-II diabetes, which inhibits the activity of α-glucosidase and α-amylase enzymes. However, it might cause diarrhoea and gastric discomfort [216].

Several *in vivo* studies were performed to evaluate the anti-diabetic potential of seaweeds using animal models [217 - 220]. One study reported the clinical trials conducted in humans with type-II diabetes using *Sacchariz apolyschides* (previously known as *Gigantea bulbosa*) and *Undaria pinnatifida* as a daily dietary supplement. The results indicated an increase in high-density lipoprotein cholesterol (HDL-C), a decline in serum triglyceride levels, and ultimately balancing of blood glucose levels [221]. Another study reported the effects of a commercially available seaweed extract (InSea2) on 11 men and 12 women (age between: 12-59, average body mass index (BMI): 24.9 kg/m^2) to determine postprandial plasma insulin and glucose levels [222]. The InSea2 contains naturally available α-glucosidase along with α-amylase inhibitors and extracts of

Ascophyllum nodosum as well as *Fucus vesiculosus*. These seaweed extracts have been previously reported for their *in vitro* anti-diabetic potential by inhibiting the above-mentioned intestinal digestive enzymes, which break down polysaccharides into simple sugars and thus reduce the levels of glucose in the blood [218]. Capsules with InSea2 powder (500 mg) were administered to 12 individuals before giving white bread (50 g) as a meal for 30 min, whereas the control group was administered with placebo capsules. After 3 h of meal consumption, glucose levels in plasma were decreased by 9% in the InSea2 administered group (1081 mmol L^{-1} min^{-1}) compared to control group (1188 mmol L^{-1} min^{-1}). Even though the difference between the treated and control groups was not significant statistically ($p>0.05$), a considerable amount of reduction in plasma insulin (around 12%) was observed in InSea2 treated group. Additionally, muscular glucose uptake and peripheral insulin sensitivity (Cederholm index) have risen by around 8%. Due to the improvements observed in insulin and glucose profiles in the seaweed extract administered group indicated that they can be employed to maintain insulin homeostasis in type-II diabetes individuals.

Another study done using the similar InSea2 seaweed extract reported a significant increase in postprandial cognitive performance in test individuals (27 men and 33 women) who normally experience severe postprandial drowsiness [223]. Factors such as choice reaction time, digit vigilance, simple reaction time and immediate word recall were recorded after and before a high-carbohydrate meal to study the error rates, performance and attention during the cognitive tests. InSea2 extract (500 mg) was given to individuals before 30 min of high carbohydrate meal containing pure maple syrup and waffles, later, the individuals were asked to repeat the cognitive tests. The seaweed extract group showed higher accuracy and a decrease in error rates by 15% in comparison to the control group. The results indicated the modulation of glucose metabolism by high polyphenol and fibre content (equivalent to 20% chlorogenic acid) in InSea2 group. Similar polyphenols found in terrestrial plants (such as grapes and cocoa) increase cerebral blood flow.

One study reported the postprandial levels of blood glucose in 12 healthy individuals (4 women and 8 men, average age: 25) after a meal without and with the addition of freshly cooked sporophylls of *U. pinnatifida* (mekabu) or entire *U. pinnatifida* (wakame) [224]. An average breakfast prepared in Japan with potatoes, broccoli, boiled soya beans and white rice (200 g) was supplemented with either 70 g of fresh mekabu or wakame. In the breakfast of the control contains 87.5 g carbohydrate and 455 kcal of energy, whereas the breakfast with mekabu contains 90 g carbohydrate and 463 kcal and wakame contains 91.5 g of carbohydrate and 466 kcal. The objective of the study was to quantify the decrease in blood sugar levels due to the presence of seaweeds because white rice is a high glycaemic index food. After 30 minutes of taking the meals, glucose levels were reduced considerably ($p<0.05$) in the mekabu group compared to the control group. The study concluded that the polysaccharide-rich content and fucoxanthin fraction of the mekabu caused the antiglycaemic effect. When cooked, the smaller shoots of mekabu, which were grown from the parent wakame seaweed thallus, were more viscous. This is because of the high soluble viscous fibre content of mekabu, which delays gastric emptying and affects carbohydrate metabolism [224]. In conclusion, the supplementation of freshly cooked brown seaweed in meals can be helpful in regulating the blood glucose levels in type-II diabetes individuals.

The seaweeds anti-diabetic potential was due to the occurrence of bio-active compounds such as polysaccharides, polyphenolics, fucoxanthin and phlorotannins [225, 226], which inhibits the activity of digestive enzymes such as aldose reductase, lipase, α-glucosidase and α-amylase as well as inhibition of hepatic gluconeogenesis [227]. Seaweed extracts might ease the health epidemic caused due to type-II diabetes, which could lead to premature death, peripheral neuropathy with loss of limb extremities, blindness and nephropathy [228].

Cardiovascular Diseases

Cardiovascular diseases (CVDs) deal with the disorders of heart and its blood vessels such as pulmonary embolism, deep vein thrombosis, congenital heart disease, rheumatic heart disease, peripheral arterial disease, cerebrovascular disease and coronary heart disease [229]. Among these, stroke and coronary heart disease are responsible for the majority of deaths. As per the World Health Organization (WHO), stroke and coronary heart disease will cause the majority of heart-related deaths in the coming years, and a population of around 23.6 million will die due to CVDs in 2030 [230].

In CVDs, dyslipidemia is considered a potential risk factor, which is generally characterized by low levels of apolipoprotein B (apo B) plasmatic and HDL-C as well as high levels of LDL-C in combination with increased fasting, postprandial total triglycerides (TG) levels along with free fatty acids [231]. Most importantly, HDL-C imparts cardio-protective effects by the cholesterol uptake from the peripheral tissues and liver arterial wall, thus resulting in the prevention of LDL oxidation. On the other hand, hypertriglyceridemia causes increased production of small dense LDL, resulting in the delay of their metabolism and consequentially atherogenicity [232] as well as in combination with damage to endothelial tissue are considered as major events in the most regular pathological mechanisms seen in CVDs [233, 234].

In an *in vivo* study, hypercholesterolemic Wistar rat's feed was supplemented with *Gigartina pistillata* (23%) or *Himanthalia elongate* (21%) [235]. The results indicated an increase in the levels of HDL-C (20%) and a decrease in the TG plasmatic levels (28%) in rats treated with *H. elongate*, whereas rats treated with *G. pistillata* showed a considerable decrease in LDL-C (16%), total cholesterol (TC) (18%) and TG (31%) [235]. Similar results were reported in studies carried out by using tropical green seaweed (5% *Derbesia tenuissinma*-dried) in diet supplements. The *D. tenuissinma* did not affect the total body fat mass, however, it reduced the TC (17%) and TG (38%) plasmatic levels [236]. A study reported the potential *in vitro* inhibition activity of ACE inhibitors by the presence of peptides extracted from red seaweed (*Gracilariopsis lemaneiformis*). In general, peptides can bound with the active site of the angiotensin-I converting enzyme, thus resulting in lower blood pressure [237]. Bio-active compounds such as sulfated polysaccharides extracted from green seaweed (*Ulva prolifera*) decrease hyperlipidaemia in *in vivo* animal studies [238, 239].

Obesity

According to Organisation for Economic Co-operation and Development, in 2015, about 38% of adults in the United States were observed to be suffering from obesity, contrastingly, Japan and Korea have only 4 and 5%, respectively [240]. Obesity in India is increasing more rapidly than the world average. More than 135 million people are suffering from obesity in India. Factors such as socio-economic status, geographical environment, gender and age were among the deciding factors for obesity. In the last two decades, obesity prevailed due to high-calorie food consumption and a sedentary lifestyle [241]. Obesity, in combination with vitamin and mineral deficiency, has become a health epidemic in several nations around the world [242]. Consumption of dietary foods rich in fibres such as seaweeds and seaweed extracts (such as alginate and carotenoids) has been associated with positive impacts on cholesterol, blood glucose, satiety and appetite [221, 243].

The occurrence of fucoxanthin primarily in brown seaweeds and several microalgae showed the anti-obesity potential. One study reported the fucoxanthin activity on 50 men and women (age ranging from 20 to 59) with BMI of 26-30 kg/m^{2}as well as a waist size of ≥85 cm for men and ≥90 cm for women [244, 245]. The results showed that the individuals were healthy at the end of the study apart from BMI more than recommended (18-28 kg/m^2). Fucoxanthin of concentrations 1 mg and 3 mg per day were given to two different groups, and placebos were administered to the control group for 4 weeks. In the 1 mg group, right thigh and waist circumference, subcutaneous fat area and total fat mass were considerably reduced. Whereas, in 3 mg group, a considerable reduction in circumferences of the thigh, arm and neck, abdominal fat, basal metabolic rate, visceral fat and BMI was observed in comparison to the control group. Additionally, no harmful health effects such as a change in blood analysis, pulse rate, blood pressure and urinalysis parameters were observed during the study duration.

Another study reported the administration of fucoxanthin capsules to 115 non-diabetic, premenopausal, obese women with a liver fat composition >11% for over a period of 16 weeks [246]. One capsule with dried, brown seaweed extract (300 mg) with fucoxanthin (2.4 mg) in combination with pomegranate seed oil (300 mg) was given daily. Whereas, for the control group, placebo capsules containing olive oil were given. Considerable weight reduction with an average weight of 4.9 kg and a significant increase in resting energy expenditure was reported in the treated group. Seaweed extracts, in combination with positive modifications in the lifestyle such as dietary changes and exercise, can be employed as a strategy in treating obesity.

Fucoxanthin, which is a major carotenoid observed in brown seaweeds, exhibited anti-obesity potential, particularly in obese individuals with non-alcoholic fatty liver disease and increased levels of chronic inflammation [246]. The fucoxanthin aids in the decrease of body weight through (1) improving resistance to insulin [247], (2) reduction of plasma insulin and blood glucose [248], (3) production of heat in white adipose tissue (WAT20) [247, 249], (4) fatty acids oxidation, (5) causing increased resting energy expenditure, by uncoupling step during cell metabolism [250] and (6) up-regulation of mitochondrial uncoupling protein 1 (UCP1) gene expression in WAT [249]. The seaweeds with omega-3 fatty acids show excellent TG reduction, and homeostasis effects [251] and aid in normalizing inflammatory factors such as C-reactive protein (CRP), which is positively associated with central adiposity [252].

Few *Sargassum sp.* (such as *S. singgildianum, S. thunbergii* and *S. hemiphyllum*) inhibit prolyl endopeptidase enzyme [253], whereas *Spatoglossum schroederi* and *Caulerpa racemosa* inhibit α-amylase [254]. Enzymes such as Phospholipase A2, 5-lipoxygenase and lipase were inhibited by *Caulerpa* sp [255]. Other seaweed species, such as *Eiseniabicyclis* and *Ecklonia cava,* inhibits human muscle aldose reductase enzyme [256].

Contraceptive Activity Applications

Various research groups have studied the effects of red seaweed extracts for contraceptive purposes [257, 258]. Seaweed extracts were given to the female rats with pregnancy in the first seven days. Based on the results, it was observed that the extracts showed post-coital contraceptive protection additionally, no relative adverse health effects were seen in the rats.

CONCLUDING REMARKS

In conclusion, seaweeds appear to be a ray of hope for humans as the diet and lifestyle related diseases are increasing day by day. Consuming seaweeds as a part of diet could make lives healthier as they contain more insoluble dietary fibre, indigestible polysaccharides and huge amounts of proteins, vitamins and minerals. Hence there is a need to concentrate on seaweeds as a functional food and invest more research interest in it to discover many more useful seaweeds to improve human health and life span.

CONSENT FOR PUBLICATION

Not applicable.

CONFLICT OF INTEREST

The author(s) declare no conflict of interest, financial or otherwise.

ACKNOWLEDGEMENT

Declared none.

REFERENCES

[1] Raghuveer C. Review Article Consumption of Functional Food and Our Health Concerns 2009; 5(1): 76-83.

[2] Hasler CM. Functional foods: benefits, concerns and challenges-a position paper from the american council on science and health. J Nutr 2002; 132(12): 3772-81. [http://dx.doi.org/10.1093/jn/132.12.3772] [PMID: 12468622]

[3] Siró I, Kápolna E, Kápolna B, Lugasi A. Functional food. Product development, marketing and consumer acceptance—A review. Appetite 2008; 51(3): 456-67. [http://dx.doi.org/10.1016/j.appet.2008.05.060] [PMID: 18582508]

[4] Doyon M, Labrecque J. Functional foods: a conceptual definition. Br Food J 2008; 110(11): 1133-49. [http://dx.doi.org/10.1108/00070700810918036]

[5] Mendis E, Kim SK. Present and future prospects of seaweeds in developing functional foods [Internet]. 1st ed. Vol. 64, Advances in Food and Nutrition Research. Elsevier Inc.; 2011; 1-15. [http://dx.doi.org/10.1016/B978-0-12-387669-0.00001-6]

[6] Lordan S, Ross RP, Stanton C. Marine bioactives as functional food ingredients: potential to reduce the incidence of chronic diseases. Mar Drugs 2011; 9(6): 1056-100. [http://dx.doi.org/10.3390/md9061056] [PMID: 21747748]

[7] Plaza M, Cifuentes A, Ibáñez E. In the search of new functional food ingredients from algae. Trends Food Sci Technol 2008; 19(1): 31-9. [http://dx.doi.org/10.1016/j.tifs.2007.07.012]

[8] Bharti SK, Neeraj Kumar S, Krishna M, Awanish K. Funct-ional Aspe-cts of Dai-ry Fo-ods in Hum-an Hea-lth: An Over-view. Int J Phar-macol Ther [Inter-net] 2012; 2(1): 22-31.

https://www.yumpu.com/en/document/read/32870060/functional-aspects--f-dairy-foods-in-human-earthjournalsorg

[9] Clarke SF, Foster JR. A history of blood glucose meters and their role in self-monitoring of diabetes mellitus. Br J Biomed Sci 2012; 69(2): 83-93. [http://dx.doi.org/10.1080/09674845.2012.12002443] [PMID: 22872934]

[10] Guthrie R, Susi A. a simple phenylalanine method for detecting phenylketonuria in large populations of newborn infants. pediatrics 1963; 32(3): 338-43. [http://dx.doi.org/10.1542/peds.32.3.338] [PMID: 14063511]

[11] Bonanno G, Orlando-Bonaca M. Chemical elements in Mediterranean macroalgae. A review. Ecotoxicol Environ Saf 2018; 148(148): 44-71. [http://dx.doi.org/10.1016/j.ecoenv.2017.10.013] [PMID: 29031118]

[12] Tanna B, Mishra A. metabolites unravel nutraceutical potential of edible seaweeds: an emerging source of functional food. Compr Rev Food Sci Food Saf 2018; 17(6): 1613-24. [http://dx.doi.org/10.1111/1541-4337.12396] [PMID: 33350143]

[13] Zerrifi S, El Khalloufi F, Oudra B, Vasconcelos V. Seaweed bioactive compounds against pathogens and microalgae: Potential uses on pharmacology and harmful algae bloom control. Mar Drugs 2018; 16(2): 55. [http://dx.doi.org/10.3390/md16020055] [PMID: 29425153]

[14] Davis GDJ, Vasanthi AHR. Seaweed metabolite database (SWMD): A database of natural compounds from marine algae. Bioinformation 2011; 5(8): 361-4. [http://dx.doi.org/10.6026/97320630005361] [PMID: 21423723]

[15] WHO. Double burden of malnutrition [Internet]. World Health Organization. 2014. Available from:. 2014. https://www.who.int/nutrition/double-burden-malnutrition/en/

[16] Nanri A, Mizoue T, Shimazu T, *et al.* Dietary patterns and all-cause, cancer, and cardiovascular disease mortality in Japanese men and women: The Japan public health center-based prospective study. PLoS One 2017; 12(4): e0174848. [http://dx.doi.org/10.1371/journal.pone.0174848] [PMID: 28445513]

[17] Iso H. Lifestyle and cardiovascular disease in Japan. J Atheroscler Thromb 2011; 18(2): 83-8. [http://dx.doi.org/10.5551/jat.6866] [PMID: 21307610]

[18] Shannon E, Abu-Ghannam N. Seaweeds as nutraceuticals for health and nutrition. Phycologia 2019; 58(5): 563-77. [Internet]. [http://dx.doi.org/10.1080/00318884.2019.1640533]

[19] Davis MC, Fiehn O, Durnford DG. Metabolic acclimation to excess light intensity in *Chlamydomonas reinhardtii*. Plant Cell Environ 2013; 36(7): 1391-405. [http://dx.doi.org/10.1111/pce.12071] [PMID: 23346954]

[20] Dunn WB, Ellis DI. Metabolomics: Current analytical platforms and methodologies. Trends Analyt Chem 2005; 24(4): 285-94. [http://dx.doi.org/10.1016/j.trac.2004.11.021]

[21] Ho L, Tang T, Hoefel D, Vigneswaran B. Determination of rate constants and half-lives for the simultaneous biodegradation of several cyanobacterial metabolites in Australian source waters. Water Res 2012; 46(17): 5735-46. [http://dx.doi.org/10.1016/j.watres.2012.08.003] [PMID: 22921397]

[22] Lee YJ, Perdian DC, Song Z, Yeung ES, Nikolau BJ. Use of mass spectrometry for imaging metabolites in plants. Plant J 2012; 70(1): 81-95. [http://dx.doi.org/10.1111/j.1365-313X.2012.04899.x] [PMID: 22449044]

[23] Gupta V, Thakur RS, Baghel RS, Reddy CRK, Jha B. Seaweed metabolomics: A new facet of functional genomics [Internet]. Vol. 71, Advances in Botanical Research. Elsevier; 2014. 2014; 31-52. [http://dx.doi.org/10.1016/B978-0-12-408062-1.00002-0]

[24] Bart HJ, Pilz S. Industrial Scale Natural Products Extraction.Wiley-VCH Verlag GmbH & Co. KGaA 2011.
[http://dx.doi.org/10.1002/9783527635122]

[25] Mushtaq MY, Choi YH, Verpoorte R, Wilson EG. Extraction for metabolomics: access to the metabolome. Phytochem Anal 2014; 25(4): 291-306.
[http://dx.doi.org/10.1002/pca.2505] [PMID: 24523261]

[26] Kim HK, Choi YH, Verpoorte R. NMR-based metabolomic analysis of plants. Nat Protoc 2010; 5(3): 536-49.
[http://dx.doi.org/10.1038/nprot.2009.237] [PMID: 20203669]

[27] Schripsema J. Application of NMR in plant metabolomics: techniques, problems and prospects. Phytochem Anal 2010; 21(1): 14-21.
[http://dx.doi.org/10.1002/pca.1185] [PMID: 19904731]

[28] Kruger NJ, Troncoso-Ponce MA, Ratcliffe RG. 1H NMR metabolite fingerprinting and metabolomic analysis of perchloric acid extracts from plant tissues. Nat Protoc 2008; 3(6): 1001-12.
[http://dx.doi.org/10.1038/nprot.2008.64] [PMID: 18536647]

[29] Tanna B, Mishra A. metabolomics of seaweeds: tools and techniques [internet]. plant metabolites and regulation under environmental stress. elsevier inc.; 2018; 37-52.
[http://dx.doi.org/10.1016/B978-0-12-812689-9.00002-9]

[30] Zhang A, Sun H, Wang P, Han Y, Wang X. Modern analytical techniques in metabolomics analysis. Analyst (Lond) 2012; 137(2): 293-300.
[http://dx.doi.org/10.1039/C1AN15605E] [PMID: 22102985]

[31] Goulitquer S, Potin P, Tonon T. Mass spectrometry-based metabolomics to elucidate functions in marine organisms and ecosystems. Mar Drugs 2012; 10(4): 849-80.
[http://dx.doi.org/10.3390/md10040849]

[32] Hall RD, De Vos CHR, Verhoeven HA, Bino RJ. Metabolomics for the assessment of functional diversity and quality traits in plants non-targeted biochemical analyses. In: Vaidyanathan S, Harrigan GG, Goodacre R (Ed.) Metabolome Analyses: Strategies for Systems Biology. Springer, Boston, MA 2005; 31-44.

[33] Yuliana ND, Khatib A, Verpoorte R, Choi YH. Comprehensive extraction method integrated with NMR metabolomics: a new bioactivity screening method for plants, adenosine A1 receptor binding compounds in Orthosiphon stamineus Benth. Anal Chem 2011; 83(17): 6902-6.
[http://dx.doi.org/10.1021/ac201458n] [PMID: 21800886]

[34] Smith CA, Want EJ, O'Maille G, Abagyan R, Siuzdak G. XCMS: processing mass spectrometry data for metabolite profiling using nonlinear peak alignment, matching, and identification. Anal Chem 2006; 78(3): 779-87.
[http://dx.doi.org/10.1021/ac051437y] [PMID: 16448051]

[35] Alsufyani T, Weiss A, Wichard T. Time course exo-metabolomic profiling in the green marine macroalga Ulva (Chlorophyta) for identification of growth phase-dependent biomarkers. Mar Drugs 2017; 15(1): 14.
[http://dx.doi.org/10.3390/md15010014] [PMID: 28075408]

[36] Stuart KA, Welsh K, Walker MC, Edrada-Ebel R. Metabolomic tools used in marine natural product drug discovery. Expert Opin Drug Discov 2020; 15(4): 499-522.
[http://dx.doi.org/10.1080/17460441.2020.1722636] [PMID: 32026730]

[37] Hamid SS, Wakayama M, Soga T, Tomita M. Drying and extraction effects on three edible brown seaweeds for metabolomics. J Appl Phycol 2018; 30(6): 3335-50.
[http://dx.doi.org/10.1007/s10811-018-1614-z]

[38] Obata T, Fernie AR. The use of metabolomics to dissect plant responses to abiotic stresses. Cell Mol Life Sci 2012; 69(19): 3225-43.

[http://dx.doi.org/10.1007/s00018-012-1091-5] [PMID: 22885821]

[39] Chester T. Recent developments in high-performance liquid chromatography stationary phases. Anal Chem. 2013; 85(2): 579-89. https://pubmed.ncbi.nlm.nih.gov/23121288/

[40] Holčapek M, Jirásko R, Lísa M. Recent developments in liquid chromatography–mass spectrometry and related techniques. J Chromatogr A 2012; 1259: 3-15. [http://dx.doi.org/10.1016/j.chroma.2012.08.072] [PMID: 22959775]

[41] Schiffmann C, Hansen R, Baumann S, *et al.* Comparison of targeted peptide quantification assays for reductive dehalogenases by selective reaction monitoring (SRM) and precursor reaction monitoring (PRM). Anal Bioanal Chem 2014; 406(1): 283-91. [http://dx.doi.org/10.1007/s00216-013-7451-7] [PMID: 24220761]

[42] Wilson ID, Brinkman UAT. Hyphenation and hypernation. J Chromatogr A 2003; 1000(1-2): 325-56. [http://dx.doi.org/10.1016/S0021-9673(03)00504-1] [PMID: 12877178]

[43] Trufelli H, Palma P, Famiglini G, Cappiello A. An overview of matrix effects in liquid chromatography-mass spectrometry. Mass Spectrom Rev 2011; 30(3): 491-509. https://apps.who.int/iris/bitstream/handle/10665/254610WHO-MSD-MER-2017.2-eng.pdf?seq-uence=1%0Ahttp://apps.who.int/iris/bitstream/handle/10665/254610/WHO-MSD-MER-2017.2-eng.pdf;jsessi-onid=F9A3B620D10ECFAF8-498AC0E08EFCBC8?seq-uence=1%0A http://apps.who.int/ [http://dx.doi.org/10.1002/mas.20298] [PMID: 21500246]

[44] Robert L L. A Daniel J, Yair Shachar H. Towards the plant metabolome and beyond. Nat Rev | Mol CELL Biol [Internet]. 2007; 8(1): 167-74. http://dx.doi.org/10.1038/s41598-017-139-23-x%0A http://dx.doi.org/10.1038/s41598-017-09690-4

[45] Gaubert J, Greff S, Thomas OP, Payri CE. Metabolomic variability of four macroalgal species of the genus Lobophora using diverse approaches. Phytochemistry 2019; 162(February): 165-72. [http://dx.doi.org/10.1016/j.phytochem.2019.03.002] [PMID: 30925377]

[46] Desbrosses GG, Kopka J, Udvardi MK. Lotus japonicus metabolic profiling. Development of gas chromatography-mass spectrometry resources for the study of plant-microbe interactions. Plant Physiol 2005; 137(4): 1302-18. [http://dx.doi.org/10.1104/pp.104.054957] [PMID: 15749991]

[47] Kopka J, Schauer N, Krueger S, *et al.* GMD@CSB.DB: The Golm metabolome database. Bioinformatics 2005; 21(8): 1635-8. [http://dx.doi.org/10.1093/bioinformatics/bti236] [PMID: 15613389]

[48] Orata F. [48] Orata F. Derivatization Reactions and Reagents for Gas Chromatography Analysis, Advanced Gas Chromatography - Progress in Agricultural, Biomedical and Industrial Applications. InTech [Internet]. 2012; 11(1_suppl): 1-27. http://journals.sagepub.com/doi/10.1038/jcbfm.1991.34

[49] Kopka J. Current challenges and developments in GC–MS based metabolite profiling technology. J Biotechnol 2006; 124(1): 312-22. [http://dx.doi.org/10.1016/j.jbiotec.2005.12.012] [PMID: 16434119]

[50] Villas-Bôas SG, Mas S, Åkesson M, Smedsgaard J, Nielsen J. Mass spectrometry in metabolome analysis. Mass Spectrom Rev 2005; 24(5): 613-46. [http://dx.doi.org/10.1002/mas.20032] [PMID: 15389842]

[51] Lei Z, Huhman DV, Sumner LW. Mass spectrometry strategies in metabolomics. J Biol Chem 2011; 286(29): 25435-42. [http://dx.doi.org/10.1074/jbc.R111.238691] [PMID: 21632543]

[52] Nilsson SL, Bylund D, Jörntén-Karlsson M, Petersson P, Markides KE. A chemometric study of active parameters and their interaction effects in a nebulized sheath-liquid electrospray interface for capillary electrophoresis-mass spectrometry. Electrophoresis 2004; 25(13): 2100-7. [http://dx.doi.org/10.1002/elps.200305937] [PMID: 15237411]

[53] Ramautar R, Somsen GW, de Jong GJ. CE-MS in metabolomics. Electrophoresis 2009; 30(1): 276-91.

[http://dx.doi.org/10.1002/elps.200800512] [PMID: 19107702]

[54] Ohta D, Kanaya S, Suzuki H. Application of Fourier-transform ion cyclotron resonance mass spectrometry to metabolic profiling and metabolite identification. Curr Opin Biotechnol 2010; 21(1): 35-44.
[http://dx.doi.org/10.1016/j.copbio.2010.01.012] [PMID: 20171870]

[55] Hu Q, Noll RJ, Li H, Makarov A, Hardman M, Graham Cooks R. The Orbitrap: a new mass spectrometer. J Mass Spectrom 2005; 40(4): 430-43.
[http://dx.doi.org/10.1002/jms.856] [PMID: 15838939]

[56] Sumner LW, Mendes P, Dixon RA. Plant metabolomics: large-scale phytochemistry in the functional genomics era. Phytochemistry 2003; 62(6): 817-36.
[http://dx.doi.org/10.1016/S0031-9422(02)00708-2] [PMID: 12590110]

[57] Barrow MP, Burkitt WI, Derrick PJ. Principles of Fourier transform ion cyclotron resonance mass spectrometry and its application in structural biology. Analyst (Lond) 2005; 130(1): 18-28.
[http://dx.doi.org/10.1039/b403880k] [PMID: 15614347]

[58] Tiago FJ, Joao AR, Camila C, *et al.* Mass spectrometry-based plant metabolomics: Metabolite responses to abiotic stress. Mass Spectrom Rev 2016; 35(5): 620-49.

[59] Wani TA, Zargar S. New ultra-performance liquid chromatography-tandem mass spectrometry method for the determination of irbesartan in human plasma. J Food Drug Anal 2015; 23(3): 569-76.
[http://dx.doi.org/10.1016/j.jfda.2015.02.008] [PMID: 28911717]

[60] Gaubert J, Payri CE, Vieira C, Solanki H, Thomas OP. High metabolic variation for seaweeds in response to environmental changes: a case study of the brown algae Lobophora in coral reefs. Sci Rep 2019; 9(1): 993.
[http://dx.doi.org/10.1038/s41598-018-38177-z] [PMID: 30700781]

[61] Emwas AH, Roy R, McKay RT, *et al.* Nmr spectroscopy for metabolomics research. Metabolites 2019; 9(7): 123.
[http://dx.doi.org/10.3390/metabo9070123] [PMID: 31252628]

[62] Wu J, An Y, Yao J, Wang Y, Tang H. An optimised sample preparation method for NMR-based faecal metabonomic analysis. Analyst (Lond) 2010; 135(5): 1023-30.
[http://dx.doi.org/10.1039/b927543f] [PMID: 20419252]

[63] Fan TWM, Lane AN. NMR-based stable isotope resolved metabolomics in systems biochemistry. J Biomol NMR 2011; 49(3-4): 267-80.
[http://dx.doi.org/10.1007/s10858-011-9484-6] [PMID: 21350847]

[64] Booth D, Provan J, Maggs CA. Molecular approaches to the study of invasive seaweeds.In: Johnson CR (Ed.) Seaweed Invasions: A Synthesis of Ecological, Economic and Legal Imperatives. Berlin, New York: De Gruyter, 2008; 65-76.
[http://dx.doi.org/10.1515/9783110211344.2.65]

[65] Hamid SS, Wakayama M, Ichihara K, *et al.* Metabolome profiling of various seaweed species discriminates between brown, red, and green algae. Planta 2019; 249(6): 1921-47.
[http://dx.doi.org/10.1007/s00425-019-03134-1] [PMID: 30891648]

[66] Gomez-Zavaglia A, Prieto Lage MA, Jimenez-Lopez C, Mejuto JC, Simal-Gandara J. The potential of seaweeds as a source of functional ingredients of prebiotic and antioxidant value. Antioxidants 2019; 8(9): 406.
[http://dx.doi.org/10.3390/antiox8090406] [PMID: 31533320]

[67] Denis C, Morançais M, Li M, *et al.* Study of the chemical composition of edible red macroalgae Grateloupia turuturu from Brittany (France). Food Chem 2010; 119(3): 913-7. [Internet].
[http://dx.doi.org/10.1016/j.foodchem.2009.07.047]

[68] Peña-Rodríguez A, Mawhinney TP, Ricque-Marie D, Cruz-Suárez LE. Chemical composition of cultivated seaweed Ulva clathrata (Roth) C. Agardh. Food Chem 2011; 129(2): 491-8.

[http://dx.doi.org/10.1016/j.foodchem.2011.04.104] [PMID: 30634256]

[69] Yaich H, Garna H, Besbes S, Paquot M, Blecker C, Attia H. Chemical composition and functional properties of *Ulva lactuca* seaweed collected in Tunisia. Food Chem 2011; 128(4): 895-901. [Internet]. [http://dx.doi.org/10.1016/j.foodchem.2011.03.114]

[70] Ortiz J, Romero N, Robert P, *et al.* Dietary fiber, amino acid, fatty acid and tocopherol contents of the edible seaweeds *Ulva lactuca* and Durvillaea antarctica. Food Chem 2006; 99(1): 98-104. [http://dx.doi.org/10.1016/j.foodchem.2005.07.027]

[71] Gómez-Ordóñez E, Jiménez-Escrig A, Rupérez P. Dietary fibre and physicochemical properties of several edible seaweeds from the northwestern Spanish coast. Food Res Int 2010; 43(9): 2289-94. [Internet]. [http://dx.doi.org/10.1016/j.foodres.2010.08.005]

[72] Dawczynski C, Schubert R, Jahreis G. Amino acids, fatty acids, and dietary fibre in edible seaweed products. Food Chem 2007; 103(3): 891-9. [http://dx.doi.org/10.1016/j.foodchem.2006.09.041] [PMID: 26065750]

[73] Pereira L. Nutritional Composition of the Main Edible Algae.In: Pereira L. Therapeutic and Nutritional Uses of Algae. Boca Raton: CRC Press 2018; 98-160. [http://dx.doi.org/10.1201/9781315152844-2]

[74] Peñalver R, Lorenzo JM, Ros G, Amarowicz R, Pateiro M, Nieto G. Seaweeds as a functional ingredient for a healthy diet. Mar Drugs 2020; 18(6): 301. [http://dx.doi.org/10.3390/md18060301] [PMID: 32517092]

[75] Okuzumi J, Nishino H, Murakoshi M, *et al.* Inhibitory effects of fucoxanthin, a natural carotenoid, on N-myc expression and cell cycle progression in human malignant tumor cells. Cancer Lett 1990; 55(1): 75-81. [http://dx.doi.org/10.1016/0304-3835(90)90068-9] [PMID: 2245414]

[76] Khalid S, Abbas M, Saeed F, Bader-Ul-Ain H, Ansar Rasul Suleria H. Therapeutic Potential of Seaweed Bioactive Compounds.In: Maiti S (Ed.) Seaweed Biomaterials. London: IntechOpen; 2018. [http://dx.doi.org/10.5772/intechopen.74060]

[77] Martins RM, Nedel F, Guimarães VBS, *et al.* Macroalgae extracts from Antarctica have antimicrobial and anticancer potential. Front Microbiol 2018; 9(MAR): 412. [http://dx.doi.org/10.3389/fmicb.2018.00412] [PMID: 29568291]

[78] Rosa GP, Tavares WR, Sousa PMC, Pagès AK, Seca AML, Pinto DCGA. Seaweed secondary metabolites with beneficial health effects: An overview of successes in in vivo studies and clinical trials. Mar Drugs 2019; 18(1): 8. [http://dx.doi.org/10.3390/md18010008] [PMID: 31861879]

[79] Belghit I, Rasinger JD, Heesch S, *et al.* In-depth metabolic profiling of marine macroalgae confirms strong biochemical differences between brown, red and green algae. Algal Res 2017; 26(September): 240-9. [Internet]. [http://dx.doi.org/10.1016/j.algal.2017.08.001]

[80] Rajapakse N, Kim SK. Nutritional and digestive health benefits of seaweed [Internet]. 1st ed. Vol. 64, Advances in Food and Nutrition Research. Elsevier Inc. 2011; 17-28. [http://dx.doi.org/10.1016/B978-0-12-387669-0.00002-8]

[81] Usman A, Khalid S, Usman A, Hussain Z, Wang Y. Chapter 5 - Algal Polysaccharides, Novel Application, and Outlook. In: Zia KM, Zuber M, Ali Blends, and Composites MBT-ABP, editors. Elsevier 2017; 115-53. http://www.sciencedirect.com/science/article/pii/B9780128123607000057

[82] Fu X, Cao C, Ren B, Zhang B, Huang Q, Li C. Structural characterization and *in vitro* fermentation of a novel polysaccharide from Sargassum thunbergii and its impact on gut microbiota. Carbohydr Polym 2018; 183: 230-9. http://www.sciencedirect.com/science/article/pii/S0144861717314546 [http://dx.doi.org/10.1016/j.carbpol.2017.12.048] [PMID: 29352879]

[83] Charoensiddhi S, Conlon MA, Vuaran MS, Franco CMM, Zhang W. Impact of extraction processes on pre-biotic potential of the brown sea-weed Ecklonia radiata by *in vitro* human gut bacteria fermentation. J Funct Foods 2016; 24: 221-30. http://www.sciencedirect.com/ science/article/pii/ S1756464616300718 [Internet]. [http://dx.doi.org/10.1016/j.jff.2016.04.016]

[84] Charoensiddhi S, Conlon MA, Vuaran MS, Franco CMM, Zhang W. Polysaccharide and phlorotannin-enriched extracts of the brown seaweed Ecklonia radiata influence human gut microbiota and fermentation *in vitro*. J Appl Phycol 2017; 29(5): 2407-16. [Internet]. [http://dx.doi.org/10.1007/s10811-017-1146-y]

[85] García-Ríos V, Ríos-Leal E, Robledo D, Freile-Pelegrin Y. Polysaccharides composition from tropical brown seaweeds. Phycol Res 2012; 60(4): 305-15. [Internet]. [http://dx.doi.org/10.1111/j.1440-1835.2012.00661.x]

[86] Maeda-Yamamoto M. Development of functional agricultural products and use of a new health claim system in Japan. Trends Food Sci Technol 2017; 69: 324-32. http://www.sciencedirect.com/science/article/pii/S0924224416303806 [Internet]. [http://dx.doi.org/10.1016/j.tifs.2017.08.011]

[87] Terada A, Hara H, Mitsuoka T. Effect of Dietary Alginate on the Faecal Microbiota and Faecal Metabolic Activity in Humans. Microb Ecol Health Dis 1995; 8(6): 259-66. [Internet]. [http://dx.doi.org/10.3109/08910609509140105]

[88] Brownlee IA, Havler ME, Dettmar PW, Allen A, Pearson JP. Colonic mucus: secretion and turnover in relation to dietary fibre intake.Proc Nutr Soc 2003; 62(1): 245-9. https://www.cambridge.org/core/article/colonic-mucus-secreti-n-and-turnover-in-relation-to-dietary-fibre-intake/0C63794FD258E16811942D604A56C2C7

[89] Desai MS, Seekatz AM, Koropatkin NM, *et al.* A Dietary Fiber-Deprived Gut Microbiota Degrades the Colonic Mucus Barrier and Enhances Pathogen Susceptibility. Cell 2016; 167(5): 1339-1353.e21. [http://dx.doi.org/10.1016/j.cell.2016.10.043] [PMID: 27863247]

[90] Kong Q, Dong S, Gao J, Jiang C. *in vitro* fermentation of sulfated polysaccharides from E. prolifera and L. japonica by human fecal microbiota. Int J Biol Macromol 2016; 91: 867-71. http://www.sciencedirect.com/science/article/pii/S0141813016305748 [http://dx.doi.org/10.1016/j.ijbiomac.2016.06.036] [PMID: 27316763]

[91] Shang Q, Shan X, Cai C, Hao J, Li G, Yu G. Dietary fucoidan modulates the gut microbiota in mice by increasing the abundance of *Lactobacillus* and *Ruminococcaceae*. Food Funct 2016; 7(7): 3224-32. [http://dx.doi.org/10.1039/C6FO00309E] [PMID: 27334000]

[92] Lahaye M, Rochas C. Chemical structure and physico-chemical properties of agar. Hydrobiologia 1991; 221(1): 137-48. [Internet]. [http://dx.doi.org/10.1007/BF00028370]

[93] Hehemann JH, Correc G, Barbeyron T, Helbert W, Czjzek M, Michel G. Transfer of carbohydrate-active enzymes from marine bacteria to Japanese gut microbiota. Nature 2010; 464(7290): 908-12. [http://dx.doi.org/10.1038/nature08937] [PMID: 20376150]

[94] Weiner ML. Food additive carrageenan: Part II: A critical review of carrageenan *in vivo* safety studies. Crit Rev Toxicol 2014; 44(3): 244-69. [http://dx.doi.org/10.3109/10408444.2013.861798] [PMID: 24467586]

[95] Lahaye M, Michel C, Barry JL. Chemical, physicochemical and in-vitro fermentation characteristics of dietary fibres from Palmaria palmata (L.) Kuntze. Food Chem [Internet]. 1993; 47(1): 29-36. http://www.sciencedirect.com/science/article/pii/030881469390298T

[96] Mirande C, Mosoni P, Béra-Maillet C, Bernalier-Donadille A, Forano E. Characterization of Xyn10A, a highly active xylanase from the human gut bacterium Bacteroides xylanisolvens XB1A. Appl Microbiol Biotechnol 2010; 87(6): 2097-105.

[http://dx.doi.org/10.1007/s00253-010-2694-0] [PMID: 20532756]

[97] Lahaye M, Robic A. Structure and functional properties of ulvan, a polysaccharide from green seaweeds. Biomacromolecules 2007; 8(6): 1765-74. [http://dx.doi.org/10.1021/bm061185q] [PMID: 17458931]

[98] Ren X, Liu L, Gamallat Y, Zhang B, Xin Y. Enteromorpha and polysaccharides from enteromorpha ameliorate loperamide-induced constipation in mice. Biomed Pharmacother 2017; 96: 1075-81. http://www.sciencedirect.com/science/article/pii/S0753332217344050 [http://dx.doi.org/10.1016/j.biopha.2017.11.119] [PMID: 29198923]

[99] de Jesus Raposo M, de Morais A, de Morais R. Marine polysaccharides from algae with potential biomedical applications. Mar Drugs 2015; 13(5): 2967-3028. [http://dx.doi.org/10.3390/md13052967] [PMID: 25988519]

[100] Vera J, Castro J, Gonzalez A, Moenne A. Seaweed polysaccharides and derived oligosaccharides stimulate defense responses and protection against pathogens in plants. Mar Drugs 2011; 9(12): 2514-25. [http://dx.doi.org/10.3390/md9122514] [PMID: 22363237]

[101] Strain CR, Collins KC, Naughton V, *et al.* Effects of a polysaccharide-rich extract derived from Irish-sourced Laminaria digitata on the composition and metabolic activity of the human gut microbiota using an *in vitro* colonic model. Eur J Nutr 2020; 59(1): 309-25. [http://dx.doi.org/10.1007/s00394-019-01909-6] [PMID: 30805695]

[102] Bai S, Chen H, Zhu L, *et al.* Comparative study on the *in vitro* effects of Pseudomonas aeruginosa and seaweed alginates on human gut microbiota. PLoS One 2017; 12(2): e0171576. [http://dx.doi.org/10.1371/journal.pone.0171576] [PMID: 28170428]

[103] Chen L, Xu W, Chen D, *et al.* Digestibility of sulfated polysaccharide from the brown seaweed Ascophyllum nodosum and its effect on the human gut microbiota *in vitro*. Int J Biol Macromol 2018; 112: 1055-61. http://www.sciencedirect.com/science/article/pii/S0141813017348511 [http://dx.doi.org/10.1016/j.ijbiomac.2018.01.183] [PMID: 29425873]

[104] Rodrigues D, Walton G, Sousa S, *et al. in vitro* fermentation and prebiotic potential of selected extracts from seaweeds and mushrooms. Lebensm Wiss Technol 2016; 73: 131-9. http://www.sciencedirect.com/science/article/pii/S002364381630336X [Internet]. [http://dx.doi.org/10.1016/j.lwt.2016.06.004]

[105] Bajury DM, Rawi MH, Sazali IH, Abdullah A, Sarbini SR. Prebiotic evaluation of red seaweed (*Kappaphycus alvarezii*) using *in vitro* colon model. Int J Food Sci Nutr 2017; 68(7): 821-8. [http://dx.doi.org/10.1080/09637486.2017.1309522] [PMID: 28393631]

[106] Di T, Chen G, Sun Y, Ou S, Zeng X, Ye H. *in vitro* digestion by saliva, simulated gastric and small intestinal juices and fermentation by human fecal microbiota of sulfated polysaccharides from Gracilaria rubra. J Funct Foods 2018; 40: 18-27. http://www.sciencedirect.com/science/article/pii/S1756464617306515 [Internet]. [http://dx.doi.org/10.1016/j.jff.2017.10.040]

[107] Seong H, Bae JH, Seo JS, Kim SA, Kim TJ, Han NS. Comparative analysis of prebiotic effects of seaweed polysaccharides laminaran, porphyran, and ulvan using *in vitro* human fecal fermentation. J Funct Foods 2019; 57: 408-16. http://www.sciencedirect.com/science/article/pii/S1756464619301914 [Internet]. [http://dx.doi.org/10.1016/j.jff.2019.04.014]

[108] Maehre HK, Malde MK, Eilertsen KE, Elvevoll EO. Characterization of protein, lipid and mineral contents in common Norwegian seaweeds and evaluation of their potential as food and feed. J Sci Food Agric 2014; 94(15): 3281-90. [http://dx.doi.org/10.1002/jsfa.6681] [PMID: 24700148]

[109] Mohamed S, Hashim SN, Rahman HA. Seaweeds: A sustainable functional food for complementary and alternative therapy. Trends Food Sci Technol 2012; 23(2): 83-96.

[http://dx.doi.org/10.1016/j.tifs.2011.09.001]

[110] Holdt SL, Kraan S. Bioactive compounds in seaweed: functional food applications and legislation. J Appl Phycol 2011; 23(3): 543-97. [http://dx.doi.org/10.1007/s10811-010-9632-5]

[111] Dhargalkar VK, Verlecar XN. Southern Ocean seaweeds: A resource for exploration in food and drugs. Aquaculture 2009; 287(3-4): 229-42. [Internet]. [http://dx.doi.org/10.1016/j.aquaculture.2008.11.013]

[112] Probst Y. A review of the nutrient composition of selected *Rubus* berries. Nutr Food Sci 2015; 45(2): 242-54. [http://dx.doi.org/10.1108/NFS-07-2014-0063]

[113] Conde E, Balboa EM, Parada M, Falqué E. Algal proteins, peptides and amino acids.Functional Ingredients from Algae for Foods and Nutraceuticals. Woodhead Publishing Series in Food Science, Technology and Nutrition 2013; 135-80. [http://dx.doi.org/10.1533/9780857098689.1.135]

[114] Quitral R V, Morales G C, Sepúlveda L M, Schwartz M M. Propiedades nutritivas y saludables de algas marinas y su potencialidad como ingrediente funcional. Rev Chil Nutr 2012; 39(4): 196-202. [http://dx.doi.org/10.4067/S0717-75182012000400014]

[115] El Gamal AA. Biological importance of marine algae. Saudi Pharm J 2010; 18(1): 1-25. [http://dx.doi.org/10.1016/j.jsps.2009.12.001] [PMID: 23960716]

[116] Lopez-Huertas E. Health effects of oleic acid and long chain omega-3 fatty acids (EPA and DHA) enriched milks. A review of intervention studies. Pharmacol Res 2010; 61(3): 200-7. [http://dx.doi.org/10.1016/j.phrs.2009.10.007] [PMID: 19897038]

[117] Fleurence J, Gutbier G, Mabeau S, Leray C. Fatty acids from 11 marine macroalgae of the French Brittany coast. J Appl Phycol 1994; 6(5-6): 527-32. [http://dx.doi.org/10.1007/BF02182406]

[118] Kumari P, Kumar M, Gupta V, Reddy CRK, Jha B. Tropical marine macroalgae as potential sources of nutritionally important PUFAs. Food Chem 2010; 120(3): 749-57. [Internet]. [http://dx.doi.org/10.1016/j.foodchem.2009.11.006]

[119] Sánchez-Machado DI, López-Cervantes J, López-Hernández J, Paseiro-Losada P. Fatty acids, total lipid, protein and ash contents of processed edible seaweeds. Food Chem 2004; 85(3): 439-44. [http://dx.doi.org/10.1016/j.foodchem.2003.08.001]

[120] Mabeau S, Fleurence J. Seaweed in food products: biochemical and nutritional aspects. Trends Food Sci Technol 1993; 4(4): 103-7. [http://dx.doi.org/10.1016/0924-2244(93)90091-N]

[121] Elleuch M, Bedigian D, Roiseux O, Besbes S, Blecker C, Attia H. Dietary fibre and fibre-rich by-products of food processing: Characterisation, technological functionality and commercial applications: A review. Food Chem 2011; 124(2): 411-21. [Internet]. [http://dx.doi.org/10.1016/j.foodchem.2010.06.077]

[122] Jiménez-Escrig A, Goñi Cambrodón I. Nutritional evaluation and physiological effects of edible seaweeds. Arch Latinoam Nutr 1999; 49(2): 114-20. [PMID: 10488389]

[123] Rasmussen RS, Morrissey MT. Marine biotechnology for production of food ingredients. Adv Food Nutr Res 2007; 52(06): 237-92. [http://dx.doi.org/10.1016/S1043-4526(06)52005-4] [PMID: 17425947]

[124] Wells ML, Potin P, Craigie JS, *et al.* Algae as nutritional and functional food sources: revisiting our understanding. J Appl Phycol 2017; 29(2): 949-82. [http://dx.doi.org/10.1007/s10811-016-0974-5] [PMID: 28458464]

[125] McDermid KJ, Stuercke B. Nutritional composition of edible Hawaiian seaweeds. J Appl Phycol 2003; 15(6): 513-24.
[http://dx.doi.org/10.1023/B:JAPH.0000004345.31686.7f]

[126] Norziah MH, Ching CY. Nutritional composition of edible seaweed *Gracilaria changgi*. Food Chem 2000; 68(1): 69-76.
[http://dx.doi.org/10.1016/S0308-8146(99)00161-2]

[127] Ferraces-Casais P, Lage-Yusty MA, Rodríguez-Bernaldo de Quirós A, López-Hernández J. Evaluation of Bioactive Compounds in Fresh Edible Seaweeds. Food Anal Methods 2012; 5(4): 828-34.
[http://dx.doi.org/10.1007/s12161-011-9321-2]

[128] Hernández-Carmona G, Carrillo-Domínguez S, Arvizu-Higuera DL, *et al.* Monthly variation in the chemical composition of *Eisenia arborea* J.E. Areschoug. J Appl Phycol 2009; 21(5): 607-16.
[http://dx.doi.org/10.1007/s10811-009-9454-5]

[129] Haddad EH, Berk LS, Kettering JD, Hubbard RW, Peters WR. Dietary intake and biochemical, hematologic, and immune status of vegans compared with nonvegetarians. Am J Clin Nutr 1999; 70(3) (Suppl.): 586s-93s.
[http://dx.doi.org/10.1093/ajcn/70.3.586s] [PMID: 10479236]

[130] Waldmann A, Koschizke JW, Leitzmann C, Hahn A. Homocysteine and cobalamin status in *German vegans*. Public Health Nutr 2004; 7(3): 467-72.
[http://dx.doi.org/10.1079/PHN2003540] [PMID: 15153278]

[131] Allen LH. Causes of vitamin B 12 and folate deficiency Causes of vitamin B 12 deficiency Overview of vitamin B 12 absorption and requirements. United Nations Univ [Internet]. 2008; 29(2): 20-34. http://journals.sagepub.com/doi/pdf/10.1177/15648265080292S105

[132] Lorenzo J, Agregán R, Munekata P, *et al.* Proximate composition and nutritional value of three macroalgae: Ascophyllum nodosum, Fucus vesiculosus and Bifurcaria bifurcata. Mar Drugs 2017; 15(11): 360.
[http://dx.doi.org/10.3390/md15110360] [PMID: 29140261]

[133] Rupérez P, Saura-Calixto F. Dietary fibre and physicochemical properties of edible Spanish seaweeds. Eur Food Res Technol 2001; 212(3): 349-54.
[http://dx.doi.org/10.1007/s002170000264]

[134] Wijesinghe WAJP, Jeon YJ. Enzyme-assistant extraction (EAE) of bioactive components: A useful approach for recovery of industrially important metabolites from seaweeds: A review. Fitoterapia 2012; 83(1): 6-12.
[http://dx.doi.org/10.1016/j.fitote.2011.10.016] [PMID: 22061659]

[135] Parys S, Rosenbaum A, Kehraus S, Reher G, Glombitza KW, König GM. Evaluation of quantitative methods for the determination of polyphenols in algal extracts. J Nat Prod 2007; 70(12): 1865-70.
[http://dx.doi.org/10.1021/np070302f] [PMID: 18052031]

[136] Li Y-X, Wijesekara I, Li Y, Kim S-K. Phlorotannins as bioactive agents from brown algae. Process Biochem 2011; 46(12): 2219-24. [Internet].
[http://dx.doi.org/10.1016/j.procbio.2011.09.015]

[137] Ford L, Stratakos AC, Theodoridou K, *et al.* Polyphenols from Brown Seaweeds as a Potential Antimicrobial Agent in Animal Feeds. ACS Omega 2020; 5(16): 9093-103.
[http://dx.doi.org/10.1021/acsomega.9b03687] [PMID: 32363261]

[138] Kumar CS, Ganesan P, Suresh PV, Bhaskar N. Seaweeds as a source of nutritionally beneficial compounds - A review. J Food Sci Technol 2008; 45(1): 1-13.

[139] Wang T, Jónsdóttir R, Kristinsson HG, *et al.* Enzyme-enhanced extraction of antioxidant ingredients from red algae Palmaria palmata. Lebensm Wiss Technol 2010; 43(9): 1387-93. [Internet].
[http://dx.doi.org/10.1016/j.lwt.2010.05.010]

[140] Souza BWS, Cerqueira MA, Martins JT, *et al.* Antioxidant potential of two red seaweeds from the Brazilian coasts. J Agric Food Chem 2011; 59(10): 5589-94. [http://dx.doi.org/10.1021/jf200999n] [PMID: 21491929]

[141] Cho M, Lee HS, Kang IJ, Won MH, You S. Antioxidant properties of extract and fractions from Enteromorpha prolifera, a type of green seaweed. Food Chem 2011; 127(3): 999-1006. [http://dx.doi.org/10.1016/j.foodchem.2011.01.072] [PMID: 25214089]

[142] Peter I C, Matthew D W, David H, Jeffrey P P. The role of seaweed bioactives in the control of digestion: Implications for obesity treatments Food Funct 2015; 6(11): 3420-7.

[143] Pangestuti R, Kim SK. Biological activities and health benefit effects of natural pigments derived from marine algae. J Funct Foods 2011; 3(4): 255-66. [Internet]. [http://dx.doi.org/10.1016/j.jff.2011.07.001]

[144] Bohn T. Norma española AENOR 6885_2016.pdf. Curr Nutr Food Sci 2008; 4(4): 240-58. [http://dx.doi.org/10.2174/157340108786263685]

[145] Sasaki K, Ishihara K, Oyamada C, *et al.* effects of fucoxanthin addition to ground chicken breast meat on lipid and colour stability during chilled storage, before and after cooking. asian-australas J anim sci 2008; 21(7): 1067-72. [http://dx.doi.org/10.5713/ajas.2008.70670]

[146] Heo SJ, Yoon WJ, Kim KN, *et al.* Evaluation of anti-inflammatory effect of fucoxanthin isolated from brown algae in lipopolysaccharide-stimulated RAW 264.7 macrophages. Food Chem Toxicol 2010; 48(8-9): 2045-51. [http://dx.doi.org/10.1016/j.fct.2010.05.003] [PMID: 20457205]

[147] Das SK, Ren R, Hashimoto T, Kanazawa K. Fucoxanthin induces apoptosis in osteoclast-like cells differentiated from RAW264.7 cells. J Agric Food Chem 2010; 58(10): 6090-5. [http://dx.doi.org/10.1021/jf100303k] [PMID: 20420432]

[148] Shimoda H, Tanaka J, Shan SJ, Maoka T. Anti-pigmentary activity of fucoxanthin and its influence on skin mRNA expression of melanogenic molecules. J Pharm Pharmacol 2010; 62(9): 1137-45. [http://dx.doi.org/10.1111/j.2042-7158.2010.01139.x] [PMID: 20796192]

[149] Bleakley S, Hayes M. algal proteins: extraction, application, and challenges concerning production. foods 2017; 6(5): 33. [http://dx.doi.org/10.3390/foods6050033] [PMID: 28445408]

[150] Vasconcelos MA, Arruda FVS, Carneiro VA, *et al.* Effect of algae and plant lectins on planktonic growth and biofilm formation in clinically relevant bacteria and yeasts. Biomed Res Int 2014; 2014; 365272. [http://dx.doi.org/10.1155/2014/365272]

[151] Frikha F, Kammoun M, Hammami N, *et al.* Composición química y algunas actividades biológicas de algas marinas recolectadas en túnez. Cienc Mar 2011; 37(2): 113-24. [http://dx.doi.org/10.7773/cm.v37i2.1712]

[152] Samarakoon K, Jeon YJ. Bio-functionalities of proteins derived from marine algae — A review. Food Res Int 2012; 48(2): 948-60. [Internet]. [http://dx.doi.org/10.1016/j.foodres.2012.03.013]

[153] Roca I, Akova M, Baquero F, *et al.* The global threat of antimicrobial resistance: science for intervention. New Microbes New Infect 2015; 6(6): 22-9. [http://dx.doi.org/10.1016/j.nmni.2015.02.007] [PMID: 26029375]

[154] Vonthron-Sénécheau C, Fleurence J, Levine IA. Medicinal Properties: Antibiotic, Tonic, and Antiparasitic PropertiesIn: Fleurence J, Levine I (Eds.) Seaweed in health and disease prevention. Academic Press 2016; 369-88. https://search.ebscohost.com/ login.aspx?direct=true&scope=site& db=nlebk&db=nlabk& AN=1198061

[155] Carlucci MJ, Scolaro LA, Damonte EB. Inhibitory action of natural carrageenans on Herpes simplex virus infection of mouse astrocytes. Chemotherapy 1999; 45(6): 429-36. [http://dx.doi.org/10.1159/000007236] [PMID: 10567773]

[156] Liu L, Heinrich M, Myers S, Dworjanyn SA. Towards a better understanding of medicinal uses of the brown seaweed *Sargassum* in Traditional Chinese Medicine: A phytochemical and pharmacological review. J Ethnopharmacol 2012; 142(3): 591-619. [http://dx.doi.org/10.1016/j.jep.2012.05.046] [PMID: 22683660]

[157] Shiney E, Reginald M, Wilsy JI. antibacterial activity and phytochemical screening of marine macro algae amphiroa anceps using three solvent extracts. int J pharmacogn 2014; 1(9): 605-8.

[158] Alghazeer R, Whida F, Abduelrhman E, Gammoudi F, Azwai S. Screening of antibacterial activity in marine green and brown macroalgae from the coast of Morocco. Nat Sci 2013; 5(1): 7-14.

[159] Basir A, Desniar , Ristyanti WK, Tarman K. Physical treatments to induce the antibacterial and antioxidant activities of green algae Halimeda sp. from Seribu Islands, North Jakarta, Indonesia. IOP Conf Ser Earth Environ Sci 2020; 414(1): 012002. [http://dx.doi.org/10.1088/1755-1315/414/1/012002]

[160] Rozema J, Björn LO, Bornman JF, *et al.* The role of UV-B radiation in aquatic and terrestrial ecosystems—an experimental and functional analysis of the evolution of UV-absorbing compounds. J Photochem Photobiol B 2002; 66(1): 2-12. [http://dx.doi.org/10.1016/S1011-1344(01)00269-X] [PMID: 11849977]

[161] Matsukawa R, Dubinsky Z, Kishimoto E, *et al.* A comparison of screening methods for antioxidant activity in seaweeds. J Appl Phycol 1997; 9(1): 29-35. [http://dx.doi.org/10.1023/A:1007935218120]

[162] Boonchum W, Peerapornpisal Y, Kanjanapothi D, *et al.* Antioxidant activity of some seaweed from the Gulf of Thailand. Int J Agric Biol 2011; 13(1): 95-9.

[163] Jayalakshmi J, Subramanian V, Anantharaman P. Evaluation of biochemical composition and *in vitro* antioxidant properties of selected seaweeds from Kanyakumari coast, Tamil Nadu, India. Adv Appl Sci Res 2014; 5(4): 74-81.

[164] LinYuan J. Effects of Grateloupia filicina polysaccharide on immune-related gene expression in Litopenaeus vannamei. Nanfang Nongye Xuebao 2014; 45(4): 682-6.

[165] Hermund DB, Plaza M, Turner C, *et al.* Structure dependent antioxidant capacity of phlorotannins from Icelandic Fucus vesiculosus by UHPLC-DAD-ECD-QTOFMS. Food Chem 2018; 240(240): 904-9. [http://dx.doi.org/10.1016/j.foodchem.2017.08.032] [PMID: 28946360]

[166] Wang T, Jónsdóttir R, Ólafsdóttir G. Total phenolic compounds, radical scavenging and metal chelation of extracts from Icelandic seaweeds. Food Chem 2009; 116(1): 240-8. [Internet]. [http://dx.doi.org/10.1016/j.foodchem.2009.02.041]

[167] Rodrigo R, Miranda A, Vergara L. Modulation of endogenous antioxidant system by wine polyphenols in human disease. Clin Chim Acta 2011; 412(5-6): 410-24. [http://dx.doi.org/10.1016/j.cca.2010.11.034] [PMID: 21130758]

[168] Keyrouz R, Abasq ML, Bourvellec CL, *et al.* Total phenolic contents, radical scavenging and cyclic voltammetry of seaweeds from Brittany. Food Chem 2011; 126(3): 831-6. [Internet]. [http://dx.doi.org/10.1016/j.foodchem.2010.10.061]

[169] Stagos D, Amoutzias GD, Matakos A, Spyrou A, Tsatsakis AM, Kouretas D. Chemoprevention of liver cancer by plant polyphenols. Food Chem Toxicol 2012; 50(6): 2155-70. [http://dx.doi.org/10.1016/j.fct.2012.04.002] [PMID: 22521445]

[170] Frikha F, Kammoun M, Hammami N, *et al.* Chemical composition and some biological activities of marine algae collected in Tunisia. Cienc Mar 2011; 37(2): 113-24. [Internet].

[http://dx.doi.org/10.7773/cm.v37i2.1712]

[171] Wijesekara I, Pangestuti R, Kim SK. Biological activities and potential health benefits of sulfated polysaccharides derived from marine algae. Carbohydr Polym 2011; 84(1): 14-21. [Internet]. [http://dx.doi.org/10.1016/j.carbpol.2010.10.062]

[172] Koutsaviti A, Ioannou E, Roussis V. Bioactive Seaweed Substances In: Qin Y (Ed) Bioactive Seaweeds for Food Applications Academic Press 2018; 25-52. http://www.sciencedirect.com/science/article/pii/B9780128133125000029 [http://dx.doi.org/10.1016/B978-0-12-813312-5.00002-9]

[173] Sun L, Wang C, Shi Q, Ma C. Preparation of different molecular weight polysaccharides from Porphyridium cruentum and their antioxidant activities. Int J Biol Macromol 2009; 45(1): 42-7. [http://dx.doi.org/10.1016/j.ijbiomac.2009.03.013] [PMID: 19447258]

[174] Wang J, Zhang Q, Zhang Z, Song H, Li P. Potential antioxidant and anticoagulant capacity of low molecular weight fucoidan fractions extracted from Laminaria japonica. Int J Biol Macromol 2010; 46(1): 6-12. [http://dx.doi.org/10.1016/j.ijbiomac.2009.10.015] [PMID: 19883681]

[175] Kim JA, Ahn BN, Kong CS, Kim SK. The chromene sargachromanol E inhibits ultraviolet A-induced ageing of skin in human dermal fibroblasts. Br J Dermatol 2013; 168(5): 968-76. [http://dx.doi.org/10.1111/bjd.12187] [PMID: 23278330]

[176] Jiao G, Yu G, Zhang J, Ewart H. Chemical structures and bioactivities of sulfated polysaccharides from marine algae. Mar Drugs 2011; 9(2): 196-223. [http://dx.doi.org/10.3390/md9020196] [PMID: 21566795]

[177] Patel S. Therapeutic importance of sulfated polysaccharides from seaweeds: updating the recent findings. 3 Biotech. 2012; 2(3): 171-85.

[178] Pati MP, Sharma SD, Nayak L, Panda CR. Uses of seaweed and its application to human welfare: A review. Int J Pharm Pharm Sci 2016; 8(10): 12-20. [http://dx.doi.org/10.22159/ijpps.2016v8i10.12740]

[179] Trinchero J, Ponce NMA, Córdoba OL, *et al.* Antiretroviral activity of fucoidans extracted from the brown seaweed *Adenocystis utricularis*. Phytother Res 2009; 23(5): 707-12. [http://dx.doi.org/10.1002/ptr.2723] [PMID: 19107862]

[180] Vlachos V, Critchley AT, von Holy A. Differential antibacterial activity of extracts from selected southern African macroalgal thalli. Bot Mar 1999; 42(2): 165-73. [http://dx.doi.org/10.1515/BOT.1999.019]

[181] Chattopadhyay K, Mateu CG, Mandal P, Pujol CA, Damonte EB, Ray B. Galactan sulfate of Grateloupia indica: Isolation, structural features and antiviral activity. Phytochemistry 2007; 68(10): 1428-35. [http://dx.doi.org/10.1016/j.phytochem.2007.02.008] [PMID: 17451760]

[182] Abu Bakar NA, Anyanji VU, Mustapha NM, Lim SL, Mohamed S. Seaweed (*Eucheuma cottonii*) reduced inflammation, mucin synthesis, eosinophil infiltration and MMP-9 expressions in asthma-induced rats compared to Loratadine. J Funct Foods 2015; 19: 710-22. http://www.sciencedirect.com/science/article/pii/S1756464615005083 [Internet]. [http://dx.doi.org/10.1016/j.jff.2015.10.011]

[183] Namvar F, Mohamed S, Fard SG, *et al.* Polyphenol-rich seaweed (*Eucheuma cottonii*) extract suppresses breast tumour via hormone modulation and apoptosis induction. Food Chem 2012; 130(2): 376-82. http://www.sciencedirect.com/science/article/pii/S0308814611010156 [http://dx.doi.org/10.1016/j.foodchem.2011.07.054] [PMID: 23122073]

[184] Colliec S, Boisson-vidal C, Jozefonvicz J. A low molecular weight fucoidan fraction from the brown seaweed *Pelvetia canaliculata*. Phytochemistry 1994; 35(3): 697-700. [http://dx.doi.org/10.1016/S0031-9422(00)90590-9]

[185] Chevolot L, Foucault A, Chaubet F, *et al.* Further data on the structure of brown seaweed fucans: relationships with anticoagulant activity. Carbohydr Res 1999; 319(1-4): 154-65. [http://dx.doi.org/10.1016/S0008-6215(99)00127-5] [PMID: 10520264]

[186] Melo FR, Pereira MS, Foguel D, Mourão PAS. Antithrombin-mediated anticoagulant activity of sulfated polysaccharides: different mechanisms for heparin and sulfated galactans. J Biol Chem 2004; 279(20): 20824-35. [http://dx.doi.org/10.1074/jbc.M308688200] [PMID: 14996843]

[187] Silva FRF, Dore CMPG, Marques CT, *et al.* Anticoagulant activity, paw edema and pleurisy induced carrageenan: Action of major types of commercial carrageenans. Carbohydr Polym 2010; 79(1): 26-33. [Internet]. [http://dx.doi.org/10.1016/j.carbpol.2009.07.010]

[188] Bilal M, Iqbal HMN. Marine seaweed polysaccharides-based engineered cues for the modern biomedical sector. Mar Drugs 2019; 18(1): 7. [http://dx.doi.org/10.3390/md18010007] [PMID: 31861644]

[189] Gupta S, Abu-Ghannam N. Bioactive potential and possible health effects of edible brown seaweeds. Trends Food Sci Technol 2011; 22(6): 315-26. [Internet]. [http://dx.doi.org/10.1016/j.tifs.2011.03.011]

[190] Sugiura Y, Matsuda K, Yamada Y, *et al.* Anti-allergic phlorotannins from the edible brown alga, *Eisenia Arborea*. Food Sci Technol Res 2007; 13(1): 54-60. [http://dx.doi.org/10.3136/fstr.13.54]

[191] Esmaeili A, Khakpoor M. Biological activities and chemical composition of solvent extracts of *Stoechospermum marginatum* (*C. Agardh*). Acta Biochim Pol 2012; 59(4): 581-5. [http://dx.doi.org/10.18388/abp.2012_2095] [PMID: 23162803]

[192] Raman M, Doble M. κ-Carrageenan from marine red algae, *Kappaphycus alvarezii* – A functional food to prevent colon carcinogenesis. J Funct Foods 2015; 15: 354-64. http://www.sciencedirect.com/science/article/pii/S1756464615001541 [Internet]. [http://dx.doi.org/10.1016/j.jff.2015.03.037]

[193] Yen Y-H, Farooqi AA, Li K-T, *et al.* Methanolic extracts of Solieria robusta inhibits proliferation of oral cancer Ca9-22 cells *via* apoptosis and oxidative stress. Molecules 2014; 19(11): 18721-32. http://europepmc.org/abstract/MED/25405289

[194] Ruqqia K, Sultana V, Ara J, Ehteshamul-Haque S, Athar M. Hypolipidaemic potential of seaweeds in normal, triton-induced and high-fat diet-induced hyperlipidaemic rats. J Appl Phycol 2015; 27(1): 571-9. [http://dx.doi.org/10.1007/s10811-014-0321-7]

[195] Sun X, Jin H, Zhang L, Hu W, Li Y, Xu N. Screening and isolation of the algicidal compounds from marine green alga *Ulva intestinalis*. Chin J Oceanology Limnol 2016; 34(4): 781-8. [http://dx.doi.org/10.1007/s00343-016-4383-z]

[196] Bhadury P, Wright P. Exploitation of marine algae: biogenic compounds for potential antifouling applications. Planta 2004; 219(4): 561-78. [http://dx.doi.org/10.1007/s00425-004-1307-5] [PMID: 15221382]

[197] Shin T, Ahn M, Hyun JW, Kim SH, Moon C. Antioxidant marine algae phlorotannins and radioprotection: A review of experimental evidence. Acta Histochem 2014; 116(5): 669-74. [http://dx.doi.org/10.1016/j.acthis.2014.03.008] [PMID: 24751171]

[198] Kwak JY. Fucoidan as a marine anticancer agent in preclinical development. Mar Drugs 2014; 12(2): 851-70. [http://dx.doi.org/10.3390/md12020851] [PMID: 24477286]

[199] Van Weelden G, Bobiński M, Okła K, Van Weelden WJ, Romano A, Pijnenborg JMA. Fucoidan structure and activity in relation to anti-cancer mechanisms. Mar Drugs 2019; 17(1): 32.

[http://dx.doi.org/10.3390/md17010032] [PMID: 30621045]

[200] P A, K A, L S, M M, K M. Anticancer effect of fucoidan on cell proliferation, cell cycle progression, genetic damage and apoptotic cell death in HepG2 cancer cells. Toxicol Rep 2019; 6(6): 556-63. [Internet].
[http://dx.doi.org/10.1016/j.toxrep.2019.06.005]

[201] Abudabbus A, Badmus JA, Shalaweh S, Bauer R, Hiss D. effects of fucoidan and chemotherapeutic agent combinations on malignant and non-malignant breast cell lines. curr pharm biotechnol 2017; 18(9): 748-57.
[PMID: 29141543]

[202] Vaikundamoorthy R, Krishnamoorthy V, Vilwanathan R, Rajendran R. Structural characterization and anticancer activity (MCF7 and MDA-MB-231) of polysaccharides fractionated from brown seaweed Sargassum wightii. Int J Biol Macromol 2018; 111: 1229-37.
[http://dx.doi.org/10.1016/j.ijbiomac.2018.01.125] [PMID: 29415413]

[203] Zubia M, Fabre MS, Kerjean V, *et al.* Antioxidant and antitumoural activities of some Phaeophyta from Brittany coasts. Food Chem 2009; 116(3): 693-701. [Internet].
[http://dx.doi.org/10.1016/j.foodchem.2009.03.025]

[204] Barbosa M, Valentão P, Andrade P. Bioactive compounds from macroalgae in the new millennium: implications for neurodegenerative diseases. Mar Drugs 2014; 12(9): 4934-72.
[http://dx.doi.org/10.3390/md12094934] [PMID: 25257784]

[205] Silva J, Alves C, Freitas R, *et al.* Antioxidant and neuroprotective potential of the brown seaweed *Bifurcaria bifurcata* in an *in vitro* Parkinson's disease model. Mar Drugs 2019; 17(2): 85.
[http://dx.doi.org/10.3390/md17020085] [PMID: 30717087]

[206] Liu J, Banskota A, Critchley A, Hafting J, Prithiviraj B. Neuroprotective effects of the cultivated *Chondrus crispus* in a *C. elegans* model of Parkinson's disease. Mar Drugs 2015; 13(4): 2250-66.
[http://dx.doi.org/10.3390/md13042250] [PMID: 25874922]

[207] Sevevirathne M, Lee KH, Ahn CB, Park PJ, Je JY. Evaluation of antioxidant, anti-alzheimer's and anti-inflammatory activities of enzymatic hydrolysates from edible brown seaweed *Laminaria japonica*). J Food Biochem 2012; 36(2): 207-16.
[http://dx.doi.org/10.1111/j.1745-4514.2010.00527.x]

[208] Leyton A, Pezoa-Conte R, Barriga A, *et al.* Identification and efficient extraction method of phlorotannins from the brown seaweed *Macrocystis pyrifera* using an orthogonal experimental design. Algal Res 2016; 16: 201-8. [Internet].
[http://dx.doi.org/10.1016/j.algal.2016.03.019]

[209] Cha SH, Heo SJ, Jeon YJ, Park SM. Dieckol, an edible seaweed polyphenol, retards rotenone-induced neurotoxicity and α-synuclein aggregation in human dopaminergic neuronal cells. RSC Advances 2016; 6(111): 110040-6. [Internet].
[http://dx.doi.org/10.1039/C6RA21697H]

[210] Ikeda K, Kitamura A, Machida H, *et al.* Effect of *Undaria pinnatifida* (Wakame) on the development of cerebrovascular diseases in stroke-prone spontaneously hypertensive rats. Clin Exp Pharmacol Physiol 2003; 30(1-2): 44-8.
[http://dx.doi.org/10.1046/j.1440-1681.2003.03786.x] [PMID: 12542452]

[211] Peng J, Yuan JP, Wu CF, Wang JH. Fucoxanthin, a marine carotenoid present in brown seaweeds and diatoms: metabolism and bioactivities relevant to human health. Mar Drugs 2011; 9(10): 1806-28.
[http://dx.doi.org/10.3390/md9101806] [PMID: 22072997]

[212] Zhou B, Lu Y, Hajifathalian K, *et al.* Worldwide trends in diabetes since 1980: a pooled analysis of 751 population-based studies with 4·4 million participants. Lancet 2016; 387(10027): 1513-30.
[http://dx.doi.org/10.1016/S0140-6736(16)00618-8] [PMID: 27061677]

[213] World Health Organization [Internet]. WHO. 2016. https://www.who.int/diabetes /country-profiles/

usa_en.pdf?ua=1

[214] World Health Organization [Internet]. WHO. 2016. https://www.who.int/diabetes/ country-profiles/ gbr_en.pdf?ua=1

[215] Centers for Disease Control and Prevention (CDC). [Internet]. Chronic disease prevention and health promotion publications 2016–2017. At a glance fact sheets - Diabetes. . https://www.cdc.gov/chronicdisease/ resources/publications/ aag/diabetes.htm

[216] Moore LE. Moore LE. Oral Hypoglycemic Agents. In: Diabetes in Pregnancy The complete guide to management. Springer Cham 2016; 103-10. https://link.springer.com/content/pdf/10.1007/978-3-3-9-65518-5.pdf%0Ahttp://link.springer.com/10.1007/978-3-319-65518-5

[217] Gabbia D, Dall'Acqua S, Di Gangi I, *et al.* The Phytocomplex from Fucus vesiculosus and ascophyllum nodosum controls postprandial plasma glucose levels: An *in vitro* and in Vivo study in a mouse model of NASH. Mar Drugs 2017; 15(2): 41.
[http://dx.doi.org/10.3390/md15020041] [PMID: 28212301]

[218] Roy MC, Anguenot R, Fillion C, Beaulieu M, Bérubé J, Richard D. Effect of a commercially-available algal phlorotannins extract on digestive enzymes and carbohydrate absorption in vivo. Food Res Int 2011; 44(9): 3026-9.
[http://dx.doi.org/10.1016/j.foodres.2011.07.023]

[219] Song W, Wang Z, Zhang X, Li Y. ethanol extract from *Ulva prolifera* prevents high-fat diet-induced insulin resistance, oxidative stress, and inflammation response in mice. biomed res int. 2018 2018.

[220] Zhao C, Yang C, Chen M, *et al.* regulatory efficacy of brown seaweed *lessonia nigrescens* extract on the gene expression profile and intestinal microflora in type 2 diabetic mice. Mol Nutr Food Res 2018; 62(4): 1700730.
[http://dx.doi.org/10.1002/mnfr.201700730] [PMID: 29210521]

[221] Kim MS, Kim JY, Choi WH, Lee SS. Effects of seaweed supplementation on blood glucose concentration, lipid profile, and antioxidant enzyme activities in patients with type 2 diabetes mellitus. Nutr Res Pract 2008; 2(2): 62-7.
[http://dx.doi.org/10.4162/nrp.2008.2.2.62] [PMID: 20126367]

[222] Paradis ME, Couture P, Lamarche B. A randomised crossover placebo-controlled trial investigating the effect of brown seaweed (*Ascophyllum nodosum* and *Fucus vesiculosus*) on postchallenge plasma glucose and insulin levels in men and women. Appl Physiol Nutr Metab 2011; 36(6): 913-9.
[http://dx.doi.org/10.1139/h11-115] [PMID: 22087795]

[223] Haskell-Ramsay C, Jackson P, Dodd F, *et al.* Acute post-prandial cognitive effects of Brown seaweed extract in humans. Nutrients 2018; 10(1): 85.
[http://dx.doi.org/10.3390/nu10010085] [PMID: 29342865]

[224] Tanemura Y, Yamanaka-Okumura H, Sakuma M, Nii Y, Taketani Y, Takeda E. Effects of the intake of *Undaria pinnatifida* (Wakame) and its *Undaria pinnatifida* (Mekabu) on postprandial glucose and insulin metabolism. J Med Invest 2014; 61(3.4): 291-7.
[http://dx.doi.org/10.2152/jmi.61.291] [PMID: 25264047]

[225] Kellogg J, Grace M, Lila M. Phlorotannins from Alaskan seaweed inhibit carbolytic enzyme activity. Mar Drugs 2014; 12(10): 5277-94.
[http://dx.doi.org/10.3390/md12105277] [PMID: 25341030]

[226] Murray M, Dordevic A, Ryan L, Bonham M. The impact of a single dose of a polyphenol-rich seaweed extract on postprandial glycaemic control in healthy adults: A randomised cross-over trial. Nutrients 2018; 10(3): 270.
[http://dx.doi.org/10.3390/nu10030270] [PMID: 29495492]

[227] Sharifuddin Y, Chin YX, Lim PE, Phang SM. Potential bioactive compounds from seaweed for diabetes management. Mar Drugs 2015; 13(8): 5447-91.
[http://dx.doi.org/10.3390/md13085447] [PMID: 26308010]

[228] Yamazaki D, Hitomi H, Nishiyama A. Hypertension with diabetes mellitus complications. Hypertens Res 2018; 41(3): 147-56.
[http://dx.doi.org/10.1038/s41440-017-0008-y] [PMID: 29353881]

[229] WHO. World Health Organisation [Internet]. Cardiovascular Diseases (CVDs). Fact Sheet No 317 2015. https://www.who.int/en/news-room/fact-sheets/detail/cardiovascular-diseases-(cvds)

[230] WHO. Worl Health Organization [Internet]. Cardiovascular Diseases. [cited 2020 Dec 4]. https://www.who.int/cardiovascular_diseases/about_cvd/en/

[231] Klop B, Elte J, Cabezas M. Dyslipidemia in obesity: mechanisms and potential targets. Nutrients 2013; 5(4): 1218-40.
[http://dx.doi.org/10.3390/nu5041218] [PMID: 23584084]

[232] Jellinger P. Smith DA, Mehta AE. American Association of Clinical Endocrinologists' Guidelines for Management of Dyslipidemia and Prevention of Atherosclerosis. Endocr Pract 2012; 18(Suppl 1): 1-78. http://aace.metapress.com/content/t7g5335740165v13/fulltext.pdf

[233] Halperin RO, Sesso HD, Ma J, Buring JE, Stampfer MJ, Michael Gaziano J. Dyslipidemia and the risk of incident hypertension in men. Hypertension 2006; 47(1): 45-50.
[http://dx.doi.org/10.1161/01.HYP.0000196306.42418.0e] [PMID: 16344375]

[234] Hadi HAR, Carr CS, Al Suwaidi J. Endothelial dysfunction: cardiovascular risk factors, therapy, and outcome. Vasc Health Risk Manag 2005; 1(3): 183-98. https://pubmed.ncbi.nlm.nih.gov/17319104 [Internet].
[PMID: 17319104]

[235] Villanueva MJ, Morcillo M, Tenorio MD, Mateos-Aparicio I, Andrés V, Redondo-Cuenca A. Health-promoting effects in the gut and influence on lipid metabolism of Himanthalia elongata and Gigartina pistillata in hypercholesterolaemic Wistar rats. Eur Food Res Technol 2014; 238(3): 409-16.
[http://dx.doi.org/10.1007/s00217-013-2116-5]

[236] Kumar S, Magnusson M, Ward L, Paul N, Brown L. Seaweed supplements normalise metabolic, cardiovascular and liver responses in high-carbohydrate, high-fat fed rats. Mar Drugs 2015; 13(2): 788-805.
[http://dx.doi.org/10.3390/md13020788] [PMID: 25648511]

[237] Cao D, Lv X, Xu X, Yu H, Sun X, Xu N. Purification and identification of a novel ACE inhibitory peptide from marine alga Gracilariopsis lemaneiformis protein hydrolysate. Eur Food Res Technol 2017; 243(10): 1829-37.
[http://dx.doi.org/10.1007/s00217-017-2886-2]

[238] Ren R, Gong J, Zhao Y, Zhuang X, Ye Y, Lin W. Sulfated polysaccharides from Enteromorpha prolifera suppress SREBP-2 and HMG-CoA reductase expression and attenuate non-alcoholic fatty liver disease induced by a high-fat diet. Food Funct 2017; 8(5): 1899-904.
[http://dx.doi.org/10.1039/C7FO00103G] [PMID: 28429814]

[239] Ren R, Gong J, Zhao Y, *et al.* Sulf-ated poly-sacchar-ide from Enter-omorpha proli-fera supp-resses SREBP-1c and ACC expre-ssion to lower ser-um trigly-cerides in high-fat diet-induced hyperli-pida--mic rats. J Fun-ct Fo-ods 2018; 40: 722-8. http://www.sciencedirect.com/science/article/ pii/ S17564-64617307351 [Internet].
[http://dx.doi.org/10.1016/j.jff.2017.12.010]

[240] Organisation for Economic Co-operation and Development (OECD). Organisation for Economic Co-operation and Development (OECD). [Internet]. Obesity Update 2017 2017.Available from:. www.oecd.org/health/obesity-update.htm

[241] Ahirwar R, Mondal PR. Prevalence of obesity in India: A systematic review. Diabetes Metab Syndr 2019; 13(1): 318-21.
[http://dx.doi.org/10.1016/j.dsx.2018.08.032] [PMID: 30641719]

[242] Medina-Remón A, Kirwan R, Lamuela-Raventós RM, Estruch R. Dietary patterns and the risk of

obesity, type 2 diabetes mellitus, cardiovascular diseases, asthma, and neurodegenerative diseases. Crit Rev Food Sci Nutr 2018; 58(2): 262-96.
[http://dx.doi.org/10.1080/10408398.2016.1158690] [PMID: 27127938]

[243] Brown EM, Allsopp PJ, Magee PJ, *et al.* Seaweed and human health. Nutr Rev 2014; 72(3): 205-16. [http://dx.doi.org/10.1111/nure.12091] [PMID: 24697280]

[244] Shannon E, Abu-Ghannam N. Optimisation of fucoxanthin extraction from Irish seaweeds by response surface methodology. J Appl Phycol 2017; 29(2): 1027-36. [Internet]. [http://dx.doi.org/10.1007/s10811-016-0983-4]

[245] Hitoe S, Shimoda H. seaweed fcoxanthin supplementation improves obesity parameters in mild obese japanese subjects. funct food health dis 2017; 7(4): 246. [http://dx.doi.org/10.31989/ffhd.v7i4.333]

[246] Abidov M, Ramazanov Z, Seifulla R, Grachev S. The effects of Xanthigen™ in the weight management of obese premenopausal women with non-alcoholic fatty liver disease and normal liver fat. Diabetes Obes Metab 2010; 12(1): 72-81. [http://dx.doi.org/10.1111/j.1463-1326.2009.01132.x] [PMID: 19840063]

[247] Miyashita K. The carotenoid fucoxanthin from brown seaweed affects obesity. Lipid Technol 2009; 21(8-9): 186-90. [http://dx.doi.org/10.1002/lite.200900040]

[248] Maeda H, Tsukui T, Sashima T, Hosokawa M, Miyashita K. Seaweed carotenoid, fucoxanthin, as a multi-functional nutrient. Asia Pac J Clin Nutr 2008; 17 (Suppl. 1): 196-9. [PMID: 18296336]

[249] Maeda H, Hosokawa M, Sashima T, Funayama K, Miyashita K. Fucoxanthin from edible seaweed, Undaria pinnatifida, shows antiobesity effect through UCP1 expression in white adipose tissues. Biochem Biophys Res Commun 2005; 332(2): 392-7. [http://dx.doi.org/10.1016/j.bbrc.2005.05.002] [PMID: 15896707]

[250] Miyashita K. Seaweed carotenoid, fucoxanthin, with highly bioactive and nutritional activities.J Mar Biosci biotech 2006; 1: 48-58.

[251] Skulas-Ray AC, West SG, Davidson MH, Kris-Etherton PM. Omega-3 fatty acid concentrates in the treatment of moderate hypertriglyceridemia. Expert Opin Pharmacother 2008; 9(7): 1237-48. [http://dx.doi.org/10.1517/14656566.9.7.1237] [PMID: 18422480]

[252] Schenck HV, Wallentin L, Lennmarken C, Larsson J. Lipoprotein metabolism following gastroplasty in obese women. Scand J Clin Lab Invest 1992; 52(4): 269-74. [http://dx.doi.org/10.3109/00365519209088358] [PMID: 1439513]

[253] Hyun-jin L, Jin-Hui K, Chu-Hyun L, *et al.* Inhibitory activites of seaweeds on prolyl endopeptidase, tyrosinase and coagulation. Korean J Pharmacogn 1999; 30(3): 231-7. https://ocean.kisti.re.kr/downfile/volume/kspmc/HKSOBF/1999/v30n3/HKSOBF_1999_v30n3_231.pdf [Internet].

[254] Teixeira VL, Rocha FD, Houghton PJ, Kaplan MAC, Pereira RC. α-Amylase inhibitors from Brazilian seaweeds and their hypoglycemic potential. Fitoterapia 2007; 78(1): 35-6. [http://dx.doi.org/10.1016/j.fitote.2006.09.017] [PMID: 17067759]

[255] Bitou N, Ninomiya M, Tsujita T, Okuda H. Screening of lipase inhibitors from marine algae. Lipids 1999; 34(5): 441-5. [http://dx.doi.org/10.1007/s11745-999-0383-7] [PMID: 10380115]

[256] Iwahori Y, Enomoto S, Okada Y, Tanaka J, Okuyama T. Naturally Occurring Substances for Prevention of Complications of Diabetes. IV: Screening of Seavegetables for Inhibitory Effect on Aldose Reductase. In Nat Med 1999; 53: 138-40.

[257] de Almeida CLF, Falcão H de S, Lima GR, *et al.* Bioactivities from marine algae of the genus Gracilaria. Int J Mol Sci 2011; 12(7): 4550-73.

[http://dx.doi.org/10.3390/ijms12074550] [PMID: 21845096]

[258] Ratnasooriya W, Premakumara GAS, Tillekeratne LMV. Post-coital contraceptive activity of crude extracts of Sri Lankan marine red algae. Contraception 1994; 50(3): 291-9. [http://dx.doi.org/10.1016/0010-7824(94)90074-4] [PMID: 7805379]

CHAPTER 14

Probiotics and Algal Functional Food

Mona M. Ismail[1,*]

[1] *National Institute of Oceanography and Fisheries, NIOF, Cairo, Egypt*

Abstract: The aim of this work is to explain the probiotic activity of algae and the role of their bioactive compounds as a dietary supplement for promoting growth and disease resistance. Micro- and macroalgae are good alternatives to chemical probiotics because they contain important and safe functional components such as polysaccharides, fatty acids, pigment and proteins, which are considered an extraordinary source of prebiotics with health benefits and enhance healthy colonic microbiota composition. Recently, algae and their extracts have been employed in ingestion systems, simulating the human colon, or animal models. Especially, macroalgae and their polysaccharides have been used for decades to improve human health depending on saccharolytic fermentation by the gut microbiota. Generally, the addition of algae leads to enhance the viability of probiotic bacteria, the acidity of food, and storage quality and, consequently, prevent certain human diseases. Therefore, further investigations should be done for screening and identifying putative prebiotic compounds from algae *via* the host organism's identification, which utilizes algal bioactive metabolites produced. There will be a need for increasing algae production to replace or supplement the intake of plant foods of terrestrial origin.

Keywords: Bioactive Compounds, Cyanobacteria, Dietary Supplement, Health Food, Seaweed.

INTRODUCTION

Probiotics are live harmless microorganisms that encourage health when consumed in adequate amounts, generally by restoring or improving the gut flora, which are frequently present in the digestive tract of biota [1]. Generally, they improve many of the host's physiological processes, including a reduction in mutagenic enzymes like nitroreductase, β-glucuronida and choloylglycine [2]. In addition, inhibiting the invasion of the pathogen in the gut encourages the proliferation of probiotic microbes, enhances the immune function, protects colon cancer [3], prevents diarrhea, irritable bowel syndrome, and *Helicobacter pylori* infection; lactose intolerance, and decreases the levels of cholesterol in blood [4].

* **Corresponding author Mona M. Ismail:** National Institute of Oceanography and Fisheries, NIOF, Egypt; Tel: +01225786197; Fax: +2027921341; E-mails: mona_es5@yahoo.com; mm.esmail@niof.sci.eg

Probiotics are now emerging as an important class of food additives, so we can consider them functional foods since they provide health benefits in addition to their traditional nutritional purpose [3]. Probiotic foods comprise 60% and 70% of the total functional food market [5]. Functional foods (probiotics and prebiotics) have been shown to be useful in the treatment or management of a variety of disorders [6]. Functional foods are made by combining a chemical or microbiological element with a dietary base, such as milk or milk products.

Both probiotics and prebiotics are of current interest because they offer alternatives for the management of different gastrointestinal disorders, multiple antibiotic resistance and overwhelming infections in hospitalized patients [7]. If prebiotics are combined with probiotic microorganisms, they will form symbiotic that activate the growth and the metabolism of one or a limited number of beneficial bacteria species [8]. The synbiotic action has a greater effect compared to the effect of the probiotic or prebiotic alone [9]. The consumption of probiotics, prebiotics, or synbiotics is favorable for the intestinal microbiota and human health [8].

The probiotic organisms are characterized by resistance to acid and bile, attachment to human epithelial cells, colonization of the human intestine, production of antimicrobial substances, good growth characteristics and beneficial effects on human health [6]. Global demand for macroalgal and microalgal foods is growing beyond the traditional considerations of nutrition and health [10].

Algae are photoautotrophic organisms, which have simple reproductive structures. These organisms constitute about 30,000 to over a million species with a great diversity of morphology and sizes, and that can exist from unicellular microscopic organisms (microalgae) to multicellular (macroalgae). They act as a very interesting natural source of bioactive compounds like proteins, minerals, vitamins, amino acids, lipids, phenolic, fatty acids, polysaccharides, nucleic acid, and carotenoids with biological ability that could be utilized as nutritional supplements (Fig. **1**) [11 - 14]. They have a high ability to live and survive under extreme conditions because they have a high adaptation to any environmental changes by producing different secondary substances that cannot be found in other organisms [15]. Moreover, they are characterized by their rapidly growing and easy cultivation, which leads to controlling the production of some bioactive compounds by manipulating the cultivation conditions. Micro- and macroalgae are rich sources of biocompounds, which have been reported to exhibit prebiotic properties [16].

Lately, microalgae have gained popularity as a novel source of nutraceutical and pharmaceutical products (Table **1**). Actually, there are some microalgae cultivated for foods and food additives [17]. Microalgae can increase the capability of probiotic bacteria, especially acid production [18]. They contain different biologically active components like unsaturated fatty acids, amino acids, Fe and Ca and many types of vitamins (A, B2, B6, B8, B12, E, and K) [19].

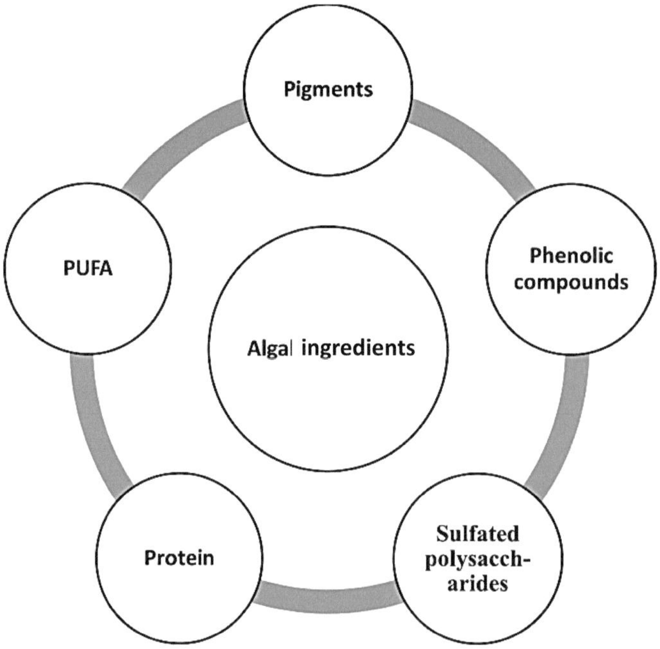

Fig. (1). Common algal prebiotic ingredients.

Co-addition of microalgae species stimulates growth and increases the viability and acid production of the probiotic bacteria [20]. They decrease blood lipid profile, blood sugar, body weight, and wound healing time [21]. Feeding microalgae to elderly people or animals have been demonstrated to protect against age-dependent diseases, particularly cardiac hypertension or hyperlipidemia [22].

The microalgae (*Chlorella vulgaris* and *Spirulina platiness*) are considered as functional food due to their diverse bioactive content [23]. The green alga "*Chlorella*" contains about 61.6% proteins, 12.5% fat, 13.7% carbohydrates and vitamins like ascorbic acid, thiamine, B1, B2, B6, D, E, and K. In addition to 38.94% polyunsaturated fatty acids (PUFA) and elements, for example, Zn, Ca, Mg, Pb, and AL [24]. *Chlorella* is a good source of nutrients such as valuable protein, calories, fat, and vitamins [25]. Numerous references on the health benefits of consuming *Chlorella* spp. have shown that their ingested extracts can lower blood sugar levels, and increase hemoglobin concentrations. Plus, it acts as hypocholesterolemic and hepatoprotective agent during malnutrition and reduces ethionine intoxication [26]. Intake of *Chlorella* was beneficial to the improvement of insulin sensitivity in normal Wistar rats [27]. *C. vulgaris* has also been proved to have immune-modulating and anticancer properties [22]. While, the cyanobacterium *Spirulina platensis* has been consumed as safe human food and has been approved for human nutrition in many countries [28]. It contains approximately 78% proteins, 13.6% carbohydrates, 4-7% lipids, pigments, minerals and vitamins (A, B2, B6, B12, E, H, and K), which can promote bacteria growth such as *Streptococcus thermophiles*, *Lactobacillus acidophilus*, and *L. casei* [29], while inhibiting the growth of pathogenic bacteria like *Bacillus subtilis*, *B. pumulis*, and *Proteus vulgaris* [30].

According to the WHO/FAO/UNU, *Chlorella* and *Spirulina* have well-balanced essential amino acid profiles which are similar to other conventional protein sources, such as eggs and soybean, so they are recommended for the nutrition of humans [19]. The addition of microalgae "*S. platensis* and *C. vulgaris*" into milk for the production of fermented milk to enhance the viability of probiotics has been the subject of recent research [23].

Particularly cyanobacteria species (blue-green algae) have received great attention since they are an excellent source of high-value compounds [31] and can stimulate the thermophilic dairy starter bacteria growth, so they are suitable for the cost-effective manufacture of novel functional fermented dairy foods [32]. From several years ago, *Spirulina* spp. has been used as a food source due to its nutritional value [33], while it contains 18 of the 20 known amino acids, more calcium than milk, more vitamin B12 than cow liver, vitamins A, B2, B 6, E, H, and K, and all essential minerals, as well as enzymes [34]. It is used as a supplement in the treatment of pernicious anemia by encouraging the growth of helpful bacteria species [35].

Many studies have tested the addition effect of *Spirulina* to yoghurt, cheese, and fermented milk and have demonstrated positive results, including an increase in the number of lactic acid bacteria, such as *L. acidophilus* and improvement of the nutritional quality of the fermented product during storage [23, 36, 37]. *Spirulina* nutritional properties could be improved through lactic acid fermentation, which could increase the digestibility of the biomass and the availability of nutritious substances, as well as provide probiotic bacteria [38].

Previously, Varga *et al.* [18] stated the addition of cyanobacterial biomass into fermented milk in order to increase the nutritional values act as a new trend. For example, *Spirulina* spp. can increase the viability of probiotic bacteria in fermented milk by promoting their proliferation. The effect of algal-free amino acids, adenine, peptone, and hypoxanthine from *S. platensis* on the viable counts of *L. bulgaricus* yoghurt samples because these nitrogenous chemicals stimulated the bacterial growth [35]. *Spirulina* has been shown to have a variety of health-promoting effects, including the alleviation of hyperlipidemia, the suppression of hypertension, the protection against renal failure, the growth promotion of intestinal *Lactobacillus*, and the suppression of elevated serum glucose levels [39]. The extract of *S. platensis* was the best algal source for prebiotics due to its stimulation effect on the growth of all three probiotic bacteria (*L. lactis, L. bulgaricus,* and *B. longum*) [40]. The aqueous extracts of different microalgae like *S. platensis*, *Chlorococcum*, *Dunaliella salina*, and *Chlorella* species are potential sources for prebiotics [33].

SEAWEEDS

Marine seaweeds are a traditional food in many countries [41] due to their diverse health-promoting bioactive compounds that have beneficial effects *via* lower incidence of cancers, preventing cardiovascular diseases, reduced blood pressure and blood sugar, antiviral, anti-inflammatory, immunomodulatory activities (Table **1**) [41 - 43]. They constitute approximately 25,000–30,000 different species, with a great diversity of forms and sizes [44]. They show rapid growth and do not require arable land, freshwater or fertilizers [45]. Moreover, they had different biological activities like cardiovascular diseases, antihypertensive, anti-obesity and anti-diabetic, anti-cancer or anti-oxidant properties [11, 43, 45, 46].

Sargassum vulgare, *Undaria pinnatifida*, *Himanthalia elongata*, *Chondrus crispus*, *Porphyra sp.*, *Cystoseira spp.* and *Ulva spp.* are the common macroalgal species recommended as natural sources of novel functional components [47]. The diet supplemented with red macroalgae *Chondrus crispus* improved the beneficial bacteria and the levels of the acetic, propionic and butyric SCFAs, increasing the colon histomorphology, immune system and immunoglobulins A and G [48].

Many studies demonstrated the prebiotic effect of red seaweed such as *Acanthopora*, *Gracilaria*, *Kappaphycus*, *Euchema*, *Grateloupia*, *Chondrus*, *Gelidium* and *Osmundea* [45, 49, 50].

The green macroalgae species " *Halimeda tuna*, *H. macroloba*, *Enteromorpha* sp., and *Acetabularia acetabulum*" have significant anti-oxidant activity for future uses in food and medicine products [51]. Kim *et al.* [52] tested the effects of supplementation of two brown algae (*U. pinnatifida and L. japonica*) on the gut microbiota and body status of laboratory rats. In both instances, the animals' body weight was reduced, which was thought to be mediated by the influence of the seaweed on the composition of the intestinal microbial communities associated with obesity, reducing the proportion of Firmicutes to Bacteroidetes, and the populations of pathogenic bacteria, such as *Clostridium*, *Escherichia* and *Enterobacter* [52]. Similarly, *L. japonica* increased beneficial bacteria and short-chain fatty acids (SCFA) while lowering pH [53]. Dierick *et al.* [54] demonstrated that dried whole brown seaweed *Ascophyllum nodosum* decreased potential pathogens (*E. coli*, *Streptococci*) and total anaerobes in batch systems inoculated with either porcine small intestinal or cecal suspensions.

Table 1. Examples of algal species as probiotic.

Algal spp.	Beneficial Effects	References
Chlorella spp., *Scenedesmus* spp., *Dunaliella salina*, *Chlamydocapsa* spp.	Anti-oxidant & anticancer activity, cardiovascular health, food colorant and prevent liver fibrosis	[100, 101]
Cyanobacteria spp.	Inhibition of the Lachnospiraceae, *Lactobacillus* , *Mollicutes* and *Mucispirillum* , during induction of the Prevotellaceae and Rickenellaceae growth.	[93]
Enteromorpha prolifera	Increasing the proportion of *Alistipes* , Lachnospiraceae and Odoribacter	[102]
Ulva rigida	Anti-oxidant activity, Growth performance, and immune stimulation.	[103]
Macrocystis pyrifera	Improving the growth performance.	[105]
Sargassum wightii; S. microcystum; *S. polycystum*	Improving the growth, immune response, and disease resistance.	[106]
Pyropia pheroplasts	Increasing the body weight, nutrient utilization, digestibility analysis of protein and lipid.	[107]
Gracilaria fisheri *Gracilaria lemaneiformis*	Enhancing the immune parameters, and disease resistance.	[108]

According to Ruperez [55], the brown alga *U. pinnatifida* and the red alga *C. crispus* could be utilized as a food supplement to help meet the recommended daily intake of some minerals, macro elements (Na, K, Ca, Mg, ranging from 8.1 to 17.9 mg/100 g), and trace elements (Fe, Zn, Mn, Cu ranging from 5.1 to 15.2 mg/100 g). *Porphyra* sp. and *C. crispus* from red algae, as well as *H. elongata* and *U. pinnatifida* from brown algae, are very interesting to consumers and the food industry due to their low-calorie content and high vitamins, minerals and dietetic fiber content [47]. Ismail *et al.* [13, 56] stated that the anti-oxidant potential of different Egyptian seaweeds reveals their potential for future applications in dietary supplementation and medicine products.

PREBIOTIC PROPERTIES OF THE ALGAL COMPOUNDS

A prebiotic is a selectively fermented ingredient that allows specific changes, both in the composition and/or activity of the gastrointestinal, resulting in positive effects on the host's wellbeing and health [57]. Moreover, it has satisfied many characters; firstly, it must be resistant to digestion in the upper gastrointestinal tract (GIT) and therefore resistant to acid and enzymatic hydrolysis; secondly, it must be a selective substrate for the growth of beneficial bacteria and therefore, result in a shift in the profile of the microflora and finally, it must induce luminal or systemic effects that are beneficial to host health [57]. Prebiotics derived from several algae species are attractive alternative source for promoting the growth of probiotic strains of lactic acid bacteria, *e.g.*, *Lactobacillus* and *Bifidobacterium* spp. (Table **2**) [8].

Table 2. Some algal compositions as a dietary supplement.

Bioactive Compounds	Beneficial Effects	References
PUFAs	Cardiac hypertension	[22]
Sulfated Polysaccharides *e.g.* Alginic acid, xylofucans, porphyrans, fucoidan	Anti-oxidant, antiviral, antimicrobial, Apoptotic activities. Regulate the bioactivity of growth factors and cytokines. Increasing in beneficial bacteria as *Lactobacillus* and *Bifidobacterium*. Decreasing the pH and increasing in lactic acid and SCFA, including acetic and butyic acids	[47, 53]
Soluble fiber	Decreasing total and LDL cholesterol.	[59]
Fucoxanthin	Preventive effect on cerebrovascular diseases. Encourage the metabolism.	[47]
Phenolic compounds	Decreasing total and LDL cholesterol. Anti-oxidant, antimicrobial, and anti-obesity activities	[44]
Carotenoids (a & β carotene; Astaxanthin)	Decreasing risk of premature death, anti-oxidant, and anticancer activity.	[100, 101]
Amino acid *e.g.* glutathione	Antioxdiant activity, detoxify the metals; Lowering the heart attack	[102]
Vitamins *e.g.* B, C, E	Antioxidant & antidiabetic properties, Anti-alzheimer's protection,	[103, 104]

Polysaccharides

Marine algae are one of the most effective sources of polysaccharides which serve as a natural source of nontoxic dietary fibers; they produce about half of the organic substances in the world and possess a significant potential to be utilized as functional food components or nutriceuticals [58]. Algal polysaccharides have unique biochemical and fermenting characteristics, which provide different biological activities such as anticancer, immunostimulating, antibacterial and antiviral [59]. They have extraordinary potential for emergent prebiotics to be used directly, and they can also be incorporated into feed or administered as pills [60] *via* normalization of intestinal microflora and selective stimulation of the growth of colonic bacteria, so prospects for inclusion of seaweed polysaccharides in the composition of functional nutrition products to correct intestinal bacterial disorders and inflammatory processes and normalize the immune and metabolic status [61].

Xylose and galactose compounds from different microalgal species are used for the stimulation of probiotic bacteria [40]. Polysaccharide complexes from *Chlorella* species, such as galactose, rhamnose, mannose, and arabinose, displayed immunostimulant characteristics and inhibited *Listeria monocytogenes* and *Candida albicans* proliferation [62].

Seaweeds contain 2.97–71.4% complex sulfated polysaccharides (SP) [46, 63], such as alginate, fucoidan (l-fucose and sulphated ester groups), and laminarin in brown seaweeds; xylan and sulphated galactans, such as agar, carrageenan (sulphated glucuronoxylorhamnan), and porphyran in red seaweeds; whilst ulvan (sulphated glucuronoxylorhamnan) and xylan are found in green seaweeds (Fig. **2**) [58]. Polysaccharides from marine algae, such as alginate, agarose oligosaccharides and κ-carrageenan oligosaccharides, have a higher prebiotic activity than fructo-oligosaccharides (FOS) *in vitro*. Sulphated polysaccharides from marine algae, in particular, exhibit anticoagulant, antiviral, anticancer, anti-inflammatory, antibacterial, immunological, anti-oxidant, and a variety of other pharmacological and physiological properties [64]. Seaweeds or their polysaccharides are considered valuable additives in the food industry because of their rheological properties as gelling and thickening agents [42]. They are the majority of seaweed biomass (up to 76% of dry weight in some species) [44]. Depending on the taxonomic classification of algae, polysaccharides can vary greatly in their composition [44]. Macroalgal polysaccharides differ in their properties and compositions from one type of algae to another, so their effects on the human granulocyte-macrophage gut microbiota (GM) will also differ [65]. They are characterized by hypolipidemic activity, which increase fecal excretion of bile acids, to stimulate lipolysis in adipocytes and suppression of adipogenesis in 3T3-1 preadipocytes *via* the MAPK signalling pathway [52]. Other beneficial actions were reported, including decreased lipopolysaccharide-binding protein in female mice [66], diminished histopathological lesions of inflammatory infiltrations in the distal colon, or modulating diabetes-related genes expression in diabetic mice [45]. These polysaccharides are not fully fermented by the human

GM [67]. Foodborne pathogens such as *Listeria monocytogenes*, *Enterococcus faecalis*, *Escherichia coli*, *Staphylococcus aureus*, and *Salmonella enterica* are inhibited by the SP from macroalgae [68].

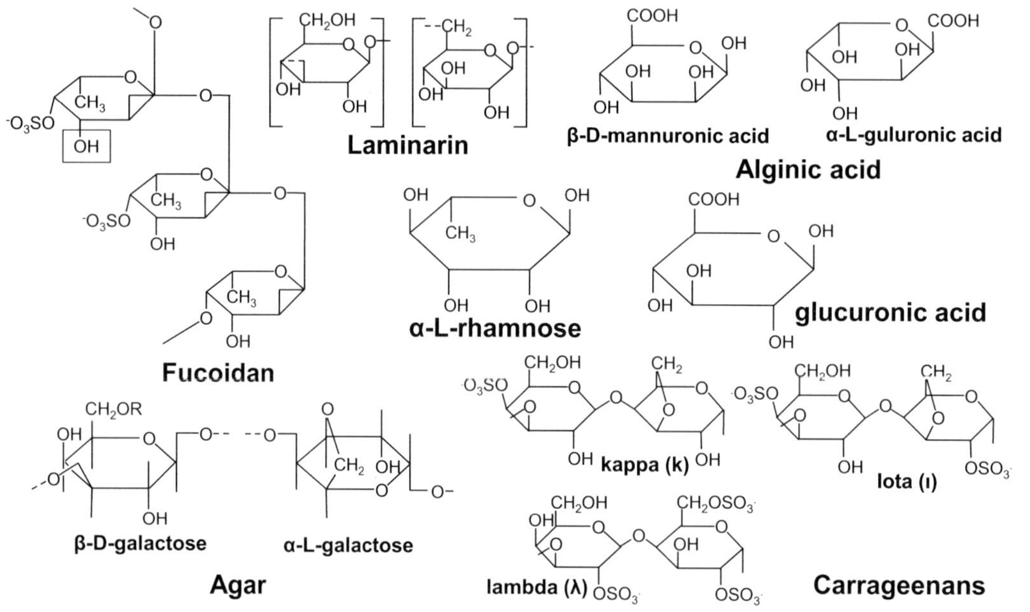

Fig. (2). Chemical structure of the common sulfated polysaccharide [58].

Sulphated structural polysaccharides, such as ulvans, which account for 8–29% of dry weight, as well as sulphated galactans, xylans and mannans, are common in green macroalgae. These polymers are composed mainly of rhamnose, xylose, glucose, glucuronic acid and sulphates, with smaller amounts of mannose, arabinose and galactose [69]. The human GM does not fully ferment these polysaccharides [58]. While β-glucans extracted from *L. digitata* increased *Bifidobacterium* and propionic and butyric acids *in vitro*, in addition to lowering pH [70]. Besides, the administration of green seaweeds or seaweed-extracted polysaccharides also resulted in an increase in SCFA production, stimulating beneficial bacteria grown such as *Lactobacillus*, *Bifidobacterium*, or *Akkermansia* [66], whereas inhibited the growth of potentially pathogen bacteria [71]. It has recently been observed that a polysaccharide of *E. prolifera* could be used as a novel agent for treating obesity and hyperlipidaemia [72].

Especially, polysaccharides from red and brown seaweeds increased SCFA production, stimulating beneficial bacteria grown like *Lactobacillus* [73] or *Bifidobacterium* [50], whereas inhibited the growth of potentially pathogen bacteria [73]. It was also reported red seaweed activity in the prevention of naproxen-induced gastrointestinal damage [74]. These sulfated polysaccharides and their derivatives have a prebiotic activity that has been utilized for decades to improve human and animal health [67]. Seong *et al.* [64] found that polysaccharides from marine algae, such as alginate, agarose oligosaccharides and carrageenan oligosaccharides, have a higher prebiotic activity than FOS *in vitro*.

Polysaccharides from "*Grateloupia filicina*, *Eucheuma spinosum*, *Ulva pertusa*, and *Ascophyllum nodosum*" affect beneficial and harmful bacteria *in vitro* [75]. These polysaccharides often stimulate the growth and activity of beneficial bacteria by acting as substrates for fermentation in the large intestine, leading to the production of SCFA, with multiple functions that help maintain health [67].

Another type of polysaccharide with prebiotic function found in red algae is the group of carrageenans, which are derived from D-galactose, and approved as food additives [76]. Carrageenans from *C. crispus* could act as a fermentable substrate for probiotic bacteria present in the gastrointestinal tract, increasing probiotic growth while suppressing pathogenic bacteria [48].

The beneficial effects of brown seaweed and their polysaccharides are not strictly related to action on GM, such as reducing serum inflammatory markers [77], reducing serum levels of lipopolysaccharide-binding protein, increasing CAZymes, reducing the activity of fecal bile salt hydrolase activity, or reduced the expression or diabetes-related genes [78, 79]. Kaewmanee *et al.* [80] conducted a study in rats, which were administered polysaccharides extracted from *A. nodosum*, and they have seen an increase in both acetate, propionate and butyrate SCFAs.

Fucoidan from brown macroalgal *F. evanescens* decreases the level of total cholesterol, the atherogenic coefficient, and stabilizes the level of products of lipid peroxidation [81]. Fucoidan induced the growth of lactobacteria in the large intestine, so it is used as a food ingredient [61]. Okolie *et al.* [65] demonstrated the prebiotic activity of a mixture of fucoidans and alginates obtained from *A. nodosum in vitro*, leading to an increase in the growth rate of *Lactobacillus delbrueckii* and *L. casei* to levels similar to those observed after administration of inulin, a standard commercial prebiotic.

Laminins stimulate the proportion of Bifidobacterium, which generates a prebiotic potential. In other research, laminarin has been shown to promote an immune response [82], and could be effective in preventing the production of putrefactive substances from undigested proteins [83]. Laminaran from brown algae "*Laminaria* and *Saccharina*" plays a role in regulating the gut metabolism by interfering with gut goblet cells of the epithelial layer and stimulating the release of mucus [84].

Many macroalgal polysaccharides are known for their prebiotic activity *in vitro*, like (*Grateloupia filicina* (GFP), *Eucheuma spinosum* (ESP), *Ulva pertusa*, and *Ascophyllum nodosum* (ANP)) [85].

Fibers

Consumption of dietetic fiber has a positive influence on several aspects related to health, reducing the risk of suffering colon cancer, constipation, hypercholesterolemia, obesity and diabetes. Besides, many constituents of the dietetic fiber show anti-oxidant ability [86].

Fibers from algae represent about 33-500 g/100 g wet weight which is higher than that in most fruits and vegetables with different chemical and physical properties. They are necessary for the growth of lactic acid bacteria, so they act as a natural nutrition source [59]. Nevertheless, many seaweed fibres are high-molecula--weight polymers that need to be transformed into oligosaccharides to increase their fermentability by the GM; the human GM did not acquire the same efficacy to degrade seaweed polysaccharides. Thus, although humans possess the enzymes necessary to degrade some algal polysaccharides, such as starches, they are unable to digest the most complex polysaccharides [42]. De Jesus Raposo *et al.* [60] suggested that most seaweed polysaccharides can be regarded as dietary fiber because they are resistant to breakdown by enzymes found in the human gastrointestinal tract and reach the distal gut. In this allocation, polysaccharides are fermented and become food for the commensal bacteria, stimulating their growth. For this reason, great efforts have been placed on developing efficient methods for seaweed polysaccharides extraction, purification and structural characteristics elucidation to improve their bioavailability, especially for insoluble fibre [42]. Due to its high amount of both soluble and insoluble fibres, the marine microalga "*Isochrysis galbana*" acts as a prebiotic [87].

Lipid

Lipids are essential for all biota since they are components of membranes, energy storage compounds and cell signaling molecules [88]. Although these algae had a low lipid content, they possessed a high level of polyunsaturated fatty acids (PUFA), which include eicosapentaenoic acid (EPA), that demonstrated their effect on decreasing coronary diseases [89].

According to Le Tutour *et al.* [90], the use of an extract containing soluble lipids from *H. elongata* improved synergically the anti-oxidant effect of vitamin E, in a percentage as high as 45%. Seaweed contains n-3 PUFAs such eicosapentanoic acid (EPA) and docosahexaenoic acid (DHA) in amounts ranging from 1–5% dry weight [91]. The n-3 PUFA is associated with the anti-inflammatory activity to reduce cardiovascular disease risk and may also exert beneficial effects on brain function and behaviour, as mediated by the microbiota-gut-brain axis [92]. In both people and animals, dietary EPA and DHA intake has been shown to promote microbial diversity, lower the Firmicutes/Bacteroidetes ratio, decrease Lipopolysaccharide (LPS) producing bacteria, and boost populations of *Bifidobacterium*, Lachnospiraceae, and LPS-suppressing bacteria in both humans and animal models [93]. Although the evidence to date has focused on fish-derived n-3 PUFA, great scope exists to evaluate the prebiotic effect of n-3 PUFA obtained from seaweeds [93].

Protein

The low molecular weight proteins obtained from seaweeds were further fermented by intestinal (gut) bacteria, indicating that they could be used as a novel source of prebiotics [94].

Seaweed proteins and phenolic compounds have also attracted interest as potential functional ingredients [44]. Protein-derived bioactive peptides and phlorotannins can prevent hypertension by inhibiting angiotensin-I converting enzyme activity [95]. Barbarino and Lourenco [96] isolated proteins of *S. vulgare* that had a high nutritional value since they contained all the essential amino acids in significant amounts.

Phenolic Compounds

Macroalgal phlorotannins inhibited adipocyte differentiation or fucosterol, decreasing the expression of the adipocyte marker proteins PPARγ and CCAAT/enhancer-binding protein alpha. Seaweeds can also prevent obesity through the modification of the relative quantities of phyla in the GM. Polysaccharides from seaweeds can reduce obesity by repairing the intestinal barrier and reducing inflammation [97]. These compounds can also reach the large intestine, which can be converted into helpful bioactive metabolites by GM. Phlorotannins are highly hydrophilic components formed by the polymerization of monomeric units of phloroglucinol (1,3,5-trihydroxybenzene). Fucols, floretoles, fucofloretols, fuhalols, isofuhalols and eckols are the six main groups and all display strong anti-oxidant properties and protect against oxidative stress [98]. Charoensiddhi *et al.* [44] found an increase in the levels of beneficial bacteria, such as *Bifidobacterium*, *Lactobacillus* and *C. coccoides* associated with the phlorotannins enriched fermentation of *E. radiate*.

Pigments

Carotenoids are utilized as food colouring additives, while fucoxanthin is recommended as a functional food ingredient, owing to putative anti-oxidant, anti-inflammatory, anti-cancer, anti-obesity, and anti-diabetic bioactivities [99]. *Spirulina* contains phycocyanin, a nontoxic blue pigment that can operate as a free radical scavenger and potent anti-oxidant [23].

CONCLUDING REMARKS

This work recommended the algae and their compounds, particularly polysaccharides, phenolic compounds and PUFA, as prebiotics which can be consumed as dietary supplements due to their beneficial effects either for the host or the gut microbiota through enhancing the immune system and, consequently, improving the human health. Microalgae as marine organisms have recently become popular as a new source for both nutraceutical and pharmaceutical products. The addition of microalgae such as *Spirulina* and *Chlorella* genera could raise the viability of probiotics in fermented dairy products like yogurt. The decrease in terrestrial agriculture and disposable water is likely to increase the consumption of algae by humans in the near future. Meanwhile, there is a profound need for more in-depth investigations into the potential prebiotic effects of marine seaweeds and their derived polysaccharides on the human GM. Additionally, other factors should be considered, such as the possible effect of other secondary compounds contained in seaweeds on the GM composition, or the potential to transfer genes from marine bacteria to human GM bacteria by coding for enzymes that could degrade seaweed polysaccharides.

CONSENT FOR PUBLICATION

Not applicable.

CONFLICT OF INTEREST

The author(s) declare no conflict of interest, financial or otherwise.

ACKNOWLEDGEMENT

Declared none.

REFERENCES

[1] Nayak SK. Probiotics and immunity: A fish perspective. Fish Shellfish Immunol 2010; 29(1): 2-14. [http://dx.doi.org/10.1016/j.fsi.2010.02.017] [PMID: 20219683]

[2] de Roos NM, Katan MB. Effects of probiotic bacteria on diarrhea, lipid metabolism, and carcinogenesis: a review of papers published between 1988 and 1998. Am J Clin Nutr 2000; 71(2): 405-11. [http://dx.doi.org/10.1093/ajcn/71.2.405] [PMID: 10648252]

[3] David C. Lin, Ph D, Facn, CN. 2004. Probiotics as functional foods. Nutr Clin Pract 2004; 18: 497-506.

[4] Lee NK, Park JS, Park E, Paik HD. Adherence and anticarcinogenic effects of Bacillus polyfermenticus SCD in the large intestine. Lett Appl Microbiol 2007; 44(3): 274-8. [http://dx.doi.org/10.1111/j.1472-765X.2006.02078.x] [PMID: 17309504]

[5] Kołożyn-Krajewska D, Dolatowski ZJ. Probiotic meat products and human nutrition. Process Biochem 2012; 47(12): 1761-72. [http://dx.doi.org/10.1016/j.procbio.2012.09.017]

[6] Zubillaga M, Weill R, Postaire E, *et al.* The use of functional foods (probiotics and prebiotics) has been demonstrated to be effective for the treatment or control of several diseases. Nutr Res 2001; 21: 569-79. [http://dx.doi.org/10.1016/S0271-5317(01)00281-0]

[7] Levy J. Immunonutrition: the pediatric experience. Nutrition 1998; 14(7-8): 641-7. [http://dx.doi.org/10.1016/S0899-9007(98)00007-0] [PMID: 9684269]

[8] Omar HH, Dighriri Kh A, Gashgary RM. The Benefit Roles of Micro- and Macro-Algae in Probiotics. Nat Sci 2019; 17(11): 258-79.

[9] Panesar PS, Kaur G, Panesar R, Bera MB. Synbiotics: potential dietary supplements in functional foods. Berkshire, UK: IFIS, 2009. [http://dx.doi.org/10.1136/gut.2005.074971]

[10] Wells ML, Potin P, Craigie JS, *et al.* Algae: anatomy, biochemistry, and biotechnology. New York, N.Y: CRC Press 2006.

[11] El Zokm GM, Ismail MM, El-Said GF. Halogen content relative to the chemical and biochemical. composition of fifteen marine macro and micro algae: nutritional value, energy supply, antioxidant potency, and health risk assessment. Environ Sci Pollut Res Int 2021; 28(12): 14893-908. [http://dx.doi.org/10.1007/s11356-020-11596-0] [PMID: 33222067]

[12] Ismail MM, El-Komy R, Diab MH. Algal Bioactive Compounds and Their biological activities. International Journal of Pharmaceutical Research 2021; 13(2): 1440-51.

[13] Ismail MM, El Zokm GM, El-Sayed AAM. Variation in biochemical constituents and master elements in common seaweeds from Alexandria Coast, Egypt, with special reference to their antioxidant activity and potential food uses: prospective equations. Environ Monit Assess 2017; 189(12): 648. [http://dx.doi.org/10.1007/s10661-017-6366-8] [PMID: 29177576]

[14] Priyadarshani I, Rath B. Commercial and industrial applications of micro algae-A review. J Algal Biomass Util 2012; 3(4): 89.

[15] Carlucci MJ, Scolaro LA, Damonte EB. Inhibitory action of natural carrageenans on *Herpes simplex* virus infection of mouse astrocytes. Chemotherapy 1999; 45(6): 429-36. [http://dx.doi.org/10.1159/000007236] [PMID: 10567773]

[16] Zhang W, Gu Y, Chen Y, *et al.* Intestinal flora imbalance results in altered bacterial translocation and liver function in rats with experimental cirrhosis. Eur J Gastroenterol Hepatol 2010; 22(12): 1. [http://dx.doi.org/10.1097/MEG.0b013e32833eb8b0] [PMID: 20739895]

[17] Summary FDA. Summary FDA. Substances Generally Regarded As Safe (Final Rule). U.S. Food and Drug Administration (HHS). http: //www. fda.gov/ AboutFDA/

[18] Varga L, Szigeti J, Kovács R, Földes T, Buti S. Influence of a *Spirulina platensis* biomass on the microflora of fermented ABT milks during storage (R1). J Dairy Sci 2002; 85(5): 1031-8. [http://dx.doi.org/10.3168/jds.S0022-0302(02)74163-5] [PMID: 12086036]

[19] Caporgno MP, Mathys A. Trends in microalgae incorporation into innovative food products with potential health benefits. Front Nutr 2018; 5: 58. [http://dx.doi.org/10.3389/fnut.2018.00058] [PMID: 30109233]

[20] Webb LE. Detection by Warburg manometry of compounds stimulatory to lactic acid bacteria. J Dairy Res 1982; 49(3): 479-86. [http://dx.doi.org/10.1017/S0022029900022615]

[21] Gyenis B, Szigeti J, Moln'ar N, Varga L. Use of dried microalgal biomasses to stimulate acid production and growth of *Lactobacillus plantarum* and *Enterococcus faecium* in milk. Acta Agrar Kvár 2005; 9(2): 53-9.

[22] Janczyk P, Franke H, Souffrant WB. Nutritional value of *Chlorella vulgaris*: Effects of ultrasonication and electroporation on digestibility in rats. Anim Feed Sci Technol 2007; 132(1-2): 163-9. [http://dx.doi.org/10.1016/j.anifeedsci.2006.03.007]

[23] Beheshtipour H, Mortazavian AM, Mohammadi R, Sohrabvandi S, Khosravi-Darani K. Supplementation of *Spirulina platensis* and *Chlorella vulgaris* algae into probiotic fermented milks. Compr Rev Food Sci Food Saf 2013; 12(2): 144-54. [http://dx.doi.org/10.1111/1541-4337.12004]

[24] Blas-Valdivia V, Ortiz-Butrón R, Pineda-Reynoso M, Hernández-Garcia A, Cano-Europa E. *Chlorella vulgaris* administration prevents HgCl2-caused oxidative stress and cellular damage in the kidney. J Appl Phycol 2011; 23(1): 53-8. [http://dx.doi.org/10.1007/s10811-010-9534-6]

[25] Belasco W. Algae burgers for a hungry world? The rise and fall of *Chlorella cuisine*. Technol Cult 1997; 38(3): 608-34. [http://dx.doi.org/10.2307/3106856]

[26] Barrow C, Shahidi F. Marine nutraceuticals and functional foods. Boca Raton, Fla.: CRC Press.2008; 367-403.

[27] Jeong H, Kwon HJ, Kim MK. Hypoglycemic effect of *Chlorella vulgaris* intake in type 2 diabetic Goto-Kakizaki and normal Wistar rats. Nutr Res Pract 2009; 3(1): 23-30. [http://dx.doi.org/10.4162/nrp.2009.3.1.23] [PMID: 20016698]

[28] Kavimandan A. Incorporation of *Spirulina platensis* into probiotic fermented dairy products. Int J Dairy Sci 2014; 10(1): 1-11. [http://dx.doi.org/10.3923/ijds.2015.1.11]

[29] Parada J, Zulpa de Caire G, Zaccaro de Mulé MC, Storni de Cano MM. Lactic acid bacteria growth promoters from *Spirulina platensis.* Int J Food Microbiol 1998; 45(3): 225-8. [http://dx.doi.org/10.1016/S0168-1605(98)00151-2] [PMID: 9927000]

[30] Bhowmik D, Dubey J, Mehra S. Probiotic efficiency of *Spirulina platensis*-stimulating growth of lactic acid bacteria. World J. Dairy Food Sci 2009; 160-3.

[31] Barsanti L, Gualtieri P. Barsanti L, Gualtieri P. Algae: Anatomy, Biochemistry, and Biotechnology. 2006; CRC Press, Taylor & Francis Group, Boca Raton, FL, USA, 301.

[32] Varga L, Szigeti J, Ord'og V. Effect of a *Spirulina platensis* biomass and that of its active components on single strains of dairy starter cultures. Milchwissenschaft 1999; 54: 187-90.

[33] Gupta C, Gupta C, Garg AP, Prakash D. prebiotic efficiency of blue green algae on probiotics microorganisms. J microbiol exp 2017; 4(4): 00120.

[http://dx.doi.org/10.15406/jmen.2017.04.00120]

[34] Fox RD. Fox RD. Algaculture: la *Spirulina*, 1986; UN espoir pour le monde de la faim, Edisude, France 1986.

[35] Ásványi-Molnár N, Gyenis B, Varga L. Influence of a powdered ′ *Spirulina platensis* biomass on acid production of lactococci in milk. Milchwissenschaft 2005; 60(4): 380-92.

[36] Beheshtipour H, Mortazavian AM, Haratian P, Darani KK. Effects of *Chlorella vulgaris* and *Arthrospira platensis* addition on viability of probiotic bacteria in yogurt and its biochemical properties. Eur Food Res Technol 2012; 235(4): 719-28. [http://dx.doi.org/10.1007/s00217-012-1798-4]

[37] Mazinani S, Fadaei V, Khosravi-Darani K. Impact of *Spirulina platensis* on physicochemical properties and viability of *Lactobacillus acidophilus* of probiotic UF feta cheese. J Food Process Preserv 2016; 40(6): 1318-24. [http://dx.doi.org/10.1111/jfpp.12717]

[38] Niccolai A, Shannon E, Abu-Ghannam N, Biondi N, Rodolfi L, Tredici MR. Lactic acid fermentation of *Arthrospira platensis* (*spirulina*) biomass for probiotic-based products. J Appl Phycol 2019; 31(2): 1077-83. [http://dx.doi.org/10.1007/s10811-018-1602-3]

[39] Spolaore P, Joannis-Cassan C, Duran E, Isambert A. Commercial applications of microalgae. J Biosci Bioeng 2006; 101(2): 87-96. [http://dx.doi.org/10.1263/jbb.101.87] [PMID: 16569602]

[40] Gourbeyre P, Denery S, Bodinier M. Probiotics, prebiotics, and synbiotics: impact on the gut immune system and allergic reactions. J Leukoc Biol 2011; 89(5): 685-95. [http://dx.doi.org/10.1189/jlb.1109753] [PMID: 21233408]

[41] Cian R, Drago S, de Medina F, Martínez-Augustin O. Proteins and carbohydrates from red seaweeds: Evidence for beneficial effects on gut function and microbiota. Mar Drugs 2015; 13(8): 5358-83. [http://dx.doi.org/10.3390/md13085358] [PMID: 26308006]

[42] Gurpilhares DB, Cinelli LP, Simas NK, Pessoa A Jr, Sette LD. Marine prebiotics: Polysaccharides and oligosaccharides obtained by using microbial enzymes. Food Chem 2019; 280: 175-86. [http://dx.doi.org/10.1016/j.foodchem.2018.12.023] [PMID: 30642484]

[43] Ismail MM, Alotaibi BS, EL-Sheekh MM. Therapeutic Uses of Red Macroalgae. Molecules 2020; 25(19): 4411. [http://dx.doi.org/10.3390/molecules25194411]

[44] Charoensiddhi S, Conlon MA, Franco CMM, Zhang W. The development of seaweed-derived bioactive compounds for use as prebiotics and nutraceuticals using enzyme technologies. Trends Food Sci Technol 2017; 70: 20-33. [http://dx.doi.org/10.1016/j.tifs.2017.10.002]

[45] Lopez-Santamarina A, Miranda JM, Mondragon AC, *et al.* Potential Use of Marine Seaweeds as Prebiotics: A Review. Molecules 2020; 25(4): 1004. [http://dx.doi.org/10.3390/molecules25041004] [PMID: 32102343]

[46] Ismail MM, Amer MS. Characterization and biological properties of sulfated polysaccharides of *Corallina officinalis* and *Pterocladia capillacea.* Acta Bot Bras 2020; 34(4): 623-32. [http://dx.doi.org/10.1590/0102-33062020abb0121]

[47] Plaza M, Cifuentes A, Ibáñez E. In the search of new functional food ingredients from algae. Trends Food Sci Technol 2008; 19(1): 31-9. [http://dx.doi.org/10.1016/j.tifs.2007.07.012]

[48] Liu J, Kandasamy S, Zhang J, *et al.* Prebiotic effects of diet supplemented with the cultivated red seaweed Chondrus crispus or with fructo-oligo-saccharide on host immunity, colonic microbiota and gut microbial metabolites. BMC Complement Altern Med 2015; 15(1): 279.

[http://dx.doi.org/10.1186/s12906-015-0802-5] [PMID: 26271359]

[49] Di T, Chen G, Sun Y, Ou S, Zeng X, Ye H. *in vitro* digestion by saliva, simulated gastric and small intestinal juices and fermentation by human fecal microbiota of sulfated polysaccharides from *Gracilaria rubra.* J Funct Foods 2018; 40: 18-27.
[http://dx.doi.org/10.1016/j.jff.2017.10.040]

[50] Bajury DM, Rawi MH, Sazali IH, Abdullah A, Sarbini SR. Prebiotic evaluation of red seaweed (*Kappaphycus alvarezii*) using *in vitro* colon model. Int J Food Sci Nutr 2017; 68(7): 821-8.
[http://dx.doi.org/10.1080/09637486.2017.1309522] [PMID: 28393631]

[51] Sivaramakrishnan T, Swain S, Saravanan K, *et al. in vitro* anti-oxidant and free radical scavenging activity and chemometric approach to reveal their variability in green macroalgae from south Andaman Coast of India. Turk J Fish Aquat Sci 2017; 17(3): 641-51.
[http://dx.doi.org/10.4194/1303-2712-v17_3_20]

[52] Kim J, Yu D, Kim J, *et al.* Effects of *Undaria linnatifida* and *Laminaria japonica* on rat's intestinal microbiota and metabolite. J Nutr Food Sci 2016; 6.

[53] Kong Q, Dong S, Gao J, Jiang C. *In vitro* fermentation of sulfated polysaccharides from *E. prolifera* and *L. japonica* by human fecal microbiota. Int J Biol Macromol 2016; 91: 867-71.
[http://dx.doi.org/10.1016/j.ijbiomac.2016.06.036] [PMID: 27316763]

[54] Dierick N, Ovyn A, De Smet S. Effect of feeding intact brown seaweed *Ascophyllum nodosum* on some digestive parameters and on iodine content in edible tissues in pigs. J Sci Food Agric 2009; 89(4): 584-94.
[http://dx.doi.org/10.1002/jsfa.3480]

[55] Ruperez P. Mineral content of edible marine seaweeds. Food Chem 2002; 79(1): 23-6.
[http://dx.doi.org/10.1016/S0308-8146(02)00171-1]

[56] Ismail MM, Gheda SF, Pereira L. Variation in bioactive compounds in some seaweeds from Abo Qir bay, Alexandria, Egypt. Rend Lincei Sci Fis Nat 2016; 27(2): 269-79.
[http://dx.doi.org/10.1007/s12210-015-0472-8]

[57] Roberfroid M. Prebiotics: Concept, definition, criteria, methodologies, and products. In: Gibson GR, Roberfroid M, Eds. Handbook of Prebiotics 2008; 39-69.
[http://dx.doi.org/10.1201/9780849381829.ch3]

[58] Jiao G, Yu G, Zhang J, Ewart H. Chemical structures and bioactivities of sulfated polysaccharides from marine algae. Mar Drugs 2011; 9(2): 196-223.
[http://dx.doi.org/10.3390/md9020196] [PMID: 21566795]

[59] Holdt SL, Kraan S. Bioactive compounds in seaweed: functional food applications and legislation. J Appl Phycol 2011; 23(3): 543-97.
[http://dx.doi.org/10.1007/s10811-010-9632-5]

[60] de Jesus Raposo M, de Morais A, de Morais R. Emergent sources of prebiotics: seaweeds and microalgae. Mar Drugs 2016; 14(2): 27.
[http://dx.doi.org/10.3390/md14020027] [PMID: 26828501]

[61] Zaporozhets TS, Besednova NN, Kuznetsova TA, *et al.* The prebiotic potential of polysaccharides and extracts of seaweeds. Russ J Mar Biol 2014; 40(1): 1-9.
[http://dx.doi.org/10.1134/S1063074014010106]

[62] Mata TM, Martins AA, Caetano NS. Microalgae for biodiesel production and other applications: A review. Renew Sustain Energy Rev 2010; 14(1): 217-32.
[http://dx.doi.org/10.1016/j.rser.2009.07.020]

[63] Thursby E, Juge N. Introduction to the human gut microbiota. Biochem J 2017; 474(11): 1823-36.
[http://dx.doi.org/10.1042/BCJ20160510] [PMID: 28512250]

[64] Seong H, Bae JH, Seo JS, Kim S-A, Kim T-J, Han NS. Comparative analysis of prebiotic effects of

seaweed polysaccharides laminaran, porphyran, and ulvan using *in vitro* human fecal fermentation. J Funct Foods 2019; 57: 408-16.
[http://dx.doi.org/10.1016/j.jff.2019.04.014]

[65] Okolie CL, Rajendran CK Sr, Udenigwe CC, *et al.* Prospects of brown seaweed polysaccharides (BSP) as prebiotics and potential immunomodulators. J Food Biochem 2017; 41: 12392.
[http://dx.doi.org/10.1111/jfbc.12392]

[66] Shang Q, Wang Y, Pan L, *et al.* Dietary polysaccharide from *Enteromorpha clathrata* modulates gut microbiota and promotes the growth of *Akkermansia muciniphila, Bifidobacterium* spp. and *Lactobacillus* spp. Mar Drugs 2018; 16(5): 167.
[http://dx.doi.org/10.3390/md16050167] [PMID: 29772753]

[67] O'Sullivan L, Murphy B, McLoughlin P, *et al.* Prebiotics from marine macroalgae for human and animal health applications. Mar Drugs 2010; 8(7): 2038-64.
[http://dx.doi.org/10.3390/md8072038] [PMID: 20714423]

[68] Pérez M, Falqué E, Domínguez H. Antimicrobial action of compounds from marine seaweed. Mar Drugs 2016; 14(3): 52.
[http://dx.doi.org/10.3390/md14030052] [PMID: 27005637]

[69] Cherry P, Yadav S, Strain CR, *et al.* Prebiotics from seaweeds: An ocean of opportunity? Mar Drugs 2019; 17(6): 327.
[http://dx.doi.org/10.3390/md17060327] [PMID: 31159359]

[70] Zhao D, Zhuo RX, Cheng SX. Alginate modified nanostructured calcium carbonate with enhanced delivery efficiency for gene and drug delivery. Mol Biosyst 2012; 8(3): 753-9.
[http://dx.doi.org/10.1039/C1MB05337J] [PMID: 22159070]

[71] Zmora N, Suez J, Elinav E. You are what you eat: diet, health and the gut microbiota. Nat Rev Gastroenterol Hepatol 2019; 16(1): 35-56.
[http://dx.doi.org/10.1038/s41575-018-0061-2] [PMID: 30262901]

[72] Rubiano-Buitrago P, Duque F, Puyana M, Ramos FA, Castellanos L. Bacterial biofilm inhibitor diterpenes from *Dictyota pinnatifida* collected from the Colombian Caribbean. Phytochem Lett 2019; 30: 74-80.
[http://dx.doi.org/10.1016/j.phytol.2019.01.021]

[73] Praveen A, Parvathy KRK, Balasubramanian P, Rasu Jayabalan R. An overview of extraction and purification techniques of seaweed dietary fibers for immunomodulation on gut microbiota. Trends Food Sci Technol 2020; 92(9): 46-64.
[http://dx.doi.org/10.1016/j.tifs.2019.08.011]

[74] Silva R, Santana A, Carvalho N, *et al.* A sulfated-polysaccharide fraction from seaweed *Gracilaria birdiae* prevents naproxen-induced gastrointestinal damage in rats. Mar Drugs 2012; 10(12): 2618-33.
[http://dx.doi.org/10.3390/md10122618] [PMID: 23342384]

[75] Chen X, Sun Y, Hu L, *et al. in vitro* prebiotic effects of seaweed polysaccharides. J Oceanol Limnol 2018; 36(3): 926-32.
[http://dx.doi.org/10.1007/s00343-018-6330-7]

[76] Sardari RR. Nordberg Karlsson, E. Marine poly-and oligosaccharides as prebiotics. J Agric Food Chem 2018; 66: 11544-9.
[http://dx.doi.org/10.1021/acs.jafc.8b04418] [PMID: 30350987]

[77] Lean QY, Eri RD, Fitton JH, Patel RP, Gueven N. Fucoidan extracts ameliorateacutecolitis. PLoS One 2015; 10(6): e0128453.
[http://dx.doi.org/10.1371/journal.pone.0128453]

[78] Huebbe P, Nikolai S, Schloesser A, *et al.* An extract from the Atlantic brown algae *Saccorhiza polyschides* counteracts diet-induced obesity in mice *via* a gut related multi-factorial mechanisms. Oncotarget 2017; 8(43): 73501-15.

[http://dx.doi.org/10.18632/oncotarget.18113] [PMID: 29088722]

[79] Yang C, Lai S, Chen Y, *et al.* Anti-diabetic effect of oligosaccharides from seaweed *Sargassum confusum via* JNK-IRS1/PI3K signalling pathways and regulation of gut microbiota. Food Chem Toxicol 2019; 131: 110562.
[http://dx.doi.org/10.1016/j.fct.2019.110562] [PMID: 31181236]

[80] Kaewmanee W, Suwannaporn P, Huang TC, Al-Ghazzewi F, Tester RF. *In vivo* prebiotic properties of *Ascophyllum nodosum* polysaccharide hydrolysates from lactic acid fermentation. J Appl Phycol 2019; 31(5): 3153-62.
[http://dx.doi.org/10.1007/s10811-019-01794-w]

[81] Maistrovsky KV, Zaporozhets TS, Fedyanina LN, *et al.* The effect of fucoidan immunomodulator from the brown algae *Fucus evanescens* on parameters of the anti-oxidant system and lipid and carbohydrate exchange in mice. Tikhookean Med J 2009; 3: 97-100.

[82] Shih YL, Hsueh SC, Chen YL, *et al.* Laminarin promotes immune responses and reduces lactate dehydrogenase but increases glutamic pyruvic transaminase in normal mice *in vivo.* In Vivo 2018; 32(3): 523-9.
[PMID: 29695555]

[83] Nakata T, Kyoui D, Takahashi H, Kimura B, Kuda T. Inhibitory effects of laminaran and alginate on production of putrefactive compounds from soy protein by intestinal microbiota *in vitro* and in rats. Carbohydr Polym 2016; 143: 61-9.
[http://dx.doi.org/10.1016/j.carbpol.2016.01.064] [PMID: 27083344]

[84] Devillé C, Damas J, Forget P, Dandrifosse G, Peulen O. Laminarin in the dietary fibre concept. J Sci Food Agric 2004; 84(9): 1030-8.
[http://dx.doi.org/10.1002/jsfa.1754]

[85] Xiaolin C, Yuhao S, Linfeng HU. Laminarin in the dietary fiber concept. J Sci. Food Agricul. 2018; 84(9): 1030-.

[86] Suzuki N, Fujimura A, Nagai T, *et al.* Antioxidative activity of animal and vegetable dietary fibers. Biofactors 2004; 21(1-4): 329-33.
[http://dx.doi.org/10.1002/biof.552210164] [PMID: 15630221]

[87] Nuño K, Villarruel-López A, Puebla-Pérez AM, Romero-Velarde E, Puebla-Mora AG, Ascencio F. Effects of the marine microalgae *Isochrysis galbana* and *Nannochloropsis oculata* in diabetic rats. J Funct Foods 2013; 5(1): 106-15.
[http://dx.doi.org/10.1016/j.jff.2012.08.011]

[88] Eyster KM. The membrane and lipids as integral participants in signal transduction: lipid signal transduction for the non-lipid biochemist. Adv Physiol Educ 2007; 31(1): 5-16.
[http://dx.doi.org/10.1152/advan.00088.2006] [PMID: 17327576]

[89] Simopoulos AP. Omega-3 essential fatty acid ratio and chronic diseases. Food Rev Int 2004; 20(1): 77-90.
[http://dx.doi.org/10.1081/FRI-120028831]

[90] Le Tutour B, Benslimane F, Gouleau MP, Gouygou JP, Saadan B, Quemeneur F. Anti-oxidant and pro-oxidant activities of the brown algae, *Laminaria digitata, Himanthalia elongata, Fucus vesiculosus, Fucus serratus* and *Ascophyllum nodosum.* J Appl Phycol 1998; 10(2): 121-9.
[http://dx.doi.org/10.1023/A:1008007313731]

[91] Robertson R, Guihéneuf F, Bahar B, *et al.* The anti-inflammatory effect of algae-derived lipid extracts on lipopolysaccharide (LPS)-stimulated human THP-1 macrophages. Mar Drugs 2015; 13(8): 5402-24.
[http://dx.doi.org/10.3390/md13085402] [PMID: 26308008]

[92] Costantini L, Molinari R, Farinon B, Merendino N. Impact of Omega-3 fatty acids on the gut microbiota. Int J Mol Sci 2017; 18(12): 2645.

[http://dx.doi.org/10.3390/ijms18122645] [PMID: 29215589]

[93] Robertson RC, Kaliannan K, Strain CR, Ross RP, Stanton C, Kang JX. Maternal omega-3 fatty acids regulate offspring obesity through persistent modulation of gut microbiota. Microbiome 2018; 6(1): 95.
[http://dx.doi.org/10.1186/s40168-018-0476-6] [PMID: 29793531]

[94] Ramnani P, Chitarrari R, Tuohy K, *et al. in vitro* fermentation and prebiotic potential of novel low molecular weight polysaccharides derived from agar and alginate seaweeds. Anaerobe 2012; 18(1): 1-6.
[http://dx.doi.org/10.1016/j.anaerobe.2011.08.003] [PMID: 21924371]

[95] Seca A, Pinto D. Overview of the antihypertensive and anti-obesity effects of secondary metabolites from seaweeds. Mar Drugs 2018; 16(7): 237.
[http://dx.doi.org/10.3390/md16070237] [PMID: 30011911]

[96] Barbarino E, Lourenço SO. An evaluation of methods for extraction and quantification of protein from marine macro- and microalgae. J Appl Phycol 2005; 17(5): 447-60.
[http://dx.doi.org/10.1007/s10811-005-1641-4]

[97] You L, Gong Y, Li L, *et al.* Beneficial effects of three brown seaweed polysaccharides on gut microbiota and their structural characteristics: An overview. Int J Food Sci Technol 2019; 55(3): 1199-206.

[98] Lee DH, Park MY, Shim BJ, *et al.* Effects of *Ecklonia cava* polyphenol in individuals with hypercholesterolemia: a pilot study. J Med Food 2012; 15(11): 1038-44.
[http://dx.doi.org/10.1089/jmf.2011.1996] [PMID: 23126663]

[99] Kulczyński B, Gramza-Michałowska A, Kobus-Cisowska J, Kmiecik D. The role of carotenoids in the prevention and treatment of cardiovascular disease – Current state of knowledge. J Funct Foods 2017; 38: 45-65.
[http://dx.doi.org/10.1016/j.jff.2017.09.001]

[100] Wu Z, Dejtisakdi W, Kermanee P, Ma C, *et al.* Outdoor cultivation of *Dunaliella salina* KU 11 using brine and saline lake water with raceway ponds in northeastern Thailand. Biotechnol Appl Biochem 2016; 1-18.
[PMID: 27696529]

[101] Chu WL. 2012. Biotechnological applications of microalgae. Int. e-J. Sci Med Edu 2006; 6: 24-37.

[102] Li Y, Wei G, Chen J. Glutathione: a review on biotechnological production. Appl Microbiol Biotechnol 2004; 66(3): 233-42.
[http://dx.doi.org/10.1007/s00253-004-1751-y] [PMID: 15480625]

[103] Akbary P, Aminikhoei Z. Effect of water-soluble polysaccharide extract from the green alga *Ulva rigida* on growth performance, anti-oxidant enzyme activity, and immune stimulation of grey mullet Mugil cephalus. J Appl Phycol 2017; 9: 1-9.

[104] Becker W. Microalgae in human and animal nutrition. In: Richmond A, Ed. Microalgal Culture Handbook Becker W. Microalgae in human and animal nutrition. In: Richmond, A. (Ed.), Microalgal Culture. Handbook, Blackwell, Oxford, In: Richmond A, Ed. 2004; 312-51.

[105] Díaz AC, Espino ML, Arzoz NS, *et al.* Free radical scavenging activity of extracts from seaweeds *Macrocystis pyrifera* and *Undaria pinnatifida*: applications as functional food in the diet of prawn *Artemesia longinaris.* Lat Am J Aquat Res 2017; 45(1): 104-12.
[http://dx.doi.org/10.3856/vol45-issue1-fulltext-10]

[106] Sivagnanavelmurugan M, Marudhupandi T, Palavesam A, Immanuel G. Antiviral effect of fucoidan extracted from the brown seaweed, *Sargassum wightii*, on shrimp *Penaeus monodon* postlarvae against white spot syndrome virus. J World Aquacult Soc 2012; 43(5): 697-706.
[http://dx.doi.org/10.1111/j.1749-7345.2012.00596.x]

[107] Shahabuddin AM, Khan MND, Mikami K, Araki T, Yoshimatsu T. Dietary supplementation of red

alga *Pyropia* spheroplasts on growth, feed utilization and body composition of sea cucumber, *Apostichopus japonicus* (Selenka). Aquacult Res 2017; 48(10): 5363-72. [http://dx.doi.org/10.1111/are.13350]

[108] Wongprasert K, Rudtanatip T, Praiboon J. Immunostimulatory activity of sulfated galactans isolated from the red seaweed *Gracilaria fisheri* and development of resistance against white spot syndrome virus (WSSV) in shrimp. Fish Shellfish Immunol 2014; 36(1): 52-60. [http://dx.doi.org/10.1016/j.fsi.2013.10.010] [PMID: 24161778]

CHAPTER 15

Seaweeds Metabolites: Characterization and Applications as Phyco-Nutraceuticals

Surabhi Agrawal[1,2], **Darshita K. Pandya**[1], **Shahrukh A. Siddiqui**[1,2] and **Mangal S. Rathore**[1,2,*]

[1] *Division of Applied Phycology and Biotechnology, CSIR-Central Salt and Marine Chemicals Research Institute, Bhavnagar, Gujarat, India*

[2] *Academy of Scientific and Innovative Research (AcSIR), Ghaziabad, India*

Abstract: For decades, it has been evidenced that seaweeds are used as part of the human diet in many parts of the world. Across the globe, it has been investigated and accepted that seaweeds are rich source of diverse bioactive compounds and can be used as functional foods. These bioactive compounds or metabolites are derived in the form of peptides, polysaccharides, fatty acids, pigments and polyphenols. They have various bioactive properties like anti-cancerous, anti-inflammatory, antibacterial, antioxidants and anti-diabetic. Further, these findings were noticed by pharmaceutical industries that seaweeds can be employed as nutraceuticals for various human ailments. Environmental factors contribute to maximum diversification in the production of metabolites; therefore, there is a need to develop a better understanding of the functional importance of novel metabolites through screening. Several modern analytical tools have been used to elucidate the functionality and characterisation of the metabolic network. This study of metabolic network regulation forms the newly emerging field regarded as metabolomics. This chapter highlights various types of metabolites in seaweeds as phyco-nutraceuticals. Also, various techniques have been discussed to characterize and profile seaweed-derived metabolites. The chapter also emphasises the seaweed-derived bioactive compounds' potential and economic importance.

Keywords: Characterization, Functional Food, Metabolites, Proximate Analysis, Phyco-Nutraceuticals, Seaweed-Proteins, Seagrass.

INTRODUCTION

Over many years, seaweeds have been continuously discussed for their socio-economic benefits among researchers. Seaweeds are considered for abundant biomass originating from the marine source. These benthic marine algae are

* **Corresponding author Mangal S. Rathore:** Division of Applied Phycology and Biotechnology, CSIR-Central Salt and Marine Chemicals Research Institute, Bhavnagar, Gujarat, India; Tel: +91 278 2567760 (Ext. 6240); Fax: +91-27--2565762; E-mail: mangalrathore@csmcri.res.in

highly diverse, simple and multicellular autotrophic organisms. They are primitive plants lacking actual stem, root or leaves and can be free-floating or attached to any hard rocky surface. Based on pigmentation, their diversity is categorised into three broad groups of red algae (Rhodophyta), brown algae (Phaeophyceae) and green algae (Chlorophyta) [1]. Due to the continuous over-growing human population, the increasing demand for food cannot be overlooked. In addition, modernisation leads to a change in lifestyle patterns due to which many health issues have been raised. It urges the discovery of new sources of functional food other than the traditional food sources. These functional foods are called nutraceuticals that can provide health and medicinal benefits. Today demand for nutraceuticals has increased tremendously around the globe as people are getting aware about the health benefit of consuming nutraceuticals. Seaweeds are well-identified for their unique biochemical composition and can be directly consumed as a food source. Seaweeds are highly enriched with proteins, carbohydrates, amino acids, minerals, vitamins, dietary fibers, and other secondary metabolites. In many Asian countries like Japan, Korea, Philippines, and China, seaweeds are directly consumed as food sources [2]. Nowadays, many European countries have gained a key interest in seaweed-derived nutraceuticals. Their distinct composition can be a naturally abundant source of numerous bioactive substances used in the nutraceutical aspects. They are also identified as one of the key organisms for producing various novel bioactive compounds [3]. These metabolites have broad-spectrum bioactive functionality that can be applied as anticancer, antioxidant, anti-inflammatory or in various nutraceutical, pharmaceutical or other industrial applications. In some countries like Scotland and United States, seaweeds are used as animal feed and additives. Many species of seaweeds have been valorised as hydrocolloids, animal feed additives, and food colorants [4]. It has been discovered that marine flora is more diverse than terrestrial, and seaweeds are considered an excellent reservoir for producing novel metabolites. However, seaweeds are an excellent source of primary protein, carbohydrates, vitamins, minerals and essential fatty acids. They are also a diverse source of antioxidants, pigments and polyphenols [5]. Seaweeds have a distinct type of carbohydrates like sulphated polysaccharides (fucoidan, agar, ulvan, and carrageenan). Unlike terrestrial plants, they are marine-originated, producing distinct or novel metabolites that can have various bioactive properties. Seaweeds residing in the tidal zones face extreme environmental conditions due to various biotic and abiotic stresses. These factors create rapid fluctuations in the seaweed growth environment and force them to adapt towards these extremities. These sessile organisms, during the adaptations, produce various primary and secondary metabolites [6]. Hence, seaweeds can be considered promising feedstocks in producing novel nutraceutical products. Even though, marine flora has immense diversity and potential but it is still less explored; hence attempts have to be made to understand its structural complexity and economic viability.

SEAWEEDS AS A SOURCE OF METABOLITES

Primary Metabolites

Seaweeds comprise primary metabolites, constituting carbohydrates, proteins, lipids, minerals and vitamins. These compounds require during the primary metabolism, which plays a role in growth and survival. These primary metabolites and their derived compounds can be of much nutraceutical importance. Some of the primary metabolites of nutraceutical value are further described.

Carbohydrates

Seaweeds are mainly composed of carbohydrates, ranging from around 4%-75% of dry biomass. It forms structural organisation and also as energy source. In seaweeds, carbohydrates are mainly present in the form of polysaccharides which primarily form the cell wall. Some of the carbohydrates that mammalian enzymes cannot degrade are called dietary fibres, which are also beneficial in nutrition.

Polysaccharides

Seaweeds are abundant source of sulphated polysaccharides, mainly present in their cell wall. Polysaccharides are chains of simple sugars linked by glycosidic bonds. All different classes of seaweeds have distinct cell wall compositions. Some colloidal polysaccharides are quite different from terrestrial plants and have various bioactive properties. In seaweeds, most of their cell walls comprise cellulose, a polymer of glucose subunits. Ulvan, a water-soluble sulphated polysaccharide, is found in ulvales species like *Ulva fasciata, Ulva lactuca*, and Ulva ohnii [7].

It occupies around 8- 29% of dry weight in green seaweeds and comprises the repeating subunits of rhamnose, xylose, iduronic acid, and glucuronic acid. It has been suggested that ulvans can be a better source of rhamnose and iduronic acids than terrestrial plants. Ulvan has many bioactive properties and can serve as an antioxidant, anticoagulant, and antibacterial agent [6, 8]. It can also control cholesterol and blood sugar levels in the blood.

In red algae, the most predominant polysaccharide is agar and carrageenan. Agar is one of the most abundant sulphated polysaccharides present in the red algae. It mainly forms the structural part of red seaweeds and is composed of the polymer of agarose and agaropectin. Both subunits are polymers of galactose with some attached sulphated groups. The agar content ranges from around 21- 40% of dry biomass in red seaweeds. Agar is mainly found in *Gelidiaceae* species and *Gracilariaceae* species. Agar has much economic importance and wide industrial applications. It is commonly used as gelling, thickening agent and stabilizer in food and beverages. Additionally, agar has bioactivities in cytotoxicity, anticoagulant, collagen supplement, and antioxidant. *Gracilaria gracilis* has the highest agar content in red seaweeds and other preceding species are *G. corticata, G. Dura* and *G. edulis* [7].

Carrageenan is a sulphated water-soluble seaweed galactan, which is a polymer of α-D-galactopyranose and β-D-galactopyranose. It is used as a thickening and stabilizing agent as the agar-agar. It is of five types: κ-carrageenan, ι-carrageenan, λ-carrageenan, μ-carrageenan and ν-carrageenan. It is mainly found in *Eucheuma* sp., *Kappaphycus* sp., and *Chondrus crispus.* Carrageenan is a highly bioactive compound used in both medical and nutraceutical aspects. Carrageenan has potent immunomodulation, anticoagulant, antiviral, anti-hypercholesteremic and anti-tumor activity. Other than these polysaccharides, porphyran and xylan polysaccharides are also found in the red species [9].

Some colloidal polysaccharides in brown seaweeds are found, like alginates and laminarin. Alginate is the colloidal polysaccharide commonly found in brown seaweeds, mainly composed of the subunits of guluronic and mannuronic acids. It also has thickening, stabilizing and bioactive properties. Alginates are abundantly found in *Fucus, Laminaria, Saccharina, Macrocystis pyrifera,* and *Sargassum.* They have many bioactive properties like anti-tumor, anti-bacterial, anti-oxidant and tissue healing [10].

Laminarin is another predominant storage polysaccharide found in brown seaweeds and mainly comprises glucose subunits. In some cases, mannitol and other uronic acid residues may also be present. Laminarin is a highly bioactive compound with anti-tumor, anticancer, anticoagulant and antibacterial properties. Laminarin is abundantly found in the *Laminaria, Sargassum and Fucus* species.

Fucoidan is another sulphated polysaccharide found in brown seaweeds. Fucoidans are mainly composed of the subunits of the fucose and sulphate groups. In brown algae, the most predominant species for the fucoidan are *Ascophyllum nodosum, Undaria pinnatifida,* and *Sargassum* [11]. They exhibit many bioactive properties like anti-tumor, anti-cancer, anti-bacterial, anti-viral, anti-inflammatory and anti-coagulant. Fucoidan also possesses anti-aging properties and can reduce wrinkles, blemishes and freckles.

Dietary Fibers

Some seaweed polysaccharides are also consumed as dietary fibres as they may not degrade with the mammalian enzymes. These types of polysaccharides are significant in the anti-cholesterolemic and anti-hyperglycaemic effects. The dietary fibres can be water-soluble like agar, carrageenan, alginic acid, laminarin and porphyran or insoluble like cellulose, xylan and mannans. The soluble fibres form viscous gels in the intestinal tract with water, while the insoluble fibres retain water and regulate the intestinal effect [12]. These fibres have properties like antibacterial, anticoagulant, anti-inflammatory and prebiotic [13].

Proteins

Another most abundant primary metabolites are proteins required to regulate, maintain, and transport the molecules and ions. These form the antibodies, hormones and enzymes which have various regulatory roles. Seaweeds can be considered a good source of proteins, an alternative to terrestrial plant-derived proteins. Proteins found in seaweeds constitute essential amino acids beneficial for human consumption. Red and green algae constitute the highest range of proteins [2], around 10%–47% of dry biomass, while brown algae precede them from 5%-24% of dry weight [12]. In brown seaweeds, *Undaria pinnatifida* is reported to have high protein content, which could be around 24%, while in red seaweeds, *Porphyra* species can have around 47%. The protein content in seaweeds largely depends on environmental conditions. Some seaweed like *Undaria pinnatifida, Pyropia* and *Porphyra*-derived proteins scores equivalent to the plant-based or egg proteins [14]. Seaweeds are also reported to contain essential amino acids whose proportions were almost similar in all three phyla [15]. Some reports concluded that amino acid levels in species like *Ulva, Kappaphycus, Hypnea and Gracilaria* are higher than the WHO-recommended milk or egg proteins [16]. Seaweeds can be used for animal feed due to the presence of protein constituting a good proportion of essential amino acids. Seaweed-derived peptides and extracts are highly bioactive and occur as enzymes and lectins. These can be used for therapeutic purposes like antioxidant, cardioprotective, antibacterial, or antidiabetic [17, 18]. Lectins derived from red seaweed *Eucheuma* were reported to have high hemagglutination activity. Seaweed-derived peptides have cosmeceutical applications as they are rich in arginine, glycine, and lysine and stimulate collagen biosynthesis [19].

Pigments

Seaweeds are a rich source of pigments, and it is one of the factors for their characteristic differentiation. Chlorophyta or green seaweeds possess chlorophyll and carotenoids. Red seaweed or Rhodophyta contains phycoerythrin, phycocyanin and phycobilin, whereas brown seaweeds or Phaeophyta contain fucoxanthin as their major pigment [18]. Seaweed pigments have colorant and bioactive properties. Red seaweed pigments like phycocyanin and phycoerythrin can be used in the food industry as colouring agents in ice cream and jellies. Also, some reports suggested phycoerythrin can be used in photodynamic therapy. It also has antihypertensive, antioxidant and immunosuppressive biological activities [19]. In green seaweeds, chlorophyll and carotenoids are used as food colorants in food industries. Some reports also suggested that these can be bioactive as antioxidants, immune-modulatory and anti-inflammatory [20]. Fucoxanthin found in brown seaweeds like *Undaria* is used as an anticancer and anti-tumour. Phycobilin proteins found in red seaweeds are also antioxidants and anti-inflammatory [21].

Lipids and Fatty Acids

Seaweeds contain lipids as another important class of primary metabolite, but their content only ranges between 0.3% - 5% of their dry weight. These compounds comprised fatty acids, waxes, fat-soluble vitamins and sterols. These molecules are an important constituent in the cell membrane and function for energy storage and cell signalling. Seaweeds are rich in essential fatty acids such as polyunsaturated acid (PUFA) and monounsaturated fatty acid (MUFA). Among PUFA, eicosapentaenoic acid (n-3) and docosahexaenoic acid (ω-3) and among MUFA, linoleic acid (n-6) and arachidonic acids (ω -6) are primarily found in seaweeds. Both PUFA and MUFA are essential from the dietary point of view found in seaweeds [22]. Red seaweeds contain high eicosapentaenoic acid, oleic acid, arachidonic acid, and palmitic acid, while brown seaweeds have higher linolenic and oleic acid content. In green seaweeds, a higher amount of DHA with palmitic, oleic and linolenic acids is found. A higher amount of lipid is found in seaweeds in species like *Ulva, Sargassum, Caulerpa,* and *Porphyra* [23].

Other than fatty acids, seaweeds also contain sterols and vitamins with bioactive properties. Seaweeds are rich in many essential vitamins, which constitute the class of sterols. Vitamin E or α-tocopherol and phytol are found highest in *Sargassum wightii*, a brown seaweed and *Gelidiella acerosa,* red seaweed. Vitamin A or β-carotene are higher in red seaweed like *Porphyra* and green seaweeds like *Ulva* sp. Another sterol type is fucosterol, abundant in brown seaweeds: Fucosterol has high health benefits as an antioxidant, anti-inflammatory, and antiphotoaging. *Undaria, Laminaria and Sargassum* species were reported for high levels of Fucosterols in brown seaweeds.

Secondary Metabolites

Unlike primary metabolites, secondary metabolites are not directly involved in growth and development. These metabolites are mainly produced during extreme conditions, including biotic or abiotic stress. Seaweeds produce a diverse range of secondary metabolites that can be novel and highly bioactive. Secondary metabolite isolated from seaweeds, such as lectins (carbohydrates binding proteins), is known for flavour enhancer, antibacterial and antifungal properties [24]. It is found in the species of *Ulva, Euchema* and *Gracilaria*. Green seaweeds are also a rich source of alkaloids with antimicrobial, antifungal, and anti-inflammatory properties. Bis-indole alkaloids like racemosins A and B are majorly found in *Caulerpa racemosa*, while Caulerpin and Caulerpic acid were found in *C. racemosa, C. lamourouxii, C. sertularioides* and *C. serrulata* [25, 26]. Other novel prenylated compounds like Caulerprenylols A and B found in *Caulerpa* species have antifungal activity. Some other secondary metabolites like terpenes, sesquiterpenes, diterpenoids, triterpenoids and sterols were also found [26]. In brown algae, more diverse secondary metabolites were found than the green algae and characterised by the presence of phlorotannin and phenolic compounds [27]. Other than these, many types of terpenoids and acetogenins are found. Brown seaweeds are also found to be rich in fucosterol [28]. Red seaweeds are found to have the most diverse class of secondary metabolites. Some reports suggested that red seaweeds are rich in cholesterol [18]. Seaweeds like *Laurencia* are rich in halogenated secondary metabolites like terpenoids and acetogenins. These types of secondary metabolites have wide biotechnological and pharmacological applications. These could be used as antifouling, antiviral, antibacterial and anticarcinogenic agents [29].

Mineral Content of Seaweeds

Seaweeds are well reported as the source of minerals, and variations in the content of these minerals depend on the seaweed type. Recently seaweeds are in the latest trend for producing nutraceutical supplements; hence can be used as a source of vitamin supplements. A report stated that brown seaweeds were used to produce soda ash, potash, and iodine production. Some species were reported to have a higher value than the recommended by (Recommended dietary intake) RDIs [30]. Some seaweeds were reported to have higher iodine, iron, manganese and zinc. In seaweed species like *Laminaria, Sargassum, Undaria, Ascophyllum, Palmaria, Gelidium, Gracilaria, Ulva, Codium* and *Monostroma* [1]. Seaweeds are also well-known contributors to iron and many species were reported for a higher range of iron. Species like *Codium fragile, Monostroma, Gelidium, Porphyra, Sargassum* and *Laminaria.*

Proximate Analysis

In addition to primary and secondary metabolites, some other components like ash contents and moisture contents are estimated as a proximate analysis. Proximate analysis of seaweed samples is performed to assess the percentage of moisture, volatile contents, fixed carbon and ash content. A proximate analysis comprises a combination of the standard test; a brief summary is as follows:

Moisture Content Analysis

It is determined by the modified Association of Official Analytical Collaboration AOAC method [31]. In this method, samples were placed in a crucible and dried under an oven overnight at a temperature around 105°C- 110°C till constant weights are obtained [32].

Ash Content

Ash content is determined with the modified AOAC method [31], in which, after moisture content determination, dried samples were used. These samples were burnt in a muffle furnace overnight at 550°C temperature till biomass is converted to ash [33].

Crude Protein and Total Amino Acid Analysis

Crude protein is analysed by the modified method described by the AOAC [31]. The most commonly used method is the Kjeldahl nitrogen method. This protein percentage is calculated by the multiplication of the conversion factor of 6.25. The amount of the protein is further expressed as the gram of proteins per 100 gram samples [34]. Total amino acids are estimated by the ninhydrin method and in which first the extraction is done, and then the ninhydrin is used to visualize the amino acids [35].

Crude Lipid Content Analysis

It is done by the gravimetric method, the modified Folch method described by the AOAC [31]. Crude lipids extraction is done by mixing the chloroform and methanol and then evaporating to dryness using nitrogen gas [32, 36].

Crude Fibre Content Analysis

The crude fibre content is determined by the gravimetric method [37]. Extraction of seaweed samples is done using the enzyme amylase and absolute ethanol. The undigested residue forms the precipitation collected, dried, and weighed, giving total dietary fibers [38].

Carbohydrate Content

Carbohydrate content is calculated by difference calculation [%Carbohydrate =100% - (%moisture + %ash + %crude fibre + %crude protein + %fat)]. Estimation of carbohydrates is done by the phenol sulphuric method [36, 38].

Determination of Total Phenolic Content

To determine the total phenolic content, extraction of the samples is performed. Total phenolic content is determined with the colorimetric method using the Folin- Ciocalteu reagent, and phloroglucinol is used as a standard. Absorbance is observed using a UV-Visible Spectrometer at the wavelength of 765 nm [39].

IDENTIFICATION AND CHARACTERIZATION OF METABOLITES

Seaweeds are large multicellular photoautotrophic organisms that are found in a marine environment [40]. They synthesize many compounds that enable them to survive in such a harsh environment. Low molecular mass compounds that an organism synthesizes (or end product of cellular function) are known as metabolites. The science that deals with identifying and characterizing these metabolites is known as metabolomics [41, 42]. It has emerged as an advanced analytical tool to study the relative concentration of metabolites. To study the metabolome's dynamics and analyse fluxes in metabolic pathways, the identification of the participating metabolites should be as unambiguous as possible [43]. Metabolomic studies can be done through a targeted or non-targeted approach. The targeted metabolomics approach is used for the identification and characterization of particular metabolites or the limited number of metabolites based on prior information.

In contrast, the non-targeted approach focuses on the identification and quantification of global metabolites that are synthesised by seaweeds/ organisms [23, 44]. Therefore, the identification and quantification of metabolites must be made with the help of sophisticated instruments. The commonly used chromatographic tools coupled to MS are GC-MS (gas chromatography mass spectrometry), LC-MS (liquid chromatography-mass spectrometry), and CE-MS (capillary electrophoresis mass spectrometry). Nuclear magnetic resonance (NMR) spectroscopy and Matrix-assisted laser desorption and ionisation time of flight mass spectroscopy (MALDI-TOF MS) also used whereas, FT-ICR-MS and orbitrap MS are direct MS based techniques applied for metabolic profiling of a large number of samples. MS based chromatographic techniques are most preferred because of their high reproducibility and sensitivity [23, 45].

Gas Chromatography-Mass Spectroscopy (GCMS) Based Metabolite Profiling

GC-MS is a robust analytical tool that can be used for the identification and quantification of a large number of metabolites from a single seaweed extract. It is rapid, reproducible and comprehensively covers pathways involved in primary metabolism [45, 47]. GC-MS is the most popular global method used for metabolite quantification, even at low concentrations [23]. Identification of the whole metabolome of seaweeds cannot be obtained by using a single analytical tool, but GC-MS has a relatively broad coverage of metabolite classes. This method is utilised for the identification of volatile and thermostable compounds. Metabolites involved in primary metabolism, such as sugar, sugar alcohols, organic and amino acids (non-volatile), can be identified using this technique after converting them to volatile compounds [45]. In routine, non-volatile compounds are converted to volatile compounds through derivatization, which is the most important step. In the derivatization procedure, one or more functional groups of chemical compounds are modified to produce a chemically related derivative using derivatizing reagent [46]. A standard derivatizing procedure for GC-MS involves two steps, *i.e.*, methoxyamination reaction followed by silylation reaction.

In the first step, functional groups of metabolites are converted to oxides by using methoxyamine hydrochloride. The other reaction involves the derivatization of active hydrogens of -COOH, -NH and is converted to more volatile and thermostable TMS derivatives by the use of N-methyl-trimethy-lsilyltrif-luroa-cetamide (MSTFA) or N, O-bis-trimethylsilyltrifluoroacetamide (BSTFA) [23, 48]. The spectrum obta-ined by the GC-MS is complex and cont-ains numerous overlapping peaks that need to decon-volute during the analysis. Soft-wares that can be util-ised for deconv-olution are Auto-mated Mass Spectral Decon-volution and Identi-fication System (http:// chemdata.nist.gov/mass-spc/ amdis/) and metabolite detector (http:// md.tu-bs.de /) [47, 49]. GC-MS has been used for the metabolites profiling in members of Chlorophyta and Phaeophyta members to reveal their nutraceutical potential [50 - 52].

Liquid Chromatography Mass Spectrometry (LCMS) Based Metabolite Profiling

LC-MS is another analytical platform that is widely used for metabolic profiling. The accuracy and resolution of LC-MS-based metabolomics have been improved over the last decades but covering the complexity of metabolome remains a challenge. An individual sample can contain thousands of metabolites, but far fewer detected metabolites. The detection of metabolites largely depends on the experimental design, such as sample extraction procedures and the choice of LC-MS methods [53]. It separates the metabolites based on their interaction with the stationary phase, after which they pass through a mass analyser [54]. This technique does not involve any derivatization, and metabolites can be extracted directly in simple alcohol solvents. LC-MS is preferably used to identify semi-polar metabolites and secondary metabolites but depending upon the type of column, primary metabolites such as amino acids and organic acids can also be analysed [55]. It utilizes a soft ionization technique such as electrospray ionisation (ESI) or atmospheric pressure chemical ionisation (APCI) for ionisation and produces protonated (+ mode) or deprotonated (-mode) m/z. This technique is highly reproducible in m/z values, but significant differences in retention time can be observed. Therefore, there is a need to run a standard quality control sample with the experimental sample. The spectra generated through LC-MS can be processed and analysed through instrument manufacturer software or through publically available tools like XCMS, MZmine and MetAlign [56]. These platforms identify chromatographic peaks, aligns them and statistically assess the comparative data based on intensity and fold change. Differentially regulated metabolites can be identified using databases such as METLIN (http://metlin.scripps.edu) and LIPID MAPS (http://www.lipidmaps.org/) [57].

Capillary Electrophoresis Mass Spectrometry (CEMS) Based Metabolic Profiling

CE-MS is a complementary technique to liquid chromatography as it also separates the metabolites on the basis of their charge to mass ratio. It is a potential technique that is used for metabolomic studies. As the name suggests, capillary electrophoresis is done in a capillary having fused silica whose ends are dipped in buffer reservoirs and voltage is applied. The electrical current must be maintained in the running buffer and MS-compatible background electrolytes are needed. The separation of different metabolites takes place upon their electrophoretic mobility [23, 58]. This technique suffers from poor sensitivity as it requires only a small sample. Using platinum needle for sample spray, change of ammonium acetate with triethylamine in background electrolyte and sheath liquid enhanced the metabolic ionisation and coverage. Earlier, only a few research groups used CE-MS for metabolomic studies due to a lack of standard operating procedures. Soga *et al.* [59] introduced the first CE-MS method for metabolomics, profiled more than 8000 samples, and provided the absolute quantification of 94 polar metabolites in plasma. Li *et al.* [60] performed metabolic profiling in tobacco leaves in a bare-fused silica capillary and detected highly polar and charged metabolites such as sugar phosphates and polyamines.

Nuclear Magnetic Resonance Spectroscopy (NMR) Based Metabolic Profiling

The metabolome of an organism is very complex as from one part of tobacco leaves, about 3000 metabolites have been identified. These metabolites are different in their stability, polarity, concentration and chemical behavior. Qualitative and quantitative identification of all these metabolites in a single experiment is very difficult. In this context, Nuclear Magnetic Resonance (NMR) is an analytical technique that allows the identification of primary metabolites (sugars, organic acids, amino acids) and secondary metabolites (alkaloids, flavonoids and terpenoids) in a single experiment [61]. The intensity of the spectral peak obtained from NMR is directly proportional to the metabolite concentration; hence it provides a scope for comparative quantification of metabolites [47, 61]. The major drawback of this technique is that it is less sensitive than MS-based techniques and also pH sensitive [62].

NMR involves the use of one-dimensional and two-dimensional techniques for metabolite profiling. 1H NMR is simple, fast, have high quantitative accuracy and does not require a standard for quantification. Peak overlapping in 1D NMR is a major problem in identifying metabolites; therefore, 1D spectra are dispersed to 2D experiments. The most commonly used 2D experiments are ^{1}H–^{1}H *J* resolved spectroscopy, HSQC (hetero-nuclear single quantum correlation), TOCSY (total correlation spectroscopy) and HMBC (hetero-nuclear multiple bonds correlation) [47]. ^{1}H–^{1}H *J* spectrum is useful for the signals in a crowded area, and it provides information about the plot of the chemical shift and the coupling constants metabolites [63].

COSY shows the correlation of protons with mutual spin-spin couplings, whereas TOCSY shows spin systems of coupled protons in the same molecules. TOCSY is useful in carbohydrate and amino acid detection due to the accuracy it ensures in correlation [64]. The HMBC, on the other hand, helps in assigning long-range correlations between carbon and protons, which leads to the identification of resonance from the same molecules. HSQC has recently found remarkable application in metabolomics, as certain metabolites in the sample extract can be discerned by comparison with the available reference compounds. It is preferably utilized for metabolic profiling, specifically for determining particular compounds [65]. Complete identification of metabolites even through 2D NMR is not accurate. Further, the coupling of NMR with HPLC is employed for metabolite identification as it provides the NMR chromatogram after separation through HPLC [66].

MALDI-TOF-MS

Nowadays, Matrix-assisted laser desorption/ionization-time of flight mass spectrometry (MALDI TOF MS) is another widely used technique in the metabolomics approach. It has high sensitivity, utilizes a soft ionization technique and allows the rapid detection of metabolites in a sample [10]. It requires a very small sample (0.1-1 ul) with matrix and analysis time depending on the amount of laser shots required for the experiment [67]. The most commonly used matrices for MALDI are α-cyano-4-hydroxycinnamic acid (CHCA) and 2,5-dihydroxybenzoic acid (DHB). Ye *et al.* [68] combined the novel matrix 1,8–bis (dimethyl-amino) naphthalene, DMAN with DHB, allowing the detection of sugars, lipids, amino acids, organic acids, flavonoids and their conjugates. The major drawback of this technique is the poor reproducibility of the signal intensity, which is largely determined by the matrices used [67]. MALDI hybrid systems such as MALDI-IT-Orbitrap and MALDI-FT-ICR can be used to image the metabolites in tissue. It is used to understand the spatial distribution of the metabolites present within a cell. For this, sample extraction is not required, but cryo-sectioning is required with the help of cryo-microtome. The analysis is usually performed with ionization techniques that ionize and desorbs the analytes directly from the tissue [23].

ECONOMIC IMPORTANCE

Riverine, estuarian and coastal populations always utilise seaweeds for various home remedies and food purposes for a long time. However, macroalgae have taken over the major industries like food, pharmaceuticals, nutraceuticals, cosmeceuticals, biofuels, and fertilizers [23, 69 - 72] in many maritime countries. In countries like Japan, Korea, and China, seaweed cultivation has become a major industry. Apart from these, diverse countries, namely Indonesia, Tanzania, Vietnam, and India, have also validated the positive correlation of socio-economic impacts with seaweed cultivation. It is a relatively simple technique with a minimum initial capital requirement, short life span and high return of investment [73, 74].

Well-known seaweed commercial products include sulphated polysaccharides like carrageenan, agar-agar, and alginate. Seaweed farming for hydrocolloids sufficiently corroborated it as a profitable project for upholding the socio-economic status of coastal communities [75, 76]. Populations living along coastal areas are closely associated with oceanic resources, particularly fisheries. But intensive fishing and current climatic conditions largely affect the breeding and population of fishes, causing a remarkable depletion of this resource, and almost the rest of the fishing grounds are replaced by big fishing operations. This makes the living of coastal populations quite hard [77]. In such a scenario seaweed cultivation stands as an excellent proposition to improve the living status of coastal populations [78, 79]. Nevertheless, a fixed income from seaweed cannot be defined as it is also affected by the logistical capabilities of the particular countries and the lack of a world market price.

Global production of seaweeds aquaculture production increased from 10.6 million tonnes in 2000 to approximately 32 million tonnes in 2015. Seaweed's cultivation increased triple times in 16 years, but production stabilized due to the slowing growth of farming in South Asia [80]. For instance, an average farmer in countries like the Philippines and Indonesia normally earns between USD 0.60 and USD 1.40/kg, whereas in remote countries like Tanzania and the Pacific Islands, it is less than USD 0.50/kg [73].

Behind the economic impact of seaweed farming is its multifarious applications in biomedical areas. Seaweeds have been identified to produce metabolites that are therapeutic, nutraceutical and food supplements, and cosmeceuticals. Therapeutic properties comprise of antioxidant [81 - 83], anti-inflammatory [84, 85], anticoagulant [86], antimicrobial [87, 88], antiviral [89], antitumor [90] and hyper cholesterolaemia activities [91, 92]. It has been perceived that marine macroalgae have been utilised since the seventeenth century due to their potential bioactive properties [93]. Fucoxanthins [94], phycoerythrobilin [95], sulphated polysaccharides [96 - 98], phlorotannins [99 - 101], phloroglucinol [102], catechols, and hesperidin [103] include some of the bioactive compounds isolated from various seaweeds.

Seaweed-derived nutraceuticals and food products are well appreciated all over the world. The foremost member of this category is always agar which is used in puddings, jellies, cakes, and desserts. Major producers of agar are *Gracilaria* and *Gelidium* sp. Furthermore, carrageenan, alginate, ulvan, fucoidan, and laminarin are the other important seaweed polysaccharides widely available in the market [16]. Additionally, bioactive peptides with potential applications in nutraceuticals are also obtained from seaweeds [17].

The term cosmeceutical broadly comprises biological potentials such as anti-melanogenesis activity, anti-tyrosinase activity, anti-aging and anti-photoaging activity, anti-wrinkle, anti-oxidant, anti-acne, anti-inflammatory activities and wound healing properties [104, 105]. Yang *et al.* [106] reported that Phlorotannins from brown algae could recover radiation dermatitis. Extract of *Grateloupia turuturu,* a red alga, has Antioxidant, UV absorbance, photoprotection anti-elastase, and anti-hyaluronidase properties, as reported by [107]. Also, sulfated polysaccharides from *Ecklonia maxima* (brown alga) have been reported to exhibit anti-oxidant, anti-melanogenesis and photoprotective as per the studies of Wang *et al.* [108].

According to the report of Chopin and Tacon, seaweeds (32.4 million tones) represent 51.3% of the total production of marine and coastal aquaculture [109]. Loss due to epiphytes, diseases, environmental stresses, transportation issues, *etc.* create a barrier in seaweed farming, demanding more attention from the government to carry out successful farming, harvesting, and marketing and thereby contribute to the elevated socio-economic status of coastal populations.

APPLICATIONS OF SEAWEED METABOLITES

Seaweeds are also used as food in Asian countries like Korea, Japan and China because of their nutritional value. With the increasing demand for natural alternatives for various nutraceutical and cosmeceutical products, seaweed metabolites have started gaining much attention. These metabolites have very less or almost negligible side effects on humans and have a very wide variety of applicability [110]. Seaweeds are considered one of the major sources of biomolecules such as polysaccharides, phenolic, lipids, vitamins and even proteins with specific activities against cancer, diabetes, microbial diseases, *etc* [111]. Bioactive phyco-metabolites also show antibacterial and antiviral properties, making them a potential source for pharmaceutical industries. Terpenes, phenol, lipids, alkaloids, *etc.*, extracted from macroalgae have shown microbial properties [112]. Macroalgae are known to survive in continuously altering environments, making them adaptable to environmental stress like fluctuating oxygen levels, salinity, humidity, *etc*. Their ability to survive makes them a vibrant source of antioxidants and polyphenols. Their cell walls also vary concerning the environment, which brings wide structural diversity to their bioactive products [113]. Mineral ions associated with the sulphation in the cell wall also play an important role as dietary supplements [114].

Macroalgal Metabolites as Nutraceutical

Polysaccharides are one of the major components of the algal cell wall. These biomolecules can be derived and used in many different industries. Polysaccharides like agar, alginate, and carrageenan have jelling properties and retain moisture. This moisture retaining property of seaweed is exploited by many cosmetic industries. Alginate is also selectively antimicrobial as it can stimulate the growth of gut microflora and inhibit the pathogenic bacteria like *Salmonella* [115]. Some of these polysaccharides are also used for the production of capsules or drug delivery vehicles in the case of lipophilic drugs [116]. Nano particles of alginate, carrageenan and fucoidan that form complexes with chitosan are used to deliver drugs against cancer and certain microbial infections [117]. Ulvan derived from *Ulva lactuca* is used as a packaging material in pharmaceutical industries. It is also used in food industries as it can aid in the colloidal formulation of food items and also behaves as a prebiotic [118]. *Gracilaria* sp. produces agar used as a gelling agent and can thus produce a vegan alternative of gelatin [119, 120]. Some of the sulfated polysaccharides are also known to have beneficial effects on fibroblast and help reform collagen molecules. Galactofucan, derived from the brown seaweed *Saccharina longicruris,* was reported to influence fibroblast growth rate [121].

Red alga is an excellent source of agar and carrageenan. Carrageenan derived from *Kappaphycus alvarezii* and *Chondrus crispus* is also a food additive [11]. Commercially produced high molecular weight carrageenan (HMWF) have great physical properties and is thus a boon to food industries [122]. It is widely used as a thickening agent and a stabilizer in various food products. Industries rely on *Gracilaria* and *Gelidium* for the production of agar. As agar is a potent gelling agent and emulsifier, it is used in various food products [123].

Brown alga is again a source of another polysaccharide known as fucoidan. The composition of fucoidan largely depends on the habitat of the seaweeds. *Fucus vesiculosus* and *Sargassum mcclurei* are very good sources [124]. Reports show that fucoidans possess anti-cancerous properties and thus can be used as drugs. Synthetic low molecular weight fucoidans (LMWF) have shown cytotoxic effects against various human cancer cell lines [125].

Seaweeds are also rich in various essential amino acids. Even though macroalga is very low in lipid content, they are a rich source of MUFA and PUFA. As the human body cannot synthesize PUFA, seaweeds can very well be exploited to produce PUFAs for the vegan and vegetarian population [30]. *Ulva* sp., *T. atomaria, C. spongiosus* and *B. secundiflora* produced α-linolenic acid (ALA) and eicosapentaenoic acid (EPA) in considerate amounts [126]. Macroalgal lipids can also be used against autoimmune and inflammatory diseases. All three classes of seaweeds hold equal importance in terms of different types of fatty acids. PUFA is also a potent drug delivery vehicle for cancer cells because of its lipophilic properties [127].

Laminarin derived from *Laminaria* and *Saccharina* produces butyric acid through fermentation. Short chain fatty acids like butyric acid help in the apoptosis of mainly colon cells with DNA damage. Seaweed-based dietary fibres help in digestion and also reduce gastro-intestinal inflammation. It is also used as a prebiotic [128]. Macroalga also produces various vitamins and minerals and can be used to manufacture dietary supplements. *Porphyra umbilicalis, Pyropia tenera, Grateloupia turuturu* have shown the presence of vitamin C, B12 and vitamin E, respectively. *Porphyra umbilicalis* and *Crassiphycus changii* have also shown considerable amounts of vitamin C. The *Codium fragile,* and *Gracilaria chilensis* contain higher vitamin A than carrot. In addition, *G. asiatica* and *G. livida* are major carotenoid sources like fucoxanthin and zeazanthin, respectively [2]. The presence of iron, iodine, and manganese have been reported in *Palmaria palmata*, whereas *Chondrus crispus*, *Porphyra* spp. and *Gracilaria* spp. have shown the presence of zinc in considerable amounts [129, 130].

Seaweeds are also a good source of essential and non-essential amino acids. Red alga possesses the maximum quantity of protein which is approximately 35-20% of their dry weights, followed by green and brown alga [2]. Reports suggest that *Ulva reticulata, Osmundea pinnatifida Ulva lactuca*, Chondrus crispus, Ulva fasciata and species of *Enteromorpha, Gracilaria, Porphyra, etc.* contain a considerable amount of essential amino acids. Therefore, these seaweeds can replace commonly consumed animal and plant proteins [131].

Metabolites with Anti-Cancer Activities

We are very well aware of the adverse effects of anti-cancer drugs and chemotherapy. According to reports, certain polyphenols and polysaccharides produced by seaweeds are used to replace conventional anti-cancer drugs. Brown algae, *Padina gymnospora,* and *Sargassum wightii* have shown cytotoxic effects on human cancer cell lines A549, HCT-15 and PC3. Flavonoids derived from macroalgae can neutralize the free radicals and can interact with enzymes like superoxide dismutase and catalase, thus blocking apoptosis [132]. Brown alga is also a rich source of sulphated polysaccharides, alginate and carotenoids like fucoxanthin. Sulphated polysaccharides show anti-tumor properties and inhibit the proliferation of cancerous cells. Thus, they play a major role in immunomodulation. Alginates derived from *Sargassum vulgare* also showed similar results [133].

Administration of ulvan has also shown a decrease in the inflammation caused due to cytokines released from cancerous cells. It shows that cytotoxic effects on cancer cells and the viability of various cancer cell lines like cervical cancer and human breast cancer have reportedly decreased [134]. Ulvan can also prevent the formation of tumor cells by upregulation of p53 protein that aids in apoptosis [135]. Jesus *et al.* [136] reported the presence of around 184 haloaryl compounds from the extracts of macro algae belonging to all three families. Lithothamnin A, derived from *Lithothamnion fragilissimum* (red algae), could inhibit the proliferation of various human cancer cell lines like melanoma, astrocytoma, ovarian serous adenocarcinoma, colon adenocarcinoma, *etc.*

Phycobiliproteins like phycoerythrin and phycocyanin, which are derived from the red alga, also show anti-cancerous properties by acting as photo stabilizers for tumor cells. *Gracilaria* is the most used seaweed to extract phycobiliproteins [119]. Terpenes and their derivatives, obtained from red algae like *Portieria hornemannii, Plocamium suhrii and Plocamium cornutum* have also shown anti-proliferative properties towards cancer cell lines. *Porphyra* produces porphyran, a type of polysaccharide, which is cytotoxic to cancerous cells and is known to induce cell death. Apart from this, it shows anti-inflammatory properties. Red seaweeds also produce certain phenolic and polyphenolic compounds, which have also been reported to show anti-cancer properties [137].

Metabolites with Anti-Oxidant Activities

The reactive oxygen species produced in the body during the metabolic process can cause skin aging and generate oxidative stress. Phenolics and some sulphated polysaccharides derived from macroalga, show cytoprotective activity by scavenging and modulating the free radicals generated in the body [138]. Zhang *et al.* [139] reported the inhibition of lipid peroxidation in mice using a sulphated galactan extracted from *Porphyra haitanensis* (Rhodophyta). Bioactive peptides derived from brown algae like *Undaria pinnatifida, Sargassum fusiformis, Ecklonia cava* and red alga like *Porphyra yezoensis, Palmaria palmata, Porphyra columbina, Porphyra dioica, etc.* also possess antioxidant activities [140]. Ulvan, a bioactive polysaccharide obtained from *Ulva* (Chlorophyta), shows antihyperlipidemic and antioxidant properties, and is also known to maintain cholesterol levels in mice [11].

Phlorotannin, a secondary metabolite found in brown algae like *Fucus spiralis* is also known to have antioxidant as well as antiallergic activities as it can inhibit the effect of the hyaluronidase enzyme [141]. Islam *et al.* [142] derived haloaryls - odonthadione and odonthalol from red alga *Odonthalia corymbifera.* Both the compounds showed antioxidant activity and could strongly inhibit the tyrosinase enzyme. Caulerpin isolated from *Caulerpa racemosa* shows anti-inflammatory activities by inhibiting the production of reactive oxygen species like cyclooxygenase and lipooxygenase that can cause inflammation [143]. Like many other algal polysaccharides, ulvan also shows antioxidant properties and helps scavenge free radicals by preventing lipid peroxidation. It is also proved that ulvan can increase antioxidant enzymes like catalase, superoxide dismutase, *etc.*, and reduce overall oxidative stress [144].

Application of Metabolites as Nutricosmetic

Algal polysaccharides like alginate, carrageenan and agar can be extracted and used as emulsifiers and thickeners in products like soaps, creams, shampoos, lotions, and shaving creams, sunburn creams, deodorants, *etc*. They help in water regulation on the skin and can thus provide a moisturizing effect. Alginate can be derived from *Isochrysis* spp., *Postelsiapa maeformis, Laminaria digilata* belonging to phaeophyceae [145]. Some species from Chlorophyta have been reported to show anti-aging properties. *Ulva lactuca* and *Codium tomentosum* (Chlorophyta) possess anti-aging, moisturising and lipolytic properties [146]. Ulvan possesses an amphipathic nature, which is very important for cosmetic industries [118].

Many naturally growing macroalgae can synthesize carotenoids which can counteract the damaging effects of ultraviolet rays. These compounds are commonly termed as mycosporins-like amino acids. Red macro alga *Chondrus crispus* and *Porphyra umbilicalis* have been reported to produce MAA and can be used to manufacture sunscreen lotions [147]. Brown algae also produce an additional pigment called fucoxanthin which enables the formation of new collagen and can thus help in maintaining skin pigmentation and also reduce inflammation. Fucoidans, derived from brown algae like *Chnoospora minima* and *Sargassum polycystum* possess antioxidant, skin-whitening, anti-inflammatory, and anti-wrinkle activities [148]. Phlorotannins like phloroeckol and phloroglucinol derived from *Macrocyctis pyrifera* can be used as a nutricosmetic supplement as it helps prevent skin aging [149].

Metabolites Showing Activities Against Various Lifestyle Disorders

Various seaweed metabolites inhibit certain enzymes, including Angiotensin-converting-enzyme (ACE). These compounds are effective against lifestyle disorders like hypertension, obesity, diabetes, *etc*. Clinical trials showed that nori peptides isolated from *Pyropia yezoensis* showed ACE I inhibition in rats as well as humans [150]. Dipeptides extracted from *Undaria pinnatifida* also showed ACE I inhibition. These peptides can be used to cure hypertension as they affect vasodilation. Phlorotannins like dieckol, isolated from *Ecklonia stolonifera* (brown algae), also showed similar properties [151]. Kang *et al.* [152] suggested that dieckol isolated from *E. cava* could prevent diabetes mellitus type II. Various applications of different seaweed metabolites have been summarized in Table **1**.

Table 1. Summary of applications of seaweed-derived products and metabolites.

Metabolite	Source	Application	References
Polysaccharide			
Agar	*Gracilaria*	Anti-cancer, nutricosmetic, food additive	[119, 120]

(Table 1) cont.....

Metabolite	Source	Application	References
Alginate	*Isochrysis, Postelsiapa, maeformis, Laminaria digilata*	Gelling agent, drug delivery vehicle, anti-bacterial	[115]
Carrageenan	*Kappaphycus alvarezii, Chondrus crispus*	Food additive	[11]
Ulvan	*Ulva lactuca*	Pharma industry, food additive, anti-cancer	[118, 134]
Porphyran	*Porphyra*	Anti-cancer	[137]
Galactofucan	*Saccharina longicruris*	Stimulate the growth of fibroblast	[121]
laminarin	*Laminaria, Saccharina*	Production of SCFA	[128]
Fucoidan	*Chnoospora minima, Sargassum polycystum*	Anti-oxidants, cosmetics	[148]
Fatty acids			
PUFA	*Ulva* sp., *T. atomaria, C. spongiosus, B. secundiflora*	Nutritional supplement	[126]
Proteins			
Bioactive peptides	*Pyropia yezoensis Undaria pinnatifida*	ACE I inhibitor	[150]
Myco-sporines like amino acids			
MAA	*Chondrus crispus, Porphyra umbilicalis*	Anti-oxidant, UV protectant	[147]
Pigments			
Caulerpin	*Caulerpa racemosa*	Anti-oxidant	[143]
Phycobiliproteins: Phycocyanin phycoerythrin	*Gracilaria*	Anti-tumor	[119]
Phlorotannin: phloroeckol phloroglucinol	*Macrocyctis pyrifera, Fucus spiralis*	Nutricosmetic supplement, anti-oxidant	[141]
Phlorotannin: dieckol	*Ecklonia stolonifera, E. cava*	ACE I inhibitor Antidiabetic	[151, 152]
Phenolic compounds			
odonthadione odonthalol	*Odonthalia corymbifera*	Anti-oxidant	[142]
Lithothamnin A	*Lithothamnion fragilissimum*	Anti-cancer	[136]
Vitamins			

(Table 1) cont.....

Metabolite	Source	Application	References
Vitamin C	*Porphyra umbilicalis*	Vitamin supplements	[18]
Vitamin B12	*Pyropia tenera*		
Vitamin E	*Grateloupia turuturu*		
Vitamin C	*Porphyra umbilicalis and Crassiphycus changii*		
Minerals			
Iron	*Palmaria palmate*	Dietary supplements	[18]
Iodine	*Palmaria palmate*		
manganese	*Palmaria palmate, C. crispus, Gracillaria* sp.		
Zinc	*Chondrus crispus, Porphyra spp. and Gracilaria spp.*		[130]

FUTURE PERSPECTIVE

Macro-algae do not just play a major role in balancing the environment but have also become very important in commercial aspects. The biological potential of these bioactive compounds has been exclusively reported by scientists worldwide. These compounds can be used to discover drugs that can be used for non-curable diseases. Phyco-polysaccharides and proteins can also work effectively against various cancers, so we might get drugs with minimum cytotoxic effects. Bioactive peptides can also prove to be an antibiotic against a broad spectrum of microbes, replacing the manufacturing of chemical drugs. Even though they are already being used as nutritional supplements in some cases, more metabolites can be ingested to meet several deficiencies. These supplements can benefit the vegan and vegetarian population as the seaweeds can replace meat or a non-vegetarian diet. Seaweeds metabolite can also benefit many nutricosmetic industries and can be exploited to manufacture health supplements for hair, skin and nails. Seaweeds are also a good source of many elements and minerals; their consumption can fulfill the nutrient requirements.

For obtaining more metabolites, the production of seaweeds through cultivation should also be increased. Seaweeds like *Gracilaria, Kappaphycus, Ulva* species, *etc.*, which provide maximum production and have selective commercial values, should be given more attention. Over exploitation of seaweed from the natural sources might cause a bottleneck effect. So, in order to avoid this, we can apply bioprocess engineering, and tissue culture approaches. With the help of metabolic engineering, the production of such metabolites can be increased. Another approach can also be taken into consideration where all the by-products are used and the required metabolites. The methods for the extraction of these metabolites should be standardized such that there is a minimum waste and maximum production. With the establishment of zero-waste, cost-effective set-ups like bio-refineries, the consumer and the manufacturers of this seaweed industry can benefit.

Although the traditional sources for nutraceuticals have been used for several decades, but the global burden of lifestyle-related health disorders has urged us to find new sources of functional foods. Several reports stated that seaweeds are rich in nutrients beneficial for human health. There is a vast potential of seaweeds in consideration for producing bioactive compounds used as dietary supplements, however, many challenges barricades. First, more research has to be conducted to conclude the isolated compound's biological potential. Secondly, biotransformation of the isolated compound is difficult due to its low structural elucidation. Thirdly, collection and extraction have to be performed on a large scale to ensure the sufficient availability of the bioactive compound. Hence new biological techniques must be adapted for the extraction and purification process. Also, metabolic engineering and genomic study can improve the production of these bioactive compounds.

CONCLUSION

Macro algae are a complete package of a variety of biomolecules, having applications in different areas, especially in health, nutraceuticals, and functional foods. Seaweeds can serve as a major reserve for proteins, carbohydrates, and other bioactive compounds, including polyphenols. The metabolites can be a potent weapon against tumors, certain diseases, lifestyle disorders, *etc*. Due to this, macro-algal metabolites have become a centre of attention for various commercial bodies like cosmetics industries, food industries, pharmaceutical units, *etc*. In spite of all the advantages provided by these seaweeds, there are still problems in the cultivation and extraction of bioactive products. Thus, more attention can be given to expanding the cultivation and optimizing these compounds' extraction. This chapter described various metabolites and their uses in various applications. With the expansion of cultivation and improvement of processing of biomolecules, the seaweeds can contribute enormously to the development of the coastal populace through blue economy approaches.

CONSENT FOR PUBLICATION

Not applicable.

CONFLICT OF INTEREST

The author(s) declare no conflict of interest, financial or otherwise.

ACKNOWLEDGEMENTS

CSIR-CSMCRI Communication No.: PRIS- 147/2021. The authors acknowledge the financial support from the Council of Scientific and Industrial Research (CSIR; www.csir.res.in), Government of India, New Delhi for infrastructure creation and in the form of different R&D projects. Authors thankfully acknowledge AcSIR, New Delhi, for registration in the Ph.D. program.

REFERENCES

[1] Shannon E, Abu-Ghannam N. Seaweeds as nutraceuticals for health and nutrition. Phycologia 2019; 58(5): 563-77.
[http://dx.doi.org/10.1080/00318884.2019.1640533]

[2] Ganesan AR, Tiwari U, Rajauria G. Seaweed nutraceuticals and their therapeutic role in disease prevention. Food Sci Hum Wellness 2019; 8(3): 252-63.
[http://dx.doi.org/10.1016/j.fshw.2019.08.001]

[3] Jiménez-Escrig A, Gómez-Ordóñez E, Rupérez P. Seaweed as a source of novel nutraceuticals: sulfated polysaccharides and peptides. Adv Food Nutr Res 2011; 64: 325-37.
[http://dx.doi.org/10.1016/B978-0-12-387669-0.00026-0] [PMID: 22054959]

[4] Fleurence J. Seaweeds as food. Seaweed in health and disease prevention. 2016; 1: 149-67.
[http://dx.doi.org/10.1016/B978-0-12-802772-1.00005-1]

[5] Suleria H, Osborne S, Masci P, Gobe G. Marine-based nutraceuticals: An innovative trend in the food and supplement industries. Mar Drugs 2015; 13(10): 6336-51.
[http://dx.doi.org/10.3390/md13106336] [PMID: 26473889]

[6] Jiménez-Escrig A, Gómez-Ordóñez E, Rupérez P. Brown and red seaweeds as potential sources of antioxidant nutraceuticals. J Appl Phycol 2012; 24(5): 1123-32.
[http://dx.doi.org/10.1007/s10811-011-9742-8]

[7] Ruocco N, Costantini S, Guariniello S, Costantini M. Polysaccharides from the marine environment with pharmacological, cosmeceutical and nutraceutical potential. Molecules 2016; 21(5): 551.
[http://dx.doi.org/10.3390/molecules21050551] [PMID: 27128892]

[8] Lopez-Santamarina A, Miranda JM, Mondragon AC, *et al.* Potential use of marine seaweeds as prebiotics: A review. Molecules 2020; 25(4): 1004.
[http://dx.doi.org/10.3390/molecules25041004] [PMID: 32102343]

[9] Salehi B, Sharifi-Rad J, Seca AML, *et al.* Current trends on seaweeds: Looking at chemical composition, phytopharmacology, and cosmetic applications. Molecules 2019; 24(22): 4182.
[http://dx.doi.org/10.3390/molecules24224182] [PMID: 31752200]

[10] Liu Y, Xue A, Ding L, *et al.* Direct identification and metabolomic analysis of Huanglongbing associated with Candidatus Liberibacter spp. in navel orange by MALDI-TOF-MS. Anal Bioanal Chem 2020; 412(13): 3091-101.
[http://dx.doi.org/10.1007/s00216-020-02555-2] [PMID: 32291516]

[11] Wijesekara I, Karunarathna WK. Usage of seaweed polysaccharides as nutraceuticals. In Seaweed Polysaccharides: Isolation, Biological and Biomedical Applications. Venkatesan J, Anil S, Kim SK. Elsevier 2017; pp. 341-8.
[http://dx.doi.org/10.1016/B978-0-12-809816-5.00018-9]

[12] Peñalver R, Lorenzo JM, Ros G, Amarowicz R, Pateiro M, Nieto G. Seaweeds as a functional ingredient for a healthy diet. Mar Drugs 2020; 18(6): 301. [http://dx.doi.org/10.3390/md18060301] [PMID: 32517092]

[13] Gomez-Zavaglia A, Prieto Lage MA, Jimenez-Lopez C, Mejuto JC, Simal-Gandara J. The potential of seaweeds as a source of functional ingredients of prebiotic and antioxidant value. Antioxidants 2019; 8(9): 406. [http://dx.doi.org/10.3390/antiox8090406] [PMID: 31533320]

[14] Nadeeshani H, Hassouna A, Lu J. Proteins extracted from seaweed *Undaria pinnatifida* and their potential uses as foods and nutraceuticals. Crit Rev Food Sci Nutr 2022; 62(22): 6187-203. [PMID: 33703974]

[15] Dumay J, Morançais M. Proteins and pigments. Seaweed in health and disease prevention 2016; 275-318. [http://dx.doi.org/10.1016/B978-0-12-802772-1.00009-9]

[16] Tanna B, Mishra A. Nutraceutical potential of seaweed polysaccharides: Structure, bioactivity, safety, and toxicity. Compr Rev Food Sci Food Saf 2019; 18(3): 817-31. [http://dx.doi.org/10.1111/1541-4337.12441] [PMID: 33336929]

[17] Admassu H, Gasmalla MAA, Yang R, Zhao W. Bioactive peptides derived from seaweed protein and their health benefits: antihypertensive, antioxidant, and antidiabetic properties. J Food Sci 2018; 83(1): 6-16. [http://dx.doi.org/10.1111/1750-3841.14011] [PMID: 29227526]

[18] Cotas J, Leandro A, Pacheco D, Gonçalves AMM, Pereira L. A comprehensive review of the nutraceutical and therapeutic applications of red seaweeds (Rhodophyta). Life (Basel) 2020; 10(3): 19. [http://dx.doi.org/10.3390/life10030019] [PMID: 32110890]

[19] Dumay J, Morancais M, Munier M, Le Guillard C, Fleurence J. Phycoerythrins: Valuable proteinic pigments in red seaweeds. InAdvances in Botanical Research 2014; 71: 321-43.

[20] Aryee ANA, Agyei D, Akanbi TO. Recovery and utilization of seaweed pigments in food processing. Curr Opin Food Sci 2018; 19(19): 113-9. [http://dx.doi.org/10.1016/j.cofs.2018.03.013]

[21] Kraan S. Kraan S. Pigments and minor compounds in algae. In: Dominguez H, Ed. Functional ingredients from algae for foods and nutraceuticals. Woodhead Publishing 2013; 205-51. [http://dx.doi.org/10.1533/9780857098689.1.205]

[22] Rocha CP, Pacheco D, Cotas J, Marques JC, Pereira L, Gonçalves AMM. Seaweeds as Valuable Sources of Essential Fatty Acids for Human Nutrition. Int J Environ Res Public Health 2021; 18(9): 4968. [http://dx.doi.org/10.3390/ijerph18094968] [PMID: 34067088]

[23] Tanna B, Mishra A. Metabolomics of seaweeds: Tools and techniques. InPlant metabolites and regulation under environmental stress 2018; 37-52.

[24] Besednova N, Zaporozhets T, Kuznetsova T, *et al.* Metabolites of seaweeds as potential agents for the prevention and therapy of influenza infection. Mar Drugs 2019; 17(6): 373. [http://dx.doi.org/10.3390/md17060373] [PMID: 31234532]

[25] Nathani NM, Mootapally C, Gadhvi IR, Maitreya B, Joshi CG, Eds. Marine Niche: Applications in Pharmaceutical Sciences: Translational Research. Springer Singapore Nature; 2020. [http://dx.doi.org/10.1007/978-981-15-5017-1]

[26] Shah SAA, Hassan SS, Bungau S, *et al.* Chemically Diverse and Biologically Active Secondary Metabolites from Marine *Phylum chlorophyta.* Mar Drugs 2020; 18(10): 493. [http://dx.doi.org/10.3390/md18100493] [PMID: 32993146]

[27] Ford L, Theodoridou K, Sheldrake GN, Walsh PJ. A critical review of analytical methods used for the

chemical characterisation and quantification of phlorotannin compounds in brown seaweeds. Phytochem Anal 2019; 30(6): 587-99.
[http://dx.doi.org/10.1002/pca.2851] [PMID: 31250488]

[28] Miyashita K, Beppu F, Hosokawa M, Liu X, Wang S. Nutraceutical characteristics of the brown seaweed carotenoid fucoxanthin. Arch Biochem Biophys 2020; 686: 108364.
[http://dx.doi.org/10.1016/j.abb.2020.108364] [PMID: 32315653]

[29] de Oliveira L, Tschoeke D, de Oliveira A, *et al.* New Insights on the terpenome of the red seaweed Laurencia dendroidea (Florideophyceae, Rhodophyta). Mar Drugs 2015; 13(2): 879-902.
[http://dx.doi.org/10.3390/md13020879] [PMID: 25675000]

[30] Mišurcová L, Machů L, Orsavová J. Seaweed minerals as nutraceuticals. Adv Food Nutr Res 2011; 64: 371-90.
[http://dx.doi.org/10.1016/B978-0-12-387669-0.00029-6] [PMID: 22054962]

[31] Horwitz W. Official methods of analysis of AOAC international 2000.

[32] Ahmad F, Sulaiman MR, Saimon W, Yee CF, Matanjun P. Proximate compositions and total phenolic contents of selected edible seaweed from Semporna, Sabah, Malaysia. Borneo science. 2016; 31: 85-96.

[33] Rupérez P. Mineral content of edible marine seaweeds. Food Chem 2002; 79(1): 23-6.
[http://dx.doi.org/10.1016/S0308-8146(02)00171-1]

[34] Rupérez P, Saura-Calixto F. Dietary fibre and physicochemical properties of edible Spanish seaweeds. Eur Food Res Technol 2001; 212(3): 349-54.
[http://dx.doi.org/10.1007/s002170000264]

[35] Friedman M. Applications of the ninhydrin reaction for analysis of amino acids, peptides, and proteins to agricultural and biomedical sciences. J Agric Food Chem 2004; 52(3): 385-406.
[http://dx.doi.org/10.1021/jf030490p] [PMID: 14759124]

[36] McDermid KJ, Stuercke B. Nutritional composition of edible Hawaiian seaweeds. J Appl Phycol 2003; 15(6): 513-24.
[http://dx.doi.org/10.1023/B:JAPH.0000004345.31686.7f]

[37] Cunniff P. Official methods of analysis of AOAC international 1995.

[38] Wong KH, Cheung PCK. Nutritional evaluation of some subtropical red and green seaweeds. Food Chem 2000; 71(4): 475-82.
[http://dx.doi.org/10.1016/S0308-8146(00)00175-8]

[39] Tibbetts SM, Milley JE, Lall SP. Nutritional quality of some wild and cultivated seaweeds: Nutrient composition, total phenolic content and *in vitro* digestibility. J Appl Phycol 2016; 28(6): 3575-85.
[http://dx.doi.org/10.1007/s10811-016-0863-y]

[40] Siddiqui SA, Agrawal S, Brahmbhatt H, Rathore MS. Metabolite expression changes in Kappaphycus alvarezii (a red alga) under hypo- and hyper-saline conditions. Algal Res 2022; 63: 102650.
[http://dx.doi.org/10.1016/j.algal.2022.102650]

[41] Lankadurai BP, Nagato EG, Simpson MJ. Environmental metabolomics: an emerging approach to study organism responses to environmental stressors. Environ Rev 2013; 21(3): 180-205.
[http://dx.doi.org/10.1139/er-2013-0011]

[42] Kumar M, Kuzhiumparambil U, Pernice M, Jiang Z, Ralph PJ. Metabolomics: an emerging frontier of systems biology in marine macrophytes. Algal Res 2016; 16: 76-92.
[http://dx.doi.org/10.1016/j.algal.2016.02.033]

[43] Moco S, Bino RJ, Vorst O, *et al.* A liquid chromatography-mass spectrometry-based metabolome database for tomato. Plant Physiol 2006; 141(4): 1205-18.
[http://dx.doi.org/10.1104/pp.106.078428] [PMID: 16896233]

[44] Roberts LD, Souza AL, Gerszten RE, Clish CB. Targeted metabolomics. Curr Protoc Mol Biol 2012;

Chapter 30(1): 2.1-4.
[PMID: 22470063]

[45] Lisec J, Schauer N, Kopka J, Willmitzer L, Fernie AR. Gas chromatography mass spectrometry–based metabolite profiling in plants. Nat Protoc 2006; 1(1): 387-96.
[http://dx.doi.org/10.1038/nprot.2006.59] [PMID: 17406261]

[46] Orata F. Derivatization reactions and reagents for gas chromatography analysis. Advanced gas chromatography-Progress in agricultural, biomedical and industrial applications. 2012; 91: 83-106.
[http://dx.doi.org/10.5772/33098]

[47] Gupta V, Thakur RS, Baghel RS, Reddy CRK, Jha B. Seaweed Metabolomics. Adv Bot Res 2014; 71: 31-52.
[http://dx.doi.org/10.1016/B978-0-12-408062-1.00002-0]

[48] Gullberg J, Jonsson P, Nordström A, Sjöström M, Moritz T. Design of experiments: an efficient strategy to identify factors influencing extraction and derivatization of Arabidopsis thaliana samples in metabolomic studies with gas chromatography/mass spectrometry. Anal Biochem 2004; 331(2): 283-95.
[http://dx.doi.org/10.1016/j.ab.2004.04.037] [PMID: 15265734]

[49] Fiehn O, Wohlgemuth G, Scholz M, *et al.* Quality control for plant metabolomics: reporting MSI-compliant studies. Plant J 2008; 53(4): 691-704.
[http://dx.doi.org/10.1111/j.1365-313X.2007.03387.x] [PMID: 18269577]

[50] Tanna B, Choudhary B, Mishra A, *et al.* Antioxidant, Scavenging, Reducing, and Anti-Proliferative Activities of Selected Tropical Brown Seaweeds Confirm the Nutraceutical Potential of *Spatoglossum asperum*. Foods 2021; 10(10): 2482.
[http://dx.doi.org/10.3390/foods10102482] [PMID: 34681531]

[51] Tanna B, Choudhary B, Mishra A. Metabolite profiling, antioxidant, scavenging and anti-proliferative activities of selected tropical green seaweeds reveal the nutraceutical potential of Caulerpa spp. Algal Res 2018; 36(36): 96-105.
[http://dx.doi.org/10.1016/j.algal.2018.10.019]

[52] Deepak P, Josebin MPD, Kasthuridevi R, *et al.* GC-MS Metabolite Profiling, Antibacterial, Antidiabetic and Antioxidant Activities of Brown Seaweeds, Sargassum wightii Greville Ex J. Agardh, 1848 and Stoechospermum marginatum (C. Agardh) Kützing 1843. Pharmacology, Toxicology and Biomedical Reports 2017; 3(2): 27-34.
[http://dx.doi.org/10.5530/PTB.2017.3.5]

[53] Chaleckis R, Meister I, Zhang P, Wheelock CE. Challenges, progress and promises of metabolite annotation for LC-MS-based metabolomics. Curr Opin Biotechnol 2019; 55(55): 44-50.
[PMID: 30138778]

[54] Okazaki Y, Saito K. Recent advances of metabolomics in plant biotechnology. Plant Biotechnol Rep 2012; 6(1): 1-15.
[http://dx.doi.org/10.1007/s11816-011-0191-2] [PMID: 22308170]

[55] Tolstikov VV, Fiehn O. Analysis of highly polar compounds of plant origin: combination of hydrophilic interaction chromatography and electrospray ion trap mass spectrometry. Anal Biochem 2002; 301(2): 298-307.
[http://dx.doi.org/10.1006/abio.2001.5513] [PMID: 11814300]

[56] De Vos RCH, Moco S, Lommen A, Keurentjes JJB, Bino RJ, Hall RD. Untargeted large-scale plant metabolomics using liquid chromatography coupled to mass spectrometry. Nat Protoc 2007; 2(4): 778-91.
[http://dx.doi.org/10.1038/nprot.2007.95] [PMID: 17446877]

[57] Guijas C, Montenegro-Burke JR, Warth B, Spilker ME, Siuzdak G. Metabolomics activity screening for identifying metabolites that modulate phenotype. Nat Biotechnol 2018; 36(4): 316-20.
[http://dx.doi.org/10.1038/nbt.4101] [PMID: 29621222]

[58] Nilsson SL, Bylund D, Jörntén-Karlsson M, Petersson P, Markides KE. A chemometric study of active parameters and their interaction effects in a nebulized sheath□liquid electrospray interface for capillary electrophoresis□mass spectrometry. Electrophoresis 2004; 25(13): 2100-7.
[http://dx.doi.org/10.1002/elps.200305937] [PMID: 15237411]

[59] Soga T, Igarashi K, Ito C, Mizobuchi K, Zimmermann HP, Tomita M. Metabolomic profiling of anionic metabolites by capillary electrophoresis mass spectrometry. Anal Chem 2009; 81(15): 6165-74.
[http://dx.doi.org/10.1021/ac900675k] [PMID: 19522513]

[60] Li L, Zhao J, Zhao Y, *et al.* Comprehensive investigation of tobacco leaves during natural early senescence via multi-platform metabolomics analyses. Sci Rep 2016; 6(1): 37976.
[http://dx.doi.org/10.1038/srep37976] [PMID: 27897248]

[61] Kim HK, Choi YH, Verpoorte R. NMR-based metabolomic analysis of plants. Nat Protoc 2010; 5(3): 536-49.
[http://dx.doi.org/10.1038/nprot.2009.237] [PMID: 20203669]

[62] Wu J, An Y, Yao J, Wang Y, Tang H. An optimised sample preparation method for NMR-based faecal metabonomic analysis. Analyst (Lond) 2010; 135(5): 1023-30.
[http://dx.doi.org/10.1039/b927543f] [PMID: 20419252]

[63] Viant MR. Improved methods for the acquisition and interpretation of NMR metabolomic data. Biochem Biophys Res Commun 2003; 310(3): 943-8.
[http://dx.doi.org/10.1016/j.bbrc.2003.09.092] [PMID: 14550295]

[64] Fan TWM. Metabolite profiling by one- and two-dimensional NMR analysis of complex mixtures. Prog Nucl Magn Reson Spectrosc 1996; 28(2): 161-219.
[http://dx.doi.org/10.1016/0079-6565(95)01017-3]

[65] Xi Y, de Ropp JS, Viant MR, Woodruff DL, Yu P. Improved identification of metabolites in complex mixtures using HSQC NMR spectroscopy. Anal Chim Acta 2008; 614(2): 127-33.
[http://dx.doi.org/10.1016/j.aca.2008.03.024] [PMID: 18420042]

[66] Bobzin SC, Yang S, Kasten TP. Application of liquid chromatography–nuclear magnetic resonance spectroscopy to the identification of natural products. J Chromatogr, Biomed Appl 2000; 748(1): 259-67.
[http://dx.doi.org/10.1016/S0378-4347(00)00289-9] [PMID: 11092603]

[67] Ren JL, Zhang AH, Kong L, Wang XJ. Advances in mass spectrometry-based metabolomics for investigation of metabolites. RSC Advances 2018; 8(40): 22335-50.
[http://dx.doi.org/10.1039/C8RA01574K] [PMID: 35539746]

[68] Ye H, Gemperline E, Venkateshwaran M, *et al.* MALDI mass spectrometry-assisted molecular imaging of metabolites during nitrogen fixation in the *Medicago truncatula-Sinorhizobium meliloti* symbiosis. Plant J 2013; 75(1): 130-45.
[http://dx.doi.org/10.1111/tpj.12191] [PMID: 23551619]

[69] Schneider G, Figueroa FL, Vega J, *et al.* Photoprotection properties of marine photosynthetic organisms grown in high ultraviolet exposure areas: Cosmeceutical applications. Algal Res 2020; 49: 101956.
[http://dx.doi.org/10.1016/j.algal.2020.101956]

[70] Sanjeewa KKA, Fernando IPS, Kim EA, Ahn G, Jee Y, Jeon YJ. Anti-inflammatory activity of a sulfated polysaccharide isolated from an enzymatic digest of brown seaweed *Sargassum horneri* in RAW 264.7 cells. Nutr Res Pract 2017; 11(1): 3-10.
[http://dx.doi.org/10.4162/nrp.2017.11.1.3] [PMID: 28194259]

[71] Elshobary ME, El-Shenody RA, Abomohra AEF. Sequential biofuel production from seaweeds enhances the energy recovery: A case study for biodiesel and bioethanol production. Int J Energy Res 2021; 45(4): 6457-67.

[http://dx.doi.org/10.1002/cr.6181]

[72] Vijayakumar S, Durgadevi S, Arulmozhi P, Rajalakshmi S, Gopalakrishnan T, Parameswari N. Effect of seaweed liquid fertilizer on yield and quality of Capsicum annum L. Acta Ecol Sin 2019; 39(5): 406-10.
[http://dx.doi.org/10.1016/j.chnaes.2018.10.001]

[73] Valderrama D, Cai J, Hishamunda N, *et al.* The economics of Kappaphycus seaweed cultivation in developing countries: a comparative analysis of farming systems. Aquac Econ Manag 2015; 19(2): 251-77.
[http://dx.doi.org/10.1080/13657305.2015.1024348]

[74] Mariño M, Breckwoldt A, Teichberg M, Kase A, Reuter H. Livelihood aspects of seaweed farming in Rote Island, Indonesia. Mar Policy 2019; 107: 103600.
[http://dx.doi.org/10.1016/j.marpol.2019.103600]

[75] Porse H, Rudolph B. The seaweed hydrocolloid industry: 2016 updates, requirements, and outlook. J Appl Phycol 2017; 29(5): 2187-200.
[http://dx.doi.org/10.1007/s10811-017-1144-0]

[76] Glicksman M. Utilization of seaweed hydrocolloids in the food industry. InTwelfth international seaweed symposium. Dordrecht Springer 1987; 31-47.

[77] Trono GC Jr. Seaweed resources in the developing countries of Asia: Production and socio-economic implications. InCulture and use of algae in Southeast Asia. Proceedings of the Symposium on Culture and Utilization of Algae in Southeast Asia. 1-7.

[78] Mantri VA, Eswaran K, Shanmugam M, *et al.* An appraisal on commercial farming of Kappaphycus alvarezii in India: success in diversification of livelihood and prospects. J Appl Phycol 2017; 29(1): 335-57.
[http://dx.doi.org/10.1007/s10811-016-0948-7]

[79] Hurtado Anicia Q, Renato F. Hurtado Anicia Q., and Renato F. Agbayani. "Deep-sea farming of Kappaphycus using the multiple raft, long-line method." 2002; 438-44.

[80] Chopin T, Tacon AG. Importance of Seaweeds and Extractive Species in Global Aquaculture Production. Rev Fish Sci Aquacult 2021; 29(2): 139-48.

[81] Yap WF, Tay V, Tan SH, Yow YY, Chew J. Decoding antioxidant and antibacterial potentials of Malaysian green seaweeds: Caulerpa racemosa and Caulerpa lentillifera. Antibiotics (Basel) 2019; 8(3): 152.
[http://dx.doi.org/10.3390/antibiotics8030152] [PMID: 31533237]

[82] Ismail GA. Biochemical composition of some Egyptian seaweeds with potent nutritive and antioxidant properties. Food Sci Technol (Campinas) 2017; 37(2): 294-302.
[http://dx.doi.org/10.1590/1678-457x.20316]

[83] Venkatesan M, Arumugam V, Pugalendi R, *et al.* Antioxidant, anticoagulant and mosquitocidal properties of water soluble polysaccharides (WSPs) from Indian seaweeds. Process Biochem 2019; 84: 196-204.
[http://dx.doi.org/10.1016/j.procbio.2019.05.029]

[84] Cumashi A, Ushakova NA, Preobrazhenskaya ME, *et al.* A comparative study of the anti-inflammatory, anticoagulant, antiangiogenic, and antiadhesive activities of nine different fucoidans from brown seaweeds. Glycobiology 2007; 17(5): 541-52.
[http://dx.doi.org/10.1093/glycob/cwm014] [PMID: 17296677]

[85] Saraswati , Giriwono PE, Iskandriati D, Tan CP, Andarwulan N. Saraswati, Giriwono PE, Iskandriati D, Tan CP, Andarwulan N. Sargassum seaweed as a source of anti-inflammatory substances and the potential insight of the tropical species: a review. Mar Drugs 2019; 17(10): 590.
[http://dx.doi.org/10.3390/md17100590]

[86] Ciancia M, Quintana I, Cerezo AS. Overview of anticoagulant activity of sulfated polysaccharides

from seaweeds in relation to their structures, focusing on those of green seaweeds. Curr Med Chem 2010; 17(23): 2503-29.
[http://dx.doi.org/10.2174/092986710791556069] [PMID: 20491645]

[87] Ford L, Stratakos AC, Theodoridou K, *et al.* Polyphenols from brown seaweeds as a potential antimicrobial agent in animal feeds. ACS Omega 2020; 5(16): 9093-103.
[http://dx.doi.org/10.1021/acsomega.9b03687] [PMID: 32363261]

[88] Ristivojević P, Jovanović V, Opsenica DM, Park J, Rollinger JM, Velicković TĆ. Rapid analytical approach for bioprofiling compounds with radical scavenging and antimicrobial activities from seaweeds. Food Chem 2021; 334: 127562.
[http://dx.doi.org/10.1016/j.foodchem.2020.127562] [PMID: 32707368]

[89] Shi Q, Wang A, Lu Z, Qin C, Hu J, Yin J. Overview on the antiviral activities and mechanisms of marine polysaccharides from seaweeds. Carbohydr Res 2017; 453-454: 1-9.
[http://dx.doi.org/10.1016/j.carres.2017.10.020] [PMID: 29102716]

[90] Gutiérrez-Rodríguez AG, Juárez-Portilla C, Olivares-Bañuelos T, Zepeda RC. Anticancer activity of seaweeds. Drug Discov Today 2018; 23(2): 434-47.
[http://dx.doi.org/10.1016/j.drudis.2017.10.019] [PMID: 29107095]

[91] Valado A, Pereira M, Caseiro A, *et al.* Effect of carrageenans on vegetable jelly in humans with hypercholesterolemia. Mar Drugs 2019; 18(1): 19.
[http://dx.doi.org/10.3390/md18010019] [PMID: 31878353]

[92] André R, Pacheco R, Bourbon M, Serralheiro ML. Brown Algae Potential as a Functional Food against Hypercholesterolemia: Review. Foods 2021; 10(2): 234.
[http://dx.doi.org/10.3390/foods10020234] [PMID: 33498824]

[93] Khalid S, Abbas M, Saeed F, Bader-Ul-Ain H, Suleria HA. Therapeutic potential of seaweed bioactive compounds. IntechOpen; 2018.Nov 5.
[http://dx.doi.org/10.5772/intechopen.74060]

[94] Yan X, Chuda Y, Suzuki M, Nagata T. Fucoxanthin as the major antioxidant in Hijikia fusiformis, a common edible seaweed. Biosci Biotechnol Biochem 1999; 63(3): 605-7.
[http://dx.doi.org/10.1271/bbb.63.605] [PMID: 10227153]

[95] Yabuta Y, Fujimura H, Kwak CS, Enomoto T, Watanabe F. Antioxidant activity of the phycoerythrobilin compound formed from a dried Korean purple laver (Porphyra sp.) during *in vitro* digestion. Food Sci Technol Res 2010; 16(4): 347-52.
[http://dx.doi.org/10.3136/fstr.16.347]

[96] Kolender AA, Pujol CA, Damonte EB, Matulewicz MC, Cerezo AS. The system of sulfated α-(1 → 3)-linked d-mannans from the red seaweed Nothogenia fastigiata: Structures, antiherpetic and anticoagulant properties. Carbohydr Res 1997; 304(1): 53-60.
[http://dx.doi.org/10.1016/S0008-6215(97)00201-2] [PMID: 9403994]

[97] Jurd KM, Rogers DJ, Blunden G, McLellan DS. Anticoagulant properties of sulphated polysaccharides and a proteoglycan from Codium fragile ssp. atlanticum. J Appl Phycol 1995; 7(4): 339-45.
[http://dx.doi.org/10.1007/BF00003790]

[98] Leiro JM, Castro R, Arranz JA, Lamas J. Immunomodulating activities of acidic sulphated polysaccharides obtained from the seaweed Ulva rigida C. Agardh. Int Immunopharmacol 2007; 7(7): 879-88.
[http://dx.doi.org/10.1016/j.intimp.2007.02.007] [PMID: 17499190]

[99] Shibata T, Fujimoto K, Nagayama K, Yamaguchi K, Nakamura T. Inhibitory activity of brown algal phlorotannins against hyaluronidase. Int J Food Sci Technol 2002; 37(6): 703-9.
[http://dx.doi.org/10.1046/j.1365-2621.2002.00603.x]

[100] Heo SJ, Ko SC, Cha SH, *et al.* Effect of phlorotannins isolated from Ecklonia cava on melanogenesis and their protective effect against photo-oxidative stress induced by UV-B radiation. Toxicol in vitro

2009; 23(6): 1123-30.
[http://dx.doi.org/10.1016/j.tiv.2009.05.013] [PMID: 19490939]

[101] Xiaojun Y, Xiancui L, Chengxu Z, Xiao F. Prevention of fish oil rancidity by phlorotannins fromSargassum kjellmanianum. J Appl Phycol 1996; 8(3): 201-3.
[http://dx.doi.org/10.1007/BF02184972]

[102] Bae JS. Antithrombotic and profibrinolytic activities of phloroglucinol. Food Chem Toxicol 2011; 49(7): 1572-7.
[http://dx.doi.org/10.1016/j.fct.2011.04.003] [PMID: 21501649]

[103] Kazłowska K, Hsu T, Hou CC, Yang WC, Tsai GJ. Anti-inflammatory properties of phenolic compounds and crude extract from Porphyra dentata. J Ethnopharmacol 2010; 128(1): 123-30.
[http://dx.doi.org/10.1016/j.jep.2009.12.037] [PMID: 20051261]

[104] Pangestuti R, Shin KH, Kim SK. Anti-photoaging and potential skin health benefits of seaweeds. Mar Drugs 2021; 19(3): 172.
[http://dx.doi.org/10.3390/md19030172] [PMID: 33809936]

[105] Alves A, Sousa E, Kijjoa A, Pinto M. Marine-derived compounds with potential use as cosmeceuticals and nutricosmetics. Molecules 2020; 25(11): 2536.
[http://dx.doi.org/10.3390/molecules25112536] [PMID: 32486036]

[106] Yang K, Kim SY, Park JH, *et al.* Topical Application of Phlorotannins from Brown Seaweed Mitigates Radiation Dermatitis in a Mouse Model. Mar Drugs 2020; 18(8): 377.
[http://dx.doi.org/10.3390/md18080377] [PMID: 32707897]

[107] Félix C, Félix R, Carmona AM, *et al.* Cosmeceutical Potential of Grateloupia turuturu: Using Low-Cost Extraction Methodologies to Obtain Added-Value Extracts. Appl Sci (Basel) 2021; 11(4): 1650.
[http://dx.doi.org/10.3390/app11041650]

[108] Wang L, Jayawardena TU, Yang HW, Lee HG, Jeon YJ. The Potential of Sulfated Polysaccharides Isolated from the Brown Seaweed *Ecklonia maxima* in Cosmetics: Antioxidant, Anti-melanogenesis, and Photoprotective Activities. Antioxidants 2020; 9(8): 724.
[http://dx.doi.org/10.3390/antiox9080724] [PMID: 32784879]

[109] Chopin T. Seaweeds: the world's largest mariculture crop. Int Aqua Feed 2018; 21(8): 14-5.

[110] Thiyagarasaiyar K, Goh BH, Jeon YJ, Yow YY. Algae metabolites in cosmeceutical: an overview of current applications and challenges. Mar Drugs 2020; 18(6): 323.
[http://dx.doi.org/10.3390/md18060323] [PMID: 32575468]

[111] Gullón B, Gagaoua M, Barba FJ, Gullón P, Zhang W, Lorenzo JM. Seaweeds as promising resource of bioactive compounds: Overview of novel extraction strategies and design of tailored meat products. Trends Food Sci Technol 2020; 100: 1-18.
[http://dx.doi.org/10.1016/j.tifs.2020.03.039]

[112] Falkenberg M, Nakano E, Zambotti-Villela L, *et al.* Bioactive compounds against neglected diseases isolated from macroalgae: a review. J Appl Phycol 2019; 31(2): 797-823.
[http://dx.doi.org/10.1007/s10811-018-1572-5]

[113] Cardozo KHM, Guaratini T, Barros MP, *et al.* Metabolites from algae with economical impact. Comp Biochem Physiol C Toxicol Pharmacol 2007; 146(1-2): 60-78.
[http://dx.doi.org/10.1016/j.cbpc.2006.05.007] [PMID: 16901759]

[114] Gómez-Ordóñez E, Jiménez-Escrig A, Rupérez P. Dietary fibre and physicochemical properties of several edible seaweeds from the northwestern Spanish coast. Food Res Int 2010; 43(9): 2289-94.
[http://dx.doi.org/10.1016/j.foodres.2010.08.005]

[115] Charoensiddhi S, Abraham RE, Su P, Zhang W. Seaweed and seaweed-derived metabolites as prebiotics. Adv Food Nutr Res 2020; 91: 97-156.
[http://dx.doi.org/10.1016/bs.afnr.2019.10.001] [PMID: 32035602]

[116] Wasupalli GK, Verma D. Polysaccharides as biomaterials. InFundamental Biomaterials. Polymers (Basel) Woodhead Publishing 2018; 37-70.

[117] Venkatesan J, Anil S, Kim SK, Shim M. Seaweed polysaccharide-based nanoparticles: preparation and applications for drug delivery. Polymers (Basel) 2016; 8(2): 30.
[http://dx.doi.org/10.3390/polym8020030] [PMID: 30979124]

[118] Dominguez H, Loret EP. *Ulva lactuca*, a source of troubles and potential riches. Mar Drugs 2019; 17(6): 357.
[http://dx.doi.org/10.3390/md17060357] [PMID: 31207947]

[119] Torres P, Santos JP, Chow F, dos Santos DYAC. A comprehensive review of traditional uses, bioactivity potential, and chemical diversity of the genus Gracilaria (Gracilariales, Rhodophyta). Algal Res 2019; 37: 288-306.
[http://dx.doi.org/10.1016/j.algal.2018.12.009]

[120] Griffiths M, Harrison ST, Smit M, Maharajh D. Major commercial products from micro-and macroalgae. InAlgae biotechnology. Cham Springer 2016; 269-300.

[121] Rioux LE, Moulin V, Beaulieu M, Turgeon SL. Human skin fibroblast response is differentially regulated by galactofucan and low molecular weight galactofucan. Bioactive Carbohydrates and Dietary Fibre 2013; 1(2): 105-10.
[http://dx.doi.org/10.1016/j.bcdf.2013.03.004]

[122] Prasetyaningrum A, Praptyana IR. Prasetyaningrum A, Praptyana IR. Carrageenan: Nutraceutical and functional food as future food. In: IOP Conference Series: Earth and Environmental Science 2019 Jun 1 (Vol. 292, No. 1, p. 012068). IOP Publishing.

[123] de Jesus Raposo M, de Morais A, de Morais R. Marine polysaccharides from algae with potential biomedical applications. Mar Drugs 2015; 13(5): 2967-3028.
[http://dx.doi.org/10.3390/md13052967] [PMID: 25988519]

[124] Chevolot L, Mulloy B, Ratiskol J, Foucault A, Colliec-Jouault S. A disaccharide repeat unit is the major structure in fucoidans from two species of brown algae. Carbohydr Res 2001; 330(4): 529-35.
[http://dx.doi.org/10.1016/S0008-6215(00)00314-1] [PMID: 11269406]

[125] Sanjeewa KKA, Lee JS, Kim WS, Jeon YJ. The potential of brown-algae polysaccharides for the development of anticancer agents: An update on anticancer effects reported for fucoidan and laminaran. Carbohydr Polym 2017; 177: 451-9.
[http://dx.doi.org/10.1016/j.carbpol.2017.09.005] [PMID: 28962791]

[126] Pereira H, Barreira L, Figueiredo F, *et al.* Polyunsaturated Fatty acids of marine macroalgae: potential for nutritional and pharmaceutical applications. Mar Drugs 2012; 10(12): 1920-35.
[http://dx.doi.org/10.3390/md10091920] [PMID: 23118712]

[127] Wang J, Luo T, Li S, Zhao J. The powerful applications of polyunsaturated fatty acids in improving the therapeutic efficacy of anticancer drugs. Expert Opin Drug Deliv 2012; 9(1): 1-7.
[http://dx.doi.org/10.1517/17425247.2011.618183] [PMID: 22171694]

[128] Rajapakse N, Kim SK. Nutritional and digestive health benefits of seaweed. Adv Food Nutr Res 2011; 64: 17-28.
[http://dx.doi.org/10.1016/B978-0-12-387669-0.00002-8] [PMID: 22054935]

[129] Choudhary B, Chauhan OP, Mishra A. Edible seaweeds: a potential novel source of bioactive metabolites and nutraceuticals with human health benefits. Front Mar Sci 2021; 8: 740054.
[http://dx.doi.org/10.3389/fmars.2021.740054]

[130] Circuncisão A, Catarino M, Cardoso S, Silva A. Minerals from macroalgae origin: Health benefits and risks for consumers. Mar Drugs 2018; 16(11): 400.
[http://dx.doi.org/10.3390/md16110400] [PMID: 30360515]

[131] Abirami RG, Kowsalya S. Phytochemical screening, microbial load and antimicrobial activity of

underexploited seaweeds. Int Res J Microbiol 2012; 3(10): 328-32.

[132] Murugan K, Iyer VV. Differential growth inhibition of cancer cell lines and antioxidant activity of extracts of red, brown, and green marine algae. in vitro Cell Dev Biol Anim 2013; 49(5): 324-34. [http://dx.doi.org/10.1007/s11626-013-9603-7] [PMID: 23645467]

[133] Gupta S, Abu-Ghannam N. Bioactive potential and possible health effects of edible brown seaweeds. Trends Food Sci Technol 2011; 22(6): 315-26. [http://dx.doi.org/10.1016/j.tifs.2011.03.011]

[134] Thanh TTT, Quach TMT, Nguyen TN, Vu Luong D, Bui ML, Tran TTV. Structure and cytotoxic activity of ulvan extracted from green seaweed *Ulva lactuca*. Int J Biol Macromol 2016; 93(Pt A): 695-702. [http://dx.doi.org/10.1016/j.ijbiomac.2016.09.040] [PMID: 27637450]

[135] M Ahmed O, Ahmed RR. Anti-proliferative and apoptotic efficacies of ulvan polysaccharides against different types of carcinoma cells *in vitro* and *in vivo*. J Cancer Sci Ther 2014; 6(6): 202-8. [http://dx.doi.org/10.4172/1948-5956.1000272]

[136] Jesus A, Correia-da-Silva M, Afonso C, Pinto M, Cidade H. Isolation and potential biological applications of haloaryl secondary metabolites from macroalgae. Mar Drugs 2019; 17(2): 73. [http://dx.doi.org/10.3390/md17020073] [PMID: 30678253]

[137] Liu Z, Gao T, Yang Y, *et al.* Anti-cancer activity of porphyran and carrageenan from red seaweeds. Molecules 2019; 24(23): 4286. [http://dx.doi.org/10.3390/molecules24234286] [PMID: 31775255]

[138] Bedoux G, Hardouin K, Burlot AS, Bourgougnon N. Bioactive components from seaweeds: Cosmetic applications and future development. Adv Bot Res 2014; 71: 345-78. [http://dx.doi.org/10.1016/B978-0-12-408062-1.00012-3]

[139] Zhang Q, Li N, Liu X, Zhao Z, Li Z, Xu Z. The structure of a sulfated galactan from Porphyra haitanensis and its *in vivo* antioxidant activity. Carbohydr Res 2004; 339(1): 105-11. [http://dx.doi.org/10.1016/j.carres.2003.09.015] [PMID: 14659676]

[140] Beaulieu L. Insights into the regulation of algal proteins and bioactive peptides using proteomic and transcriptomic approaches. Molecules 2019; 24(9): 1708. [http://dx.doi.org/10.3390/molecules24091708] [PMID: 31052532]

[141] Gómez I, Huovinen P. Brown algal phlorotannins: an overview of their functional roles. In: Gómez I, Huovinen P, Eds. Antarctic Seaweeds. Springer, Cham 2020; 365-88. [http://dx.doi.org/10.1007/978-3-030-39448-6_18]

[142] Islam MR, Mikami D, Kurihara H. Two new algal bromophenols from Odonthalia corymbifera. Tetrahedron Lett 2017; 58(43): 4119-21. [http://dx.doi.org/10.1016/j.tetlet.2017.09.044]

[143] De Souza ÉT, Pereira de Lira D, Cavalcanti de Queiroz A, *et al.* The antinociceptive and anti-inflammatory activities of caulerpin, a bisindole alkaloid isolated from seaweeds of the genus Caulerpa. Mar Drugs 2009; 7(4): 689-704. [http://dx.doi.org/10.3390/md7040689] [PMID: 20098607]

[144] Cindana Mo'o FR, Wilar G, Devkota HP, Wathoni N. Ulvan, a Polysaccharide from Macroalga Ulva sp.: A Review of Chemistry, Biological Activities and Potential for Food and Biomedical Applications. Appl Sci (Basel) 2020; 10(16): 5488. [http://dx.doi.org/10.3390/app10165488]

[145] Joshi S, Kumari R, Upasani VN. Applications of algae in cosmetics: An overview. Int J Innov Res Sci Eng Technol 2018; 7(2): 1269. [http://dx.doi.org/10.15680/IJIRSET.2018.0702038]

[146] Harnedy PA, Fitzgerald RJ. Bioactive proteins and peptides from macroalgae, fish, shellfish and marine processing waste. Marine proteins and peptides: Biological activities and applications.

2013.Apr 19:5-39.
[http://dx.doi.org/10.1002/9781118375082.ch2]

[147] Rosic N. Mycosporine-like amino acids: making the foundation for organic personalised sunscreens. Mar Drugs 2019; 17(11): 638.
[http://dx.doi.org/10.3390/md17110638] [PMID: 31726795]

[148] Shanura Fernando IP, Asanka Sanjeewa KK, Samarakoon KW, *et al.* The potential of fucoidans from Chnoospora minima and Sargassum polycystum in cosmetics: antioxidant, anti-inflammatory, skin-whitening, and antiwrinkle activities. J Appl Phycol 2018; 30(6): 3223-32.
[http://dx.doi.org/10.1007/s10811-018-1415-4]

[149] Purcell-Meyerink D, Packer MA, Wheeler TT, Hayes M. Aquaculture Production of the Brown Seaweeds *Laminaria digitata* and *Macrocystis pyrifera*: Applications in Food and Pharmaceuticals. Molecules 2021; 26(5): 1306.
[http://dx.doi.org/10.3390/molecules26051306] [PMID: 33671085]

[150] Venkatraman KL, Mehta A. Health benefits and pharmacological effects of Porphyra species. Plant Foods Hum Nutr 2019; 74(1): 10-7.
[http://dx.doi.org/10.1007/s11130-018-0707-9] [PMID: 30543042]

[151] Seca A, Pinto D. Overview on the antihypertensive and anti-obesity effects of secondary metabolites from seaweeds. Mar Drugs 2018; 16(7): 237.
[http://dx.doi.org/10.3390/md16070237] [PMID: 30011911]

[152] Kang MC, Wijesinghe WAJP, Lee SH, *et al.* Dieckol isolated from brown seaweed Ecklonia cava attenuates type II diabetes in db/db mouse model. Food Chem Toxicol 2013; 53: 294-8.
[http://dx.doi.org/10.1016/j.fct.2012.12.012] [PMID: 23261675]

CHAPTER 16

Role of Algal Metabolites and Pigments: Characterization and Application

G. Kavitha Aravind[1] and **D. Inbakandan**[2,*]

[1] *Providence College for Women, Bandishola, Spring Field, Coonoor, The Nilgiris, Tamil Nadu 643104, India*

[2] *Centre for Ocean Research (DST-FIST Sponsored Centre), MoES - Earth Science & Technology Cell (Marine Biotechnological Studies), Col. Dr. Jeppiaar Research Park, Sathyabama Institute of Science and Technology, Chennai 600 119, India*

Abstract: Algae establish a dissimilar group owning huge morphological and metabolic variety. Algae are an exclusive source of beneficial constituents, holding exciting biological activity. They consist of several groups, including unicellular, colonial or filamentous, photosynthetic or heterotrophic micro-organisms containing chlorophyll and other pigments. The photosynthetic pigment composition of algae depends on the quantity and quality of light received by them. In addition to chlorophyll, algal cells comprise a numeral of accessory pigments, such as phycobiliproteins and carotenoids, that aid in photosynthesis and guard the cells against photo blanching. Pigments are light-absorbing molecules in algae. The conventional botanists categorize the algae based on their colour, more specifically based on the distribution of the pigments, which has great taxonomic significance in algal systematics. More than 40,000 algal species have been classified under prokaryotic algae (cyanobacteria) and several eukaryotes, including green algae, diatoms, yellow-green algae, golden algae, red algae, brown algae, dinoflagellates and others. These intense metabolites from algae have numerous biotechnological and industrial applications in cosmetic, food, diagnostic and pharmacological industries owing to their nontoxic, noncarcinogenic environment and additional health-promoting properties like antioxidative and immune-boosting effects. Cumulative trials of clinical research on health aids of algal metabolites and pigments have made them a harmless and viable replacement for chemically synthesized drugs. Additional perception into the biological properties of these molecules and their method of action will aid in the improvement of effective pharmaceutical products.

Keywords: Algae, Antioxidants, Metabolites, Pharmaceutical, Pigments.

* **Corresponding author D. Inbakandan:** Centre for Ocean Research (DST-FIST Sponsored Centre), MoES- Earth Science & Technology Cell (Marine Biotechnological Studies), Col. Dr. Jeppiaar Research Park, Sathyabama Institute of Science and Technology, Chennai, India; Tel: +91 99655 40310; E-mail: inbakandan@gmail.com

Avinash Mishra (Ed.)

INTRODUCTION

Algae contribute 50% of the planet's primary productivity, which uses solar radiation as a principal energy source to drive physiological processes, such as photosynthesis, for their growth and development. The aquatic ecosystem balances the terrestrial ones in biomass production and is assumed to incorporate a large amount of atmospheric inorganic carbon into organic carbon annually; thus, they are the most important sink for CO_2 [1].

Microalgae are a potential source of high-value products for biotechnological exploitation, which include polyunsaturated fatty acids (PUFA), carotenoids (β-carotene), phycobiliproteins, polysaccharides, and lipids such as triglycerides and hydrocarbons. A mass of compounds is formed by microalgae as a guard against stress conditions. A benefit of using algae for bioactive molecules is that they can be refined on a huge scale for the making of desired chemicals. Amongst microalgae, the green alga *Dunaliella salina* has been mass cultured in open ponds for the production of β-carotene [2]. The interest in commercializing microalgae [3], specifically secondary metabolites of Cyanobacteria, has been increasing recently in different fields, including environment, nutrition, medicine and pharmacology [4]. Many of the constituents display a multitude of bioactivities with application in food, pigment source, natural sunscreens and biofuel production. The biotechnological application of microalgae for the production of fine chemicals has created much importance [5].

MICROALGAL METABOLITES

Microalgae are important producer which dominates terrestrial and aquatic ecosystems. They form the basis of the food web and account for up to 60% of the total global oxygen production and about 50% of the global CO_2 fixation (Fig. **1**) [6]. Microalgae have been exploited for millennia and have several applications, from human to animal nutrition to cosmetics and the production of high-value molecules (*e.g.,* fatty acids, pigments, stable isotope biochemical) [7]. Microscopic algae constitute the original source of fossil carbon found in crude oil and natural gas [8]. Besides, Cyanobacteria and microalgae are engaged in a wide variety of biotechnological applications as a health food, food for organisms and source of antioxidants or antimicrobial agents [9, 10]. Still, the world is facing several threats, such as soaring oil prices, diminishing world oil reserves and the environmental worsening affiliated with fossil fuel consumption. All these have risen and renewed the interest in using microalgae as an alternate and renewable feedstock for biofuel production [11].

Microalgae play a key role in food/feed additives and feed supplements for humans and animals [12]. It was reported that besides serving as a valuable model in biological sciences, *Chamydomonas reinhardtii* had been successfully used during the last decade in the biotechnology area to establish alga-based platforms for biopharmaceutical production at low costs [13]. This promises the exploitation of algae species, such as *Dunaliella salina*, *Nannochloropsis* spp. and *Phaeodactylum tricornutum* for biopharmaceuticals production as a key trend in the field [13].

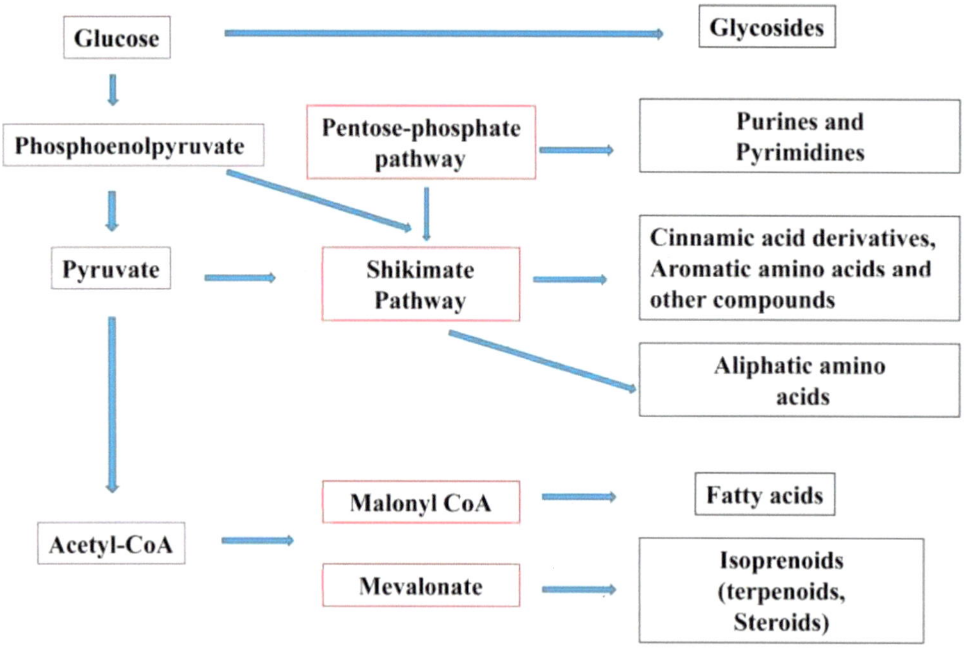

Fig. (1). Algal metabolites.

There is a growing interest in exploring potential microalgae for various industrial applications from nutraceutical to biodiesel feedstock; there is an increasing search worldwide. The green colonial lipid-rich microalga, *Botryococcus,* is widespread in freshwater, brackish lakes, reservoirs and ponds [14]. *Botryococcus* is characterized by its ability to synthesise and accumulate a high level of lipids. These lipid substances include numerous hydrocarbons, *i.e.*, highly reduced compounds comprising only carbon and hydrogen as elements [15]. Being a photosynthetic organism, it has been reported to reduce CO_2 emissions by 1.5 x 10^5 tons/yr/8.4 x 10^3 ha. Thus, it offers an eco-friendly process for producing lipid and other bioactive compounds along with its CO_2 mitigation credits [16].

BIOFUELS

Biofuels are one of the most likely substitute energy sources for fossil fuels because they are renewable and help to reduce greenhouse gas releases. The lipid composition of microalga, *Botryococcus braunii* (Chlorophyceae), cultured outdoors and used in biofuel production, is due to the high content of saturated fatty acids. The production of defensible renewable energy is being discussed universally since it is gradually increasing; it needs to be understood that first-generation biofuels primarily produced from food crops and mostly oil seeds, compete for arable land, freshwater or biodiverse natural landscapes and are limited in their ability to achieve targets for biofuel production, climate change mitigation and economic growth. These alarms have amplified the interest in emerging second and third-generation biofuels such as lignocelluloses and microalgae, correspondingly, which possibly bid inordinate occasions in the longer term, and there is no need to contend for arable land and valuable freshwater [17].

Fig. (2). Cultivation and Extraction of lipids of *Botryococcus*.

A number of biofuel candidates are proposed to displace fossil fuels in order to eliminate the vulnerability of the energy sector. Also, algal biomass productivity is several folds higher than terrestrial plant crops [18]. The algae biomass cultivation, from producing renewable biomass, biofuels and bioenergy, helps in cycling CO_2 as huge magnitudes of CO_2 are utilized during the algal growth. Since the rising demand for energy, microalgae are one of the most important energy sources for the future. Microalgae can be changed into energy products, such as bioethanol, biodiesel, biogas and crude oil. Algal cultivation by means of sunlight energy can be done in open or closed photobioreactors. It was reported that *B. braunii* with 1.7 g L^{-1} dry wt. of biomass contained 17% lipid, 17% proteins, 32%

carbohydrate and 13% hydrocarbon content (Fig. **2**), confirming the mass cultivation of the alga in outdoor conditions [19]. The morphological heterogeneity of the alga, *Botryococcus* makes the identification difficult. Molecular tools like PCR play a vivacious part in approving the efficient place of the organism [20].

POLYUNSATURATED FATTY ACIDS

Lipids produced by microalgae can be grouped into two categories, non-polar lipids and polar lipids. Storage lipids are mainly in the form of TAG made of predominately saturated fatty acids and some unsaturated fatty acids, which can be trans esterified to yield biodiesel. Structural lipids characteristically have a high amount of polyunsaturated fatty acids (PUFA), which are also indispensable nutrients for marine animals and creatures [21]. Polar lipids and sterols are significant mechanical components of cell membranes that act as a selective penetrable fence for cells and organelles. These lipids preserve specific membrane functions, providing the matrix for a varied diversity of metabolic developments and participating directly in membrane fusion events. In addition to a structural function, some polar lipids may act as key intermediates (or precursors of intermediates) in cell signalling pathways and play a role in responding to environmental changes. Triacylglycerides (TAGs) generally serve as energy storage in microalgae that, once extracted, can be easily converted into biodiesel through transesterification reactions [22]. These neutral lipids bear a common structure of triple esters where usually three long-chain fatty acids (FA's) are attached to a glycerol molecule.

ASTAXANTHIN

Many algal biotechnologists have studied the application of photosynthetic machinery of algal cells for the production of new bioactive compounds. Microalgae have vast potential as sources for valuable pharmaceuticals, pigments and other fine chemicals. A red ketocarotenoid pigment, Astaxanthin (3, 3'-dihydroxy- β, β'-carotene-4, 4'-dione) received increasing interest in cosmetics and food industries. Astaxanthin is considered a king of antioxidants. It is widely used as a colouring agent and antioxidant compound and has the potential to cure several health problems. Astaxanthin is probably best known for eliciting the pinkish-red hue to the flesh of salmonids as well as shrimp, lobster and crayfish. As these animals cannot synthesize Astaxanthin de novo, carotenoid pigments must be supplied in their diet. *Haematococcus pluvialis* is considered the most promising producer of astaxanthin, as it accumulates up to 4% dry wt., mostly in its mono and diester forms in cytoplasmic lipid bodies as a secondary carotenoid. In the natural aquatic environment, astaxanthin is biosynthesized in the food chain within microalgae consumed by zooplankton, insects or crustaceans, which accumulate astaxanthin and, in turn, are ingested by salmonids. The desirable reddish-orange colour is imparted by feeding the astaxanthin-rich diet. Adding to its pigmentation efficiency, astaxanthin has several essential biological functions, including protection against UV light effects, immune response and reproductive behaviour in aquatic animals.

BETA CAROTENE

The unicellular wall-less green alga, *Dunaliella,* is the most extreme halotolerant eukaryotic organism. They are generally ovoid in shape, 4–15 µm in width and 10–25 µm in length. The cells have two long flagella and contain a large cup-shaped chloroplast. Beta carotene is an antioxidant molecule of profitable value that can be logically made by certain microalgae that mostly have their place in the genus *Dunaliella*. Since of its resistance to high salinity and the simplicity of its cytoarchitecture, *Dunaliella* has been widely used to investigate fundamental molecular mechanisms as an important model organism [23]. Carotenoids are the major nutraceutical constituent of *Dunaliella* sp. *D. salina* is able to grow in water with high salt content exploited for its β-carotene content that can reach 14% of its dry weight [24]. They protect the alga from damage caused by excessive irradiance by acting as a light filter. Carotenoids also prevent the formation of reactive oxygen species, quenches the triplet state chlorophyll or the formation of singlet oxygen [25]. Carotenoids are isoprenoid, synthesized by the isoprenoid pathway. Isopentenyl pyrophosphate (IPP) is the precursor of all isoprenoid compounds [26].

Dunaliella salina Teod. is the richest source of carotenoid, β-carotene. The halophilic species of *Dunaliella* also accumulate a very high concentration of glycerol [27]. *Dunaliella salina* was first proposed as a commercial source of β-carotene production [28]. The microalga, *Dunaliella,* under favourable conditions, is green in colour and contains the pigments (Chlorophylls and carotenoids) necessary for photosynthesis. In contrast, under specific inductive conditions, high amounts of β-carotene may be produced, and the colour of the algal cells change from green to rust orange [29]. Natural carotenoids have high in demand globally because of their widespread application in nutrition, medicine, food coloring agent, as well as the natural and health preference of consumers today. Some strains of *Dunaliella* are well known for their massive β-carotene accumulation. β-carotene is highest in *Dunaliella* among all the natural carotenoid sources [30].

The increasing interest in carotenoid production from microalgae is related to the commercial applications of these natural compounds and market demands for carotenoids, especially for pharmaceutical and nutritional applications [31]. Conventionally, carotenoids have been utilised commercially as food additives, counting as antioxidants, colorants and vitamins [32, 33]. Their ability to defend against oxygen free radicals has resulted in some therapeutic applications as degenerative disease preventives, anti-cancer agents and immune system stimulators [34].

Carotenoids are hydrocarbons containing C_{40} polyene chain backbones. The linear chain is cyclized at terminals, the molecule with hydrocarbon terminal is known as carotene, and that of oxygenated terminals is known as xanthophylls. Those carotenes with at least half of β-carotene molecules, *i.e.*, an ionone ring having 11 carbon polyene side chain,s are classified as metabolic precursors of vitamin A. β-carotene is a $C_{40}H_{56}$ molecule containing hydrocarbon cycle at the terminal. It consists of eight isoprenoid units joined in such a manner that the joining of isoprenoid units at the middle is reversed at the centre of the molcecule so that two central methyl groups are at the C_{16} position and the remaining non-terminal methyl groups are at the C_{15} position [35].

The natural β-carotene with a 9-cis isomer plays a major role in quenching free oxygen radicals and preventing oxidative damage to the cell. Moreover, free radicals are highly reactive molecules, and can cause harmful reactions in humans and plants. Antioxidants such as vitamin A play a preventive role in quenching free radicals. The transform of β-carotene is active twice as the cis form in Vitamin A formation. Pharmacological activity can be classified as Vitamin A, antioxidant, immunomodulatory, anticancer, skin protection and biological signalling molecule. β-carotene has shown anticarcinogenesis activity in the case of prostate cancer in human studies. β-carotene is known to reprogramme the oncogene through the antioxidant property through changes in phosphorylation and transcription [36].

CARBON SEQUESTRATION

Nutrients are inorganic or organic compounds other than CO_2 and water, which are used for growth whose presence in the cell is necessary for cellular function. Some algae need precise organic compounds created by other organisms. Though, numerous algae necessitate only inorganic nutrients, and it is likely that these algae could be used as feedstock for biomass fuel production. Growth limiting nutrients to algae are nitrogen, phosphorus, and iron [37].

Apart from carbon, nitrogen is the most significant nutrient causative to the production of biomass. The nitrogen content in the microalgal biomass can range from 1% to more than 10%, depending on its supply and availability, and it can vary between different groups and within specific types [38]. It was stated that the microalgae fed with a blend of urea, and sodium nitrate had the main ash-free dry biomass contented with a product of 4.15 g L^{-1} [39]. The urea molecule comprises a carbon atom as well as two nitrogen atoms. This carbon atom is released as CO_2 when urea is utilized, and presumably, it is available for photosynthetic assimilation, therefore, urea has the potential of providing both the nitrogen and carbon requirements [37].

Phosphorus is essential for the growth of many cellular processes such as energy transfer and during the biosynthesis of nucleic acids. Orthophosphate is the preferred form of algae, and its uptake is said to be energy dependent [38]. Even though algal biomass contains less than 1% phosphorous (P), it usually becomes one of the most important growth limiting factors in algal culture [37]. Sodium, magnesium, potassium, sulphur, iron, calcium and trace elements like boron, copper, manganese, zinc, molybdenum, cobalt, vanadium, and selenium are also important in algal nutrition [38].

CONCLUSION

Fresh water and marine microalgae provide a promising potential to address the climate changes associated with increasing CO_2 levels and the current global energy demands. The ability of many microalgae to photosynthesize and convert CO_2 into organic compounds, which are potential sources of metabolite production, offer these algae as prime candidates for use as experimental organisms to study the effect of climatic changes, *i.e.*, as source organisms for both biodiesel production and bioremediation of CO_2. Considering the soaring oil price in the global market, there is an urgent need to evaluate the potential of algae in the production of energy as bio-diesel. The cost-effective production of bio-diesel, antioxidants and nutraceutical substances from microalgae depends on the cultivation method used for the potential algae.

CONSENT FOR PUBLICATION

Not applicable.

CONFLICT OF INTEREST

The author(s) declare no conflict of interest, financial or otherwise.

ACKNOWLEDGEMENTS

The authors acknowledge the fellowship support from SERB (PDF/2016/002806), DST-FIST (SR/FST/ESI-145/2016) and the Management of Sathyabama Institute of Science and Technology, Chennai for their constant support.

REFERENCES

[1] Siegenthaler U, Sarmiento JL. Atmospheric carbon dioxide and the ocean. Nature 1993; 365(6442): 119-25.
[http://dx.doi.org/10.1038/365119a0]

[2] Abd El-Baky HH, El Baz FK, El-Baroty GS. Production of lipids rich in omega 3 fatty acids from the halotolerant alga *Dunaliella salina*. Biotechnology 2004; 3(1): 102-8.

[3] Stengel DB, Connan S, Popper ZA. Algal chemodiversity and bioactivity: Sources of natural variability and implications for commercial application. Biotechnol Adv 2011; 29(5): 483-501. [http://dx.doi.org/10.1016/j.biotechadv.2011.05.016] [PMID: 21672617]

[4] Rastogi RP, Sinha RP. Biotechnological and industrial significance of cyanobacterial secondary metabolites. Biotechnol Adv 2009; 27(4): 521-39.
[http://dx.doi.org/10.1016/j.biotechadv.2009.04.009] [PMID: 19393308]

[5] de la Coba F, Aguilera J, de Gálvez MV, *et al.* Prevention of the ultraviolet effects on clinical and histopathological changes, as well as the heat shock protein-70 expression in mouse skin by topical application of algal UV-absorbing compounds. J Dermatol Sci 2009; 55(3): 161-9.
[http://dx.doi.org/10.1016/j.jdermsci.2009.06.004] [PMID: 19586754]

[6] Vimalabai CPM, Kulandaivelu G. Effects of prolonged UV-B enhanced fluorescent radiation on some marine microalgae. Biol Plant 2002; 45(3): 389-94.
[http://dx.doi.org/10.1023/A:1016217700979]

[7] Tramper J, Battershill C, Brandenburg W, *et al.* What to do in marine biotechnology? Biomol Eng 2003; 20(4-6): 467-71.
[http://dx.doi.org/10.1016/S1389-0344(03)00077-7] [PMID: 12919833]

[8] Andersen RA, Ed. Algal culturing techniques. Elsevier 2005.

[9] Cohen Z. Production of polyunsaturated fatty acids by the microalga *Porphyridium cruentum*. In: Cohen Z, Ed. Taylor and Francis: London 1999; 1-24.

[10] Richmond A, Ed. Handbook of microalgal culture: biotechnology and applied phycology. John Wiley & Sons 2008.

[11] Kumar Saha S, Uma L, Subramanian G. Nitrogen stress induced changes in the marine cyanobacterium Oscillatoria willei BDU 130511. FEMS Microbiol Ecol 2003; 45(3): 263-72. [http://dx.doi.org/10.1016/S0168-6496(03)00162-4] [PMID: 19719595]

[12] Yamaguchi K. Recent advances in microalgal bioscience in Japan, with special reference to utilization of biomass and metabolites: a review. J Appl Phycol 1996; 8(6): 487-502.
[http://dx.doi.org/10.1007/BF02186327]

[13] Bañuelos-Hernández B, Beltrán-López JI, Rosales-Mendoza S. Production of biopharmaceuticals in microalgae. Academic Press 2015; 281-98.
[http://dx.doi.org/10.1016/B978-0-12-800776-1.00018-2]

[14] Metzger P, Largeau C. *Botryococcus braunii*: a rich source for hydrocarbons and related ether lipids. Appl Microbiol Biotechnol 2005; 66(5): 486-96.
[http://dx.doi.org/10.1007/s00253-004-1779-z] [PMID: 15630516]

[15] Brown AC, Knights BA, Conway E. Hydrocarbon content and its relationship to physiological state in the green alga *Botryococcus braunii*. Phytochemistry 1969; 8(3): 543-7.
[http://dx.doi.org/10.1016/S0031-9422(00)85397-2]

[16] Sawayama S, Minowa T, Yokoyama S-Y. Possibility of renewable energy production and CO2 mitigation by thermochemical liquefaction of microalgae. Biomass Bioenergy 1999; 17(1): 33-9. [http://dx.doi.org/10.1016/S0961-9534(99)00019-7]

[17] Chisti Y. Biodiesel from microalgae. Biotechnol Adv 2007; 25(3): 294-306.
[http://dx.doi.org/10.1016/j.biotechadv.2007.02.001] [PMID: 17350212]

[18] Singh A, Smyth BM, Murphy JD. A biofuel strategy for Ireland with an emphasis on production of biomethane and minimization of land-take. Renew Sustain Energy Rev 2010; 14(1): 277-88. [http://dx.doi.org/10.1016/j.rser.2009.07.004]

[19] Ashokkumar V, Rengasamy R. Mass culture of *Botryococcus braunii* Kutz. under open raceway pond for biofuel production. Bioresour Technol 2012; 104: 394-9. [http://dx.doi.org/10.1016/j.biortech.2011.10.093] [PMID: 22115530]

[20] Haddad R, Alemzadeh E, Ahmadi AR, Hosseini R, Moezzi M. Identification of Chlorophyceae based on 18S rDNA sequences from Persian Gulf. Iran J Microbiol 2014; 6(6): 437-42. [PMID: 25926963]

[21] Pohl P. Lipids and fatty acids of microalgae. CRC handbook of biosolar resources, 1982.

[22] Fukuda H, Kondo A, Noda H. Biodiesel fuel production by transesterification of oils. J Biosci Bioeng 2001; 92(5): 405-16. [http://dx.doi.org/10.1016/S1389-1723(01)80288-7] [PMID: 16233120]

[23] Chen M, Tang H, Ma H, Holland TC, Ng KYS, Salley SO. Effect of nutrients on growth and lipid accumulation in the green algae Dunaliella tertiolecta. Bioresour Technol 2011; 102(2): 1649-55. [http://dx.doi.org/10.1016/j.biortech.2010.09.062] [PMID: 20947341]

[24] Metting FB. Biodiversity and application of microalgae. J Ind Microbiol 1996; 17(5): 477-89.

[25] Telfer A. What is β–carotene doing in the photosystem II reaction centre? Philos Trans R Soc Lond B Biol Sci 2002; 357(1426): 1431-40. [http://dx.doi.org/10.1098/rstb.2002.1139] [PMID: 12437882]

[26] Fraser P, Bramley PM. The biosynthesis and nutritional uses of carotenoids. Prog Lipid Res 2004; 43(3): 228-65. [http://dx.doi.org/10.1016/j.plipres.2003.10.002] [PMID: 15003396]

[27] Borowitzka LJ, Moulton TP, Borowitzka MA. The mass culture of *Dunaliella salina* for fine chemicals: from laboratory to pilot plant. In: Eleventh international seaweed symposium. Dordrecht Springe 1984; 115-21.

[28] Massyuk NP. Mass culture of the carotene bearing alga *Dunaliella salina*. Ukr Bot Zh 1968; 23: 12-9.

[29] Orset S, Young AJ. Low-temperature-induced synthesis of α-carotene in the microalga *Dunaliella salina* (Chlorophyta). J Phycol 1999; 35(3): 520-7. [http://dx.doi.org/10.1046/j.1529-8817.1999.3530520.x]

[30] Ye ZW, Jiang JG, Wu GH. Biosynthesis and regulation of carotenoids in Dunaliella: Progresses and prospects. Biotechnol Adv 2008; 26(4): 352-60. [http://dx.doi.org/10.1016/j.biotechadv.2008.03.004] [PMID: 18486405]

[31] Nelis HJ, De Leenheer AP. Microbial sources of carotenoid pigments used in foods and feeds. J Appl Bacteriol 1991; 70(3): 181-91. [http://dx.doi.org/10.1111/j.1365-2672.1991.tb02922.x]

[32] Borowitzka MA. Algal biotechnology products and processes — matching science and economics. J Appl Phycol 1992; 4(3): 267-79. [http://dx.doi.org/10.1007/BF02161212]

[33] Armstrong GA. Genetics of eubacterial carotenoid biosynthesis: a colorful tale. Annu Rev Microbiol 1997; 51(1): 629-59. [http://dx.doi.org/10.1146/annurev.micro.51.1.629] [PMID: 9343362]

[34] Agarwal S, Rao AV. Carotenoids and chronic diseases. Drug Metabol Drug Interact 2000; 17(1-4): 189-210. [http://dx.doi.org/10.1515/DMDI.2000.17.1-4.189] [PMID: 11201295]

[35] Goodwin TW. Biochemistry of the carotenoids. London; New York: Chapman and Hall.

[36] Raja R, Anbazhagan C, Ganesan V, Rengasamy R. Efficacy of *Dunaliella salina* (Volvocales, Chlorophyta) in salt refinery effluent treatment. Asian J Chem 2004; 16(2): 1081.

[37] Neenan B. Feinberg D, Hill A, McIntosh R, Terry K, Fuels from Microalgae: Technology Status, Potential, and Research Requirements, Colorado: Solar Energy Research Institute 1986. [http://dx.doi.org/10.2172/6685301]

[38] Grobbelaar JU. Algal nutrition: mineral nutrition In: Richmond A Ed Handbook of Microalgal Culture: Biotechnology and Applied Phycology Blackwell Publishing Ltd 2004; 95-115.

[39] Lin Q, Lin J. Effects of nitrogen source and concentration on biomass and oil production of a Scenedesmus rubescens like microalga. Bioresour Technol 2011; 102(2): 1615-21. [http://dx.doi.org/10.1016/j.biortech.2010.09.008] [PMID: 20875734]

CHAPTER 17

Seaweed in Food Industries: Raw Materials, Processing, Formulations, Packaging

Marlene A. Trindade[1], Cláudia Nunes[2,3], Manuel A. Coimbra[3], Fernando J.M. Gonçalves[1,4], João C. Marques[5] and Ana M.M. Gonçalves[1,4,5,*]

[1] *Department of Biology, University of Aveiro, Campus Universitário de Santiago, 3810-193 Aveiro, Portugal*

[2] *CICECO, Department of Chemistry, University of Aveiro, Campus Universitário de Santiago, 3810-193 Aveiro, Portugal*

[3] *LAQV-REQUIMTE, Department of Chemistry, University of Aveiro, Campus Universitário de Santiago, 3810-193 Aveiro, Portugal*

[4] *CESAM, University of Aveiro, Campus Universitário de Santiago, 3810-193 Aveiro, Portugal*

[5] *University of Coimbra, MARE – Marine and Environmental Sciences Centre / ARNET - Aquatic Research Network, Department of Life Sciences, Calçada Martim de Freitas, 3000-456 Coimbra, Portugal*

Abstract: Seaweeds are an abundant source of bioactive compounds and hydrocolloids that can be used in several industries, especially in the food industry. They can be directly consumed as a whole, used as an ingredient, or a source of several compounds that can enrich the nutrient content and sensory experience of foods. The demand for more sustainable packaging options resulted in the increase in research and development of edible seaweed-based films that ensure the safety and quality of the food products, while reducing packaging waste. This chapter will make an overview of the seaweed compounds used in the food industry, their use in the formulation of edible active packaging, and how this packaging can become widely available for consumers.

Keywords: Bioactive Compounds, Food Packaging, Hydrocolloids, Macroalgae, Sustainability.

INTRODUCTION

Seaweed or marine macroalgae are multicellular photosynthetic organisms found worldwide. They are divided into three main groups according to their color: brown (phylum Ochrophyta, class Phaeophyceae), red (phylum Rhodophyta), and

[*] **Corresponding author Ana M.M. Gonçalves:** University of Coimbra, MARE – Marine and Environmental Sciences Centre ARNET - Aquatic Research Network, Department of Life Sciences, Calçada Martim de Freitas, 3000-456 Coimbra, Portugal; Department of Biology & CESAM, University of Aveiro, Campus Universitário de Santiago, 3810-193 Aveiro, Portugal; Tel: +351 239 240 700 (Ext. 262 286); E-mail: amgoncalves@uc.pt

Avinash Mishra (Ed.)

green (phylum Chlorophyta) algae. In comparison with land plants, they are different physiologically, morphologically and, consequently, chemically.

Over centuries, macroalgae have been used as medicine and food in most East Asian countries and in the island nations of the Pacific. However, its use in Europe was negligible, focused on Celtic nations and other coastal regions closely associated, and dependent on activities related to the sea [1 - 3]. The absence of lignin makes seaweeds easy to handle and process, which makes them a useful and easy-to-obtain material, not only as food and medicine but also as a source of highly valuable materials for several areas [4]. Nowadays, macroalgae are primarily used as food and for the production of hydrocolloids and bio-fertilizers, but there is a growing interest in the production of raw materials for animal feed, chemicals, and biofuels, as well as its application in wastewater treatment [5 - 8]. Seaweeds are also processed into high-value products such as cosmetics, health care, pharmaceutical, and medicinal products, as they are a source of excipients, active ingredients, and additives. Its high content in bioactive compounds with a broad spectrum of biological activities, such as antimicrobial, antioxidant, and anti-inflammatory properties, contribute to its advantageous use in these fields [9].

Consumers interest in seaweeds in Western cultures is growing due to the demand for more sustainable and healthy food sources [1, 10]. Modern problems require modern solutions, so with the exponential growth of the human world population, the climate crisis, and the decrease of arable land, rethinking the current food sources and searching for sustainable practices are needed. Even though the use of seaweeds is secular, this marine resource can be a solution to the future food shortage problem. Compared with most agricultural crops, seaweeds can be more nutritious, have higher productivity and faster growth rates, and do not require arable land, freshwater or fertilizers. Additionally, they can improve water quality and function as carbon sinks.

About 145 species of macroalgae are consumed as food worldwide, and there are no reports or proof of illnesses or deaths caused by the seaweed. Any reports are related to toxins from a dubious source or exogenous to the seaweed, so they are considered a safe ingredient and food product [11]. However, potential risks of macroalgae consumption include exposure or excess intake of toxic compounds like heavy metals, cyanotoxins, certain secondary metabolites and contamination with pathogens [12, 13]. Moreover, the environment where they are collected or farmed needs to be monitored, as toxic and harmful compounds can be accumulated by the algae. Nevertheless, to ensure consumer safety, there is a need for monitoring programs and processing technologies, and more case-by-case studies of the nutrient and toxicological profile of the species used in food production. Furthermore, there is a need for more legislation regarding the labeling of seaweed-based products to provide a safe portion size to prevent excess intakes of nutrients, minerals, or heavy metals [13].

Aquaculture is a way to standardize the nutritional value of macroalgae and the possibility of customizing its content according to its final purpose by changing the algae culture conditions, namely pH, salinity, temperature, conductivity, agitation, light, and nutrients availability. Aquaculture is the most sustainable way of seaweed production and accounts for ~96% of the annual global production (~27.3 million tonnes). However, the growing demand for seaweed-based products is putting pressure on the establishment of guidelines and regulations that ensures their sustainable status [14]. According to the Food and Agriculture Organization report of 2020 on "The State of World Fisheries and Aquaculture", in 2018, the trade in seaweed and other aquatic plants accounted for 63% of the USD 2 billion in exports. The production of marine macroalgae has grown "from 10.6 million tonnes in 2000 to 32.4 million tonnes in 2018," and some species were produced primarily for human consumption as food, like the case of *Porphyra* spp. (~2 million tonnes in 2018) and *Undaria pinnatifida* (~2.3 million tonnes in 2018). The aquaculture sector employs almost 60 million people worldwide [13] and creates a wide range of jobs and financial services, especially at a local level.

With the increase of environmental awareness, consumers' demand for environmentally friendly products is increasing and, consequentially, the growth in research and development of seaweed-based products. Therefore, this chapter will explore the uses of seaweed and its bioactive compounds in different food industries, highlighting the food packaging industry.

MACROALGAE AS A SOURCE OF BIOACTIVE COMPOUNDS AND RAW MATERIALS FOR THE FOOD INDUSTRY

Macroalgae can be consumed directly as whole or as food supplements and, indirectly, in the form of additives [2], as some hydrocolloids are used to improve the quality and sensory properties of certain food products. Macroalgae are low-caloric foods, rich in protein, carbohydrates and minerals, which are of nutritional interest. They also produce secondary metabolites, like phenolic compounds, terpenoids, halogenated compounds, sulfur derivatives, and nitrogen derivatives, that could be used to improve food quality. All these compounds have a potential biological activity and can, therefore, be considered bioactive compounds, defined "as phytochemicals, which can be extracted from foods or foods by-products, and able to regulate metabolic functions leading to beneficial effects" [15].

The chemical and biochemical composition of seaweed varies between species, season, geographic, and ecology of the harvesting location. The characterization of seaweeds composition according to these variability-inducing factors can assist to establish the optimal conditions for the isolation of specific bioactive components [2]. Additionally to their nutritional value, these bioactive compounds have several potential benefits to human health. Even though most of the studies report all these benefits, few studies address their bioavailability, the quantity released from the algae and taken by the tissues, metabolism, and its physiological effects. Moreover, most of the results are from animal or *in vitro* tests, hence the need for more human intervention studies [13]. Table **1** summarizes the bioactive components present in macroalgae and their properties used in different industries [9, 13, 16 - 20].

Table 1. Macroalgae compounds and respective bioactive properties.

Compound		Bioactive Properties
Proteins and essential amino acids		• Antioxidant • Control of hyperglycemia and hypercholesterolemia • Inhibition of calcium precipitation • Improvement of hepatic activity • Treatment of neurodegenerative diseases
Polysaccharides		• Antioxidant • Anti-inflammatory • Antitumoral • Antibacterial • Anticoagulant • Antiviral • Control of hyperglycemia • Anti-aging • Hepatoprotective • Immunoregulatory function
Lipids	**Fatty acids**	• Treatment and prevention of: ▪ some cancers ▪ cardiovascular diseases ▪ ocular diseases ▪ neurodegenerative diseases ▪ autoimmune diseases • Anti-inflammatory effects on obesity and obesity-associate--comorbidities
	Sterols	• Antioxidant • Antibacterial • Antiviral • Antifungal

(Table 1) cont.....

Compound		Bioactive Properties
Phytochemicals	**Carotenoids**	• Photoprotective • Antioxidant • Cytoprotective • Antitumoral • Anticancer • Anti-inflammatory • Neuroprotective
	Phenolic compounds	• Antioxidant • Anti-inflammatory • Antimicrobial • Anticancer • Antidiabetic • Neuroprotective • Radioprotective
Minerals		• Prevention of: o Cancer o Cardiovascular diseases o Degenerative diseases o Nutritional deficiency • Synthesis of thyroid hormones

Up to 45% of seaweeds dry weight (DW) is protein. Red seaweeds have the highest protein content (0.67-45% [13]) and the highest ratio of essential amino acids and non-essential amino acids, followed by green algae (3.4-29.8% [13]) and brown algae (5.0-19.7% [13]). Therefore, seaweed can be an alternative to animal protein [2, 12, 13, 17]. Seaweed content of essential amino acids is similar to the one found in vegetable food sources, with glutamic acid and aspartic acid being the most present. Seaweeds are an alternative lysine source, an essential amino acid, since their content in terrestrial plants is limited. Taurine is particularly abundant in the phylum Rodophyta, and even though it is not an essential amino acid, it is a bile constituent that helps to lower cholesterol levels [12, 13, 17]. Due to its antioxidant activity, the use of seaweed peptides has also been suggested as food preservatives [13].

Macroalgae lipid content is usually very low (0.4-5% DW [12,17,21]) being usually a cell membrane constituent. Most of its content is long-chain fatty acids, particularly polyunsaturated fatty acids (PUFAs), with eicosapentaenoic acid (EPA; C20:5 n-3) and arachidonic acid (C20:4 n-6) the most predominant [12]. Since consumption of whole seaweeds is unlikely to contribute significantly to dietary fat intake, they can be a sustainable source of extractable PUFAs, that may be used as dietary supplements [13]. Brown seaweeds are the richest in essential fatty acids and PUFAs, followed by green and red seaweeds [12].

Comparing with "common" vegetables and fruits, edible seaweeds have higher fiber levels (23.5-64% [12]) due to the presence of high molecular weight carbohydrates [2]. Polysaccharides are the most important macromolecules in seaweed as they are the most produced and consumed seaweed food product, being incorporated in dairy products, meat, and beverages. The most used algal polysaccharides include 1) ulvan, xylan, and cellulose from Chlorophyta; 2) laminarin, fucoidans, and alginate from Phaeophyceae; and 3) agar, carrageenan, and xylan from Rhodophyta [13]. Sulphated polysaccharides are present in the three main groups of seaweeds and have several uses in the nutraceutical and biomedical industry [17], mainly due to their antioxidant and antimicrobial activity [17]. Also, these antioxidant properties have been explored for the development of active packaging. Alginate, agar, and carrageenan are the most frequent polysaccharides from algae in the food industry due to their emulsifying, stabilizing, and thickening characteristics but have limited interest as functional dietary fibers [12, 13].

Seaweeds have many pigments that are not found in terrestrial plants (*e.g.*, phycobiliproteins) with interesting biological activities, like antioxidant, anti-inflammatory, antidiabetic, and immune modulatory [22]. These pigments are convenient for food formulation, especially in drinks [2], because of their bioactive and sensorial properties.

Vitamins are essential for metabolic functions, and algae are a very rich source, containing all essential and non-essential vitamins. For example, the vitamin A content of a portion (5 g) of dried seaweed varies from 14.5 µg in *Ulva rigida* to 70.5 µg in *Fucus spiralis*, and vitamin B9 content ranges from 7.5 µg in *Ulva* spp. to 5400 µg in *Ulva rigida* [13]. Macroalgae contain levels of vitamin C comparable to "common" vegetables, varying from 0.41 mg in *Ascophyllum nodosum* to 9.24 mg in *Undaria pinnatifida*, and kelp (brown algae) has comparable levels of vitamin E to plant oils. Seaweed is one of the few non-animal sources of vitamin B12, essential in meatless diets since higher plants do not synthesize it. *Enteromorpha* spp. is reported to contain 63.58 µg and *Porphyra* spp. from 32.26 to 133.8 µg per 100 g of dry weight [12, 13, 17]. Therefore, they are a good source of vitamins A, B-group, C, and E.

Phytochemicals, mostly carotenoids and polyphenols, have a commercial interest, especially in pharmaceutical industries [12]. Polyphenols are often bound to cell wall polysaccharides, protecting the cells from oxidative damage. Red and brown algae are rich in these compounds. Fucoxanthin, extracted from brown seaweeds, is the main carotenoid with potential application in the food industry. Its antioxidant activity may potentially be a preservative of meat products preventing lipid peroxidation [13].

Macroalgae consumption can contribute to dietary mineral intake. They are good sources of inorganic elements with nutritional benefits, like iodine and iron. Besides enhancing total dietary iron content, seaweed facilitates iron uptake from wheat, maize, and rice meal [12, 13]. The extraction of individual components from the seaweed matrix is a good strategy to produce added value to food products, since the recommended portion size of seaweed may be limited by the balance between iodine and/or heavy metal ingestion and the amount of seaweed needed to ingest a significant number of macronutrients [19].

Macroalgae Hydrocolloids in Food Industry

This section will focus mainly on seaweed hydrocolloids since they are commercially valuable, especially as functional ingredients in the food industry as stabilizers, thickening agents, emulsifiers, and texture modifiers. The extraction of hydrocolloids is a slow multistep process dependent on chemicals. Figs. **1** and **2** summarize the most used methods for phycocolloids extraction [19, 23, 24].

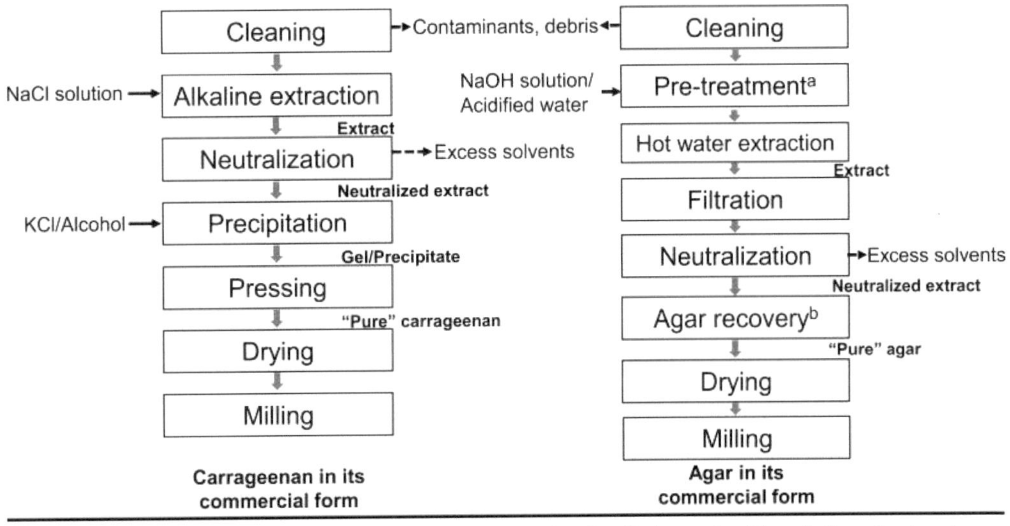

[a]Acidified water for *Gelidium* species and strong alkali for *Gracilaria* species. The alkali pre-treatment is to convert porphyran in 3,6-anhydro-galactopyranose improving the gelling capacity og agar the end mechanical properties of its resulting gels.

[b]Through a method of freeze-thaw or syneresis. The syneresis method has a higher recovery yield.

Fig. (1). General method of extraction of carrageenan and agar.

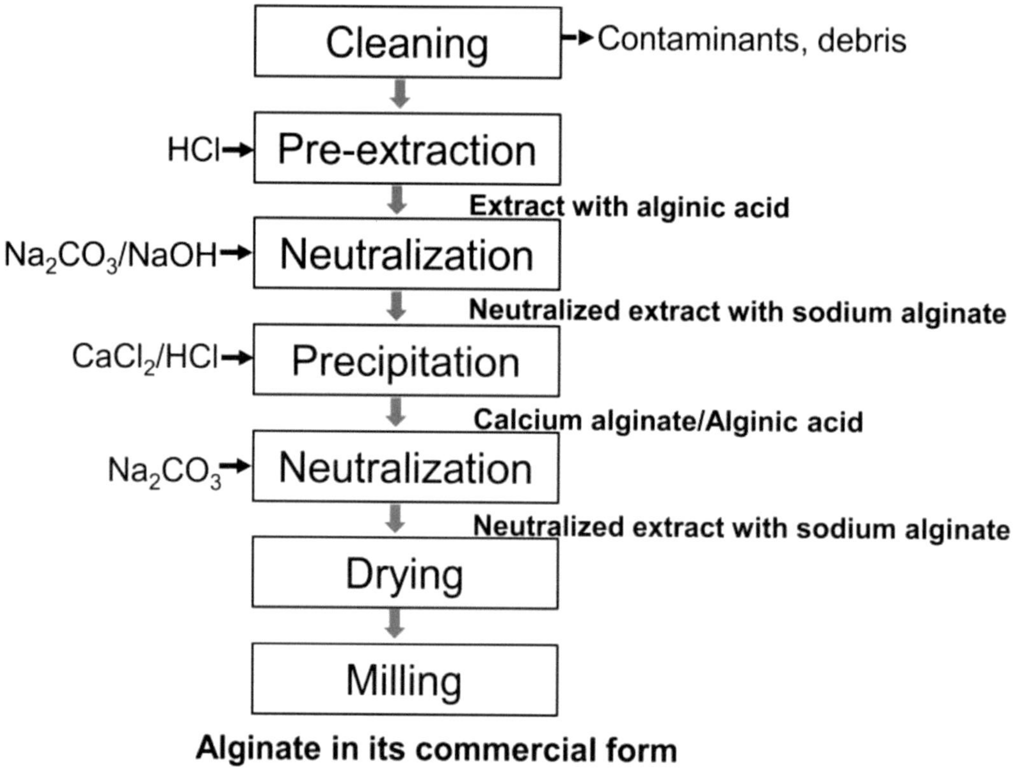

Fig. (2). General method of extraction of alginate.

Seaweed hydrocolloids are water soluble, so the solvent used for the extraction is typically hot water [19, 23]. Alginate is extracted with acid, followed by alkali solvents. Agar and carrageenan extracted with hot water have weak rheological properties and are not the most desirable for industrial applications. Therefore, the yields of production and the colloid's properties are improved through an alkali pretreatment. Nevertheless, the yields can eventually decrease due to polysaccharide degradation and cannot be recovered in the following steps [19].

Some drawbacks of the current extraction methods at an industrial level are the massive consumption of water, energy, and time. The extraction industries also use large amounts of reactants and solvents to achieve optimum yields and quality. However, some of these chemicals are hazardous to the environment and consumer's health. From an economic point of view, some solvents can be expensive, due to the high costs of the process and equipment needed to recover them. Additionally, it is difficult to optimize phycocolloids production because its properties are affected by biotic and abiotic factors, as well processing-related variables affect their yields and rheological properties [19].

Currently, economic efficiency, phycocolloids quality, and ecology are the main aspects that the extractive industry is looking for. Thus, the development of "green" extraction methods, like ultrasound-assisted and enzyme-assisted extraction, whose advantages include reducing extraction times, and avoiding or reducing the use of solvents. But there is still not much literature available on these methods due to their limited application in phycocolloids extraction [19]. Abdul Khalil *et al.* explore several "green" extraction and processing methods, describing them and their advantages and limitations [19].

The most common phycocolloids in the food industry are alginate, agar and carrageenan. Its use is not limited to functional ingredients, as they have been used in the development of new types of packaging.

"Alginate" is the term used to designate alginic acid and all its derivatives and salts [19, 23]. It is the main polysaccharide of brown algae (14-40% DW) [12], and the main species for commercial production of alginate are *Laminaria hyperborean*, *Laminaria digitata, Macrocystis pyrifera*, and *Ascophyllum nodosum* [19, 23]. Alginate is a linear polysaccharide consisting of two monomeric uronic acids: β-D-mannuronic acid (M) and α-L-guluronic acid (G), linked by 1→4 glycosidic linkages. These uronic acids are asymmetrically organized with variable proportions of MM, GG, and MG blocks depending on the source (seaweed species), harvest time, and extraction technique [2, 19, 23]. The M/G ratio and the length of the blocks in the alginate structure impact its physicochemical properties and, consequently, the gels made from alginate. Higher contents of guluronic acid result in stronger gelling properties. High M/G ratios result in more elastic gels with good freeze-thaw stability, while low ratios result in strong and brittle gels with good heat stability but show syneresis after freeze-thaw processing [19, 23].

The presence of ions in the solution, the pH and ionic strength of the solvent, will determine the solubility of alginate. The selective binding of divalent and multivalent cations determines their ability to form gels, but they also form gel following the cation-independent gelation mechanism [19, 23]. Alginate is an excellent stabilizing and thickening agent, utilized in frozen desserts, bakery products, ice creams, syrups, sauces, and beverages, due to its capacity to chelate metal ions to form highly viscous solutions [19]. In the presence of calcium ions, alginate forms heat-stable gels at room temperature. Moreover, the formation of calcium alginate fibers through the cross-link of calcium ions in the alginate matrix forms waterproof membranes, an important characteristic in the development of films for packaging purposes. Films made from alginates are interesting because of their biocompatibility, low toxicity, and insoluble film formation via ionotropic gelation. Alginate films have good oxygen barrier properties, prevent lipid oxidation, and improve food flavor and texture [25].

Carrageenan is a linear sulfated polysaccharide composed of alternating and repeating disaccharides units of β-D-galactopyranose and α-D-galactopyranose or 3,6-anhydro-α-D-galactopyranose, linked through β-(1→4) and α-(1→3) glycosidic linkages. According to the number of sulfate groups per disaccharide unit, carrageenan can be differentiated as κ-, ι-, and λ-carrageenan, the most relevant types to industry. κ- and ι-carrageenan have one and two sulfate groups, respectively, and can form gels, being mainly extracted from *Kappaphycus alvarezii* and *Euchema denticulatum*. λ-carrageenan has three sulfate groups and is a thickening agent primarily extracted from *Gigartina skottsbergi* and *Sarcothalia crispata* [2, 23, 26]. Regardless of being technically considered a dietary fiber, carrageenan has gelling, thickening and stabilizing properties, therefore, they are mainly used in dairy and meat products [23].

All different types of carrageenan are soluble in water, and factors like temperature, pH, ionic strength of the medium, and the presence of cations influence its aqueous solubility. The sulfate and hydroxyl groups cause their hydrophilic character whilst their hydrophobicity came mostly from the 3,6-anhydro-α-D-galactopyranose units [23]. The low hydrophobicity of carrageenan is a disadvantage, especially for the manufacturing of packaging, because they would be susceptible to moisture transfer and have low water resistance. However, blending carrageenan with hydrophobic compounds, reinforcing the matrix with nanomaterials, or laying it with other polymer films enhance the properties of carrageenan films, hence their potential in the development of eco-friendly, cost-effective packaging material [27].

In the formation of gels from κ- and ι-carrageenan, there are two steps involved, coil-to-helix transition upon cooling and cation-induced aggregation of helices. However, they have different properties, κ-carrageenan forms hard, strong, and brittle gels and ι-carrageenan forms soft, weak and elastic gels. The different chemical structure of carrageenan is important for their physicochemical properties [28]. Its strength depends on carrageenan concentration, type of the salt, and the concentration of the gelling cation, usually, higher concentrations imply greater gelling temperature and gel strength. The presence of co-solutes also increases the gelation and melting temperatures of carrageenan solutions [23].

Carrageenan and alginate are highly hydrophilic, hence their weak moisture barrier properties. However, they are good barriers to fats and oils. These polysaccharides can prevent microbial proliferation in meat products and fruits. Additionally, films formulated with alginate or carrageenan prevent shrinkage and surface discoloration of meat products, and delay the ripening of fruits, increasing their shelf-life [29].

Red seaweed of the genera *Gelidium* and *Gracilaria* are the major producers of commercially used agar [19]. Agar is also a linear polysaccharide composed of alternating and repeating disaccharide units of β-D-galactopyranose (agarobiose) and 3,6-anhydro-α-L-galactopyranose (neoagarobiose), linked through β-(1→4) and α-(1→3) glycosidic linkages. The hydrogen bonding between the 3,6-anhydro-α-L-galactopyranose residues stabilizes the agar gels. Agar has a fraction with high gelling capacity, the neutral, low sulphate/methoxyl substituted one, and a fraction with low gelling capacity, the charged, heterogenous, highly substituted one [23]. Usually, the non-gelling fraction is removed during its extraction to obtain a powder with a higher gelling capacity [30]. In the agar gel, a network of agarose double helices is formed, as they aggregate by their external hydroxyl groups and are stabilized by water molecules. Gels produced from agar are usually transparent and stiff, but the addition of sugars (*e.g.* sucrose or glucose) can increase its strength through the promotion of helices formation [22].

Photodegradation and fluctuations in temperature and humidity in the surrounding environment promote the formation of microfractures and embrittlement. Therefore, agar has not been used in the development of edible films. However, when mixed with film-forming emulsions, for example, it improves barrier performance [31]. Type and origin of agar, the film production method and the components incorporated in the matrix influence the needed concentration to produce an agar-based film. Agar films are biologically inert and can easily interact with different bioactive substances [30].

Comparing with carrageenan and alginate, agar has a less hygroscopic nature, which is an advantage in the production of packaging since one of its main functions is to reduce moisture transfer between the product and the surrounding environment [32]. The relatively high production cost of agar and its brittleness are drawbacks to its commercialization as food packaging, but the addition of plasticizers or the use of unpurified agar can be a way to overcome this. The plasticizers help to form softer and elastic gels, as well as the presence of proteins, polyphenols and other polysaccharides, considered as impurities in the agar extract, that additionally can give antioxidant properties to the gels [33].

SEAWEED-BASED FOOD PACKAGING

As a result of a global shift from reusable containers to disposable ones, the packaging industry creates a high content of plastic waste [34]. As it is cheap to produce, long-lasting, liquid proof, resistant to handling, and maintaining the sensory properties of the food products, petroleum-based synthetic polymers have been widely used in food packaging [28, 34–36]. Even though they are non-biodegradable, most of these packages are recyclable [36]. Moreover, the increase of the world population increases, as well as the demand for plastic products [35], it is challenging to eliminate or reduce its global use. Therefore, more eco-friendly alternatives are being researched and developed.

Bio-based plastics are an existing alternative to traditional plastic, and even though they have advantages, some are produced from crops, thus their production competes for resources that could be used for food production. Completely biodegradable, bioplastic has a minimal environmental impact, but not all bioplastics are completely biodegradable. However, they are sustainable options, since non-biodegradable bioplastics can serve as carbon sinks and integrated into sustainable and long-lasting materials and infrastructures, making bio-based plastics part of a circular economy in which plastic never becomes waste [35].

Using a marine resource, there is no need for land acquisition, deforestation and use of resources originally destined for crops to produce bioplastics. Thus, macroalgae can be an option for the production of bio-based plastics with the additional advantage of being completely biodegradable. Moreover, macroalgae can be used in the largescale production of biopolymers, since their biomass can be the substrate for bacteria and *Archaea* that produce biopolymers used in the production of bioplastics, like polyhydroxyalkanoate [37, 38]. As they are nontoxic, biodegradable, and derived from a renewable natural source, agar, carrageenan and alginate are widely used in several food industries as aforementioned, especially in the production of films for packaging [24, 29, 39, 40].

Seaweed-based packaging needs to tackle some main points to be a real alternative, namely, the aforementioned advantages in addition to keeping food crispy, fresh fruit/vegetables and meat/fish products "fresh", allowing freezing of food products, blocking germs, among others that will influence the products shelf-life.

Nevertheless, seaweed-based packaging can also add advantages to packaging that traditional plastic does not have. For example, it can produce edible and bioactive packaging that can increase the nutritious value of the product while simultaneously protect from microbial spoilage. Other challenges that need to be addressed, before edible packaging can be more widespread, include dealing with moisture and heat/cold variations.

Seaweed-based biofilms consist of a matrix of a phycocolloid or a combination of phycocolloids with other polymers and/or a hydrophobic component and/or nanoparticles forming a hybrid material, a composite [24, 31, 41]. The production of composite films is a way to overcome the poor mechanical and water barrier properties shown by seaweed-based packaging. Nevertheless, phycocolloid-based films show good oxygen vapor barrier properties and are impermeable to oils. Its mechanical and barrier properties depend on the properties of the individual polymers and their compatibility. The use of lipids in the seaweed-based matrix to produce emulsified films can also be used to encapsulate active components. Edible films can be incorporated with antioxidants and other nutraceuticals supplementing the nutritional value of the food [40, 42].

Formulation and Processing

The most common method used in research for the manufacturing of films from phycocolloids, as food packaging, is solvent casting (Fig. **3**) [24,26,31].

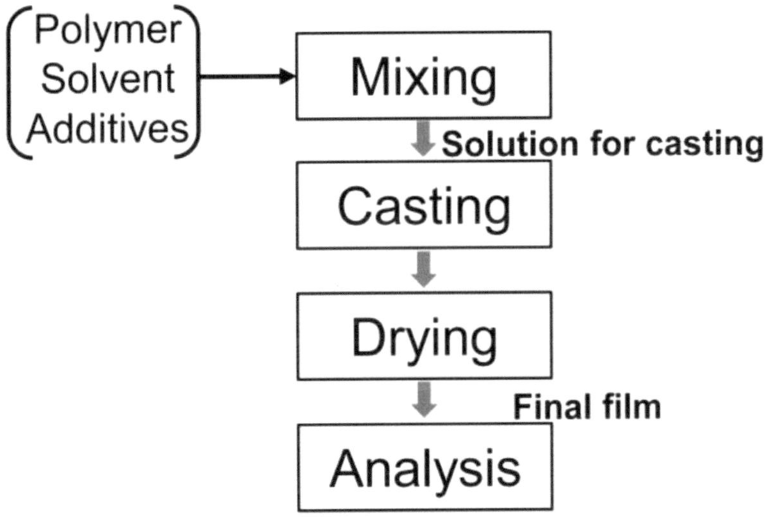

Fig. (3). General steps involved in the solvent casting method for the production of phycocolloid-based films.

Initially, the components are all mixed and stirred, to incorporate all the components, and centrifuged, to eliminate any air bubbles, only then a solution for casting is obtained. This solution is poured in casting plates and then air-dried in a ventilated oven. When the films are dried, a sample is taken for analysis to evaluate thickness, diffusion rates, antioxidant activity, and other relevant parameters.

Some important consideration of the solvent casting method is the use of water-based solvents, to avoid solvents that can be hazardous to human health and to the environment. When selecting the phycocolloid should consider that its characteristics will influence the disintegration rate, taste, and mechanical properties of the obtained film, as well as the drying conditions and the casting thickness [26, 31]. The food product itself, its designated storage conditions, and required shelf life will also influence the films properties and formulation [36]. The concentration of the polymer and additives, the spreading of the film solution, and its density, viscosity, and surface tension affect the thickness of the film, which in turn will influence the shelf life of the product [28].

The diffusion of compounds, whether water, gases, or any active compound, in the films depends on the physicochemical properties of these compounds and its interaction with the film matrix. Water resistance is important to prevent the easy disintegration of the packaging in high moisture content environments. Suitable mechanical properties are needed to prevent tears during its manufacturing, handling, and storage. Thermal properties are also important to create packaging resistant to heat fluctuations or to freezing oruse in the microwave, for example. Also important in the consumer perspective is its general aspect, taste, and smell [31].

Films made only from phycocolloids do not meet the criteria for commercial use, as they are usually brittle, and lack of flexibility and toughness. The addition of plasticizers is a common way to improve mechanical properties, where the type of plasticizer and its concentration affect its properties. The most utilized plasticizers include polyols, like glycerol and sorbitol, sugars, and water. Plasticizers intercalate between the biopolymer chains spreading them apart and disrupting hydrogen bonds, increasing flexibility, and water vapor and gas permeability [31]. However, while adding plasticizers increases the elongation at break values, meaning it increases the ability of the film to resist to changes of shape, it decreases the tensile strength, meaning the ability of the film to be stretched without breaking [25, 31].

The development of active edible packaging is an advantage of seaweed-based packages, as it can add nutritional and health benefits to the food product while improving or maintaining its organoleptic characteristics and shelf-life. Antioxidant compounds, antimicrobial agents, pre- and probiotics, vitamins and other beneficial compounds are some of the actives that can be incorporated into the phycocolloid-based matrix. This is more efficient than applying the compounds directly to the food product since its diffusion will be slowed, thus its effects are prolonged in time and increase food product shelf-life.

Some antimicrobial compounds can also act as antioxidant compounds. In the formulation of edible films, essential oils are one of the most common antimicrobial and/or antioxidant bio-preservatives with proven effects against Gram-negative and Gram-positive bacteria and antifungal activity, as well as preventing oxidative rancidity, degradation, and discoloration of the food product. Rather than its type, the concentration of the essential oils will determine its efficiency, and they can also act as plasticizers. Essential oils are usually extracted with toxic organic solvents, which adds purification steps to the process, then, the development of less toxic extraction techniques can also reduce its cost and increase the safety and sustainability of the product [25, 43 - 47].

Another way of improving the properties of phycocolloid-based films is the production of nanocomposites, meaning the development of composite film with the incorporation of nanomaterials. The most common nanomaterials used in the development of seaweed-based nanocomposite films include nanocellulose, nanoclays, organically modified nanoclays, carbon nanotubes, chitin nanofibers and silver nanoparticles [24]. There are several evidences in the literature of improved thermal, mechanical, barrier, and antimicrobial properties of phycocolloid-based nanocomposite packaging [24, 30, 40, 48]. Mechanical strength is improved, since the stress is transferred to the nanomaterials, while improving barrier properties as well by creating a denser matrix, making it "difficult" for the release of active compounds. Additionally, nanomaterials like nanoclays and silver nanoparticles have antimicrobial properties [30, 49]. However, the use of nanomaterials in the food industry raises potential environmental and health safety concerns regarding its size, allergenicity, and toxicity. It needs more studies to ensure its safety [36, 48, 50].

Regarding the quality and shelf-life of food products packed in seaweed-based packaging, especially alginate and carrageenan, the literature reports several successes in several food products, like meat and fish products, different fresh and cut fruits and vegetables, mushrooms, and cheese [25, 28 - 30, 40, 43, 44, 46, 47, 49, 51 - 55]. They report the maintenance or minimal alterations of the organoleptic characteristics (texture, color, smell, taste) of the products during the designated storage period, reduction of bacterial and fungal growth, reduction of moisture loss, efficient delivery and effect of antioxidant agents as well as reduced weight loss in fresh cut fruits and vegetables. Thus, allowing a shelf-life for these products comparable with products in traditional packaging, and improving shelf-life for fresh-cut food products.

Commercially Available Options

Even though they are not commercially available on a large scale, some products are already available for sale, namely Notpla's Ooho capsules, made from brown seaweeds, flexible edible packaging for beverages and sauces. In 2019, seaweed-based edible sachets by Evoware were distributed to runners in the London marathon instead of the traditional water bottles that serve as a replacement for disposable plastic sachets. Also, the project Mak-Pak, a cooperation project between German researchers of the Alfred-Wegener Institut and Hochschule Bremerhaven University and the fast-food chain Nordsee GmbH, creates a sustainable macroalgae-based alternative to traditional packaging for fast-food and already has their product tested by real consumers.

CONSUMERS' ACCEPTANCE

Although there is evidence that seaweed-based food and products are beneficial for human health and for the environment, one of the challenges of the food industry to disseminate new sustainable products, like seaweed-based edible packaging, is to attract consumers while introducing them to unfamiliar products. People will only consume healthy, nutritious, and sustainable food products if they taste good and do not have a weird or strong smell. Moreover, the consumer needs to get used to the idea of eating something that traditionally is thrown away. The price point is also a crucial factor for the consumer. Edible seaweed-based packaging might increase the price of a certain product. Nevertheless, as the consumer demand for more sustainable packaging options increases, the demand for the manufacturers to meet this standard increases as well, decreasing production costs, and, in this case, the only limitations will be creativity and innovation in the product development process.

A case of success of consumers' acceptance of this type of packaging is the results of consumer testing of the project Mak-Pak. The comments were very positive as the consumers praised the stability, appearance, taste, and smell of the thin packaging developed. As the ultimate goal is to develop a sustainable and commercially viable alternative to traditional packaging, consumer testing should be included in the research studies of seaweed-based packaging development. This would enrich them and give the "seal of consumer approval" which can attract investors, promoting the scale-up of the manufacturing process and allowing a great availability of seaweed-based packaging.

CONCLUDING REMARKS

Marine macroalgae are an abundant source of nutritious food products, bioactive compounds beneficial for human health, hydrocolloids useful for the food, pharmaceutical, biomedical, cosmetic, fertilizers, and other industries, and biomass for wastewater treatment, biofuel and biopolymers production. Industries related to marine macroalgae, including aquaculture, generate great profit and employability, especially at a local level. They can be integrated into a system of circular economy as the byproduct of one industry can be used as feedstock for another industry. Not forgetting all the benefits attributed to seaweed, few are the studies that address bioavailability of algal nutrients, and the quantity taken by the tissues, metabolism, and its consequential physiological effects, thus the need for more human intervention studies, since most of the results that report bioactivity and bioavailability of algal components are from animal or *in vitro* tests.

The increased environmental awareness of consumers is increasing the demand for new sustainable packaging options. While bio-based plastics exist, they are mainly produced from terrestrial plants and are also used as crops. Therefore, seaweed is a good alternative to produce completely biodegradable bio-based material that does not use resources for agricultural purposes.

The most common phycocolloids in the development of sustainable, active, and edible films for food packaging are alginate, agar and carrageenan, due to their biocompatibility, low toxicity, and bioactive properties and ability to form films. These compounds are used as thickeners, stabilizers, emulsifiers and as texture improving agents in several food products and beverages. Even though seaweed-based packaging is a sustainable product, the methods of phycocolloid extraction need to be improved by reducing the use of water, energy, and hazardous and expensive chemical solvents. Additionally, the macroalgae itself can affect the extraction, so aquaculture can be a solution by tailoring the macroalgae production which allow to optimize the yields of extraction. Moreover, for large-scale commercial availability of seaweed-based packaging, the reduction of the cost of the raw materials, the reduction of the production time, and the assurance of uniformity of the films produced, need to occur. The improvement of extraction and production methods will ensure economic efficiency, sustainability, and food safety.

Packaging solutions made only from phycocolloids do not meet the criteria for commercial use as they show poor mechanical and water barrier properties. The addition of plasticizers and the production of composites are solutions to this problem. Also, the addition of antioxidants and antibacterial agents will improve shelf life, whereas the addition of nutraceuticals increases the nutritional value of the product, with health benefits. All these additional components can be resourced from marine macroalgae as well, or from other natural and sustainable sources, maintaining the sustainability status of the packaging.

Long is the list of literature that reports the benefits of algal products, as well as the literature that successfully report the use of films made from phycocolloids to improve the shelf life of food products. Several authors reported that the seaweed-based films/composites reduced chemical and biological spoilage while keeping the organoleptic characteristics of the food products during several days of storage. Despite all the benefits and advantages reported for seaweed-based packaging, it will only be widely available if more consumer tests are included in the research and development of these products. This could also attract investment, which is important to finance research and improve the development of the technologies and methods involved in its production. It could also introduce sustainable new products to the consumer.

CONSENT FOR PUBLICATION

Not applicable.

CONFLICT OF INTEREST

The author(s) declare no conflict of interest, financial or otherwise.

ACKNOWLEDGMENTS

This work was financed by national funds through FCT – Foundation for Science and Technology, I.P., within the scope of the projects LA/P/0069/2020 granted to the Associate Laboratory ARNET, UIDB/04292/2020 granted to MARE – Marine and Environmental Sciences Centre and UIDP/50017/2020+UIDB/50017/2020 granted to CESAM – Centre for Environmental and Marine Studies. CICECO-Aveiro Institute of Materials, (FCT UIDB/50011/2020 & UIDP/50011/2020) and LAQV-REQUIMTE (UIDB/50006/2020) financed by national funds through the FCT/MEC when appropriate co-financed by FEDER under the PT2020 Partnership Agreement. CN is funded by national funds (OE), through FCT – Fundação para a Ciência e a Tecnologia, I.P., in the scope of the framework contract foreseen in the numbers 4, 5 and 6 of article 23, of the Decree-Law 57/2016, of August 29, changed by Law 57/2017, of July 19. Ana M. M. Gonçalves acknowledges the University of Coimbra for the contract IT057-18-7253. Authors thank the project MENU – Marine Macroalgae: Alternative recipes for a daily nutritional diet (FA_05_2017_011) which financed this research, funded by the Blue Fund under Public Notice No. 5 – Blue Biotechnology.

REFERENCES

[1] Mouritsen OG, Rhatigan P, Pérez-Lloréns JL. The rise of seaweed gastronomy: phycogastronomy. Bot Mar 2019; 62(3): 195-209. [http://dx.doi.org/10.1515/bot-2018-0041]

[2] Fleurence J. Seaweeds as Food. Seaweed in Health and Disease Prevention. Elsevier 2016; 149-67. [http://dx.doi.org/10.1016/B978-0-12-802772-1.00005-1]

[3] Pereira L. Nutritional Composition of the Main Edible Algae. Therapeutic and Nutritional Uses of Algae [Internet]. Boca Raton, FL: CRC Press 2018; 65-127. https://www.taylorfrancis.com/books/9781315152844 [http://dx.doi.org/10.1201/9781315152844-2]

[4] Delaney A, Frangoudes K, Ii S-A. Society and Seaweed. Seaweed in Health and Disease Prevention [Internet]. Elsevier 2016; 7-40. https://linkinghub.elsevier.com/retrieve/pii/B9780128027721000026 [http://dx.doi.org/10.1016/B978-0-12-802772-1.00002-6]

[5] Kim JK, Yarish C, Hwang EK, Park M, Kim Y. Seaweed aquaculture: cultivation technologies, challenges and its ecosystem services. Algae 2017; 32(1): 1-13. [http://dx.doi.org/10.4490/algae.2017.32.3.3]

[6] Michalak I. The application of seaweeds in environmental biotechnology [Internet]. Advances in Botanical Research. Elsevier Ltd; 2020. [http://dx.doi.org/10.1016/bs.abr.2019.11.006]

[7] Raikova S, Allen MJ, Chuck CJ. Hydrothermal liquefaction of macroalgae for the production of renewable biofuels. Biofuels Bioprod Biorefin 2019; 13(6): 1483-504. [http://dx.doi.org/10.1002/bbb.2047]

[8] Wang S, Zhao S, Uzoejinwa BB, *et al.* A state-of-the-art review on dual purpose seaweeds utilization for wastewater treatment and crude bio-oil production. Energy Conversion and Management 2020; 222: 113253. [http://dx.doi.org/10.1016/j.enconman.2020.113253]

[9] Couteau C, Coiffard L. Seaweed Application in Cosmetics. Seaweed in Health and Disease Prevention. Internet Elsevier 2016; 423-41. [http://dx.doi.org/10.1016/B978-0-12-802772-1.00014-2]

[10] Gonçalves AMM. Sustainable Premium Ready Meals for a Daily Nutritional Diet: Human Population Growing Demand. In: Filho WL, Azul AM, Brandli L, Salvia AL, Wall T, editors. Decent Work and Economic Growth Encyclopedia of the UN Sustainable Development Goals [Internet]. Springer, Cham 2020. http://link.springer.com/10.1007/978-3-319-71058-7_112-1

[11] Cheney D. Toxic and Harmful Seaweeds. Seaweed in Health and Disease Prevention [Internet]. Elsevier 2016; 407-21. [http://dx.doi.org/10.1016/B978-0-12-802772-1.00013-0]

[12] Wells ML, Potin P, Craigie JS, *et al.* Algae as nutritional and functional food sources: revisiting our understanding. J Appl Phycol 2017; 29(2): 949-82. [http://dx.doi.org/10.1007/s10811-016-0974-5] [PMID: 28458464]

[13] Cherry P, O'Hara C, Magee PJ, McSorley EM, Allsopp PJ. Risks and benefits of consuming edible seaweeds. Nutr Rev 2019; 77(5): 307-29. [http://dx.doi.org/10.1093/nutrit/nuy066] [PMID: 30840077]

[14] FAO. The State of World Fisheries and Aquaculture 2020. Rome: FAO; 2020. http://www.fao.org/documents/card/en/c/ca9229en

[15] Pateiro M, Munekata PES, Tsatsanis C, *et al.* Evaluation of the protein and bioactive compound bioaccessibility/bioavailability and cytotoxicity of the extracts obtained from aquaculture and fisheries by-products. Adv Food Nutr Res 2020; 92: 97-125. [http://dx.doi.org/10.1016/bs.afnr.2019.12.002] [PMID: 32402448]

[16] Rengasamy KR, Mahomoodally MF, Aumeeruddy MZ, Zengin G, Xiao J, Kim DH. Bioactive compounds in seaweeds: An overview of their biological properties and safety. Food and Chemical Toxicology [Internet]. Elsevier; 2020; 135: 111013.
[http://dx.doi.org/10.1016/j.fct.2019.111013]

[17] Ganesan AR, Tiwari U, Rajauria G. Seaweed nutraceuticals and their therapeutic role in disease prevention. Food Science and Human Wellness [Internet]. Beijing Academy of Food Sciences. 2019; 252-63.
[http://dx.doi.org/10.1016/j.fshw.2019.08.001]

[18] Manlusoc JKT, Hsieh CL, Hsieh CY, Salac ESN, Lee YT, Tsai PW. Pharmacologic Application Potentials of Sulfated Polysaccharide from Marine Algae. Polymers (Basel) 2019; 11(7): 1163. https://www.mdpi.com/2073-4360/11/7/1163
[http://dx.doi.org/10.3390/polym11071163] [PMID: 31288474]

[19] Abdul Khalil HPS, Lai TK, Tye YY, *et al.* A review of extractions of seaweed hydrocolloids: Properties and applications. Express Polym Lett 2018; 12(4): 296-317. http://www.expresspolymlett.com/letolt.php?file=EPL-0008729&mi=c [Internet].
[http://dx.doi.org/10.3144/expresspolymlett.2018.27]

[20] Leandro A, Pacheco D, Cotas J, Marques JC, Pereira L, Gonçalves AMM. Seaweed's Bioactive Candidate Compounds to Food Industry and Global Food Security. Life (Basel) 2020; 10(8): 140. https://www.mdpi.com/2075-1729/10/8/140
[http://dx.doi.org/10.3390/life10080140] [PMID: 32781632]

[21] Fleurence J. Seaweeds as Food [Internet]. Seaweed in Health and Disease Prevention. Elsevier I 2016.
[http://dx.doi.org/10.1016/B978-0-12-802772-1.00005-1]

[22] Aryee AN, Agyei D, Akanbi TO. Recovery and utilization of seaweed pigments in food processing.Elsevier Ltd 2018; 19: 113-9.
[http://dx.doi.org/10.1016/j.cofs.2018.03.013]

[23] Alba K, Kontogiorgos V. Seaweed polysaccharides (agar, alginate carrageenan). Encyclopedia of Food Chemistry. Elsevier 2018; 240-50.

[24] Abdul Khalil HPS, Saurabh CK, Tye YY, *et al.* Seaweed based sustainable films and composites for food and pharmaceutical applications: A review. Renewable and Sustainable Energy Reviews. Elsevier Ltd 2017; 77: 353-62.
[http://dx.doi.org/10.1016/j.rser.2017.04.025]

[25] Oussalah M, Caillet S, Salmiéri S, Saucier L, Lacroix M. Antimicrobial effects of alginate-based films containing essential oils on *Listeria monocytogenes* and *Salmonella typhimurium* present in bologna and ham. J Food Prot 2007; 70(4): 901-8. https://meridian.allenpress.com/jfp/article/70/4/901/172518/Antimicrobial-Effects-of-AlginateBased-Films
[http://dx.doi.org/10.4315/0362-028X-70.4.901] [PMID: 17477259]

[26] Ozilgen S, Bucak S. Functional Biopolymers in Food Manufacturing. Biopolymers for Food Design [Internet]. Elsevier 2018; 157-89. https://linkinghub.elsevier.com/retrieve/pii/B9780128114490000062

[27] Sedayu BB, Cran MJ, Bigger SW. A Review of Property Enhancement Techniques for Carrageenan-based Films and Coatings. Carbohydrate Polymers. Elsevier 2019; 216: 287-302.
[http://dx.doi.org/10.1016/j.carbpol.2019.04.021]

[28] Farhan A, Hani NM. Active edible films based on semi-refined κ-carrageenan: Antioxidant and color properties and application in chicken breast packaging. Food Packaging and Shelf Life [Internet]. Elsevier 2020; 24: 100476.
[http://dx.doi.org/10.1016/j.fpsl.2020.100476]

[29] Senturk Parreidt T, Müller K, Schmid M. Alginate-based edible films and coatings for food packaging applications. Foods 2018; 7(10): 170.
[http://dx.doi.org/10.3390/foods7100170] [PMID: 30336642]

[30] Mostafavi FS, Zaeim D. Agar-based edible films for food packaging applications - A review. International Journal of Biological Macromolecules. Elsevier BV 2020; 159: 1165-76.

[31] Skurtys O, Acevedo C, Pedreschi F, Enronoe J, Osorio F, Aguilera JM. Food Hydrocolloid Edible Films and Coatings. Food Hydrocolloid Edible Films and Coatings. New York: Nova Science Publishers, 2010 2010.

[32] Sousa AMM, Sereno AM, Hilliou L, Gonçalves MP. Biodegradable Agar Extracted from *Gracilaria Vermiculophylla*: Film Properties and Application to Edible Coating. Mater Sci Forum 2010; 636-637: 739-44. https://www.scientific.net/MSF.636-637.739 [Internet]. [http://dx.doi.org/10.4028/www.scientific.net/MSF.636-637.739]

[33] Martínez-Sanz M, Martínez-Abad A, López-Rubio A. Cost-efficient bio-based food packaging films from unpurified agar-based extracts. Food Packaging and Shelf Life. Elsevier 2019; 21: 100367. [http://dx.doi.org/10.1016/j.fpsl.2019.100367]

[34] Geyer R, Jambeck JR, Law KL. Production, use, and fate of all plastics ever made. Sci Adv 2017; 3(7): e1700782. [http://dx.doi.org/10.1126/sciadv.1700782] [PMID: 28776036]

[35] Karan H, Funk C, Grabert M, Oey M, Hankamer B. Green Bioplastics as Part of a Circular Bioeconomy.Trends in Plant Science. Elsevier Ltd 2019; 24: 237-49. [http://dx.doi.org/10.1016/j.tplants.2018.11.010]

[36] Silva FAGS, Dourado F, Gama M, Poças F. Nanocellulose bio-based composites for food packaging. Nanomaterials (Basel) 2020; 10(10): 2041. [http://dx.doi.org/10.3390/nano10102041] [PMID: 33081126]

[37] Ghosh S, Gnaim R, Greiserman S, Fadeev L, Gozin M, Golberg A. Macroalgal biomass subcritical hydrolysates for the production of polyhydroxyalkanoate (PHA) by Haloferax mediterranei. Bioresour Technol 2019; 271: 166-73. [http://dx.doi.org/10.1016/j.biortech.2018.09.108] [PMID: 30268011]

[38] Marques MM. From Macroalgae to Bioplastic Seaweed Hydrolysates for Polyhydroxyalkanoate Production by Marine Bacteria Biological Engineering [Internet]. 2017. https://fenix.tecnico.ulisboa.pt/downloadFile/1970719973966714/Mafalda Marques Master Thesis_ corrected_upload.pdf

[39] Kanmani P, Rhim J-W. Development and characterization of carrageenan/grapefruit seed extract composite films for active packaging. International Journal of Biological Macromolecules. Elsevier B.V. 2014; 68: 258-66. https://linkinghub.elsevier.com/retrieve/pii/S0141813014002955

[40] Tavassoli-Kafrani E, Shekarchizadeh H, Masoudpour-Behabadi M. Development of edible films and coatings from alginates and carrageenans Carbohydrate Polymers Elsevier Ltd 2016; 137: 360-74. https://linkinghub.elsevier.com/retrieve/pii/S0144861715010541

[41] Paixão LC, Lopes IA, Barros Filho AKD, Santana AA. Alginate biofilms plasticized with hydrophilic and hydrophobic plasticizers for application in food packaging. J Appl Polym Sci 2019; 136(48): 48263. [http://dx.doi.org/10.1002/app.48263]

[42] Gomaa M, Hifney AF, Fawzy MA, Abdel-Gawad KM. Use of seaweed and filamentous fungus derived polysaccharides in the development of alginate-chitosan edible films containing fucoidan: Study of moisture sorption, polyphenol release and antioxidant properties. Food Hydrocolloids [Internet]. Elsevier Ltd 2018; 82: 239-47. [http://dx.doi.org/10.1016/j.foodhyd.2018.03.056]

[43] Raybaudi-Massilia RM, Rojas-Graü MA, Mosqueda-Melgar J, Martín-Belloso O. Comparative study on essential oils incorporated into an alginate-based edible coating to assure the safety and quality of fresh-cut Fuji apples. J Food Prot 2008; 71(6): 1150-61. https://meridian.allenpress.com/jfp/article/71/6/1150/173070/Comparative-Study-on- Essential-Oil-

-Incorporated
[http://dx.doi.org/10.4315/0362-028X-71.6.1150] [PMID: 18592740]

[44] Azarakhsh N, Osman A, Ghazali HM, Tan CP, Mohd Adzahan N. Lemongrass essential oil incorporated into alginate-based edible coating for shelf-life extension and quality retention of fresh-cut pineapple. Postharvest Biol Technol 2014; 88: 1-7. [Internet].
[http://dx.doi.org/10.1016/j.postharvbio.2013.09.004]

[45] Alboofetileh M, Rezaei M, Hosseini H, Abdollahi M. Antimicrobial activity of alginate/clay nanocomposite films enriched with essential oils against three common foodborne pathogens. Food Control [Internet]. Elsevier Ltd 2014; 36: 1-7.
[http://dx.doi.org/10.1016/j.foodcont.2013.07.037]

[46] Kazemi SM, Rezaei M. Antimicrobial Effectiveness of Gelatin-Alginate Film Containing Oregano Essential Oil for Fish Preservation. J Food Saf 2015; 35(4): 482-90. http://doi.wiley.com/10.1111/jfs.12198 [Internet].
[http://dx.doi.org/10.1111/jfs.12198]

[47] Jalali N, Ariiai P, Fattahi E. Effect of alginate/carboxyl methyl cellulose composite coating incorporated with clove essential oil on the quality of silver carp fillet and *Escherichia coli* O157:H7 inhibition during refrigerated storage. J Food Sci Technol 2016; 53(1): 757-65. http://link.springer.com/10.1007/s13197-015-2060-4
[http://dx.doi.org/10.1007/s13197-015-2060-4] [PMID: 26787996]

[48] Shafiq M, Anjum S, Hano C, Anjum I, Abbasi BH. An Overview of the Applications of Nanomaterials and Nanodevices in the Food Industry. Foods 2020; 9(2): 148. https://www.mdpi.com/2304-8158/9/2/148
[http://dx.doi.org/10.3390/foods9020148] [PMID: 32028580]

[49] Jiang T, Feng L, Wang Y. Effect of alginate/nano-Ag coating on microbial and physicochemical characteristics of shiitake mushroom (Lentinus edodes) during cold storage. Food Chemistry [Internet]. Elsevier Ltd 2013; 141: 954-60.
[http://dx.doi.org/10.1016/j.foodchem.2013.03.093]

[50] Otles S, Sahyar BY. Nanomaterials: Ecotoxicity, Safety, and Public Perception [Internet]. Rai M, Biswas JK, editors. Nanomaterials: Ecotoxicity, Safety, and Public Perception. Cham: Springer International Publishing 2018.
[http://dx.doi.org/10.1007/978-3-030-05144-0_3]

[51] Sipahi RE, Castell-Perez ME, Moreira RG, Gomes C, Castillo A. Improved multilayered antimicrobial alginate-based edible coating extends the shelf life of fresh-cut watermelon (*Citrullus lanatus*). LWT - Food Science and Technology [Internet]. Elsevier Ltd 2013; 51: 9-15. https://linkinghub.elsevier.com/retrieve/pii/S0023643812004616

[52] Rojas-Graü MA, Tapia MS, Rodríguez FJ, Carmona AJ, Martin-Belloso O. Alginate and gellan-based edible coatings as carriers of antibrowning agents applied on fresh-cut Fuji apples. Food Hydrocoll 2007; 21(1): 118-27. https://linkinghub.elsevier.com/retrieve/pii/S0268005X0600066X [Internet].
[http://dx.doi.org/10.1016/j.foodhyd.2006.03.001]

[53] Reboleira J, Adão P, Guerreiro SFC, *et al.* Poultry shelf-life enhancing potential of nanofibers and nanoparticles containing porphyra dioica extracts. Coatings 2020; 10(4): 315.
[http://dx.doi.org/10.3390/coatings10040315]

[54] Surendhiran D, Cui H, Lin L. Encapsulation of Phlorotannin in Alginate/PEO blended nanofibers to preserve chicken meat from *Salmonella* contaminations. Food Packaging and Shelf Life [Internet]. Elsevier 2019; 21: 100346.
[http://dx.doi.org/10.1016/j.fpsl.2019.100346]

[55] Albertos I, Martin-Diana AB, Burón M, Rico D. Development of functional bio-based seaweed (*Himanthalia elongata* and *Palmaria palmata*) edible films for extending the shelflife of fresh fish burgers. Food Packaging and Shelf Life [Internet]. Elsevier 2019; 22: 100382. [http://dx.doi.org/10.1016/j.fpsl.2019.100382]

CHAPTER 18

Edible Seaweed-Based Biodegradable Films and Coatings for Food and Nutraceutical Applications

M.M. Jayakody[1], **K.G. Kaushani**[1] and **M.P.G. Vanniarachchy**[1,*]

Department of Food Science and Technology, Faculty of Applied Sciences, University of Sri Jayewardenepura, Gangodawila, Nugegoda, Sri Lanka

Abstract: Accumulation of non-biodegradable plastics is causing high levels of environmental pollution. Currently, edible and biodegradable films and coatings for food and nutraceutical applications are a fast emerging technology with increased attention among researchers and consumers, which acts as an alternative to these synthetic plastics. Edible coatings are most commonly developed from polysaccharides and proteins. Among polysaccharides, seaweed-based polysaccharides play a vital role. These seaweed-based polysaccharides, which are utilized in the development of edible coatings and films, include Agar, Alginate, and carrageenan. Alginates are extracted from brown algae, while agar and carrageenan are extracted from red algae varieties. These developed coatings and films are commonly applied to extend the shelf life and maintain the desired quality level in food. These films or coatings can be applied to foods such as fruits, vegetables, meat, poultry, seafood, and dairy products. They improve the quality of the product by retarding moisture loss, reducing lipid oxidation and discoloration, sealing in volatile flavors, and functioning as carriers of food additives such as antimicrobial and antioxidant agents. This book chapter discusses the application of seaweed-based biodegradable films and coatings for food and nutraceuticals.

Keywords: Agar, Active Packages, Alginate, Applications, Biodegradable Packages, Carrageenan, Coatings, Composite Films, Eco-Friendly, Edible Packages, Films, Food, Lipids, Nutraceuticals, Plasticizers, Polysaccharides, Primary Packages, Seaweeds, Stabilizers, Technology.

INTRODUCTION

At present, the accumulation of non-biodegradable conventional synthetic plastic is causing huge pollution to the environment. Thus, there is a need to promote environmentally friendly bio-degradable polymers among people. Currently, edible and biodegradable films and coatings are a fast emerging technology with

* **Corresponding author M.P.G. Vanniarachchy:** Department of Food Science and Technology, Faculty of Applied Sciences, University of Sri Jayewardenepura, Nugegoda, Sri Lanka; Tel: +94 767288380; E-mail: mihiripg@sjp.ac.lk

Avinash Mishra (Ed.)

increased attention among researchers and consumers, which acts as an alternative to these synthetic plastics.

Edible films or coatings can be introduced as thin layers of edible components which are used in the primary packaging of food and consumed with the food itself. Edible films can be introduced as self-standing structures in nature, while edible coatings adhere to the surface of the food [1]. The main difference between an edible coating and an edible film is that the edible coating is applied in liquid form on the product by dipping the product into the coating solution, while edible films are first developed as sheets and then used as wrappers [2]. Edible films and coatings are most commonly developed by polysaccharides and proteins. Among polysaccharides, cellulose derivatives, starches, alginates, pectins, chitosan, pullulan, and carrageenans are widely used. Soybean proteins, wheat gluten, corn zein, sunflower proteins, gelatin, whey, casein, and keratin are the types of proteins widely used in the development of edible films and coatings [2]. These polysaccharides and proteins are hydrophilic. Hence, lipid sources are added to enhance the water vapour barrier properties of films and coatings [3]. Different types of lipid sources that are incorporated into film-forming solutions to form emulsion-based structures include animal and plant waxes, vegetable oils, and fatty acids [3]. At present, research studies are carried out about multi-component edible films with two or more film-forming substances. These coatings and films have improved functional properties in terms of physical, mechanical, and barrier properties. Therefore, components such as plasticizers, cross-linking agents, emulsifiers, and reinforcements are incorporated to enhance the properties [3]. Other than that, films and coatings are formulated incorporating active compounds such as antimicrobials, antioxidants, colour agents, flavours, and nutraceuticals to provide additional benefits such as antibacterial effects, antifungal effects, to prolong the shelf life and to maintain the quality of the packed food [4]. This chapter describes edible Seaweed based biodegradable films and coatings for food and nutraceutical applications.

EDIBLE SEAWEEDS VARIETIES FOR FORMING FILMS AND COATINGS

Petrochemical-based packages have caused serious environmental issues due to their non-biodegradability. Therefore, substitute packages developed by natural polymers are coming to the market to replace petrochemical-based plastic packages [5]. Coatings and films can be developed by using polysaccharides. Seaweeds are also rich in polysaccharides such as alginate, agar, and carrageenan. Alginates can be extracted from brown algae, such as *Macrocystis, Laminaria, Ascophyllum*, *Ecklonia, Eisenia, Nereocystis,* and *Sargassum*. Carrageenan can be extracted from red algae, such as *Chondrus, Gigartina, Eucheuma, Hypnea,*

Iridaea, Gymnogongrus, Ahnfeltia, and *Furcellaria.* Agar can be extracted from red algae such as *Gelidium, Gracilaria,* and *Pterocladia* [6].

FORMULATION OF SEAWEED-BASED BIODEGRADABLE FILMS AND COATINGS

Films and coatings are formulated using seaweed-based extracts of alginate, carrageenan, and agar. The main property of these polysaccharides in the formulation of biodegradable films and coatings is their ability to form gels. Alginate is a hydrophilic polysaccharide that is composed of a monomeric unit of (1→4)-linked β-D-mannuronic acid (M) and α-L-guluronic acid (G) residues. The property of alginate used in the formulation of alginate-based films and coatings is its ability to form gels in the presence of divalent cations, such as calcium, magnesium, manganese, aluminium, and iron. Alginate can also form gels by lowering the Ph value of the solution below the pK_a value of guluronic residue. Calcium is the most common divalent cation used in gel formation [7]. The affinity of alginates towards divalent cation decreases in the following order: Pb > Cu > Cd > Ba > Sr > Ca > Co, Ni, Zn > Mn [8]. When these divalent ions are incorporated into an alginate solution, these ions facilitate the formation of an association between M and G blocks. The ability and selectivity to form these interactions are determined by the length of the G block. M blocks and MG blocks are almost without selectivity. During gelation, water-soluble alginate exchanges its counter-ions with the externally diffused ions into the alginate solution and undergoes a sol/gel transition [9]. When forming alginate gels, two contiguous, diaxially linked guloronic residues form a cavity that acts as a binding site for calcium ions. This forms an "eggbox" model resulting in a 3-dimensional gel network [10]. A higher amount of G-Blocks will create rigid and dense gels, while a higher amount of M-Blocks will build flexible, porous gels [11]. Among different types of alginates, sodium alginate can form films with the properties of being water-soluble, strong, glossy, tasteless, odorless, flexible, low permeable to oxygen and oils, and better spreadability on hydrophobic surfaces [12]. A higher amount of G-Blocks will create rigid and dense gels, while a higher amount of M-Blocks will build flexible, porous gels [11]. Among different types of alginates, sodium alginate can form films with the properties of being water-soluble, strong, glossy, tasteless, odorless, flexible, low permeable to oxygen and oils, and better spreadability on hydrophobic surfaces [12]. There are different proposed methods to obtain alginate films and coatings. There, an alginate solution of concentration ranging from 0.5 – 3% (*w/v*) can be taken. Glycerol is the most commonly used plasticizer. Various percentages of glycerol ranging from 2-5% (*w/v*) of alginate are incorporated into alginate-based films and coatings. Generally, to cast alginate coatings, the food product is first immersed into the coating formulation, and then spraying of $CaCl_2$ or immersion of the food product into a $CaCl_2$ solution is done.

On the other hand, films are generally produced by drying a thin layer of alginate solution at 25-40^{0}C for 20 to 24 hours and by adding 2 – 20% (*w/v*) of $CaCl_2$ for 1 – 20 minutes [13].

Carrageenan can be introduced as a water-soluble polymer with a linear chain of partially sulfonated galactans. Carrageenan can be classified into 3 major types as lambda (λ), kappa (κ), and iota (ι), considering the sulphate groups on basic repeating units. According to previous studies, κ-carrageenan contains about 25-30% of ester sulfate groups with 28-35% of 3,6-anhydrogalactose. ι-carrageenan consists of 28-30% of ester sulfate groups and 25-30% of 3,6-anhydrogalactose, while λ-carrageenan consists of 32-39% of ester sulphate groups and no 3,6-anhydrogalactose content [14]. κ-carrageenan and ι-carrageenan forms thermo-reversible sol-gels at proper conditions, but λ-carrageenan does not form gels [15]. Carrageenan can form gels through ionotropic gelation coupled to a mechanism that involves helix formation upon cooling and cross-linking in the presence of potassium or calcium ions. Gels formed by κ-carrageenan are rigid and brittle, while gels formed by ι-carrageenan are softer, elastic, and cohesive [7].

Agar is a mixture of agarose and agaropectin. Gelation of agar takes place only by the agarose content of agar which is produced exclusively by hydrogen bonds. Unlike alginate and carrageenan, the specialty of agar is that it does not need any other substance to gel [16].

During the formulation of a seaweed-based coating, different components can be incorporated into the coating formulation. Those components include plasticizers stabilizers, lipids, proteins, antimicrobial agents, antioxidants, anti-browning agents, flavors, and pigments [17 - 19]. Mechanical, functional, organoleptic, and nutritional characteristics of films and coatings can be modified by incorporating these various natural or chemical additives. Biodegradable materials and modifier agents can be blended with other food additives to improve the weak mechanical properties (tensile strength and tensile elongation) and poor barrier properties and to extend applications of edible films and coatings [20]. For example, lipids can be added to provide good moisture barrier properties [21].

Plasticizers can improve the flexibility and reduce the brittleness of films. Glycerol, sorbitol, propylene glycol, triethylene glycol, polyethylene glycol, and ethylene glycol are the commonly used plasticizers. Certain plasticizers, such as lactic acid, have the ability to inhibit fungal growth [22]. There are two types of interactions in a film formation as a coating. Those interactions include cohesion which is the attraction between film polymer molecules, and adhesion, which is the attraction force between film and substrate. Cohesive forces can impart brittleness to films. When plasticizers are added to the film, plasticizers decrease the intermolecular forces resulting from chain-to-chain interactions minimizing the brittleness of films [23].

In certain cases, stabilizers are also added to the coating solutions. Stabilizers will preserve the film's integrity and improve mechanical properties and transparency in certain cases. Most commonly, proteinaceous nature stabilizers such as gelatin, collagen, prolamine, *etc*., are preferred [22].

Lipids can also be incorporated into edible films and coatings. These lipids can be saturated or unsaturated with long carbon chains from 10 to 20 carbon atoms. Those lipids include oils (canola oil), fatty acids (palmitic acid), fatty alcohols (stearyl alcohol), monoglycerides, and triglycerides [22].

Composite materials are increasingly used to combine the favorable properties of each component [20]. Mechanical and barrier properties of composite films and coatings depend on the properties of each film-forming polymer and its compatibility [24]. Over the years, lipids, pectin, wood cellulose, locust bean gum, inorganic particles, chitosan, proteins, and gellan, as well as synthetic polymers, have been combined to improve film functionality [25].

ALGINATE-BASED FILMS AND COATINGS

Application of an alginate-based coating on hydrophobic surfaces can be performed by dipping, spraying, brushing, or vacuum impregnation [26]. Also, alginate can be made into thin films by solvent casting and extrusion [11, 27]. There are many potential uses of sodium alginate-based edible films and coatings.

Fresh-Cut Fruits and Vegetables

Fresh-cut fruits and vegetables become susceptible to enzymatic browning, undesirable volatile formation, alterations in texture, and contamination due to the growth of spoilage and pathogenic bacteria [28]. Alginate-based coatings are effective as postharvest treatments to maintain the quality of fresh-cut fruits, such as tomato, peach, sweet cherry, blueberry, watermelon, pineapples, *etc*., where the edible coatings can delay ripening life, reduce the respiration of fruits and vegetables due to their selective permeabilities to the O_2 and CO_2 gases and extend shelf-life [29]. Alginate coating incorporated with prebiotics can increase the soluble solid content, and stabilize the browning index, acidity, and firmness levels of fresh-cut apples [30]. Edible coatings made from alginate were developed for their capacity to preserve the quality of minimally processed "Gala" apples [31]. Overall, it was found that alginate coatings prolonged the shelf-life of cut "Gala" apples without causing anaerobic respiration [31]. The fresh-cut "Fuji" apples coated with alginate coating maintained the desired qualities and extended the shelf life nearly three times when compared to that of control fruits stored at a refrigerated temperature [32]. A study done by Azarakhsh *et al.* [33] has concluded that alginate-based edible coating with sunflower oil and lemongrass oil as an antimicrobial on the fresh-cut pineapples can reduce yeast, mold, and the total plate count significantly, and the shelf-life was prolonged. The sapota fruits coated with sodium alginate – pectin edible coating showed a significant reduction in changes in the physicochemical parameters at refrigerated temperature storage [34]. Alginate-coated pears had higher tensile strength, elongation, and elasticity; on the other hand, they had a lower water loss, pH increase, metabolic activities with maintained firmness, and green color [35].

Vegetables such as carrot, garlic, and lettuce can be coated with alginates to reduce discoloration, respiration rate, and ethylene synthesis [36 - 38]. Alginate coating with nanosilver is used to coat Shiitake mushrooms to prevent spoilage, improve sensory attributes, and lower weight loss [20]. According to Nussinovitch and Hershko [36], alginate coated on garlic prevented moisture loss and increased the salability of garlic.

Meats, Poultry, and Seafood

Alginate-based film coating for meat and meat products can prevent shrinkage, microbial contamination, and surface discoloration by delaying moisture transport [11]. Alginate-based coating formulation commercially known as Flavor-Tex® is used to coat pork patties to solve the flavor and texture problems of precooked meat products. Coating decreased oxidative rancidity, cooking losses, and increased meat tenderness [39]. In addition, alginate film coatings are applied to carry different functional agents to improve their applications. The application of nisin-containing alginate film for sliced beef could be effective against *Staphylococcus aureus* [20]. It could be suggested that nisin-containing edible films could control the growth of pathogens at the surface of ground beef or other meat products [20]. It has been found that the use of alginate coating with nisin and cinnamon oil could maintain the quality of northern snakehead fish fillets [40]. In alginate-coated raw and precooked pork patties wrapped in polyethylene-coated freezer paper stored at -20°C, it has been found that alginate-coated patties had improved texture, flavor, juiciness, and overall palatability over uncoated control samples [39]. Dehydration of frozen beef cuts can be reduced by coating by successive immersions into aqueous solutions of 150 g/L sodium alginate, 35 g/L $CaCl_2$, and 100 g/L glycerin before freezing [11]. Alginate has been coated on frozen pork to maintain the functional properties of pork, to decrease the thawing loss, and the concentration of Ca^{2+} influenced the tenderness of the meat and total protein solubility [41]. Alginate/polycaprolactone-based films containing essential oils have been developed to use as active packaging in food systems, mainly for meat products and broccoli to inhibit the growth of pathogens [42]. To increase the warming efficiency of microwaveable foods (*e.g.*: chicken nuggets), alginate with salt films has been used [43].

Cheese

Several microbiological and chemical defects can be developed in cheeses due to uncontrolled processing, ripening, and external environmental conditions during handling and storage conditions [44]. Loss of cheese weight and drying of cheese rind represent economic losses and poor cheese quality. Edible coatings and films are used to dispose of these defects. Low moisture mozzarella cheese, which was coated with a solution containing sodium alginate and 1% potassium sorbate, and stored under modified atmosphere packaging (MAP) at different temperatures, remarkably extended the shelf life up to 100, 40, and 11 days at 4, 8, and 14^{0}C, respectively [11]. The low moisture mozzarella cheese coated with sodium alginate presented lower values of hardness than cheese coated with chitosan and soy protein isolate-based coatings [11]. Edible coatings have been prepared from alginate with the addition of ginger essential oil to decrease the *Escherichia coli* and *Staphylococcus aureus* levels in Kashar cheese [11]. A composite of alginate-whey protein isolate coatings fortified with ginger essential oil has been used on Kashar cheese to obtain a cheese with lower acidity and higher fat level [45]. Active packaging films have been developed by incorporating natural antimicrobial agents like cinnamon essential oil into sodium alginate- calcium formulations to extend the shelf- life of paneer at refrigeration temperature (4 ± 1 °C) [46].

Other

Several investigations have been done using alginate-based films without an application on food products. Pyrogallic acid incorporated in sodium alginate-CMC matrix has increased the gas, vapor, and UV barrier properties of the edible film [4]. Fang *et al.* [47], used potassium alginate, extracted from seaweed and coupled with environmentally friendly antimicrobial polyhexamethylene guanidine phosphate (PHMGP) to build flame retardant and antimicrobial coating on cotton fabric via layer-by-layer technique. Films formed from alginate-whey protein complexes had higher tensile strength, elastic modulus, and elongation than whey protein alone [48]. Interactions between encapsulated n-hexanal (aroma compound) and alginate matrix affected the barrier properties, permeability, and surface properties of emulsified alginate edible films [11].

AGAR-BASED FILMS AND COATINGS

There are previous research studies done using agar-based films and coatings to improve the quality and storage life of food products.

Fish

Lopez *et al.* [49] developed an agar-based bioactive film incorporating green tea extract and probiotic strains, *Lactobacillus paracasei* and *Bifidobacterium lactis.* These films have shown success in delaying the growth of H_2S -producing and total viable bacteria. Applying these bioactive films has increased the count of beneficial lactic acid bacteria in fish and successfully extended the shelf -life of hake fillets for one week. Da Rocha *et al.* [50] developed an agar-based film incorporating clove essential oil and fish protein hydrolysate. The film has used to cover the fillets of *Paralichthys orbignyanus.* They have observed low total volatile bases, pH value, and a decline in H_2S, the growth of producing bacteria in the coated fillet. Vejdan *et al.* [51] developed a gelatin/agar bilayer film containing TiO_2 nanoparticles to study the oxidative stability of fish oil. According to the results, the highest amounts of eicosapentaenoic acid (EPA) and docosahexaenoic acid (DHA), as well as the lowest amount of changes in peroxide values, thiobarbituric acid, conjugated diene, fatty acids profile, and colour change were observed in oils covered with the nanocomposite.

Fruits and Vegetables

Moreno *et al.* [52] developed agar, alginate, and agar/alginate-based films incorporating *Larrea nitida* extract. These coatings have shown antiviral activity when applied to blueberries against murine norovirus (MNV). Thus, they have recorded that these coatings could be used as an alternative to reduce or eliminate foodborne viruses and protect the food against oxidative processes. Hussein *et al.* [53] applied an agar-based coating to protect banana fruits against the crown, finger, neck, and flower end rots caused by *Colletotrichum musae* and *Fusarium moniliforme.* Application of $2.0 g.l^{-1}$ agar suspension as a coating has exhibited a reduction in the incidence and severity of postharvest diseases compared to untreated fruit. They have also observed that the application of agar suspension as a coating agent at 2.0 g · l−1 significantly decreased weight loss (%), firmness loss (%), and soluble solid concentration of banana fruit for 15 days of storage at 25 ± 2°C. Thus, they have concluded that agar-based coating could be used to protect banana fruit against postharvest rot diseases and extend fruit storage life during ripening and storage. Gudadhe *et al.* [54] developed an agar-based film incorporating silver nanoparticles. The film was surface-coated over the fruit *Citrus aurantifolium* (Thornless lime) and *Pyrus malus* (Apple). The results of this study have revealed that the film possesses antimicrobial activity and also has increased the shelf life of fruits.

Cheese

The agar-based coating developed by Guitian *et al.* [55] functionalized with enterocins synthesized by *Enterococcus avium* has shown effective in controlling *L. monocytogenes* in cheese. Incoronato *et al.* [56] have studied agar hydrogel with active silver nanoparticles to prolong the shelf life of Fior di Latte cheese. Study results have revealed that the developed active packaging system could increase the shelf life of Fior di Latte cheese, due to the ability of silver cations to control microbial proliferation, without affecting the functional dairy microbiota and the sensory characteristics of the product.

CARRAGEENAN-BASED FILMS AND COATINGS

Films and coatings developed by the seaweed polysaccharide carrageenan can be applied to foods such as fresh and frozen meat, ham or sausages casings, oily foods, dry solid foods, in soft capsules and non-gelation capsules [57].

Fruits

Previous studies have proven that carrageenan-based coatings have become efficient to extend the shelf life of fresh-cut and fresh whole fruits. Hamzah *et al.* [58] applied carrageenan as a coating for papaya. They have developed the coating formulation by dissolving carrageenan (0.2–0.8%, *w/v*) in 600 mL of distilled water at 70 ◦C for 10 min and incorporating different concentrations ranging from 0-1% *w/v* of glycerol as a plasticizer. They have optimized the coating formulation and have identified the formulation with carrageenan and glycerol levels at 0.78% (*w/v*) and 0.85% (*w/v*), respectively as the best formulation considering the firmness and colour component of coated and uncoated fruits. Previous research was conducted by developing composite carrageenan-based films incorporated with antimicrobial or antioxidant agents. These films can extend the shelf-life of perishable foods by protecting against microbial growth, enzymatic browning, vitamin losses, and oxidation [59]. Falco *et al.* [60] studied the effect of κ–, ι– and λ– carrageenan-based film-forming dispersion incorporated with green tea extract for blueberries and raspberries. They have developed 1% (*w/w*) κ–, ι– and λ– carrageenan-based films forming dispersion by incorporating and without incorporating 0.7% (*w/w*) green tea extract. All these coatings have become effective in extending the shelf life of raspberries and blueberries under refrigeration, preserving their firmness to a greater extent and promoting a better appearance. They have also recorded that the addition of green tea extract in the film-forming dispersion has enhanced the antiviral activity for murine norovirus (MNV) and hepatitis A virus (HAV) at both refrigerated and ambient temperatures in blueberries and raspberries.

Dairy Products

Liu *et al.* [61] developed pH-sensitive and antioxidant-active packing films based on κ-carrageenan by incorporating mulberry polyphenolic extract. Mulberry possesses a high anthocyanin content. Thus, in this study, they have extracted polyphenolic compounds from mulberry fruit and have incorporated them into the κ-carrageenan-based film to develop an active and intelligent film. They have used these films to determine the freshness of milk. They have developed the film formation solution by incorporating 0, 1, 2, and 4 (*w/w*)% of mulberry polyphenolic extract and 30% glycerol for 1% carrageenan solution. Anthocyanins possess an intelligent pH-sensitive property. Anthocyanins can change their colour at different pH values. They have utilized that property of anthocyanins to determine the freshness of milk. According to their results, incorporation of mulberry polyphenolic extract has significantly increased the thickness, water vapor and UV–vis light barrier ability, antioxidant activity, and pH-sensitive property of κ-carrageenan film and also physical, antioxidant and pH sensitivity of carrageenan-based films have changed according to the mulberry polyphenolic extract. Thus, the film with the highest mulberry polyphenolic extract has shown the most evident colour change when applied to detect the freshness in milk. Liu *et al.* [61] have stated that the film with the highest anthocyanin content can be used as a visible pH-sensing label to monitor the freshness of other foods such as pork, shrimp, and fish and also can be used as an active packaging material to extend the shelf life of the packaged food since this film provides the antioxidant activity by releasing polyphenolic compounds from film matrix into the packaged food.

Other

Shojaee *et al.* [62] developed κ-carrageenan-based films incorporating plant essential oils to enhance the antimicrobial activity of these films to be used in food applications. Film-forming dispersion has developed by dissolving 1% κ–carrageenan (*w/v*), 50% (*w/w*) glycerol based on carrageenan weight, and different concentrations of *Zataria multiflora Boiss* (ZEO) and *Mentha pulegium* (MEO) essential oil and Tween 80. Essential oils are lipids. Thus, the incorporation of essential oils has also been shown to increase the water barrier properties of these films. They have stated that these films developed incorporating *Zataria multiflora Boiss* (ZEO) and *Mentha pulegium* (MEO) essential oils in their research can be applied in packaging a wide range of oxidative and microbial-sensitive food products [62]. Kanmani and Rhim [59] have developed an active κ-carrageenan-based composite film incorporating grapefruit seed extract. They have stated that carrageenan and grape seed extract composite films exhibited good antibacterial activity against food-borne pathogens. Therefore from their results, they have suggested that carrageenan-based composite films have a high potential for being used as antimicrobial or active food packaging applications [59].

ADVANTAGES OF USING SEAWEED-BASED COATINGS AND FILMS

Food additives, such as antioxidants, antimicrobials, probiotics, and flavors, are highly sensitive to external environmental factors such as oxygen, temperature, and light, whereas, at the same time, food processing and storage often involve harsh environmental conditions. In this situation, coating matrices are useful in offering protection for foods against the environment, improving solubility, and controlling compound release, where bioactive compounds introduced into an edible coating are protected by a solid barrier between the additive and environmental conditions [29]. One of the most attractive advantages of the use of seaweed-based coatings is the ability to control the release of bioactive compounds by parameters such as humidity, temperature, changes in pH, and mechanical rupture of the matrix. As the mode of action of an additive may be required at different times during processing, storage, or ingestion of food, controlling the release of the additive at a given point is crucial to improve quality and ensure food safety and functionality [63]. Some bioactive compounds, such as vitamins and antioxidants, are adversely affected by UV light, and the seaweed-based coatings may create a shield against UV light [63]. The major quality loss of packaged foods is caused by oxygen. Oxygen is responsible for many degradation processes in foods, such as lipid oxidation, microorganism growth, enzymatic browning, and vitamin loss [64]. The oxidation of fat results in off-flavor, off-color, and nutrient loss [64]. Bonilla *et al.* [65] concluded that the inclusion of strongly flavored antioxidants such as N-acetylcysteine and glutathione in seaweed-based coatings allows for their encapsulation and can reduce their strong aroma. In addition, oxidation could be effectively reduced and can guarantee the quality of coated foods as seaweed-based coatings have the ability to act as a semipermeable barrier to oxygen, and carbon dioxide and allow the reduction of water loss and maturation rate control is higher than that of other coatings [66]. When considering sodium alginate extracted from the brown seaweed sargassum, the coating produces a strong film that, compared to other film materials like sodium caseinate and potato starch films, presents better performances regarding water permeability and mechanical properties [67]. The capacity to cross-link, when guluronic residues bind to Ca^{2+} ions allows for the development of tailor-made structures where it is possible to modify the barrier and mechanical properties of the alginate-based coatings [68]. Sodium alginate coating resulted in better overall qualities, which can be explained by its better spreadability on hydrophobic food surfaces [26]. Another importance is that edible seaweed-based coatings can be used to overcome many obstacles involved in the marketing of foods. These functions can be specified as retarding moisture, gas, solute, and oil migration, improving structural integrity, retaining volatile flavor compounds, and conveying food additives [69]. Seaweeds-based coatings are less expensive packaging materials for industry and lower the prices of food products. In addition, they improve the aesthetic appearance by minimizing the development of physical damage, hiding scars, and improving surface shine [70]. With regard to the fresh produce industry, Rooney has described the potential benefits of using edible coatings which include, to provide moisture barrier on the surface of produce to prevent the weight loss and changes in texture, flavour, and appearance during postharvest storage, to provide sufficient gas barrier for controlling gas exchange between the fresh produce and its surrounding atmosphere, which would slow down respiration, delay deterioration, retard the enzymatic oxidation, protect the fresh produce from browning discoloration and texture softening during storage, to restrict the exchange of volatile compounds between the fresh produce and its surrounding environment which prevents the loss of natural volatile flavour and colour compounds from fresh produce and the acquisition of foreign odours, to protect from physical damage of produce and to act as carriers of other functional ingredients, such as antimicrobial and antioxidant agents, nutraceuticals, and colour and flavour ingredients for reducing microbial loads and improving quality [71]. The efficiency of alginate-based coatings can be further improved by the incorporation of antimicrobial compounds. There are many benefits that can be obtained by consuming foods with edible seaweed-based coatings. For example, the positive effects of alginates which are known as dietary fiber on intestinal absorption and colonic health have been reported [71]. Some researchers have reported the pharmaceutical activity of alginate molecules. This ability refers to alginates with high mannuronate residues, which could induce cytokine production 10 times more than alginates containing high guluronate blocks [72]. Some known positive biological effects of alginates, agar, and carrageenan, such as antitumor, antioxidant, immunomodulatory, anticoagulant, antithrombotic, and antiviral activities, have been known [73]. The quantity of packaging materials has been increasing by 8% annually. It was shown that less than 5% of all plastics are being recycled, leading to a high accumulation of plastics in the environment [11]. Besides, increasing consumer concerns about food safety led to the development of biodegradable, seaweed-based edible coatings suitable for food packaging applications. As the coating is edible, coated food products can be consumed as it is without releasing the cover, into the environment [11].

CONCLUDING REMARKS

Seaweed possesses polysaccharides, agar, alginate, and carrageenan, which can be successfully utilized to develop edible seaweed-based biodegradable films and coatings. These films and coatings can be applied to fruits, vegetables, meat, poultry, and dairy products to prolong the shelf life and maintain the quality of the food product.

CONSENT FOR PUBLICATION

Not applicable.

CONFLICT OF INTEREST

The author(s) declare no conflict of interest, financial or otherwise.

ACKNOWLEDGEMENT

The authors would like to acknowledge the University Research Grants (Grant No. ASP/ 01/RE/SCI/2019/16), University of Sri Jayewardenepura, Gangodawila, Nugegoda, Sri Lanka.

REFERENCES

[1] Hassan B, Chatha SAS, Hussain AI, Zia KM, Akhtar N. Recent advances on polysaccharides, lipids and protein based edible films and coatings: A review. Int J Biol Macromol 2018; 109: 1095-107. [http://dx.doi.org/10.1016/j.ijbiomac.2017.11.097] [PMID: 29155200]

[2] Falguera V, Quintero JP, Jiménez A, Muñoz JA, Ibarz A. Edible films and coatings: Structures, active functions and trends in their use. Trends Food Sci Technol 2011; 22(6): 292-303. [http://dx.doi.org/10.1016/j.tifs.2011.02.004]

[3] Galus S, Kadzińska J. Food applications of emulsion-based edible films and coatings. Trends Food Sci Technol 2015; 45(2): 273-83. [http://dx.doi.org/10.1016/j.tifs.2015.07.011]

[4] Salgado PR, Ortiz CM, Musso YS, Di Giorgio L, Mauri AN. Edible films and coatings containing bioactives. Curr Opin Food Sci 2015; 5: 86-92. [http://dx.doi.org/10.1016/j.cofs.2015.09.004]

[5] Praseptiangga D. Development of seaweed-based biopolymers for edible films and lectins. In: IOP Conference Series: Materials Science and Engineering. 2017.Apr 1;193:012003. [http://dx.doi.org/10.1088/1757-899X/193/1/012003]

[6] Glicksman M. Utilization of seaweed hydrocolloids in the food industry. In: Twelfth international seaweed symposium 1987; 31-47. [http://dx.doi.org/10.1007/978-94-009-4057-4_3]

[7] Khalil HP, Tye YY, Saurabh CK, *et al.* Biodegradable polymer films from seaweed polysaccharides: A review on cellulose as a reinforcement material. Express Polym Lett 2017; 11: 4.

[8] Pawar SN, Edgar KJ. Alginate derivatization: A review of chemistry, properties and applications. Biomaterials 2012; 33(11): 3279-305. [http://dx.doi.org/10.1016/j.biomaterials.2012.01.007] [PMID: 22281421]

[9] Jayakody MM, Vanniarachchy MPG, Wijesekara I. Seaweed derived alginate, agar, and carrageenan based edible coatings and films for the food industry: a review. J Food Meas Charact 2022; 16(2): 1195-227.
[http://dx.doi.org/10.1007/s11694-021-01277-y]

[10] Cuadros TR, Skurtys O, Aguilera JM. Mechanical properties of calcium alginate fibers produced with a microfluidic device. Carbohydr Polym 2012; 89(4): 1198-206.
[http://dx.doi.org/10.1016/j.carbpol.2012.03.094] [PMID: 24750932]

[11] Müller K, Schmid M. Alginate-Based Edible Films and Coatings for Food Packaging Applications. Foods (Basel, Switzerland). 2018; 17: 7(10).

[12] Xiao Q, Gu X, Tan S. Drying process of sodium alginate films studied by two-dimensional correlation ATR-FTIR spectroscopy. Food Chem 2014; 164: 179-84.
[http://dx.doi.org/10.1016/j.foodchem.2014.05.044] [PMID: 24996322]

[13] Kapetanakou AE, Manios SG, Skandamis PN. Application of edible films and coatings on food. Novel food preservation and microbial assessment techniques. 2014; 52: 237-73.

[14] Sedayu BB, Cran MJ, Bigger SW. A review of property enhancement techniques for carrageenan-based films and coatings. Carbohydr Polym 2019; 216: 287-302.
[http://dx.doi.org/10.1016/j.carbpol.2019.04.021] [PMID: 31047069]

[15] Yuguchi Y, Thu Thuy TT, Urakawa H, Kajiwara K. Structural characteristics of carrageenan gels: temperature and concentration dependence. Food Hydrocoll 2002; 16(6): 515-22.
[http://dx.doi.org/10.1016/S0268-005X(01)00131-X]

[16] Armisen R, Gaiatas F. Agar. In: Handbook of hydrocolloids. Woodhead Publishing, 2009; 82-107.

[17] Tapia MS, Rojas-Graü MA, Carmona A, Rodríguez FJ, Soliva-Fortuny R, Martin-Belloso O. Use of alginate- and gellan-based coatings for improving barrier, texture and nutritional properties of fresh-cut papaya. Food Hydrocoll 2008; 22(8): 1493-503.
[http://dx.doi.org/10.1016/j.foodhyd.2007.10.004]

[18] Parris N, Coffin DR. Composition factors affecting the water vapor permeability and tensile properties of hydrophilic zein films. J Agric Food Chem 1997; 45(5): 1596-9.
[http://dx.doi.org/10.1021/jf960809o]

[19] Quintavalla S, Vicini L. Experimental Station for the Food Preserving Industry, V. le Tanara, 31/A, 43100 Parma, Italy. Meat Sci 2002; 62: 373-80.
[http://dx.doi.org/10.1016/S0309-1740(02)00121-3] [PMID: 22061613]

[20] Tavassoli-Kafrani E, Shekarchizadeh H, Masoudpour-Behabadi M. Development of edible films and coatings from alginates and carrageenans. Carbohydr Polym 2016; 137: 360-74.
[http://dx.doi.org/10.1016/j.carbpol.2015.10.074] [PMID: 26686140]

[21] Tharanathan RN. Biodegradable films and composite coatings: past, present and future. Trends Food Sci Technol 2003; 14(3): 71-8.
[http://dx.doi.org/10.1016/S0924-2244(02)00280-7]

[22] De Azeredo HM, Rosa MF, De Sá M, Souza Filho M, Waldron KW. The use of biomass for packaging films and coatings. 2014; 819-74.
[http://dx.doi.org/10.1533/9780857097385.2.819]

[23] Sothornvit R, Krochta JM. Plasticizers in edible films and coatings. In: Innovations in food packaging 2005; 403-33.
[http://dx.doi.org/10.1016/B978-012311632-1/50055-3]

[24] Galus S, Lenart A. Development and characterization of composite edible films based on sodium alginate and pectin. J Food Eng 2013; 115(4): 459-65.
[http://dx.doi.org/10.1016/j.jfoodeng.2012.03.006]

[25] Zhang D, Zhang M, Gu X. Seaweed-derived hydrocolloids as food coating and encapsulation

agents.In: Bioactive Seaweeds for Food Applications. 2018; 153-75. [http://dx.doi.org/10.1016/B978-0-12-813312-5.00008-X]

[26] Zhong Y, Cavender G, Zhao Y. Investigation of different coating application methods on the performance of edible coatings on Mozzarella cheese. Lebensm Wiss Technol 2014; 56(1): 1-8. [http://dx.doi.org/10.1016/j.lwt.2013.11.006]

[27] Nussinovitch A. Biopolymer films and composite coatings. In: Modern biopolymer science 2009; 295-326. [http://dx.doi.org/10.1016/B978-0-12-374195-0.00010-0]

[28] Pires TCSP, Dias MI, Barros L, *et al.* Edible flowers as sources of phenolic compounds with bioactive potential. Food Res Int 2018; 105: 580-8. [http://dx.doi.org/10.1016/j.foodres.2017.11.014] [PMID: 29433250]

[29] Embuscado ME, Huber KC. Edible films and coatings for food applications. New York, NY, USA: Springer; 2009 Jun 10.

[30] Rößle C, Brunton N, Gormley RT, Wouters R, Butler F. Alginate coating as carrier of oligofructose and inulin and to maintain the quality of fresh-cut apples. J Food Sci 2011; 76(1): H19-29. [http://dx.doi.org/10.1111/j.1750-3841.2010.01902.x] [PMID: 21535683]

[31] Olivas GI, Mattinson DS, Barbosa-Cánovas GV. Alginate coatings for preservation of minimally processed 'Gala' apples. Postharvest Biol Technol 2007; 45(1): 89-96. [http://dx.doi.org/10.1016/j.postharvbio.2006.11.018]

[32] Augusto A, Simões T, Pedrosa R, Silva SFJ. Evaluation of seaweed extracts functionality as post-harvest treatment for minimally processed Fuji apples. Innov Food Sci Emerg Technol 2016; 33: 589-95. [http://dx.doi.org/10.1016/j.ifset.2015.10.004]

[33] Azarakhsh N, Osman A, Ghazali HM, Tan CP, Mohd Adzahan N. Lemongrass essential oil incorporated into alginate-based edible coating for shelf-life extension and quality retention of fresh-cut pineapple. Postharvest Biol Technol 2014; 88: 1-7. [http://dx.doi.org/10.1016/j.postharvbio.2013.09.004]

[34] Menezes J, Athmaselvi KA. Polysaccharide based edible coating on sapota fruit. Int Agrophys 2016; 30(4): 551-7. [http://dx.doi.org/10.1515/intag-2016-0019]

[35] Moraes KS, Fagundes C, Melo MC, Andreani P, Monteiro AR. Conservation of Williams pear using edible coating with alginate and carrageenan. Food Sci Technol (Campinas) 2012; 32(4): 679-84. [http://dx.doi.org/10.1590/S0101-20612012005000106]

[36] Nussinovitch A, Hershko V. Gellan and alginate vegetable coatings. Carbohydr Polym 1996; 30(2-3): 185-92. [http://dx.doi.org/10.1016/S0144-8617(96)00095-1]

[37] Tay SL, Perera CO. Effect of 1-methylcyclopropene treatment and edible coatings on the quality of minimally processed lettuce. J Food Sci 2004; 69(2): fct131-5. [http://dx.doi.org/10.1111/j.1365-2621.2004.tb15504.x]

[38] Amanatidou A, Slump RA, Gorris LGM, Smid EJ. High oxygen and high carbon dioxide modified atmospheres for shelf-life extension of minimally processed carrots. J Food Sci 2000; 65(1): 61-6. [http://dx.doi.org/10.1111/j.1365-2621.2000.tb15956.x]

[39] Wanstedt KG, Seideman SC, Donnelly LS, Quenzer NM. Sensory attributes of precooked, calcium alginate-coated pork patties. J Food Prot 1981; 44(10): 732-5. [http://dx.doi.org/10.4315/0362-028X-44.10.732] [PMID: 30856754]

[40] Lu F, Ding Y, Ye X, Liu D. Cinnamon and nisin in alginate–calcium coating maintain quality of fresh northern snakehead fish fillets. Lebensm Wiss Technol 2010; 43(9): 1331-5. [http://dx.doi.org/10.1016/j.lwt.2010.05.003]

[41] Yu XL, Li XB, Xu XL, Zhou GH. Coating with sodium alginate and its effects on the functional properties and structure of frozen pork. J Muscle Foods 2008; 19(4): 333-51.
[http://dx.doi.org/10.1111/j.1745-4573.2008.00113.x]

[42] Salmieri S, Lacroix M. Physicochemical properties of alginate/polycaprolactone-based films containing essential oils. J Agric Food Chem 2006; 54(26): 10205-14.
[http://dx.doi.org/10.1021/jf062127z] [PMID: 17177561]

[43] Albert A, Salvador A, Fiszman SM. A film of alginate plus salt as an edible susceptor in microwaveable food. Food Hydrocoll 2012; 27(2): 421-6.

[44] McSweeney PL. 197 What is enzyme-modified cheese?. Cheese problems solved. 2007; 387.

[45] Kavas N, Kavas G, Saygili D. Use of ginger essential oil-fortified edible coatings in Kashar cheese and its effects on *Escherichia coli* O157:H7 and *Staphylococcus aureus*. CYTA J Food 2016; 14(2): 317-23.
[http://dx.doi.org/10.1080/19476337.2015.1109001]

[46] Raju A, Sasikala S. Natural antimicrobial edible film for preservation of paneer. Biosci Biotechnol Res Asia 2016; 13(2): 1083-8.
[http://dx.doi.org/10.13005/bbra/2136]

[47] Fang F, Chen X, Zhang X, *et al.* Environmentally friendly assembly multilayer coating for flame retardant and antimicrobial cotton fabric. Prog Org Coat 2016; 90: 258-66.
[http://dx.doi.org/10.1016/j.porgcoat.2015.09.025]

[48] Coughlan K, Shaw NB, Kerry JF, Kerry JP. Combined effects of proteins and polysaccharides on physical properties of whey protein concentrate-based edible films. J Food Sci 2004; 69(6): E271-5.
[http://dx.doi.org/10.1111/j.1365-2621.2004.tb10997.x]

[49] López de Lacey AM, López-Caballero ME, Montero P. Agar films containing green tea extract and probiotic bacteria for extending fish shelf-life. Lebensm Wiss Technol 2014; 55(2): 559-64.
[http://dx.doi.org/10.1016/j.lwt.2013.09.028]

[50] da Rocha M, Alemán A, Romani VP, *et al.* Effects of agar films incorporated with fish protein hydrolysate or clove essential oil on flounder (*Paralichthys orbignyanus*) fillets shelf-life. Food Hydrocoll 2018; 81: 351-63.
[http://dx.doi.org/10.1016/j.foodhyd.2018.03.017]

[51] Vejdan A, Ojagh SM, Abdollahi M. Effect of gelatin/agar bilayer film incorporated with TiO_2 nanoparticles as a UV absorbent on fish oil photooxidation. Int J Food Sci Technol 2017; 52(8): 1862-8.
[http://dx.doi.org/10.1111/ijfs.13461]

[52] Moreno MA, Bojorges H, Falcó I, *et al.* Active properties of edible marine polysaccharide-based coatings containing Larrea nitida polyphenols enriched extract. Food Hydrocoll 2020; 102: 105595.
[http://dx.doi.org/10.1016/j.foodhyd.2019.105595]

[53] Ziedan ES, El Zahaby HM, Maswada HF, Zoeir HA. Agar-agar a promising edible coating agent for management of postharvest diseases and improving banana fruit quality. J Plant Prot Res 2018; 58: (3).

[54] Gudadhe JA, Yadav A, Gade A, Marcato PD, Durán N, Rai M. Preparation of an agar-silver nanoparticles (A-AgNp) film for increasing the shelf-life of fruits. IET nanobiotechnology. 2014; 8(4): 190-5.

[55] Guitián MV, Ibarguren C, Soria MC, Hovanyecz P, Banchio C, Audisio MC. Anti-Listeria monocytogenes effect of bacteriocin-incorporated agar edible coatings applied on cheese. Int Dairy J 2019; 97: 92-8.
[http://dx.doi.org/10.1016/j.idairyj.2019.05.016]

[56] Incoronato AL, Conte A, Buonocore GG, Del Nobile MA. Agar hydrogel with silver nanoparticles to

prolong the shelf life of Fior di Latte cheese. J Dairy Sci 2011; 94(4): 1697-704. [http://dx.doi.org/10.3168/jds.2010-3823] [PMID: 21426957]

[57] Karbowiak T, Debeaufort F, Champion D, Voilley A. Wetting properties at the surface of iota-carrageenan-based edible films. J Colloid Interface Sci 2006; 294(2): 400-10. [http://dx.doi.org/10.1016/j.jcis.2005.07.030] [PMID: 16259994]

[58] Hamzah HM, Osman A, Tan CP, Mohamad Ghazali F. Carrageenan as an alternative coating for papaya (Carica papaya L. cv. Eksotika). Postharvest Biol Technol 2013; 75: 142-6. [http://dx.doi.org/10.1016/j.postharvbio.2012.08.012]

[59] Kanmani P, Rhim JW. Development and characterization of carrageenan/grapefruit seed extract composite films for active packaging. Int J Biol Macromol 2014; 68: 258-66. [http://dx.doi.org/10.1016/j.ijbiomac.2014.05.011] [PMID: 24832986]

[60] Falcó I, Randazzo W, Sánchez G, López-Rubio A, Fabra MJ. On the use of carrageenan matrices for the development of antiviral edible coatings of interest in berries. Food Hydrocoll 2019; 92: 74-85. [http://dx.doi.org/10.1016/j.foodhyd.2019.01.039]

[61] Liu Y, Qin Y, Bai R, Zhang X, Yuan L, Liu J. Preparation of pH-sensitive and antioxidant packaging films based on κ-carrageenan and mulberry polyphenolic extract. Int J Biol Macromol 2019; 134: 993-1001. [http://dx.doi.org/10.1016/j.ijbiomac.2019.05.175] [PMID: 31129204]

[62] Shojaee-Aliabadi S, Hosseini H, Mohammadifar MA, *et al.* Characterization of κ-carrageenan films incorporated plant essential oils with improved antimicrobial activity. Carbohydr Polym 2014; 101: 582-91. [http://dx.doi.org/10.1016/j.carbpol.2013.09.070] [PMID: 24299814]

[63] Ayala-Zavala JF, Vega-Vega V, Rosas-Domínguez C, *et al.* Agro-industrial potential of exotic fruit byproducts as a source of food additives. Food Res Int 2011; 44(7): 1866-74. [http://dx.doi.org/10.1016/j.foodres.2011.02.021]

[64] Ayranci E, Tunc S. A method for the measurement of the oxygen permeability and the development of edible films to reduce the rate of oxidative reactions in fresh foods. Food Chem 2003; 80(3): 423-31. [http://dx.doi.org/10.1016/S0308-8146(02)00485-5]

[65] Bonilla J, Atarés L, Vargas M, Chiralt A. Edible films and coatings to prevent the detrimental effect of oxygen on food quality: Possibilities and limitations. J Food Eng 2012; 110(2): 208-13. [http://dx.doi.org/10.1016/j.jfoodeng.2011.05.034]

[66] Baldwin EA, Hagenmaier R, Bai J, Eds. editors. Edible coatings and films to improve food quality. CRC press 2011. [http://dx.doi.org/10.1201/b11082]

[67] Wang LZ, Liu L, Holmes J, Kerry JF, Kerry JP. Assessment of film-forming potential and properties of protein and polysaccharide-based biopolymer films. Int J Food Sci Technol 2007; 42(9): 1128-38. [http://dx.doi.org/10.1111/j.1365-2621.2006.01440.x]

[68] Costa MJ, Maciel LC, Teixeira JA, Vicente AA, Cerqueira MA. Use of edible films and coatings in cheese preservation: Opportunities and challenges. Food Res Int 2018; 107: 84-92. [http://dx.doi.org/10.1016/j.foodres.2018.02.013] [PMID: 29580546]

[69] Bourtoom T. Edible films and coatings: characteristics and properties. Int Food Res J 2008; 15(3): 237-48.

[70] Ncama K, Magwaza LS, Mditshwa A, Tesfay SZ. Plant-based edible coatings for managing postharvest quality of fresh horticultural produce: A review. Food Packag Shelf Life 2018; 16: 157-67. [http://dx.doi.org/10.1016/j.fpsl.2018.03.011]

[71] Lin D, Zhao Y. Innovations in the development and application of edible coatings for fresh and minimally processed fruits and vegetables. Compr Rev Food Sci Food Saf 2007; 6(3): 60-75. [http://dx.doi.org/10.1111/j.1541-4337.2007.00018.x]

[72] Lee KY, Mooney DJ. Alginate: Properties and biomedical applications. Prog Polym Sci 2012; 37(1): 106-26.
[http://dx.doi.org/10.1016/j.progpolymsci.2011.06.003] [PMID: 22125349]

[73] Campo VL, Kawano DF, Silva DB Jr, Carvalho I. Carrageenans: Biological properties, chemical modifications and structural analysis – A review. Carbohydr Polym 2009; 77(2): 167-80.
[http://dx.doi.org/10.1016/j.carbpol.2009.01.020]

CHAPTER 19

Current Applications of Seaweed-Based Polysaccharides in Edible Packaging

Sandunika Senarathna[1], Indira Wickramasinghe[1,*] and Seneviratne Navaratne[1]

[1] *Department of Food Science & Technology, Faculty of Applied Sciences, University of Sri Jayewardenepura, Nugegoda, Sri Lanka*

Abstract: The use of biopolymers instead of synthetic polymers for food packaging has become a recent trend since it successfully solves the global issue of plastic waste due to their biodegradability, biocompatibility and renewability. Moreover, edible packaging has gained the attention of the current research world. Thus, the natural polymer sources applicable in forming edible packaging materials, such as polysaccharides, proteins and lipids, are studied. Seaweed, referred to as marine macroalgae, is a rich source of polysaccharides. Different types of polysaccharides can be identified in the three main varieties of seaweed, carrageenan and agar in red algae, alginate, laminaran and fucoidan in brown algae, while ulvan is the major polysaccharide in green algae. The film-forming properties of these seaweed-based polysaccharides are enhanced due to their colloidal nature; meanwhile, the abundance and the low cost make them more applicable in edible packaging. Several modifications were carried out to achieve packaging materials with better mechanical and barrier properties. Hence, this chapter discusses the current applications of seaweed-based polysaccharides in edible packaging with improved properties in different sections such as fruits, vegetables and meat industries by analyzing recent research findings.

Keywords: Agar, Alginate, Applications of Seaweed, Biodegradability, Biopolymers, Brown Algae, Carrageenan, Edible Coating, Edible Film, Edible Packaging, Film Properties, Fucoidan, Green Algae, Laminaran, Recent Studies on Edible Packaging, Red Algae, Renewability, Seaweed-Based Packaging, Seaweed-Based Polysaccharides, Ulvan.

INTRODUCTION

Environmental accumulation of plastic waste has become a global problem with synthetic polymers due to their non-biodegradability and non-renewability. The

[*] **Corresponding author Indira Wickramasinghe:** Department of Food Science & Technology, Faculty of Applied Sciences, University of Sri Jayewardenepura, Nugegoda, Sri Lanka, Tel: +94716861908; E-mail: indiraw@sjp.ac.lk

Avinash Mishra (Ed.)

replacement of synthetic polymers with biopolymers is a successful alternative for this burning issue. Recent researchers have focused on developing biodegradable, biocompatible, renewable and edible packaging materials to reduce the environmental accumulation of synthetic polymers [1]. Edible packaging, a type of packaging suitable for human consumption, consists of two main categories: edible coatings and edible films. A thin layer formed directly on the food surface is identified as the edible coating, while edible film means a thin sheet wrapped later over the food surface. The ingredients used for edible packaging should be able to form a continuous, cohesive, and strong network while being consumable along with the contained food product making the packaging waste zero.

Furthermore, as edible packaging materials are safe for human consumption, no issue has arisen with the transmission of the compounds in packaging material to the food product [2]. However, in the development of edible packaging material, there are some factors needed to be considered, such as its mechanical strength, barrier properties, safety for human consumption, toxic compounds that can be formed, microbial stability, biodegradability, sensory attributes, the abundance of raw materials and the cost [2, 3]. Therein, abundant biopolymers that have appropriate characteristics in forming continuous matrixes from embedded constituents such as polysaccharides, proteins and lipids are focused. Edible films with better mechanical strength can be obtained from proteins but not polysaccharides. However, polysaccharides exhibit higher gas barrier properties. The water vapor permeability is high in both polysaccharide and protein-based packaging materials. Although lipids have hydrophobic characteristics, they are unable to form self-supporting film structures. Hence, the usage of lipids is limited to use as an edible coating material and as an additive in protein and polysaccharide-based packaging films, which lead to making composite films with better properties [2, 4]. In addition to making composite films, looking for more novel sources of edible packaging is another way to achieve a successful edible coating or a film.

Seaweed, biologically identified as marine macroalgae, is a part of traditional daily diets in Asia-Pacific regions [5]. In the nutritional aspects, seaweed is low in calories and lipids but high in bioactive polysaccharides, which majorly account for dietary fibers. The polysaccharide content found in seaweed species is in the range of 4-76% of the dry weight [6]. Seaweed-based polysaccharides are recently getting attentive significantly in the packaging industry due to their excellent film-forming properties, colloidal nature, abundance and low cost [1, 7]. Further, the films produced from seaweed-based polysaccharides are soluble, non-toxic, good in barrier properties (oxygen, water vapor and UV light), optical properties (transparency) and have considerably good mechanical properties such as tensile strength and elongation at break [5, 7]. Besides, seaweed-based polysaccharides become a sustainable and superior alternative to the traditional food packaging materials due to bioactive properties, such as antioxidant and antimicrobial activities, which would fulfill the active packaging characteristics too [5, 8].

Three main varieties of marine macroalgae can be found, namely red algae (Rhodophyta), brown algae (Phaeophyta) and green algae (Chlorophyta) (Fig. **1**). Red algae- the oldest division of lower plants that contains unique sulfated polysaccharides of two groups named, carrageenan and agar [9]. Brown algae, considered the second large group of seaweeds, are enriched with different bioactive compounds such as polysaccharides, carotenoids, peptides, polyphenols, omega-3 fatty acids, vitamins and minerals. The major polysaccharides obtained from brown algae are alginate, laminaran and fucoidan, accounting for 40-80% of dried, defatted brown algae biomass [10]. The polysaccharides from brown algae exhibit higher antioxidant activities than red and green algae due to a comparatively high number of polyphenols and sulphates [6, 10, 11]. Ulvan is the major polysaccharide found in the cell walls of green algae and belongs to genus *Ulva* which is considered a largely unexploited group of algae but widely distributed on the rock surfaces of the intertidal zone and used as a food ingredient that enriches the coastal residents' diets with protein, vitamins, minerals and more with dietary fiber [12 - 15].

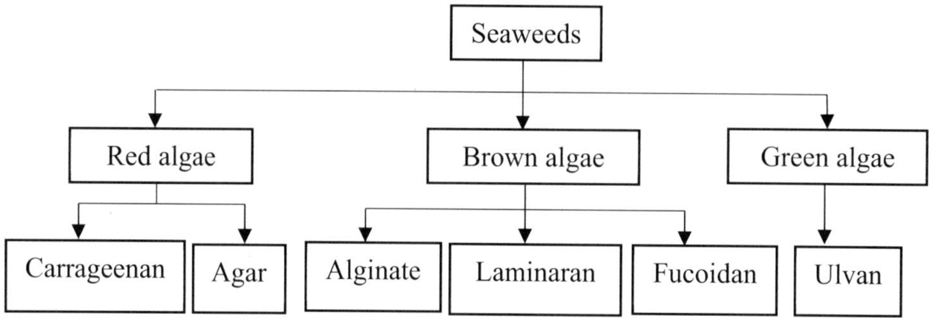

Fig. (1). Classification of seaweed-based polysaccharides [10].

CURRENT APPLICATIONS OF RED ALGAE-BASED POLYSACCHARIDES IN EDIBLE PACKAGING

Carrageenan

Carrageenans are water-soluble, sulfated polysaccharides in which the repeating disaccharide units are formed by alternating 3-linked β-D-galactopyranose (G-units) and 4-linked α-D-galactopyranose (D-units) or 4-linked 3,6-anhydro--D-galactopyranose (DA-units). Carrageenans are categorized as kappa (mono-sulfate), iota (di-sulfate) and lambda (tri-sulfate), considering their structural properties due to the existence of 3,6-anhydrobridge on the 4-linked-galactose residue, the position of sulfate groups and the number of sulfate groups. Meanwhile, the 3,6-anhydrobridges are available only in kappa (κ) and iota (ι), not in lambda (λ) carrageenans [16, 17]. Therefore, it is essential to identify these structural variances properly as they determine the physicochemical properties of carrageenans, which lead them to different applications. Among the three forms of carrageenans, kappa (κ) and iota (ι) carrageenans are used in the formation of packaging films as only these two categories have the ability to form gels. Nevertheless, κ-carrageenans form strong, hard and brittle gels, while the gels produced from ι- carrageenans are weak, soft and elastic leading to widening κ-carrageenans' applications in making packaging materials [16].

Although κ-carrageenan has gained more attention due to its possibility of forming stronger and lesser hydrophilic films, the high cost of refined κ-carrageenan obstructs its application in edible packaging. Therefore, using semi-refined κ-carrageenan, which is not as expensive as refined, is an excellent alternative for making edible packaging materials [16, 17]. Typically, carrageenans are good film-forming agents with better mechanical and barrier properties to gases such as oxygen and carbon dioxide [18]. However, to increase the application of carrageenans in the food packaging section, it is needed to enhance their functional properties in terms of active packaging. Thus, many researchers have focused on incorporating functional compounds in making edible packaging materials with carrageenans. The application of semi-refined κ-carrageenan-based film in chicken breast packaging has been studied by Farhan *et al.* [17], and results revealed that the incorporation of water extract of germinated fenugreek seeds to the film formation formula of carrageenan could increase the phenolic content, antioxidant and antimicrobial activities of the film which have caused to increase the shelf life of the food product considerably. Moreover, the food applications of carrageenan are widened by blending with the water extract of germinated fenugreek seeds as it forms stronger films by forming stronger chemical interactions with carrageenan, which makes handling and storage easy. The dark color imparted to the carrageenan film from the water extract of germinated fenugreek seeds adds an additional value; the light barrier property to the film makes it applicable for the packaging of light-sensitive food products. Sandhu *et al.* [19] and Huang *et al.* [20] suggested using carrageenan-based composite film in packaging and coating applications in the food industry as it can lower the water vapor permeability while increasing the tensile strength and elongation at break. Zhou *et al.* [21] developed konjac glucomannan/carrageenan-based edible emulsion coating by combining with camellia oil for chicken meat which is a highly deteriorative food product due to bacterial and oxidative reactions. According to the findings, the composite coating could significantly reduce weight loss, pH, microbial counts, thiobarbituric acid reactive substance and total volatile nitrogen. Further, camellia oil incorporated konjac glucomannan/carrageenan-based coating could increase the shelf life of chicken meat owing to retardation of microbial growth and the oxidation of proteins and lipids. To increase the applicability of carrageenan-based biodegradable and edible films as active packaging materials in the food industry, Martiny *et al.* [22] have developed carrageenan film blending with olive leaves extract following the casting technique. The results of the study indicated that the developed composite film was a successful alternative to antimicrobial food packaging, which can be applied to the meat industry as it exhibited *E. coli* inhibition with a 167-fold reduction in initial mesophile count with higher total phenolic content (41.40 mg GAE/g). Thus, diversified applications of carrageenan are found in fish, poultry, and frozen meat industries to prevent superficial dehydration of meat products as well as in sausage casings. Therein, for the preservation of grass carp fillets, a unidirectionally permeable film was developed by He *et al.* [23] in which the inner hydrophilic layer was based on κ-carrageenan, gelatin, and curcumin while having a zein outer layer. The mechanical properties and water resistance of carrageenan-based films could be increased owing to the hydrophobic outer layer. Whereas, the developed film could act as an intelligent packaging material in terms of freshness as its color was changed from yellow to red with the storage time of grass carp fillets. Moreover, the carrageenan-based film has been made by Ganesan *et al.* [24] as an alternative to conventional plastic packing material for edible oil storage by incorporating konjac glucomannan to increase the bioactivity of the packaging material. The developed film has resulted in a low oxidation rate, and less peroxide and iodine values of the oil with the time of storage. Further, the physical and mechanical properties of the film, such as thickness, tensile strength, and elongation at break, were increased due to the incorporation of konjac glucomannan into carrageenan. The impact of bee pollen and honey extracts incorporated in edible κ-carrageenan films on antioxidant and antibacterial properties has been reported by Velasquez *et al.* [25]. According to the results, the antioxidant capacity in terms of DPPH (2,2-diphenyl-1-picrylhydrazyl) free radicals and iron-reducing power has been improved significantly by incorporating the extracts. The carrageenan-based film combined with bee pollen extract indicated higher antibacterial activity for mesophilic aerobes. Besides, the incorporation of these extracts into carrageenan films could enhance the physical properties of the developed films. According to a study carried out by Abdillah *et al.* [26], the incorporation of ι-carrageenan to arrowroot starch-based biodegradable and edible film could improve the mechanical and barrier properties, swelling index, and water solubility. Meanwhile, the composite film could reduce weight loss and increase the shelf life of cherry tomatoes. A composite of chitosan and κ-carrageenan based coating was studied for the preservation of dragon fruit by Nguyen *et al.* [27]. The findings showed that the application of κ-carrageenan-based coating for the fruit could decrease the weight loss and protect the freshness and bract color. Moreover, an investigation of preserving strawberry fruits by a coating conducted by Wani *et al.* [28] revealed carrageenan was a successful coating material for strawberries as it could delay weight loss and maintain the anthocyanins, ascorbic acid and phenolic contents effectively. Besides, an edible film has been developed by Mahajan *et al.* [29], combining carrageenan with *Aloe vera* extract for the packaging of Kulfi, a type of Indian ice cream. The study showed the incorporation of *Aloe vera* extract could increase the antioxidant and antimicrobial activities of carrageenan film, which leads to extending the shelf lives of frozen dairy products with *E. coli* inhibition, and reduction of the total plate, psychrophilic and yeast/mould counts and the stability with lipid oxidation. According to a study carried out by Falco *et al.* [30], carrageenan-based coatings could extend the shelf life of berries, and the incorporation of green tea extract into the carrageenan coating formula could enhance its antiviral activity both at ambient and refrigerated temperatures against murine norovirus and hepatitis A virus. As well, Mahajan *et al.* [29] highlighted the health importance of *Aloe vera* extract incorporated packaging film in SARS-CoV-2 pandemic owing to its high antiviral and immune boosting properties. Accordingly, it is essential to study more applicable, cost-effective methods to enhance the properties of carrageenan-based packaging materials with the aim of making carrageenan a commercially applicable agent of edible packaging.

Agar

Agar is a polysaccharide that forms the supportive structure of the cell walls of red algae [31, 32]. Due to the esterification of agar with up to 6% (W/W) sulfate, it is considered a sulfated polysaccharide, which consists of α (1-4)-3, 6-anhydr--L-galactose and β 9(1-3)-D-galactose residues [8]. The chemical structure of agar contains two main compounds, namely agaropectin, which is identified as the non-gelling fraction, and agarose as the gelling fraction. The commercially available agar is produced from *Gelidium sp.* and *Gracilaria sp.* after removal of agaropectin fraction to get gel with higher strength [32]. This will increase the commercial applications of agar not only as a thickening, gelling, emulsifying and texture modifying agent in medical and food industries [8] but also as a film-forming agent in the packaging industry since the gelling forming property of agar determines the continuity of the film matrix [32].

The excellent film-forming capacity, moderate water resistance, thermoplasticity, transparency, biocompatibility and biodegradation of agar-based films make agar more applicable in food packaging. However, the low thermal stability, low mechanical properties and high water vapor permeability characteristics interfere with the commercialization of agar-based films. Therefore, researchers focus on improving the properties by incorporating hydrophobic components, plasticizers, nano-fillers such as clay, copper, and silver and blending them with other biopolymers [32, 33]. By considering those, Wang *et al.* [33] studied incorporating nano-bacterial cellulose into the agar-based film, which has significantly increased the mechanical properties like tensile strength, elongation at break and thermal stability. Besides, the physical, mechanical and barrier properties, pasting behavior and crystallinity of starch-based films were able to improve by incorporating agar due to the strong interactions formed by agar with starch. Furthermore, hydrophobic aggregates formation by agar reduces the water solubility and water sensitivity of films [34, 35]. Agar can produce an excellent cohesive film matrix and is an excellent agent to blend with other polysaccharides in edible film making to have better mechanical properties [36]. Zhang *et al.* [37] suggested agar/maltodextrin-based films incorporated with hydrophobic agents as a promising potential material for food packaging. The findings revealed that the hydrophobicity, mechanical strength, and barrier properties of the agar-based films could be improved with the addition of hydrophobic agents such as beeswax, shortening and liquid paraffin. Whereas a bioindicator edible film was developed by Wardana *et al.* [38] for sausages from agar, tapioca starch and red cabbage juice. The results showed that the incorporation of agar into the film caused to increase in tensile strength, and elongation at break and thickness while decreasing the water vapor transmission rate. Further, Guitian *et al.* [39] investigated the applicability of agar-based edible coating in cheese to control *Listeria monocytogenes*, in which the results revealed the agar coating was a safe, natural, cost-effective and efficient alternative accordingly. However, further research should be done on using agar in edible packaging with a view to enhance physical, mechanical and barrier properties.

CURRENT APPLICATIONS OF BROWN ALGAE-BASED POLYSACCHARIDES IN EDIBLE PACKAGING

Alginates

Alginates are binary linear copolymers that contain α-L-guluronic acid (G) and (1/4)-linked β-D-mannuronic acid residues in which the chemical composition differs with the species and harvesting period of the year. Both salt and acid forms of alginates are available. The salt form is an important component of the brown algae cell wall which accounts for 40-47% of dried biomass. The acid form comprises polyuronic acid, which is identified as alginic acid. Alginate is a widely used polysaccharide as a thickening and stabilizing agent not only in the food industry but also in the cosmetic, paper and pharmaceutical industries [40]. In recent years, it has been proven that a considerable amount of alginates is being used as an edible coating material and edible film-forming agent due to their colloidal property [41, 42]. The use of alginate in edible films and coatings can add more value to food packaging as it is a rich source of dietary fiber, which can reduce glucose uptake and cholesterol level while dealing with gastrointestinal and cardiovascular diseases [42].

Alginate-based edible coatings have been widely developed recently to extend the shelf life of fruits and vegetables due to their potential for delay in the ripening process. Moreover, the effectiveness of alginate-based films and coatings is enhanced by incorporating more ingredients such as essential oils, fibers, nanoparticles *etc*. The alginate-based edible coating has been developed by Tabassum *et al*. [42], to increase the quality and the post-harvest life of fresh-cut papaya cubes as papaya is a fruit that is highly susceptible to pathogens due to its fast ripening that increases the softening of pulp and accumulating water and sugar. However, the addition of essential oils into the alginate coating formula could increase the antimicrobial activity of the coating. Further, to overcome the chilling injury, a major issue occurred with tropical fruits, Duong *et al*. [43] studied alginate coating, which was cross-linked with calcium ions and evaluated the post-harvest quality of rose apples. The findings revealed that calcium ion cross-linking could increase mechanical strength and reduce the permeability to water vapor and oxygen. Regarding the post-harvest quality of fruits in cold storage, the respiration rate and weight loss of coated fruits have been decreased, which significantly improves the physical appearance while retarding the chilling injury. Also, the total phenolic content and antioxidant activity in coated fruits remained comparatively higher. Nevertheless, a study carried out by Joshi *et al*. [44] revealed that blackberries coated with alginate could not exhibit a significant difference in anthocyanin and polyphenolic contents between the coated and uncoated fruits at cold storage even though the respiration rate and weight loss of coated fruits were decreased. Thus, the single step to get cross-linked alginate with calcium ions could result in fruits with higher phenolic content and antioxidant activity with respect to the stand-alone alginate coating. Mahcene *et al*. [45] evaluated the characteristics of alginate-based edible film blended with essential oils in terms of using it as a food packaging material. The findings revealed that the bioactive properties such as antibacterial and antioxidant activities and the biodegradability, thermal and barrier properties of the alginate-based films could be enhanced with the incorporation of essential oils obtained from *A. herba alba* Asso, *M. pulegium* L., *O. basilicum* L. and *R. officinalis* L. Besides, table grapes, a fruit widely consumed worldwide, require proper post-harvest practices to extend their shelf life owing to their high perishability. An alginate coating composite with cashew gum, gelatin and galactomannan has been studied for table grapes in which findings revealed the applicability of the particular alginate-based coating due to its contribution to the reduction of weight loss, significant shelf-life extension and the improvement in phenolic contents and antioxidant activity [46]. Alginate/hydroxyethyl cellulose-based edible coating incorporated with asparagus waste extract has been used for extending the post-harvest life of strawberry fruits by Liu *et al.* [47], in which the research findings indicated the developed edible coating could retain the fruit quality while exhibiting an inhibition against the growth of *Penicillium italicum*. Further, the research study conducted by Aziz *et al.* [48] suggested the addition of *Aloe vera* and garlic oil in making the alginate coating, which could increase the bioactive properties of tomatoes in storage while increasing the UV shielding, thermal and mechanical properties of the edible coating.

Guava is also considered a highly perishable fruit due to its high metabolic activities. Nair *et al.* [41] developed chitosan and alginate-based coating combined with pomegranate peel extract to extend the post-harvest life of guava. The research findings of this study indicate that not only the visual attributes of the fruit but also the nutritional parameters were also able to retain by retarding the respiration, which caused to delay the senescence. Thus, introducing the plant extracts with alginate-based films and coatings can improve the antioxidant and antimicrobial properties. The use of chitosan and alginate-based coating to improve the quality and shelf life of guava has been developed up to incorporating nanoparticles [49]. Ading ZnO nanoparticles could inhibit the formation of rot appearance in guava while delaying the ripening process. Similarly, the research findings of Salama *et al.* [50] prove that incorporating nanoparticles can improve the shelf life of fruits due to the enhancement of antimicrobial properties. In addition, the incorporation of the nanoparticles will lead to an increase in the thermal and mechanical properties of the coating.

A study carried out by Shahrampour *et al.* [51] has proven that an edible film made from an alginate pectin blend could survive the probiotic cells of *Lactobacillus plantarum* KMC45 in refrigeration temperatures. Also, Fan *et al.* [52] investigated the potential of using alginate with pectin and xanthan gum in making a composite edible film. Besides, a study conducted by Silva *et al.* [53] proved that the coating produced based on alginate could act as a good matrix for the cells of lactic acid bacteria in bacteriocin production, which caused to work as a protective barrier for mesophilic bacteria and extension of the shelf life of fresh cheese. The results showed a film with better mechanical strength and water vapor transmission, which reveals the applicability in food packaging. With the aim of developing alginate-based active edible composite film, Dou *et al.* [54] incorporated tea polyphenol, which has increased not only the antioxidant properties but also the mechanical and physical properties of the film. As well this fact is proven by the findings of Biao *et al.* [55] in which the alginate and tea polyphenols composite film could increase both antioxidant and anti-inflammatory properties while resulting in a film with better mechanical properties, reduced water vapor permeability and optical transmittance. Further, enhanced mechanical properties, water vapor barrier properties, thermal stability and antimicrobial properties with significant inhibitory effects for Gram-positive bacteria can be achieved from the alginate-based novel, bioactive, edible films incorporated with castor oil [56]. In order to increase the antibacterial activity of alginate-based coatings for effective controlling of the growth of *Listeria monocytogenes* found in turkey fillets during the storage, the nano form of *Trachyspermum ammi* essential oil could be applied to the alginate formula [57]. Blending alginate with chitosan biguanidine hydrochloride in film making is another successful way to achieve better antibacterial effect, improved mechanical properties, color attributes and reduced water vapor permeability [58]. Ferulic acid has been found as a potential crosslinking agent for alginate in making more stable, transparent and homogeneous films [59]. The antioxidant properties in terms of DPPH and ferric ions can be achieved by adding ferulic acid to alginate film-forming solutions. Alginate dialdehyde, considered an ecofriendly crosslinking agent, has been used by Park *et al.* [60] to enhance the thermal stability and mechanical properties of fish gelatin-based films and to widen the its applications in food packaging. The findings exhibited a significant increase in tensile strength, Young's modulus by 400 and 600%, respectively, with the crosslinking. Further, value-added alginate films and coatings could be obtained by blending with an inclusion complex, star anise ethanol extract/hydroxypropyl-β-cyclodextrin, which could be beneficial in applying in active packaging for the food products such as fresh cut Chinese yams as revealed by Guan *et al.* [61]. According to the findings, the composite film exhibited better antioxidant capacity, antimicrobial activity against *Rhizopus stolonifera* while having good mechanical properties and physical properties such as thermal stability and color. Besides, the antiviral effect to the alginate-based film could be achieved by incorporating green tea extract and grape seed extract [62]. The incorporation of epigallocatechin gallate to alginate films could effectively increase the antioxidant activity of the films while having better mechanical strength, which could be applicable in the packaging of fatty foods [63]. The composite films exhibited antiviral activity against murine norovirus and hepatitis A virus. Moreover, the extracts could increase the antioxidant activity of alginate-based films lead to widening its applications in active food packaging. Even though alginate is a highly researched edible coating and film-forming agent, there is a gap between the research findings and the industrial applicability. Therefore, it is highly worthful to fill that gap by making alginate-based edible films and coatings commercially available. Regarding this, researching cost-effective methodologies of making alginate-based edible films and coatings is a necessity.

Laminaran

Laminarans are low molecular weight polysaccharides that are biologically active, low toxic and water-soluble. The simple structure of laminarans contains (1,3)---D-glucan and some b-(1,6)-intrachain links, make it attractive in researching on. It is reported that the dry basis content of laminarans in brown algae is up to the levels of 35%, which can be varied with the factors such as the species, habitat, harvesting period and extraction procedures [6].

Laminarans exhibit a range of bifunctional activities, including antioxidant, anti-apoptotic, anticoagulant, anti-inflammatory, and antitumor activities, which can be enhanced by applying novel processing methodologies and appropriate chemical alterations [6]. Nevertheless, the application of laminaran in edible packaging is not reported to the best of our knowledge.

Fucoidan

Fucoidans are cell wall reinforcing constituents, and fucose is rich in sulfated polysaccharides which contain substantial percentages of L-fucose (α-1,3-linked) and sulphate ester groups. Fucoidans can be found only in brown seaweeds, not in other algae or higher plants. The highest fucoidan content, which is up to 20%, is in *Fucus vesiculosus* species. Further, fucoidans account for 40% of dried algal cell walls. Fucoidans form viscous solutions of water in which viscosity decreases with the increase of concentration [40] which is beneficial in the formation of edible films and coatings. Besides, sulfate substitution of fucoidans leads to a broad range of applications with bioactive effects [64].

To extend the shelf life, Xu *et al.* [65] applied fucoidan coating for mango fruits by considering the good coating forming characteristics and the antibacterial activities that fucoidans have. The results of the study revealed that the fucoidan coating could suppress the respiration and decay of fruits with the inhibition of ascorbic acid, weight loss while increasing the total soluble solids, which lead to the effective shelf-life extension of mango fruits and that method can be applied for other fruits, and vegetables too. To have edible packaging materials with attractive properties, researchers focus on developing fucoidans based films and coatings by combining them with alginate. Alginate and fucoidan blended film was developed with calcium ion cross-linking by Gomaa *et al.* [66, 67]. The research findings indicate that the incorporation of fucoidan and Ca^{2+} could decrease the water solubility of the film while increasing thickness, oxygen and water vapor permeability. Further, the developed films exhibited good antioxidant properties. However, researching the application of fucoidans in the formation of edible films and coatings are at the basic level. Therefore, further studies should be carried out in order to find out more methods to incorporate fucoidans in films and coatings and to make them industrially applicable.

CURRENT APPLICATIONS OF GREEN ALGAE-BASED POLYSACCHARIDES IN EDIBLE PACKAGING

Ulvan

Ulvan is a sulfated heteropolysaccharide that accounts for 9-36% of the algae [68]. It is a water-soluble, anionic stable polysaccharide that contains an ionized carboxylic group. Rhamnose (12.73-45.00%), xylose (2-12%), uronic acids (6.50-25.96%) and sulfate (12.80-23.00%) are the main components of ulvan, and it is structured as disaccharide repeating units of D-glucuronic or L-iduronic acid, associated to L-rhamnose-3- sulphate. Further, the uronic acids in several ulvan extracts are replaced by sulphated xylose residues [13, 69]. Ulvan exhibits a wide range of biological and physiological activities such as an anticoagulant, antitumor, anticancer, antioxidant and antifungal activities, making ulvan applicable in active packaging.

Amin *et al.* [70] suggested ulvan silver nanoparticles composite film as a potential alternative for active food packaging to extend the shelf life of foods in processing, transportation and storage as it exhibits better physical properties as well as antimicrobial and antioxidant activities. Moreover, the incorporation of ulvan into chitosan film could increase its antioxidant and whitening ability while increasing the tensile strength revealing the potential of ulvan as a future biomaterial [71]. Although successful film formulations can be achieved by ulvan, the film properties such as structural, thermal, optical and also antioxidant properties are strongly affected by the extraction condition of ulvan and the plasticizer type and concentration [12]. The extraction conditions of ulvan can vary not only its yield but also its chemical properties and molecular weight, which may lead to an effect on the film formation and film properties [7]. The ulvan films plasticized with glycerol exhibit a better structure than with sorbitol. However, the effect of ulvan extraction conditions and the plasticizer on the formation of ulvan smart active films have not been widely explored yet [7, 12]. With regards to this, the studies in depth on the application of ulvan in edible films and coating is essential as it can be a cost-effective source for edible packaging since ulvan is from *Ulva*, which is a highly abundant alga in the world.

CONCLUDING REMARKS

By looking at the recent research findings, it can be identified that seaweed-based polysaccharides have great potential as excellent sources of edible films and coatings. More scientific studies are needed to be conducted to make these polysaccharide-based edible packaging materials commercially available at an affordable price.

CONSENT FOR PUBLICATION

Not applicable.

CONFLICT OF INTEREST

The author(s) declare no conflict of interest, financial or otherwise.

ACKNOWLEDGEMENT

The authors would like to acknowledge University Research Grants (Grant No. ASP/01/RE/SCI/2019/11), University of Sri Jayewardenepura, Nugegoda, Sri Lanka.

REFERENCES

[1] Ili Balqis AM, Nor Khaizura MAR, Russly AR, Nur Hanani ZA. Effects of plasticizers on the physicochemical properties of kappa-carrageenan films extracted from *Eucheuma cottonii.* Int J Biol Macromol 2017; 103: 721-32.
[http://dx.doi.org/10.1016/j.ijbiomac.2017.05.105] [PMID: 28528954]

[2] Jeya Jeevahan J, Chandrasekaran M, Venkatesan SP, *et al.* Scaling up difficulties and commercial aspects of edible films for food packaging: A review. Trends Food Sci Technol 2020; 100: 210-22.
[http://dx.doi.org/10.1016/j.tifs.2020.04.014]

[3] Otoni CG, Avena-Bustillos RJ, Azeredo HMC, *et al.* Recent Advances on Edible Films Based on Fruits and Vegetables-A Review. Compr Rev Food Sci Food Saf 2017; 16(5): 1151-69.
[http://dx.doi.org/10.1111/1541-4337.12281] [PMID: 33371612]

[4] Jeevahan J, Chandrasekaran M. Nanoedible films for food packaging: a review. J Mater Sci 2019; 54(19): 12290-318.
[http://dx.doi.org/10.1007/s10853-019-03742-y]

[5] Ramu Ganesan A, Shanmugam M, Bhat R. Producing novel edible films from semi refined carrageenan (SRC) and ulvan polysaccharides for potential food applications. Int J Biol Macromol 2018; 112: 1164-70.
[http://dx.doi.org/10.1016/j.ijbiomac.2018.02.089] [PMID: 29454950]

[6] Kadam SU, Tiwari BK, O'Donnell CP. Extraction, structure and biofunctional activities of laminarin from brown algae. Int J Food Sci Technol 2015; 50(1): 24-31.
[http://dx.doi.org/10.1111/ijfs.12692]

[7] Guidara M, Yaich H, Benelhadj S, *et al.* Smart ulvan films responsive to stimuli of plasticizer and extraction condition in physico-chemical, optical, barrier and mechanical properties. Int J Biol Macromol 2020; 150: 714-26.
[http://dx.doi.org/10.1016/j.ijbiomac.2020.02.111] [PMID: 32061702]

[8] Carina D, Sharma S, Jaiswal AK, Jaiswal S. Seaweeds polysaccharides in active food packaging: A review of recent progress. Trends Food Sci Technol 2021; 110: 559-72.
[http://dx.doi.org/10.1016/j.tifs.2021.02.022]

[9] Usov AI. Polysaccharides of the red algae. In: Horton D, Ed. Advances in Carbohydrate Chemistry and Biochemistry. Academic Press 2011; 115-217.

[10] Mohd Fauziee NA, Chang LS, Wan Mustapha WA, Md Nor AR, Lim SJ. Functional polysaccharides of fucoidan, laminaran and alginate from Malaysian brown seaweeds (Sargassum polycystum, Turbinaria ornata and Padina boryana). Int J Biol Macromol 2021; 167: 1135-45. [http://dx.doi.org/10.1016/j.ijbiomac.2020.11.067] [PMID: 33188815]

[11] Fernando IPS, Sanjeewa KKA, Samarakoon KW, *et al.* FTIR characterization and antioxidant activity of water soluble crude polysaccharides of Sri Lankan marine algae. Algae 2017; 32(1): 75-86. [http://dx.doi.org/10.4490/algae.2017.32.12.1]

[12] Guidara M, Yaich H, Richel A, *et al.* Effects of extraction procedures and plasticizer concentration on the optical, thermal, structural and antioxidant properties of novel ulvan films. Int J Biol Macromol 2019; 135: 647-58. [http://dx.doi.org/10.1016/j.ijbiomac.2019.05.196] [PMID: 31150672]

[13] Thanh TTT, Quach TMT, Nguyen TN, Vu Luong D, Bui ML, Tran TTV. Structure and cytotoxic activity of ulvan extracted from green seaweed *Ulva lactuca*. Int J Biol Macromol 2016; 93(Pt A): 695-702. [http://dx.doi.org/10.1016/j.ijbiomac.2016.09.040] [PMID: 27637450]

[14] Chi Y, Zhang M, Wang X, Fu X, Guan H, Wang P. Ulvan lyase assisted structural characterization of ulvan from *Ulva pertusa* and its antiviral activity against vesicular stomatitis virus. Int J Biol Macromol 2020; 157: 75-82. [http://dx.doi.org/10.1016/j.ijbiomac.2020.04.187] [PMID: 32344076]

[15] Gurpilhares DB, Cinelli LP, Simas NK, Pessoa A Jr, Sette LD. Marine prebiotics: Polysaccharides and oligosaccharides obtained by using microbial enzymes. Food Chem 2019; 280: 175-86. [http://dx.doi.org/10.1016/j.foodchem.2018.12.023] [PMID: 30642484]

[16] Farhan A, Hani NM. Characterization of edible packaging films based on semi-refined kappa-carrageenan plasticized with glycerol and sorbitol. Food Hydrocoll 2017; 64: 48-58. [http://dx.doi.org/10.1016/j.foodhyd.2016.10.034]

[17] Farhan A, Hani NM. Active edible films based on semi-refined κ-carrageenan: Antioxidant and color properties and application in chicken breast packaging. Food Packag Shelf Life 2020; 24: 100476. [http://dx.doi.org/10.1016/j.fpsl.2020.100476]

[18] Tavassoli-Kafrani E, Shekarchizadeh H, Masoudpour-Behabadi M. Development of edible films and coatings from alginates and carrageenans. Carbohydr Polym 2016; 137: 360-74. [http://dx.doi.org/10.1016/j.carbpol.2015.10.074] [PMID: 26686140]

[19] Sandhu KS, Sharma L, Kaur M, Kaur R. Physical, structural and thermal properties of composite edible films prepared from pearl millet starch and carrageenan gum: Process optimization using response surface methodology. Int J Biol Macromol 2020; 143: 704-13. [http://dx.doi.org/10.1016/j.ijbiomac.2019.09.111] [PMID: 31783070]

[20] Huang X, Luo X, Liu L, *et al.* Formation mechanism of egg white protein/κ-Carrageenan composite film and its application to oil packaging. Food Hydrocoll 2020; 105: 105780. [http://dx.doi.org/10.1016/j.foodhyd.2020.105780]

[21] Zhou X, Zong X, Zhang M, *et al.* Effect of konjac glucomannan/carrageenan-based edible emulsion coatings with camellia oil on quality and shelf-life of chicken meat. Int J Biol Macromol 2021; 183: 331-9. [http://dx.doi.org/10.1016/j.ijbiomac.2021.04.165] [PMID: 33930444]

[22] Martiny TR, Pacheco BS, Pereira CMP, *et al.* A novel biodegradable film based on κ□carrageenan activated with olive leaves extract. Food Sci Nutr 2020; 8(7): 3147-56. [http://dx.doi.org/10.1002/fsn3.1554] [PMID: 32724579]

[23] He F, Kong Q, Jin Z, Mou H. Developing a unidirectionally permeable edible film based on κ-carrageenan and gelatin for visually detecting the freshness of grass carp fillets. Carbohydr Polym 2020; 241: 116336.

[http://dx.doi.org/10.1016/j.carbpol.2020.116336] [PMID: 32507192]

[24] Ganesan AR, Shanmugam M, Ilansuriyan P, Anandhakumar R, Balasubramanian B. Composite film for edible oil packaging from carrageenan derivative and konjac glucomannan: Application and quality evaluation. Polym Test 2019; 78: 105936. [http://dx.doi.org/10.1016/j.polymertesting.2019.105936]

[25] Velásquez P, Montenegro G, Valenzuela LM, Giordano A, Cabrera-Barjas G, Martin-Belloso O. k-carrageenan edible films for beef: Honey and bee pollen phenolic compounds improve their antioxidant capacity. Food Hydrocoll 2022; 124: 107250. [http://dx.doi.org/10.1016/j.foodhyd.2021.107250]

[26] Abdillah AA, Charles AL. Characterization of a natural biodegradable edible film obtained from arrowroot starch and iota-carrageenan and application in food packaging. Int J Biol Macromol 2021; 191: 618-26. [http://dx.doi.org/10.1016/j.ijbiomac.2021.09.141] [PMID: 34582908]

[27] Nguyen HT, Boonyaritthongchai P, Buanong M, Supapvanich S, Wongs-Aree C. Chitosan- and κ-carrageenan-based composite coating on dragon fruit (*Hylocereus undatus*) pretreated with plant growth regulators maintains bract chlorophyll and fruit edibility. Sci Hortic (Amsterdam) 2021; 281: 109916. [http://dx.doi.org/10.1016/j.scienta.2021.109916]

[28] Wani SM, Gull A, Ahad T, *et al.* Effect of gum Arabic, xanthan and carrageenan coatings containing antimicrobial agent on postharvest quality of strawberry: Assessing the physicochemical, enzyme activity and bioactive properties. Int J Biol Macromol 2021; 183: 2100-8. [http://dx.doi.org/10.1016/j.ijbiomac.2021.06.008] [PMID: 34102235]

[29] Mahajan K, Kumar S, Bhat ZF, Naqvi Z, Mungure TE, Bekhit AEDA. Functionalization of carrageenan based edible film using *Aloe vera* for improved lipid oxidative and microbial stability of frozen dairy products. Food Biosci 2021; 43: 101336. [http://dx.doi.org/10.1016/j.fbio.2021.101336]

[30] Falcó I, Randazzo W, Sánchez G, López-Rubio A, Fabra MJ. On the use of carrageenan matrices for the development of antiviral edible coatings of interest in berries. Food Hydrocoll 2019; 92: 74-85. [http://dx.doi.org/10.1016/j.foodhyd.2019.01.039]

[31] Martínez-Sanz M, Gómez-Mascaraque LG, Ballester AR, Martínez-Abad A, Brodkorb A, López-Rubio A. Production of unpurified agar-based extracts from red seaweed *Gelidium sesquipedale* by means of simplified extraction protocols. Algal Res 2019; 38: 101420. [http://dx.doi.org/10.1016/j.algal.2019.101420]

[32] Mostafavi FS, Zaeim D. Agar-based edible films for food packaging applications - A review. Int J Biol Macromol 2020; 159: 1165-76. [http://dx.doi.org/10.1016/j.ijbiomac.2020.05.123] [PMID: 32442572]

[33] Wang X, Guo C, Hao W, *et al.* Development and characterization of agar-based edible films reinforced with nano-bacterial cellulose. Int J Biol Macromol 2018; 118(Pt A): 722-30. [http://dx.doi.org/10.1016/j.ijbiomac.2018.06.089] [PMID: 29944938]

[34] Nagar M, Sharanagat VS, Kumar Y, Singh L, Mani S. Influence of xanthan and agar-agar on thermo-functional, morphological, pasting and rheological properties of elephant foot yam (*Amorphophallus paeoniifolius*) starch. Int J Biol Macromol 2019; 136: 831-8. [http://dx.doi.org/10.1016/j.ijbiomac.2019.06.133] [PMID: 31228503]

[35] Wongphan P, Harnkarnsujarit N. Characterization of starch, agar and maltodextrin blends for controlled dissolution of edible films. Int J Biol Macromol 2020; 156: 80-93. [http://dx.doi.org/10.1016/j.ijbiomac.2020.04.056] [PMID: 32294497]

[36] Phan The D, Debeaufort F, Voilley A, Luu D. Biopolymer interactions affect the functional properties of edible films based on agar, cassava starch and arabinoxylan blends. J Food Eng 2009; 90(4): 548-58.

[http://dx.doi.org/10.1016/j.jfoodeng.2008.07.023]

[37] Zhang R, Wang W, Zhang H, Dai Y, Dong H, Hou H. Effects of hydrophobic agents on the physicochemical properties of edible agar/maltodextrin films. Food Hydrocoll 2019; 88: 283-90. [http://dx.doi.org/10.1016/j.foodhyd.2018.10.008]

[38] Wardana AA, Widyaningsih TD. Development of edible films from tapioca starch and agar, enriched with red cabbage (*Brassica oleracea*) as a sausage deterioration bio-indicator. IOP Conf Ser Earth Environ Sci 2017; 109: 012031. [http://dx.doi.org/10.1088/1755-1315/109/1/012031]

[39] Guitián MV, Ibarguren C, Soria MC, Hovanyecz P, Banchio C, Audisio MC. Anti-*Listeria monocytogenes* effect of bacteriocin-incorporated agar edible coatings applied on cheese. Int Dairy J 2019; 97: 92-8. [http://dx.doi.org/10.1016/j.idairyj.2019.05.016]

[40] Usman A, Khalid S, Usman A, Hussain Z, Wang Y. Algal Polysaccharides, Novel Application, and Outlook. In: Zia KM, Zuber M, Ali M, Eds. Algae Based Polymers, Blends, and Composites. Elsevier 2017; 115-53. [http://dx.doi.org/10.1016/B978-0-12-812360-7.00005-7]

[41] Nair MS, Saxena A, Kaur C. Effect of chitosan and alginate based coatings enriched with pomegranate peel extract to extend the postharvest quality of guava (Psidium guajava L.). Food Chem 2018; 240: 245-52. [http://dx.doi.org/10.1016/j.foodchem.2017.07.122] [PMID: 28946269]

[42] Tabassum N, Khan MA. Modified atmosphere packaging of fresh-cut papaya using alginate based edible coating: Quality evaluation and shelf life study. Sci Hortic (Amsterdam) 2020; 259: 108853. [http://dx.doi.org/10.1016/j.scienta.2019.108853]

[43] Duong NTC, Uthairatanakij A, Laohakunjit N, Jitareerat P, Kaisangsri N. An innovative single step of cross-linked alginate-based edible coating for maintaining postharvest quality and reducing chilling injury in rose apple cv. 'Tabtimchan' (Syzygium samarangenese). Sci Hortic (Amsterdam) 2022; 292: 110648. [http://dx.doi.org/10.1016/j.scienta.2021.110648]

[44] Joshi P, Becerra-Mora N, Vargas-Lizarazo AY, Kohli P, Fisher DJ, Choudhary R. Use of edible alginate and limonene-liposome coatings for shelf-life improvement of blackberries. Future Foods 2021; 4: 100091. [http://dx.doi.org/10.1016/j.fufo.2021.100091]

[45] Mahcene Z, Khelil A, Hasni S, *et al.* Development and characterization of sodium alginate based active edible films incorporated with essential oils of some medicinal plants. Int J Biol Macromol 2020; 145: 124-32. [http://dx.doi.org/10.1016/j.ijbiomac.2019.12.093] [PMID: 31843601]

[46] de Souza WFC, de Lucena FA, da Silva KG, Martins LP, de Castro RJS, Sato HH. Influence of edible coatings composed of alginate, galactomannans, cashew gum, and gelatin on the shelf- life of grape cultivar '*Italia*': Physicochemical and bioactive properties. Lebensm Wiss Technol 2021; 152: 112315. [http://dx.doi.org/10.1016/j.lwt.2021.112315]

[47] Liu C, Jin T, Liu W, Hao W, Yan L, Zheng L. Effects of hydroxyethyl cellulose and sodium alginate edible coating containing asparagus waste extract on postharvest quality of strawberry fruit. Lebensm Wiss Technol 2021; 148: 111770. [http://dx.doi.org/10.1016/j.lwt.2021.111770]

[48] Abdel Aziz MS, Salama HE. Developing multifunctional edible coatings based on alginate for active food packaging. Int J Biol Macromol 2021; 190: 837-44. [http://dx.doi.org/10.1016/j.ijbiomac.2021.09.031] [PMID: 34517032]

[49] Arroyo BJ, Bezerra AC, Oliveira LL, Arroyo SJ, Melo EA, Santos AMP. Antimicrobial active edible

coating of alginate and chitosan add ZnO nanoparticles applied in guavas (Psidium guajava L.). Food Chem 2020; 309: 125566.
[http://dx.doi.org/10.1016/j.foodchem.2019.125566] [PMID: 31706674]

[50] Salama HE, Abdel Aziz MS. Optimized alginate and Aloe vera gel edible coating reinforced with nTiO_2 for the shelf-life extension of tomatoes. Int J Biol Macromol 2020; 165(Pt B): 2693-701.
[http://dx.doi.org/10.1016/j.ijbiomac.2020.10.108] [PMID: 33098892]

[51] Shahrampour D, Khomeiri M, Razavi SMA, Kashiri M. Development and characterization of alginate/pectin edible films containing *Lactobacillus plantarum* KMC 45. Lebensm Wiss Technol 2020; 118: 108758.
[http://dx.doi.org/10.1016/j.lwt.2019.108758]

[52] Fan Y, Yang J, Duan A, Li X. Pectin/sodium alginate/xanthan gum edible composite films as the fresh-cut package. Int J Biol Macromol 2021; 181: 1003-9.
[http://dx.doi.org/10.1016/j.ijbiomac.2021.04.111] [PMID: 33892026]

[53] Silva SPM, Ribeiro SC, Teixeira JA, Silva CCG. Application of an alginate-based edible coating with bacteriocin-producing *Lactococcus* strains in fresh cheese preservation. Lebensm Wiss Technol 2022; 153: 112486.
[http://dx.doi.org/10.1016/j.lwt.2021.112486]

[54] Dou L, Li B, Zhang K, Chu X, Hou H. Physical properties and antioxidant activity of gelatin-sodium alginate edible films with tea polyphenols. Int J Biol Macromol 2018; 118(Pt B): 1377-83.
[http://dx.doi.org/10.1016/j.ijbiomac.2018.06.121] [PMID: 29959018]

[55] Biao Y, Yuxuan C, Qi T, *et al.* Enhanced performance and functionality of active edible films by incorporating tea polyphenols into thin calcium alginate hydrogels. Food Hydrocoll 2019; 97: 105197.
[http://dx.doi.org/10.1016/j.foodhyd.2019.105197]

[56] Abdel Aziz MS, Salama HE, Sabaa MW. Biobased alginate/castor oil edible films for active food packaging. Lebensm Wiss Technol 2018; 96: 455-60.
[http://dx.doi.org/10.1016/j.lwt.2018.05.049]

[57] Kazemeini H, Azizian A, Adib H. Inhibition of *Listeria monocytogenes* growth in turkey fillets by alginate edible coating with *Trachyspermum ammi* essential oil nano-emulsion. Int J Food Microbiol 2021; 344: 109104.
[http://dx.doi.org/10.1016/j.ijfoodmicro.2021.109104] [PMID: 33676333]

[58] Salama HE, Abdel Aziz MS, Sabaa MW. Novel biodegradable and antibacterial edible films based on alginate and chitosan biguanidine hydrochloride. Int J Biol Macromol 2018; 116: 443-50.
[http://dx.doi.org/10.1016/j.ijbiomac.2018.04.183] [PMID: 29727654]

[59] Yerramathi BB, Kola M, Annem Muniraj B, Aluru R, Thirumanyam M, Zyryanov GV. Structural studies and bioactivity of sodium alginate edible films fabricated through ferulic acid crosslinking mechanism. J Food Eng 2021; 301: 110566.
[http://dx.doi.org/10.1016/j.jfoodeng.2021.110566]

[60] Park J, Nam J, Yun H, Jin HJ, Kwak HW. Aquatic polymer-based edible films of fish gelatin crosslinked with alginate dialdehyde having enhanced physicochemical properties. Carbohydr Polym 2021; 254: 117317.
[http://dx.doi.org/10.1016/j.carbpol.2020.117317] [PMID: 33357880]

[61] Guan T, Li N, Zhang G, Xue P. Characterization and evaluation of sodium alginate-based edible films by incorporation of star anise ethanol extract/hydroxypropyl-β-cyclodextrin inclusion complex. Food Packag Shelf Life 2022; 31: 100785.
[http://dx.doi.org/10.1016/j.fpsl.2021.100785]

[62] Fabra MJ, Falcó I, Randazzo W, Sánchez G, López-Rubio A. Antiviral and antioxidant properties of active alginate edible films containing phenolic extracts. Food Hydrocoll 2018; 81: 96-103.
[http://dx.doi.org/10.1016/j.foodhyd.2018.02.026]

[63] Ruan C, Zhang Y, Wang J, *et al.* Preparation and antioxidant activity of sodium alginate and carboxymethyl cellulose edible films with epigallocatechin gallate. Int J Biol Macromol 2019; 134: 1038-44.
[http://dx.doi.org/10.1016/j.ijbiomac.2019.05.143] [PMID: 31128181]

[64] Fernando IPS, Kim D, Nah JW, Jeon YJ. Advances in functionalizing fucoidans and alginates (bio)polymers by structural modifications: A review. Chem Eng J 2019; 355: 33-48.
[http://dx.doi.org/10.1016/j.cej.2018.08.115]

[65] Xu B, Wu S. Preservation of mango fruit quality using fucoidan coatings. Lebensm Wiss Technol 2021; 143: 111150.
[http://dx.doi.org/10.1016/j.lwt.2021.111150]

[66] Gomaa M, Fawzy MA, Hifney AF, Abdel-Gawad KM. Use of the brown seaweed *Sargassum latifolium* in the design of alginate-fucoidan based films with natural antioxidant properties and kinetic modeling of moisture sorption and polyphenolic release. Food Hydrocoll 2018; 82: 64-72.
[http://dx.doi.org/10.1016/j.foodhyd.2018.03.053]

[67] Gomaa M, Hifney AF, Fawzy MA, Abdel-Gawad KM. Use of seaweed and filamentous fungus derived polysaccharides in the development of alginate-chitosan edible films containing fucoidan: Study of moisture sorption, polyphenol release and antioxidant properties. Food Hydrocoll 2018; 82: 239-47.
[http://dx.doi.org/10.1016/j.foodhyd.2018.03.056]

[68] Kidgell JT, Magnusson M, de Nys R, Glasson CRK. Ulvan: A systematic review of extraction, composition and function. Algal Res 2019; 39: 101422.
[http://dx.doi.org/10.1016/j.algal.2019.101422]

[69] Yaich H, Amira AB, Abbes F, *et al.* Effect of extraction procedures on structural, thermal and antioxidant properties of ulvan from *Ulva lactuca* collected in Monastir coast. Int J Biol Macromol 2017; 105(Pt 2): 1430-9.
[http://dx.doi.org/10.1016/j.ijbiomac.2017.07.141] [PMID: 28754621]

[70] Amin HH. Safe Ulvan Silver Nanoparticles Composite Films for Active Food Packaging. Am J Biochem Biotechnol 2021; 17(1): 28-39.
[http://dx.doi.org/10.3844/ajbbsp.2021.28.39]

[71] Don TM, Liu LM, Chen M, Huang YC. Crosslinked complex films based on chitosan and ulvan with antioxidant and whitening activities. Algal Res 2021; 58: 102423.
[http://dx.doi.org/10.1016/j.algal.2021.102423]

CHAPTER 20

Toxicity of Algal Foods With Respect to Human Health

Soraya Paz[1,*], **Carmen Rubio**[1], **Ángel J. Gutiérrez**[1], **Dailos González-Weller**[1,2] and **Arturo Hardisson**[1]

[1] *Area of Toxicology, School of Medicine, Universidad de La Laguna, 38071 Santa Cruz de Tenerife, Spain*

[2] *Health Inspection and Laboratory Service, Canary Health Service, 38006 Santa Cruz de Tenerife, Tenerife, Canary Islands, Spain*

Abstract: Since ancient times, algae have been a staple in the diet of Asian countries. Currently, due to migratory currents, globalization and new eating habits, algae are increasingly common in the markets and shops of Western countries. Edible algae can accumulate elements that, in high quantities, pose a toxic risk to consumers. Various authors have recorded high concentrations of toxic metals and trace elements in edible algae. However, in all cases, it has been shown that some species of algae, such as the *Hizikia fusiform* alga, tend to accumulate high concentrations of inorganic arsenic, making it necessary to modulate its consumption. Likewise, algae of some geographical origins are more exposed to contamination; therefore, it is recommended to avoid their consumption. Likewise, in cases of people with a susceptibility to suffering from thyroid problems, it is recommended to consume smaller amounts of algae, avoiding some species, such as some species of red algae, which can accumulate higher concentrations of iodine. In conclusion, it is recommended that the consumer varies between species, avoiding consuming a single species of algae, as well as avoiding those algae that come from places with a higher degree of contamination, as it has been shown that they are accumulators of toxic elements. Likewise, it is necessary to monitor the levels of certain toxic elements in edible algae in order to ensure the quality of these foods.

Keywords: Algal Foods, Arsenic, Benchmark Dose Level (BMDL), Cadmium, Chlorophyta, Dietary Intake, Edible Seaweeds, Environmental Pollution, Human Exposure, Iodine, Lead, Mercury, Phaeophyta, Rhodophyta, Tolerable Daily Intake (TDI), Tolerable Weekly Intake (TWI), Toxic Elements, Toxic Metals, Toxicity, Trace Elements.

* **Corresponding author Soraya Paz:** Area of Toxicology, School of Medicine, Universidad de La Laguna, Santa Cruz de Tenerife, Spain; Tel: +34 922318903; Fax: +34 922319307; E-mail: spazmont@ull.edu.es

Avinash Mishra (Ed.)

INTRODUCTION

The algae are marketed as food for human use. Since ancient times, algae have been a staple in the diet of Asian countries, such as China, Japan or Thailand. However, edible algae are also relevant in European countries, such as Galicia (Spain) or the Republic of Ireland, where algae play an important role in the Irish diet and its customs. Currently, due to migratory currents, globalization and new eating habits, algae are increasingly common in the markets and shops of Western countries [1].

Diet is the main route of exposure to pollutants for humans [2]. Toxic metals reach the algae mainly through anthropogenic sources such as the use of fertilizers, industrial activity, the use of pesticides or biocides, or wastewater, among others, but these pollutants also reach the algae through natural phenomena such as erosion or volcanic activity. These toxic metals are generally cumulative in nature and have no function in the body, being harmful to health. Likewise, algae can accumulate other elements that, in high quantities, can pose a toxic risk to consumers.

In addition, it should be noted that algae have a high capacity to absorb metals and potentially toxic elements, and are even used as a bioindicator of marine pollution [3], being necessary to establish the main derived risks of its consumption because, in some cases, the persistence of toxic metals in edible algae can counteract its nutritional value. That is why numerous research groups have determined and evaluated the content of toxic metals and toxic elements in algae, with special emphasis on determining the dietary intake of toxic elements and toxic metals from algae, studying the existence of significant differences between its content in different species of algae, as well as according to its geographical origin to detect those areas of the world with higher levels of contamination and avoid the consumption of algae from those places.

This book chapter will collect the main contaminants in the most common edible algae, their toxic effects on human health and possible recommendations to avoid or reduce dietary exposure to these toxic elements from the consumption of algae.

THE BRIEF HISTORY OF THE ALGAE

Algae is one of the staple foods in Asian countries such as China, Thailand, Japan and Korea. Seaweed has been used as food since 2700 BC in China; the first evidence of the use of seaweed as food is found in China and dates back to the 6th century BC. This evidence is collected in a few words written by Sze Tsu, who refers to algae by saying that "algae is an appropriate delicacy for the most honorable of guests". Later, in the 4th century BC, Chi Han wrote a book on edible algae. Some of the most used species in China have been the Laminaria japonica which was imported from Japan in the 5th century.

Regarding the use of algae in other regions of the world, in Europe, for example, algae from the Mediterranean Sea were commonly used in Medicine at the time of the Greeks and Romans [4]. Some species of algae, especially red algae, were used as drying agents and for the treatment of parasites. The reason why algae could be used for medicinal purposes is due to their composition. From 100 BC, the Greeks also used algae as food for animals.

In the Republic of Ireland, algae were collected and used by farmers as fertilizer for cultivated soils; this use has continued to this day being very common to see how Irish farmers collect algae from the coasts. It is precisely in Ireland where one of the most consumed seaweed is found, the "dulse seaweed" (Palmaria) or Creathnach (Iirish). This species of algae belonging to the red algae is one of the oldest varieties used as food since 961 BC.

In regions located in the Atlantic Ocean, such as the Canary Islands, algae are very abundant, being formerly used as food for farm animals. Even today, some species such as *Cystoseira abies-marina* are used to give a fresh smell to fish exposed in the fish markets of the islands [5], or as organic fertilizer [6].

Algae Classification

Algae are autotrophic organisms that are capable of oxygenic photosynthesis. Although it refers to algae as "marine plants", this concept is completely wrong since the main difference between algae and plants is the absence of roots in the algae, which have endings (in English, "holdfast") that allow them to attach to rocks. In addition, these endings are not used for the absorption of nutrients, unlike with plants, since they are only used for their fixation on a substrate.

Algae belong to the Protista kingdom and can be classified into four large groups. Table **1** shows the classification of the algae and the main edible algae belonging to each group.

Table 1. Algae classification.

Group	Common Name	Most Known Edible Specie
Cyanobacteria	Blue algae	*Spirulina spp.* (consumed for its high content of β-carotene)
Rhodophyta	Red algae	*Chondrus crispus* (irish moss), *Palmaria palmata* (dulse)
Phaeophyceae	Brown algae	*Sargassum spp., Fucus spp., Laminaria spp., Undaria pinnatifida*
Clorophyta	Green algae	*Ulva lactuca* (Sea lettuce)

Blue algae are best known for their toxicity and damage to ecosystems. Cyanobacterial proliferations or "blooms" are very common in highly eutrophic waters (due to their higher content of phosphates, nitrates, *etc.*) [7]. Cyanobacteria are organisms capable of producing toxins after metabolism, which can be cytotoxic, hepatotoxic, or neurotoxic [8, 9]. The proliferation of cyanobacterial "blooms" is favored by wastewater discharges [10].

Red algae are characterized by their particular red color due to the pigment phycoerythrin. It is a group of algae that comprises a large number of edible species currently used for human consumption. Brown or brown algae comprise one of the groups with the greatest variety of edible algae, receiving their name from the characteristic brown color due to the pigment fucoxanthin. Green algae, get their name from the compound of chlorophyte, a pigment that gives it a characteristic green color. It is necessary to consider the different groups of algae because, due to their particular components, they will absorb and accumulate toxic elements and metals to different degrees.

The life and reproductive cycle of algae are of great importance and must be considered. Algae, diploid organisms (2n), share the reproductive cycle with plants, being able to find them in two completely different phases, in its sporophyte phase, in which the algae is usually harvested for food purposes, and on the other hand, in the gametophytic phase.

The reproductive cycle of algae begins with the algae in the form of a sporophyte, which undergoes a first cell division, meiosis, in which two haploid daughter cells are obtained from a mother cell. Next, each of the spores (haploid cells) will undergo a second division, mitosis, through which two gametes will be obtained, the female and the male. These gametes will be fertilized, giving rise to the zygote (diploid cell).

It is noteworthy that the reproductive cycle of algae, which combines meiosis and mitosis, gives rise to a great genetic variability between individuals.

ALGAL FOOD AND TOXIC ELEMENTS

Despite the fact that algae are a food rich in vitamins, fiber and essential elements such as phosphorus, potassium, iron, iodine and sodium [11], it is precisely its high content of these nutrients which can cause an excessive intake of them.

Studies have found relationships between excessive consumption of algae and hyperthyroidism, due to the high content of iodine [12]. On the other hand, studies carried out by Michikawa *et al.* [13] in postmenopausal Japanese women demonstrated a direct relationship between the abusive consumption of edible algae and thyroid cancer.

Another drawback of the abusive consumption of algae is interactions with medications, especially with anticoagulants. This is due to the high content of vitamin K, which is involved in blood clotting. In addition, a high concentration of sodium can cause hypertension.

Precisely its high capacity for absorption and accumulation of elements and metals means that algae can present high concentrations of elements and toxic metals such as arsenic [14, 15]. However, practically no country has set a maximum limit for toxic elements and metals in algae. However, some manufacturers and producers of algae indicate that the packaging did not exceed the consumption of 5 grams of dehydrated algae per day.

Iodine (I)

Iodine (I), whose molecular weight is 126.9 g/mol and its atomic number is 53, is a non-metal belonging to the group of halogens. This element is widely distributed in nature, especially abundant in the marine environment, where concentrations around 50-60 µg/L of seawater can be found [16]. This element can be found in three different ionic forms due to its oxidation states. The most common ion is iodide (I^-), whose oxidation state is 1-; it is generally found by forming salts such as sodium iodide (NaI) or potassium iodide (KI). However, it is also common to find iodine in the form of iodates (IO_3^-, IO_4^-), whose oxidation states are 5+ and 7+, respectively.

Iodine is present in the environment from the erosion of soils, especially rocks and volcanic soils [17]. The iodine released from marine sediments can accumulate in marine organisms, such as algae. Emissions of iodine from anthropogenic activities are rare [18].

The iodine cycle in nature is found naturally in the marine environment, evaporating in the form of organic molecular iodine compounds (I_2), later deposited in the terrestrial environment with the rain.

In the case of iodine, it is necessary to consider that it is an essential element for humans since it is part of the thyroid hormones (thyroxine and triiodothyronine) (Fig. **1**) [19-23]. Once ingested, iodine is reduced to iodide in the intestine. Iodide is almost entirely absorbed by the small intestine. Iodide absorption is around 90% of the total ingested [17]. Once absorbed, it passes into the blood and is distributed through the tissues, mainly accumulating in the thyroid glands. In healthy adults with normal levels of iodine in the body, most of the ingested iodine is excreted in the urine.

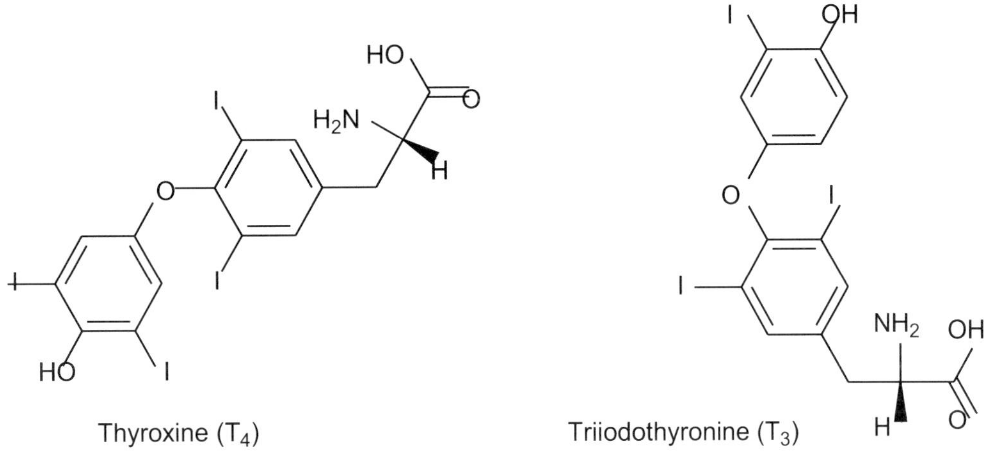

Fig. (1). Chemical structure of thyroid hormones.

Excessive intake of iodine can produce inhibition of the synthesis of thyroid hormones due to the fact that iodides (I^-), in large concentrations, act directly on the thyroid giving rise to the so-called *Wolff-Chaikoff effect* [24]. Several studies affirm that excessive intake of iodine can lead to excessive secretion of thyroid hormones, causing hyperthyroidism, goiter, hyperactivity, tachycardia, thyroid cancer, *etc* [25-29].

Such are the toxic and harmful effects derived from a high intake of iodine that organizations such as the American organization Institute of Medicine, Food and Nutrition Board have established an acceptable daily intake (ADI) value of 1100 µg/day for adults [30].

Seaweed is the main source of iodine with such high contributions that limits on seaweed intake have been suggested [31-33]. Some seaweed, such as hijiki, recommends not exceeding 5 grams of dehydrated seaweed per day. In addition, cases of hyperthyroidism associated with high consumption of algae have been detected [34].

The study carried out by González *et al.* [35], in which the iodine content in edible algae marketed in Europe was determined, registered high levels of European kombu algae being 27.7 ± 5.4 mg/kg dry weight.

Another more recent article, published by Darias-Rosales *et al.* [36], in which he studied the iodine content in red algae also commercialized in Europe, recorded levels of 3.86 ± 1.49 mg/kg dry weight in Irish moss from Europe. In this case, it can be seen that among the brown algae and the red algae from the same geographical area, the iodine levels are quite different, which is why it is advisable to vary the consumption between the different types of algae, avoiding monotonous consumption of the same species.

Aluminium (Al)

Aluminum (Al) is a metal from the p block of the periodic table, with atomic number 13 and a molecular weight of 26.9 g/mol. It is the third most abundant element in the earth's crust. However, Al concentrations have increased due to anthropogenic activities such as bauxite mining or aluminum industries, and due to natural erosion and acidification activities of soils [37, 38].

Al is a neurotoxic metal that lacks functionality in the human body. The toxicity of Al depends on the solubility of the aluminum compound and the route of exposure. This metal tends to accumulate in the brain, bones, liver, and kidneys. Prolonged exposure to high levels of Al has been linked to neurodegenerative diseases such as Alzheimer's [39, 40]. Furthermore, Al can interfere with other essential metals such as Ca [41, 42]. Due to the toxic effects of Al, the EFSA (European Food Safety Authority) has established a tolerable weekly intake (TWI) of 1 mg Al/kg body weight/week [43].

The absorption of Al by the gastrointestinal route is low, being 0.3% of Al present in drinking water and 0.1% of Al present in food and other beverages, approximately. However, an increase in acidity favors the absorption of aluminum [43]. Once absorbed, Al binds to transferrin molecules and circulates through the blood-brain barrier. The kidneys are responsible for the elimination of Al, which is expelled with urine and feces in amounts between 15-55 μg/day [37].

Algae can contain high concentrations of aluminum, especially those that grow in waters close to industrialized areas [44]. In a study carried out by Paz *et al.* (2019) [45], the Al levels recorded in wakame algae (*Undaria pinnatifida*) from Asia were 20 mg/kg, while in wakame algae from Europe, the average Al level was 31.5 mg/kg, being higher the level of Al recorded in European wakame algae compared to Asian wakame algae, despite the higher degree of industrialization in Asian countries. Likewise, it should be noted how the average content of Al has increased in a few years; in a study carried out by Rubio *et al.* [2], in which they analyzed European wakame algae, the mean level of Al was 11.7 mg/kg, much lower than that registered in later years by Paz *et al.* (2019) [45].

Cadmium (Cd)

Cadmium (Cd) is a toxic transition metal with atomic number 48 and a molecular weight of 112.4 g/mol. This metal can be found as part of minerals such as blends or sphalerites, formed by zinc sulfide, in which traces of Cd can be found which, due to their great affinity with Zn, replace it. Its persistence in the environment and its tendency to accumulate, make Cd one of the pollutants with the greatest impact. Cd is used in alloys, cells and batteries, so anthropogenic activities are the main source of Cd in the environment, which can reach drinking water and food.

Even at low concentrations, Cd has toxic effects due to its high half-life and bioaccumulation [46]. Cd^{2+} competes in the body with other divalent cations, such as Zn^{2+}, Fe^{2+} or Cu^{2+}, essential for the development of multiple physiological functions [47]. It is precisely the divalent nature of Cd that gives it the ability to form a large number of stable coordination complexes with biomolecules, altering their functioning.

This metal mainly affects the renal system, causing irreversible damage to the renal tubules, which are involved in nutrient reabsorption mechanisms [48-50]. In addition, it interferes with Ca homeostasis and affects the cardiovascular system. Cd is an inducer of metallothionein formation. The complex formed between Cd and metallothionein [Cd-MT] is stable and inert, present in the liver where it is metabolized, giving rise to low molecular weight metabolites. The [Cd-MT] complex crosses the glomerular membrane, where the lysosomes destroy the complex leaving free Cd (II) that will again induce metallothionein formation [51]. Given the serious consequences of a high intake of Cd, the EFSA (European Food Safety Authority) has set a tolerable weekly intake of 2.5 µg/kg body weight/week [52].

In non-smokers, the main route of absorption of Cd is through the gastrointestinal tract. Once absorbed, this metal is transported through the blood and attached to various molecules; Among one of them, metallothionein, plays an important role in Cd transport in humans [53, 54]. A small fraction, between 0.01 - 0.02% of the absorbed Cd is excreted with urine and feces, and the rest accumulates in the body.

Algae can absorb high Cd concentration due to its high absorption capacity. This uptake depends on the type of algae and the quality of the environment in which it is found [55, 56]. Khan *et al.* [57] recorded Cd levels in Asian wakame algae of 0.072 mg/kg, while Rubio *et al.* [2] recorded a mean Cd level in European wakame algae of 0.06 mg/kg. On the other hand, Paz *et al.* [45] recorded a mean Cd level in Asian wakame algae of 1.11 mg/kg, which is higher than that recorded in previous studies [2, 57]. Paz *et al.* [45] evaluated the dietary intake of Cd from

Asian wakame algae, being 22.7% of the TWI established for Cd (2.5 μg/kg of body weight/week [52]), being an alarming percentage of contribution since, the global intake of Cd could exceed 100% of the TWI with the consequent toxic risk. Squadrone *et al.* (2018) [52] determined the content of trace elements in algae from the Mediterranean Sea, registering the highest concentrations in green algae (*Codium bursa*) being 0.32 mg/kg. The Mediterranean is a closed sea in which there may be a greater accumulation of trace elements such as Cd, however, the level recorded is lower than that found for the Asian wakame algae analyzed by Paz *et al.* [45].

It is, therefore, necessary to avoid prolonged consumption of some algae species and, especially, to avoid some geographical origins since, as has been found in the studies indicated above, the levels of Cd in Asian algae are usually higher than the levels of Cd in algae of others places of the world. In addition, it has been found that the photosynthesis process can increase the bioaccumulation and bioabsorption of Cd [58], being another factor to consider in the harvesting stage of algae destined for human consumption, perhaps it would be convenient to take those algae that are deeper and therefore do not have as much light available for photosynthesis.

Lead (Pb)

Lead (Pb) is a heavy metal with atomic number 82 and a molecular weight of 207.2 g/mol. It is one of the best-known pollutants. Although its harmful effects have been known since the time of the Romans, who referred to the symptoms produced by Pb intoxication as "lead poisoning", this has not been prevented until relatively a few years ago, it has been one of the metals that are most commonly used. It has been used, forming an important part in the military industry, in the manufacture of pigments, or as a fuel additive [59].

This neurotoxic metal tends to accumulate in the body, causing serious damage to the central nervous system (CNS), especially in developing children and fetuses [60-63]. It can cause nephropathies, gastrointestinal tract disorders and Alzheimer's [53]. Pb, which can be found as Pb^{2+}, is capable of replacing Zn, inhibiting enzymatic activity.

Gastrointestinal absorption studies carried out with radioactive markers have shown absorption of between 37-70% of the total Pb ingested. Other studies show less absorption (15-20%). Absorption varies from one individual to another and depends on various factors, for example, Fe deficiencies increase Pb absorption, while the presence of compounds such as phytates produces a decrease [53, 64]. Absorbed Pb is distributed throughout the body in both soft and hard tissues; accumulation in bone tissue is high given the ability of Pb to interfere with calcium homeostasis. Part of it crosses the blood-brain barrier and becomes concentrated in the gray matter of the brain [64]. Finally, the elimination of this toxic metal is through urine and feces.

The EFSA (European Food Safety Authority) has established a benchmark dose level (BMDL) for Pb of 0.63 μg Pb/kg bw/day for nephrotoxic effects and 1.50 μg Pb/kg bw/day for cardiovascular effects [65]. Pb is deposited in the environment from lead particles present in the atmosphere [53]. In algae, Pb can accumulate, reaching levels that are dangerous for humans. Lead reaches the environment from various sources; once in the marine environment, it is found as Pb^{2+} and forms organic complexes that are absorbed by algae and accumulated in them.

Hwang *et al.* [66] recorded a mean Pb level in Korean edible algae of 0.70 mg/kg. Khan *et al.* [57] recorded a mean Pb level of 0.05 mg/kg in Asian wakame algae. Rubio *et al.* [2] recorded a mean Pb level of 0.07 mg/kg in European wakame algae. Paz *et al.* [45] reported a mean Pb level of 0.23 mg/kg in sea spaghetti (*Himanthalia elongata*), 0.30 mg/kg in wakame and 0.31 mg/kg in European kombu (*Laminaria ochroleuca*) and 0.38 mg/kg in Asian wakame. Again, Pb concentrations in edible algae vary considerably between geographic regions and species of algae.

Mercury (Hg)

Mercury (Hg), with atomic number 80 and molecular weight of 200.6 g/mol, is a chemical element belonging to block *d* of the periodic table. Its most remarkable characteristic is that it is the only metallic element that is in a liquid state at room temperature [53].

This metal is one of the best-known marine toxins. Widely distributed in the earth's crust, it has been mined in cinnabar (mercury sulfide) mines for its value as a material for fluorescent lamps, thermometers, barometers, pigments, *etc.*

However, this metal has high toxicity and a marked tendency to accumulate and biomagnifies along the food chain. Among the different mercury compounds that can be found, methylmercury (MeHg) is the most toxic and dangerous of all, since this organic compound has a great affinity for fats and is also capable of crossing the placental barrier [67-72].

The accumulation of mercury in algae is due to the deposition of metallic mercury in the sediments, which, after the action of various bacteria, is methylated, becoming organic mercury and accumulating in marine organisms such as algae (Fig. **2**).

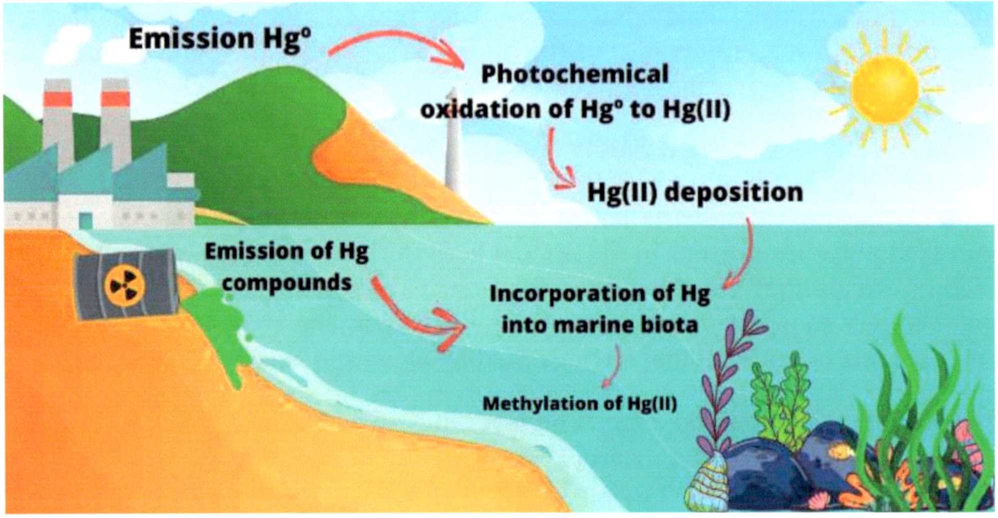

Fig. (2). Mercury cycle in the marine environment and absorption by algae.

The toxic effects of this element depend on the chemical form in which it is found and the route of entry into the body. Through the digestive tract, the mercury compound that has the greatest repercussions is methylmercury. This is a neurotoxic agent that accumulates in nerve tissues [73, 74]. Being an organic compound, this compound can cross cell membranes attacking nucleophilic groups and enzymes of the central nervous system (CNS) [75].

Gastrointestinal absorption of methylmercury is high, around 80%. While, the absorption of inorganic mercury is lower, reaching from 2 to 38% [76]. Once absorbed, MeHg crosses cell membranes and accumulates in tissues. The highest concentrations of this compound have been found in the kidney. The elimination of mercury depends on the chemical form in which it is found, therefore, inorganic mercury is eliminated with urine and feces, while methylmercury remains in the body for about 70 - 80 days and, after this time, it is eliminated as inorganic mercury with faeces [76].

Prolonged exposuretomethylmercury causes severe brain damage, deafness, visionloss,weakness, motor problems, and psychologicaldisorders. In the case of children or pregnant women, it is necessary to take into account that methylmercury can cross the placental barrier causing irreversible damage to the fetus [69, 76-78].

Given the toxic effects and the tendency of mercury to accumulate, the EFSA (European Food Safety Authority) has set tolerable weekly intake (TWI) values of

4 μg/kg body weight/week for inorganic mercury and 1.3 μg/kg body weight/week for methylmercury [76].

Besada *et al.* [79] determined the mercury content in different edible species of algae marketed in Spain, registering remarkable concentrations of 0.024-0.035 mg/kg dry weight in seaweed salad (wakame, ogonori, kombu, agar and akamodoki), 0.010-0.057 mg/kg dw in wakame algae (*U. pinnatifida*) and 0.023-0.047 mg/kg dw in arame algae (*Eisenia bicyclis*). In addition, the statistical study carried out demonstrated the existence of significant differences between the different origins of the algae analyzed [79], which reinforces what was stated in this chapter about the need to alternate different origins and avoid the consumption of algae from geographic locations more polluted.

In another more recent study carried out by Paz *et al.* [45], mean Hg concentrations of 0.012 mg/kg dw were recorded in European wakame algae, 0.015 mg/kg dw in European sea spaghetti (*Himanthalia elongata*) and 0.024 mg/kg dw in European kombu (*Laminaria ochroleuca*) algae.

Arsenic (As)

Arsenic (As) with atomic number 33 and molecular weight of 74.9 g / mol is a chemical element belonging to the group of metalloids or semi-metals. During the 19th century, it was widely used in the formulation of rodenticides, pigments, wallpapers, pesticides and even medicines. However, its use for homicidal purposes made him the king of poisons. The tasteless arsenolite (diarsenic trioxide, As_2O_3) came to be known as "succession powder." The Borgia family and the Marchioness of Brinvilliers are examples of the many historical Figs who used this compound with criminal intentions because its symptoms were confused with those of other gastrointestinal pathologies.

As is classified by the International Agency for Research on Cancer [80] within Group I as the evidence relates it to cancer in humans. The toxicity of As depends on the oxidation state in which it is found and the chemical form, with inorganic forms (iAs) being the most toxic [81-83]. Speciation is essential in risk assessment for As, since considering total arsenic (tAs), without spicing inorganic arsenic (iAs) would lead to an overestimation of risk [84, 85]. Arsine or arsenic hydride (AsH_3) is the most toxic As a compound, producing a lethal hemolytic picture.

Algae can absorb large amounts of As [85]. The soluble species of inorganic arsenic (iAS) are rapidly absorbed (around 95% of the total ingested), and the absorption of organic compounds does not exceed 70% [84]. As is transported mainly bound to the thiol groups (–SH) of proteins and low molecular weight compounds such as glutathione (GSH) and cysteine. The distribution of As reaches the entire organism and although there is an initial accumulation in organs such as the liver, kidney and lung, As also reaches the hair, nails, teeth, skin, thyroid, lens and skeleton, among others.

Arsenic is metabolized by methylation, generating methylated metabolites, dimethylarsine (DMA) and monomethylarsonic (MMA). These organoarsenical forms are less toxic and more easily eliminated [86], and therefore, the methylation of As is considered a form of detoxification. Interindividual and population differences in methylation capacity, as well as nutritional status, may be important factors in susceptibility to As toxicity. People with diets deficient in protein and other nutrients are more susceptible to cancer caused by arsenic [87]. As and its metabolites are excreted in bile and urine, although it can also be eliminated through skin, hair, nails, and sweat.

As (III) and As (V) differ in the mechanisms of toxic action. As (V) replaces the phosphate anion in various biochemical reactions, including the synthesis of glucose-6-phosphate, uncouples ATP formation, and inhibits phosphorylation (arsenolysis). As (III) reacts with cellular components with thiol groups (–SH), such as GSH and cysteine, inhibiting enzymes such as pyruvate dehydrogenase [84]. Regarding its carcinogenicity, there are several mechanisms proposed, but the chromosomal aberrations, gene amplification, alteration of DNA repair, induction of oxidative stress and alteration of DNA methylation stand out, among others, being trivalent species involved in most of them [88].

The EFSA has established the lower confidence limit of the reference dose ($BMDL_{01}$) for the risk assessment for As between 0.3 and 8 μg/kg body weight/day for lung, skin and bladder cancers, as well as for lesions cutaneous [84].

Almela *et al.* [89] determined total and inorganic arsenic content in edible algae marketed in Spain, recording concentrations of total As 0.031–149 mg/kg dw and inorganic As of <0.014–117 mg/kg dw. Furthermore, they found that the *Hizikia fusiforme* alga samples exceeded the limit concentration of inorganic As established in different countries. Besada *et al.* [79] reported elevated inorganic As concentrations of 32.1–69.5 mg/kg dw in *Hizikia fusiforme* algae, higher than the previous study [89]. Therefore, in view of the articles consulted, the *H. fusiforme* alga would not be suitable for consumption considering the high levels of inorganic As recorded.

GENERAL GUIDELINES FOR ALGAL FOODS CONSUMERS

Considering the above, the prolonged consumption of some species of algae as well as algae from contaminated geographical areas could increase the risk of ingestion of toxic elements. For this reason, (Fig. **3**) includes a series of recommendations for regular seaweed consumers. The case of healthy adults has been considered, and at no time can these recommendations be taken for individuals with certain pathologies, such as hypo or hyperthyroidism, *etc*. In this case, it is recommended to consult a doctor before consuming this type of food.

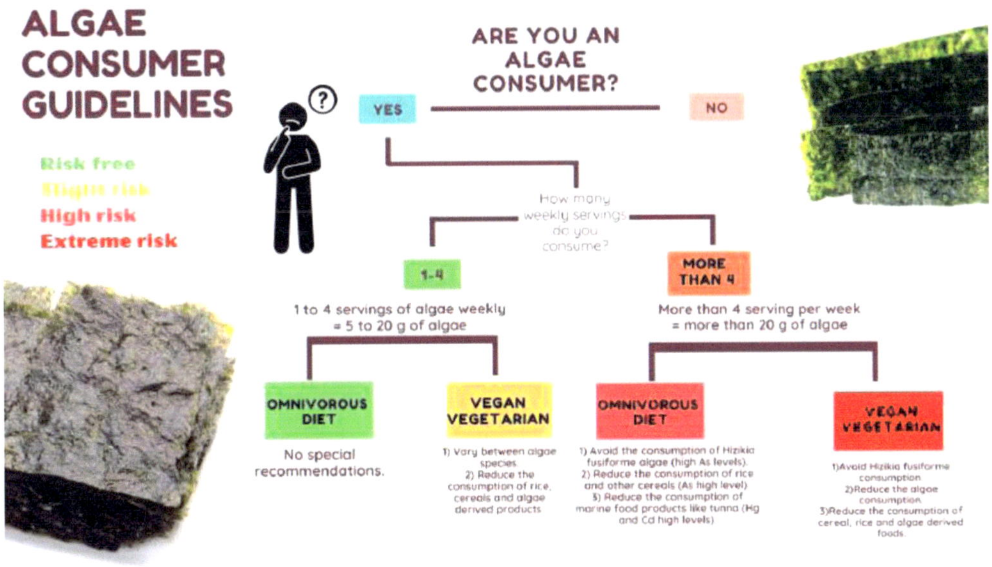

Fig. (3). Algae consumer guideline.

In the case of consumers of 1 to 4 weekly servings of dried seaweed (5 to 20 grams of seaweed a week, considering the manufacturers' recommendations not to consume more than 5 grams of dried seaweed a day), if they are individuals who they follow an omnivorous diet, no special recommendations are considered. In the case of vegans or vegetarians, it is necessary to consider a series of recommendations, such as varying the species of algae consumed and considering the place of origin of these algae to avoid overexposure to contaminants, reduce the consumption of rice, cereals and products derived from algae with the aim of avoiding a high intake of inorganic As, especially.

In the case of regular consumers who ingest more than 4 weekly servings of seaweed, the consumption would be more than 20 grams of dried seaweed per week, therefore, the risks of exposure to toxic trace elements increase.

CONCLUSION

Algae are a new food that is increasingly being consumed in Western countries. However, due to the greater anthropic pressure on the oceans as well as the environmental pollution caused by volcanic eruptions, erosion, among others, has caused these marine organisms to absorb and accumulate high concentrations of trace elements such as iodine (I), arsenic (As), cadmium (Cd), aluminum (Al), lead (Pb) and mercury (Hg), which in high concentrations can cause harmful effects on health.

Some species of algae, such as red algae, have been found to appear to accumulate higher concentrations of iodine. Likewise, Asian algae tend to contain higher levels of Cd. On the other hand, *Hizikia fusiform* algae must be consumed with extreme caution since it is the one that accumulates the highest levels of inorganic arsenic.

Taking into account all the above, it is recommended that regular algae consumers vary between species to avoid overexposure to toxic elements. Likewise, it has been shown that the place of origin of the algae directly influences the content of toxic elements in these, being advisable to avoid algae in certain geographical areas to minimize exposure. In addition, it is necessary to establish programs to monitor the content of toxic elements in algae in order to ensure the quality and safety of this food.

CONSENT FOR PUBLICATION

Not applicable.

CONFLICT OF INTEREST

The author(s) declare no conflict of interest, financial or otherwise.

ACKNOWLEDGEMENT

Declared none.

REFERENCES

[1] Paz S, Rubio C, Frías I, *et al.* Metal Concentrations in Wild-Harvested Phaeophyta Seaweed from the Atlantic Ocean (Canary Islands, Spain). J Food Prot 2018; 81(7): 1165-70. [http://dx.doi.org/10.4315/0362-028X.JFP-18-038] [PMID: 29939794]

[2] Rubio C, Napoleone G, Luis-González G, *et al.* Metals in edible seaweed. Chemosphere 2017; 173: 572-9. [http://dx.doi.org/10.1016/j.chemosphere.2017.01.064] [PMID: 28152409]

[3] Herrera-Paz DL, de la Nuez D, Valero-Rodrígeuz JM. Algae Communities as Bioindicators of Environmental Quality on a Rocky Shore of Mediterranean (S.E. Iberian Peninsula). Rev Cien 2015;

19(1): 25-40.

[4] Pooja S. Algae used as Medicine and Food-A Short Review. J Pharm Sci Res 2014; 6(1): 33-5.

[5] Haroun R, Gil-Rodríguez MC, Wildpret de la Torre W. Plantas Marinas de las Islas Canarias. Madrid: Canseco Editores S.L. 2003.

[6] Díaz Pérez SF, Hardisson de la Torre A, Real Hardisson F. La utilización del alga *C. abies-marina* como abono orgánico. Cultivo y operaciones aplicadas a las algas. Hort Rev Ind Distr Socioec Hort 1988; 41: 25-9.

[7] Andersson M, de Benoist B, Rogers L. Epidemiology of iodine deficiency: Salt iodisation and iodine status. Best Pract Res Clin Endocrinol Metab 2010; 24(1): 1-11. [http://dx.doi.org/10.1016/j.beem.2009.08.005] [PMID: 20172466]

[8] Pilotto LS, Douglas RM, Burch MD, *et al.* Health effects of exposure to cyanobacteria (blue-green algae) during recreational water-related activities. Aust N Z J Public Health 1997; 21(6): 562-6. [http://dx.doi.org/10.1111/j.1467-842X.1997.tb01755.x] [PMID: 9470258]

[9] Zurawell RW, Chen H, Burke JM, Prepas EE. Hepatotoxic cyanobacteria: a review of the biological importance of microcystins in freshwater environments. J Toxicol Environ Health B Crit Rev 2005; 8(1): 1-37. [http://dx.doi.org/10.1080/10937400590889412] [PMID: 15762553]

[10] Rahav E, Bar-Zeev E. Sewage outburst triggers Trichodesmium bloom and enhance N_2 fixation rates. Sci Rep 2017; 7(1): 4367. [http://dx.doi.org/10.1038/s41598-017-04622-8] [PMID: 28663560]

[11] Astorga-España MS, Rodríguez Galdón B, Rodríguez Rodríguez EM, Díaz Romero C. Mineral and trace element concentrations in seaweeds from the sub-Antarctic ecoregion of Magallanes (Chile). J Food Compos Anal 2015; 39: 69-76. [http://dx.doi.org/10.1016/j.jfca.2014.11.010]

[12] Bouga M, Combet E. Emergence of Seaweed and Seaweed-Containing Foods in the UK: Focus on Labeling, Iodine Content, Toxicity and Nutrition. Foods 2015; 4(4): 240-53. [http://dx.doi.org/10.3390/foods4020240] [PMID: 28231201]

[13] Michikawa T, Inoue M, Shimazu T, *et al.* Seaweed consumption and the risk of thyroid cancer in women. Eur J Cancer Prev 2012; 21(3): 254-60. [http://dx.doi.org/10.1097/CEJ.0b013e32834a8042] [PMID: 22414981]

[14] Mehta SK, Gaur JP. Use of Algae for Removing Heavy Metal Ions from Wastewater: Progress and Prospects. Off Rev Biotechn 2005; 113-52.

[15] Atici T, Obali O, Altindag A, Ahiska S, Aydin D. The accumulation of heavy metals (Cd, Pb, Hg, Cr) and their state in phytoplantonic algae and zooplanktonic organisms in Beysehir Lake and Mogan Lake, Turkey. Afr J Biotechnol 2010; 9(4): 475-87.

[16] Hetzel BS, Maberly GF. Iodine. In: Mertz W (Eds), Trace elements in human and animal nutrition. 5th edition. USA: Academic Press Inc 1986. [http://dx.doi.org/10.1016/B978-0-08-092469-4.50006-6]

[17] Scientific Opinion on Dietary Reference Values for iodine. EFSA J 2014; 12(5): 3660.

[18] Leblanc C, Colin C, Cosse A, *et al.* Iodine transfers in the coastal marine environment: the key role of brown algae and of their vanadium-dependent haloperoxidases. Biochimie 2006; 88(11): 1773-85. [http://dx.doi.org/10.1016/j.biochi.2006.09.001] [PMID: 17007992]

[19] Morreale de Escobara G, Escobar del Rey F. El yodo durante la gestación, lactancia y primera infancia. Cantidades mínimas y máximas: de microgramos a gramos. An Pediatr (Barc) 2000; 53(1): 1-5. [http://dx.doi.org/10.1016/S1695-4033(00)77404-2]

[20] Blanco A. Química Biológica. 8th edition. Madrid: Editorial El Ateneo 2008.

[21] Untoro J, Timmer A, Schultink W. The challenges of iodine supplementation: a public health programme perspective. Best Pract Res Clin Endocrinol Metab 2010; 24(1): 89-99. [http://dx.doi.org/10.1016/j.beem.2009.08.011] [PMID: 20172473]

[22] Zimmermann MB. The role of iodine in human growth and development. Semin Cell Dev Biol 2011; 22(6): 645-52. [http://dx.doi.org/10.1016/j.semcdb.2011.07.009] [PMID: 21802524]

[23] Fuge R, Johnson CC. Iodine and human health, the role of environmental geochemistry and diet, a review. Appl Geochem 2015; 63: 282-302. [http://dx.doi.org/10.1016/j.apgeochem.2015.09.013]

[24] Barrett KE, Barman SM, Boitano S, Brooks HL. Ganong, Fisiología Médica. 23rd edition. Mexico: McGraw-Hill Interamericana Editores, S.A 2010.

[25] Pennington JAT. A review of iodine toxicity reports. J Am Diet Assoc 1990; 90(11): 1571-81. [http://dx.doi.org/10.1016/S0002-8223(21)01843-5] [PMID: 2229854]

[26] Martino E, Bartalena L, Bogazzi F, Braverman LE. The effects of amiodarone on the thyroid. Endocr Rev 2001; 22(2): 240-54. [PMID: 11294826]

[27] Bürgi H. Iodine excess. Best Pract Res Clin Endocrinol Metab 2010; 24(1): 107-15. [http://dx.doi.org/10.1016/j.beem.2009.08.010] [PMID: 20172475]

[28] Luo Y, Kawashima A, Ishido Y, *et al.* Iodine excess as an environmental risk factor for autoimmune thyroid disease. Int J Mol Sci 2014; 15(7): 12895-912. [http://dx.doi.org/10.3390/ijms150712895] [PMID: 25050783]

[29] Wang C, Yatsuya H, Li Y, *et al.* Prospective study of seaweed consumption and thyroid cancer incidence in women. Eur J Cancer Prev 2016; 25(3): 239-45. [http://dx.doi.org/10.1097/CEJ.0000000000000168] [PMID: 26011104]

[30] IOM (Institute of Medicine). Food and Nutrition Board of the Institute of Medicine of the National Academies. Dietary Reference Intakes for Vitamin A, Vitamin K, Arsenic, Boron, Chromium, Copper, Iodine, Iron, Manganese, Molybdenum, Nickel, Silicon, Vanadium, and Zinc. Washington: National Academy Press 2001.

[31] Hou X, Chai C, Qian Q, Yan X, Fan X. Determination of chemical species of iodine in some seaweeds (I). Sci Total Environ 1997; 204(3): 215-21. [http://dx.doi.org/10.1016/S0048-9697(97)00182-4]

[32] Chance R, Baker AR, Küpper FC, Hughes C, Kloareg B, Malin G. Release and transformations of inorganic iodine by marine macroalgae. Estuar Coast Shelf Sci 2009; 82(3): 406-14. [http://dx.doi.org/10.1016/j.ecss.2009.02.004]

[33] Stévant P, Marfaing H, Duinker A, *et al.* Biomass soaking treatments to reduce potentially undesirable compounds in the edible seaweeds sugar kelp (Saccharina latissima) and winged kelp (Alaria esculenta) and health risk estimation for human consumption. J Appl Phycol 2018; 30(3): 2047-60. [http://dx.doi.org/10.1007/s10811-017-1343-8]

[34] Salas Coronas J, Cruz Caparrós G, Laynez Bretones F, Díez García F. Hipertiroidismo inducido por consumo de algas marinas. Med Clin (Barc) 2002; 118(20): 797-8. [http://dx.doi.org/10.1016/S0025-7753(02)72534-7]

[35] González A, Paz S, Rubio C, Gutiérrez ÁJ, Hardisson A. Human exposure to iodine from the consumption of edible seaweeds. Biol Trace Elem Res 2019; 2019: 1-6. [PMID: 31820353]

[36] Darias-Rosales J, Rubio C, Gutiérrez ÁJ, Paz S, Hardisson A. Risk assessment of iodine intake from the consumption of red seaweeds (Palmaria palmata and Chondrus crispus). Environ Sci Pollut Res Int 2020; 27(36): 45737-41.

[http://dx.doi.org/10.1007/s11356-020-10478-9] [PMID: 32803579]

[37] Sjögren B, Iregren A, Elinder CG, Yokel RA. Chapter 17: Aluminum. In: Nordberg GF, Fowler BA, Nordberg M, Friberg L (eds) Handbook on the Toxicology of Metals. 3rd Edition. Amsterdam: Academic Press
[http://dx.doi.org/10.1016/B978-012369413-3/50072-0]

[38] Hardisson A, Revert C, González-Weller D, Gutiérrez A, Paz S, Rubio C. Aluminium Exposure Through the Diet. HSOA J Food Sci Nutr 2017; 3: 1-10.

[39] Arvand M, Kermanian M. Potentiometric Determination of Aluminum in Foods, Pharmaceuticals, and Alloys by AlMCM-41-Modified Carbon Paste Electrode. Food Anal Methods 2013; 6(2): 578-86.
[http://dx.doi.org/10.1007/s12161-012-9476-5]

[40] Martinez CS, Alterman CDC, Peçanha FM, *et al.* Aluminum Exposure at Human Dietary Levels for 60 Days Reaches a Threshold Sufficient to Promote Memory Impairment in Rats. Neurotox Res 2017; 31(1): 20-30.
[http://dx.doi.org/10.1007/s12640-016-9656-y] [PMID: 27473855]

[41] Davidson T, Ke Q, Costa M. [41] Davidson T, Ke Q, Costa M. Chapter 5: Selected Molecular Mechanism of Metal Toxicity and Carcinogenicity. En: Nordberg GF, Fowler BA, Nordberg M, Friberg L (Eds.). Handbook on the Toxicology of Metals. 3ra Edición. Amsterdam: Academic Press 2007.
[http://dx.doi.org/10.1016/B978-012369413-3/50060-4]

[42] Krewski D, Yokel RA, Nieboer E, *et al.* Human health risk assessment for aluminium, aluminium oxide, and aluminium hydroxide. J Toxicol Environ Health B Crit Rev 2007; 10(sup1) (Suppl. 1): 1-269.
[http://dx.doi.org/10.1080/10937400701597766] [PMID: 18085482]

[43] Statement on the evaluation on a new study related to thee bioavailability of aluminium in food. EFSA J 2011; 9(5): 2157.

[44] Golding LA, Angel BM, Batley GE, Apte SC, Krassoi R, Doyle CJ. Derivation of a water quality guideline for aluminium in marine waters. Environ Toxicol Chem 2015; 34(1): 141-51.
[http://dx.doi.org/10.1002/*etc*.2771] [PMID: 25318392]

[45] Paz S, Rubio C, Frías I, *et al.* Toxic metals (Al, Cd, Pb and Hg) in the most consumed edible seaweeds in Europe. Chemosphere 2019; 218: 879-84.
[http://dx.doi.org/10.1016/j.chemosphere.2018.11.165] [PMID: 30609492]

[46] Barbier O, Jacquillet G, Tauc M, Cougnon M, Poujeol P. Effect of heavy metals on, and handling by, the kidney. Nephron, Physiol 2005; 99(4): p105-10.
[http://dx.doi.org/10.1159/000083981] [PMID: 15722646]

[47] Goyer RA. Toxic and essential metal interactions. Annu Rev Nutr 1997; 17(1): 37-50.
[http://dx.doi.org/10.1146/annurev.nutr.17.1.37] [PMID: 9240918]

[48] Godt J, Scheidig F, Grosse-Siestrup C, *et al.* The toxicity of cadmium and resulting hazards for human health. J Occup Med Toxicol 2006; 1(1): 22.
[http://dx.doi.org/10.1186/1745-6673-1-22] [PMID: 16961932]

[49] Liu J, Qu W, Kadiiska MB. Role of oxidative stress in cadmium toxicity and carcinogenesis. Toxicol Appl Pharmacol 2009; 238(3): 209-14.
[http://dx.doi.org/10.1016/j.taap.2009.01.029] [PMID: 19236887]

[50] Fowler BA. Monitoring of human populations for early markers of cadmium toxicity: A review. Toxicol Appl Pharmacol 2009; 238(3): 294-300.
[http://dx.doi.org/10.1016/j.taap.2009.05.004] [PMID: 19433102]

[51] Casas JS, Moreno V, Sánchez A, Sánchez JL, Sordo J. Química bioinorgánica. Biblioteca de químicas. 2002.

[52] Panel on Contaminants in the Food Chain (CONTAM). Statement on tolerable weekly intake for cadmium. EFSA J 2011; 9(2): 1975.

[53] Nordberg GF, Fowler BA, Nordberg M, Friberg L. Handbook on the Toxicology of Metals. 3rd edition. Amsterdam: Academic Press 2007.

[54] Klaassen CD, Liu J, Diwan BA. Metallothionein protection of cadmium toxicity. Toxicol Appl Pharmacol 2009; 238(3): 215-20.
[http://dx.doi.org/10.1016/j.taap.2009.03.026] [PMID: 19362100]

[55] Shams El-Din NG, Mohamedein LI, El-Moselhy KM. Seaweeds as bioindicators of heavy metals off a hot spot area on the Egyptian Mediterranean Coast during 2008–2010. Environ Monit Assess 2014; 186(9): 5865-81.
[http://dx.doi.org/10.1007/s10661-014-3825-3] [PMID: 24844431]

[56] Singh NK, Raghubanshi AS, Upadhyay AK, Rai UN. Arsenic and other heavy metal accumulation in plants and algae growing naturally in contaminated area of West Bengal, India. Ecotoxicol Environ Saf 2016; 130: 224-33.
[http://dx.doi.org/10.1016/j.ecoenv.2016.04.024] [PMID: 27131746]

[57] Khan N, Ryu KY, Choi JY, *et al.* Determination of toxic heavy metals and speciation of arsenic in seaweeds from South Korea. Food Chem 2015; 169: 464-70.
[http://dx.doi.org/10.1016/j.foodchem.2014.08.020] [PMID: 25236252]

[58] Han TW, Tseng CC, Cai M, Chen K, Cheng SY, Wang J. Effects of cadmium on bioaccumulation, bioabsorption, and photosynthesis in Sarcodia suiae. Int J Environ Res Public Health 2020; 17(4): 1294.
[http://dx.doi.org/10.3390/ijerph17041294] [PMID: 32085376]

[59] Singh RP, Tripathi RD, Sinha SK, Maheshwari R, Srivastava HS. Response of higher plants to lead contaminated environment. Chemosphere 1997; 34(11): 2467-93.
[http://dx.doi.org/10.1016/S0045-6535(97)00087-8] [PMID: 9192470]

[60] Mahaffey KR. Relation between quantities of lead ingested and health effects of lead in humans. Pediatrics 1977; 59(3): 448-56.
[http://dx.doi.org/10.1542/peds.59.3.448] [PMID: 840565]

[61] Rabinowitz MB, Kopple JD, Wetherill GW. Effect of food Intake and fasting on gastrointestinal lead absorption in humans. Am J Clin Nutr 1980; 33(8): 1784-8.
[http://dx.doi.org/10.1093/ajcn/33.8.1784] [PMID: 7405881]

[62] Carpenter DO. Effects of metals on the nervous system of humans and animals. Int J Occup Med Environ Health 2001; 14(3): 209-18.
[PMID: 11764847]

[63] Rubio C, Gutiérrez AJ, Martín-Izquierdo RE, Revert C, Lozano G, Hardisson A. El plomo como contaminante alimentario. Rev Toxicol 2004; 21: 72-80.

[64] Charlet L, Chapron Y, Faller P, Kirsch R, Stone AT, Baveye PC. Neurodegenerative diseases and exposure to the environmental metals Mn, Pb, and Hg. Coord Chem Rev 2012; 256(19-20): 2147-63.
[http://dx.doi.org/10.1016/j.ccr.2012.05.012]

[65] EFSA (European Food Safety Authority). Scientific opinion on lead in food. EFSA J 2010; 8(4): 1570.

[66] Hwang YO, Park SG, Park GY, Choi SM, Kim MY. Total arsenic, mercury, lead, and cadmium contents in edible dried seaweed in Korea. Food Addit Contam Part B Surveill 2010; 3(1): 7-13.
[http://dx.doi.org/10.1080/19440040903532079] [PMID: 24785310]

[67] Alvares AP, Leigh S, Cohn J, Kappas A. Lead and methyl mercury: effects of acute exposure on cytochrome P-450 and the mixed function oxidase system in the liver. J Exp Med 1972; 135(6): 1406-9.
[http://dx.doi.org/10.1084/jem.135.6.1406] [PMID: 5025440]

[68] Ali MB, Vajpayee P, Tripathi RD, *et al.* Mercury bioaccumulation induces oxidative stress and toxicity to submerged macrophyte Potamogeton crispus L. Bull Environ Contam Toxicol 2000; 65(5): 573-82.
[http://dx.doi.org/10.1007/s0012800162] [PMID: 11014840]

[69] Berlin M, Zalups RK, Fowler BA. Mercury. In: Nordberg GF, Fowler BA, Nordberg M, Friberg M, Friberg L, Eds. Handbook on the Toxicology of Metals 3rd ed. 2007; 675-729.
[http://dx.doi.org/10.1016/B978-012369413-3/50088-4]

[70] Rubio C, Gutiérrez Á, Burgos A, Hardisson A. Total dietary intake of mercury in the Canary Islands, Spain. Food Addit Contam Part A Chem Anal Control Expo Risk Assess 2008; 25(8): 946-52.
[http://dx.doi.org/10.1080/02652030801993597] [PMID: 18629690]

[71] Carro L, Herrero R, Barriada JL, Sastre de Vicente ME. Mercury removal: a physicochemical study of metal interaction with natural materials. J Chem Technol Biotechnol 2009; 84(11): 1688-96.
[http://dx.doi.org/10.1002/jctb.2232]

[72] Grandjean P. Mercury. In: Heggenhougen K, Ed. International Encyclopedia of Public Health 2nd ed. 2017; 110-6.
[http://dx.doi.org/10.1016/B978-0-12-803678-5.00290-3]

[73] Burbacher TM, Rodier PM, Weiss B. Methylmercury developmental neurotoxicity: A comparison of effects in humans and animals. Neurotoxicol Teratol 1990; 12(3): 191-202.
[http://dx.doi.org/10.1016/0892-0362(90)90091-P] [PMID: 2196419]

[74] van Wijngaarden E, Thurston SW, Myers GJ, *et al.* Methyl mercury exposure and neurodevelopmental outcomes in the Seychelles Child Development Study Main cohort at age 22 and 24years. Neurotoxicol Teratol 2017; 59: 35-42.
[http://dx.doi.org/10.1016/j.ntt.2016.10.011] [PMID: 27989696]

[75] Antunes dos Santos A, Appel Hort M, Culbreth M, *et al.* Methylmercury and brain development: A review of recent literature. J Trace Elem Med Biol 2016; 38: 99-107.
[http://dx.doi.org/10.1016/j.jtemb.2016.03.001] [PMID: 26987277]

[76] Scientific Opinion on the risk for public health related to the presence of mercury and methylmercury in food. EFSA J 2012; 10(12): 2985.

[77] Myers GJ, Thurston SW, Pearson AT, *et al.* Postnatal exposure to methyl mercury from fish consumption: A review and new data from the Seychelles Child Development Study. Neurotoxicology 2009; 30(3): 338-49.
[http://dx.doi.org/10.1016/j.neuro.2009.01.005] [PMID: 19442817]

[78] Fujimura M, Usuki F, Cheng J, Zhao W. Prenatal low-dose methylmercury exposure impairs neurite outgrowth and synaptic protein expression and suppresses TrkA pathway activity and eEF1A1 expression in the rat cerebellum. Toxicol Appl Pharmacol 2016; 298: 1-8.
[http://dx.doi.org/10.1016/j.taap.2016.03.002] [PMID: 26965727]

[79] Besada V, Andrade JM, Schultze F, González JJ. Heavy metals in edible seaweeds commercialised for human consumption. J Mar Syst 2009; 75(1-2): 305-13.
[http://dx.doi.org/10.1016/j.jmarsys.2008.10.010]

[80] IARC (International Agency for Research on Cancer). Arsenic, Metals, Fibres and Dusts. Volume 100C. Monographs on the evaluation of carcinogenic risks to humans; 2012.http://monographs.iarc.fr/ENG/Monographs/vol100C/

[81] González Weller D, Rubio C, Revert C, Hardisson A. El arsénico en los alimentos. Alimentaria 2003; (347): 21-7.

[82] Mohammed Abdul KS, Jayasinghe SS, Chandana EPS, Jayasumana C, De Silva PMCS. Arsenic and human health effects: A review. Environ Toxicol Pharmacol 2015; 40(3): 828-46.
[http://dx.doi.org/10.1016/j.etap.2015.09.016] [PMID: 26476885]

[83] Singh R, Singh S, Parihar P, Singh VP, Prasad SM. Arsenic contamination, consequences and remediation techniques: A review. Ecotoxicol Environ Saf 2015; 112: 247-70.
[http://dx.doi.org/10.1016/j.ecoenv.2014.10.009] [PMID: 25463877]

[84] EFSA. Scientific Opinion on Arsenic in Food. EFSA Panel on Contaminants in the Food Chain (CONTAM). EFSA J 2009; 7(10): 1351.
[http://dx.doi.org/10.2903/j.efsa.2009.1351]

[85] Brandon EFA, Janssen PJCM, de Wit-Bos L. Arsenic: bioaccessibility from seaweed and rice, dietary exposure calculations and risk assessment. Food Addit Contam Part A Chem Anal Control Expo Risk Assess 2014; 31(12): 1993-2003.
[http://dx.doi.org/10.1080/19440049.2014.974687] [PMID: 25393691]

[86] Vahter M. Mechanisms of arsenic biotransformation. Toxicology 2002; 181-182: 211-7.
[http://dx.doi.org/10.1016/S0300-483X(02)00285-8] [PMID: 12505313]

[87] Steinmaus C, Carrigan K, Kalman D, Atallah R, Yuan Y, Smith AH. Dietary intake and arsenic methylation in a U.S. population. Environ Health Perspect 2005; 113(9): 1153-9.
[http://dx.doi.org/10.1289/ehp.7907] [PMID: 16140620]

[88] Zhou Q, Xi S. A review on arsenic carcinogenesis: Epidemiology, metabolism, genotoxicity and epigenetic changes. Regul Toxicol Pharmacol 2018; 99: 78-88.
[http://dx.doi.org/10.1016/j.yrtph.2018.09.010] [PMID: 30223072]

[89] Almela C, Jesús Clemente M, Vélez D, Montoro R. Total arsenic, inorganic arsenic, lead and cadmium contents in edible seaweed sold in Spain. Food Chem Toxicol 2006; 44(11): 1901-8.
[http://dx.doi.org/10.1016/j.fct.2006.06.011] [PMID: 16901603]

SUBJECT INDEX

A

Avinash Mishra (Ed.)

B

D

E

F

G

H

I

K

L

M

N

O

P

Q

R

S

U

V

W

Z

Made in the USA
Las Vegas, NV
03 November 2026

57423617R00314